Phenomenology of Ultra-Relativistic Heavy-Ion Collisions

Phenomenology of Ultra-Relativistic Heavy-Ion Collisions

Wojciech Florkowski

Jan Kochanowski University, Kielce,
Poland, & Institute of Nuclear Physics,
Polish Academy of Sciences, Kraków, Poland

NEW JERSEY • LONDON • SINGAPORE • BEIJING • SHANGHAI • HONG KONG • TAIPEI • CHENNAI

Published by

World Scientific Publishing Co. Pte. Ltd.
5 Toh Tuck Link, Singapore 596224
USA office: 27 Warren Street, Suite 401-402, Hackensack, NJ 07601
UK office: 57 Shelton Street, Covent Garden, London WC2H 9HE

British Library Cataloguing-in-Publication Data
A catalogue record for this book is available from the British Library.

PHENOMENOLOGY OF ULTRA-RELATIVISTIC HEAVY-ION COLLISIONS

ISBN-13 978-981-4280-66-2
ISBN-10 981-4280-66-6

Printed in Singapore.

Publishers' page

To the memory of Barbara Buczek (1940–1993)

Polish composer and pianist

my Teacher of Music

Preface

The experimental studies of strongly interacting matter produced in ultra-relativistic heavy-ion collisions belong to the avant-garde of contemporary high-energy physics. This new and vastly developing field requires theoretical understanding how new experimental phenomena are related to the physical properties of the created system. This book delivers foundations of such understanding — it shows the links between basic theoretical concepts, discussed gradually from the elementary to more advanced level, and the results of experiments. In this way, I hope, experimentalists may learn more about the foundations of the models used by them to fit and interpret the data, while theoreticians may learn more about practical applications of their ideas.

The book emphasizes the role played in the interpretation of the experimental results by thermodynamics, relativistic hydrodynamics, and relativistic kinetic theory. These frameworks are used to analyze the soft hadron production, i.e., the production of hadrons with relatively small transverse momenta with respect to the collision axis. Soft hadrons contribute to more than 90% of the produced particles and their measured properties reveal information about the bulk properties of the very hot and dense system formed at the early stages of collisions — an interacting quark-gluon plasma.

A discussion of hard phenomena, i.e., the production of hadrons with large transverse momenta, has been omitted (except for the general comments concerning the jet quenching). Certainly, the role of hard processes is becoming more and more important with the increasing beam energy, however, their discussion would double the size of the present book.

The successful application of the perfect-fluid hydrodynamics in description of ultra-relativistic heavy-ion collisions allows us to establish a uniform picture of these complicated processes. In some sense we are lucky that such complex systems may be described within a concise and well-defined framework. The perfect-fluid hydrodynamics, combined with the modeling of the initial state by the Glauber model or the color glass condensate on one side, and supplemented by the kinetic simulations of the freeze-out process on the other side, forms the foundation of an approach that may be regarded as the standard model of ultra-relativistic heavy-ion collisions.

Yet, several shortcomings of the hydrodynamic approach indicate directions for many current and new investigations. To list a few: The problem of very early equilibration of matter formed in heavy-ion collisions is discussed in a very broad context including more elementary processes such as e^+e^- annihilation. Difficulties connected with the correct description of the correlations form a challenge for consistent modeling of the momentum and spacetime distributions of particles. Finally, the role of the viscosity and other dissipative effects should be elucidated. By the way, the recent developments in the field of dissipative hydrodynamics are examples of the permanent progress that is taking place in the heavy-ion physics. Again, due to the limited space, the issues of the viscous hydrodynamics have not been covered in this book. However, the perfect-fluid hydrodynamics was analyzed in greater detail, allowing all newcomers to enter this field very fast.

My intention was to write a possibly self-explanatory book, therefore, the chapters describing general formalism of the kinetic theory and hydrodynamics have been included. Moreover, the book has a modular structure where different parts and chapters may be read independently. I think this would be useful for young researches in the field who may find all the necessary information, usually scattered among various textbooks, in one volume. Nevertheless, I assume that the reader knows basic facts from the special theory of relativity, electrodynamics, quantum mechanics, statistical physics, and elementary-particle physics. Basic acquaintance with Feynman diagrams is necessary for reading the chapter about electromagnetic signals. About 50 exercises have been included, which appear after each part. They should help the reader to gain familiarity with the ideas and techniques introduced in the text.

We should be aware that the physics of ultra-relativistic heavy-ion collisions is a very broad, interdisciplinary field of physics. It is impossible to cover its all important and interesting problems in one book. Therefore, I have included many references to the original papers. They will guide the reader in further studies. I apologize for any possible omissions of the important papers in the lists of references.

This book stems from my lectures given at the Institute of Nuclear Physics in Kraków, the Jan Kochanowski University in Kielce, and the Warsaw University of Technology. I would like to thank all my collaborators, colleagues, and students for helpful comments, remarks, and questions. I am grateful to Piotr Bożek, Wojtek Broniowski, Adam Bzdak, Mikołaj Chojnacki, Marek Gaździcki, Krzysztof Golec-Biernat, Adam Kisiel, Staszek Mrówczyński, Radosław Ryblewski, and Giorgio Torrieri for critical comments concerning the manuscript. I thank Piotr Bożek for supplying me with figures showing the results of his 3+1 hydrodynamic calculations presented in Sec. 22.2.1. I also thank Mikołaj Chojnacki for his great help in preparation of my own figures.

During the work on this book, I have gained a lot from the inspiring atmosphere in the Cracow theoretical high-energy group created by Andrzej Białas, Wiesław

Czyż, and Jan Kwieciński. I hope that this book will serve as a depository of many creative ideas initialized in Cracow. I also thank Marek Pajek and Jan Pluta for their constant support and sharing enthusiasm about the idea of writing a book about relativistic heavy-ion collisions. Last but not least I thank my family for their constant support, encouragement, and patience.

I would be grateful for any comments concerning the text and ideas presented in this book. I would appreciate if they are sent to me directly to the electronic address Wojciech.Florkowski@ifj.edu.pl.

Wojciech Florkowski
Kraków–Pcim–Kielce, 2007–2009

Contents

PART IV: RELATIVISTIC HYDRODYNAMICS 193

PART I

OVERVIEW

Chapter 1

Introduction

1.1 High-energy nuclear collisions

Physics of the ultra-relativistic heavy-ion collisions is an interdisciplinary field which connects the high-energy physics of elementary particles with the nuclear physics. The name "heavy-ions" is used for heavy atomic nuclei, whereas the term "ultra-relativistic energy" denotes the energy regime where the kinetic energy exceeds significantly the rest energy. Typically, the high-energy particle physics deals with single particles (leptons, quarks, hadrons), and the interactions are derived from first principles. On the other hand, the nuclear physics deals with extended, complicated objects (nuclei), and the interactions are described by effective models. In the new field of the ultra-relativistic heavy-ion collisions one tries to analyze the properties of hot and dense nuclear/hadronic matter in terms of elementary interactions. Of the special importance are experimental searches for theoretically predicted new phases of hadronic matter, identification of the phase transitions between those phases, and a possible reconstruction of the phase diagram of strongly interacting matter in the broad range of the thermodynamic parameters such as temperature or baryon chemical potential.

In the last thirty years the nuclear physics has changed its character in a very significant way. In the 1970s and at the beginning of 1980s several accelerators used by the particle physics community were modified to accelerate heavy ions. For example, the Bevatron in Berkeley was coupled with the SuperHilac to form the Bevalac [1]. The SuperHilac was a linear accelerator where the ions of heavy elements were created and sent for further acceleration to the Bevatron. In this way the relativistic energies of 1–2 GeV per nucleon were achieved. Similarly, the Dubna Syncrophasotron was converted to accelerate heavy ions. On the other hand, in the same time a number of accelerators used in the nuclear research were developed, yielding relativistic beams of heavy ions in many places, for example, at the Gesellschaft für Schwerionenforschung (GSI) in Darmstadt.

The first experiments with the ultra-relativistic heavy ions (with energies exceeding 10 GeV per nucleon in the projectile beam) took place at the Brookhaven National Laboratory (BNL) and at the European Organization for Nuclear Re-

Table 1.1 Summary of the RHIC runs in the years 2000–2006: number of the run, date, colliding systems, and energy.

Run	Year	Species	$\sqrt{s_{NN}}$ [GeV]
01	2000	Au+Au	130
02	2001–2	Au+Au	200
		p+p	200
03	2002–3	d+Au	200
		p+p	200
04	2003–4	Au+Au	200
		Au+Au	62
05	2004–5	Cu+Cu	200
		Cu+Cu	62
		Cu+Cu	22.5
		p+p	200
06	2006	p+p	200
		p+p	62

search (CERN) in 1986. The Alternating Gradient Synchrotron (AGS) at BNL accelerated beams up to ^{28}Si at 14 GeV per nucleon. At CERN, the Super Proton Synchrotron (SPS) accelerated ^{16}O at 60 and 200 GeV per nucleon in 1986, and ^{32}S at 200 GeV per nucleon in 1987. In 1990 a long-term project on heavy-ion physics was realized at CERN with several weeks of ^{32}S beams. In the spring of 1992 the experiments with ^{197}Au beams at 11 GeV per nucleon were initiated at BNL. In 1995 the completely new experiments took place at CERN with ^{208}Pb beams at 158 GeV per nucleon. These were for the first time really ultra-relativistic "heavy" ions providing large volumes and lifetimes of the reaction zone.

In 2000 the first data from the Relativistic Heavy Ion Collider (RHIC) at BNL were collected. RHIC was designed to accelerate fully stripped Au ions to a collision center-of-mass energy of 200 GeV per nucleon pair, i.e., for $\sqrt{s_{NN}} = 200$ GeV (for the history of the construction of RHIC see [2]). The design luminosity corresponds to approximately 1400 Au+Au collisions per second. During the first run in 2000, the maximum energy of 130 GeV per nucleon pair was achieved, with 10% of the designed luminosity. In the years 2001–2004 the next three runs took place with the maximum energy of 200 GeV per nucleon pair. One of those runs was devoted to the study of the deuteron-gold collisions which were analyzed in order to get the proper reference point for the more complicated gold on gold collisions. Altogether, in the years 2000–2006 six runs took place with different colliding systems and at different beam energies, see Table 1.1.

There are four experiments at RHIC. The two smaller experiments are BRAHMS and PHOBOS, and the two larger experiments are PHENIX and STAR. The experimental aim of BRAHMS is particle identification over a broad rapidity range.

Table 1.2 *Quark Matter* Conferences

Number	Date	Place	Proceedings
1th	Aug. 24–31, 1980	Bielefeld, Germany	[3]
2nd	May 10–14, 1982	Bielefeld, Germany	[4]
3rd	Sept. 26–30, 1983	Upton, USA	[5]
4th	June 17–21, 1984	Helsinki, Finland	[6]
5th	April 13–17, 1986	Pacific Grove, USA	[7]
6th	Aug. 24–28, 1987	Nordkirchen, Germany	[8]
7th	Sept. 26–30, 1988	Lenox, USA	[9]
8th	May 7–11, 1990	Menton, France	[10]
9th	Nov. 11–15, 1991	Gatlinburg, USA	[11]
10th	June 20–24, 1993	Borlänge, Sweden	[12]
11th	Jan. 9–13, 1995	Monterey, USA	[13]
12th	May 20–24, 1996	Heidelberg, Germany	[14]
13th	Dec. 1–5, 1997	Tsukuba, Japan	[15]
14th	May 10–15, 1999	Torino, Italy	[16]
15th	Jan. 15–20, 2001	Stony Brook, USA	[17]
16th	July 18–24, 2002	Nantes, France	[18]
17th	Jan. 11–17, 2004	Oakland, USA	[19]
18th	Aug. 4–9, 2005	Budapest, Hungary	[20]
19th	Nov. 14–20, 2006	Shanghai, China	[21]
20th	Feb. 4–10, 2008	Jaipur, India	[22]
21th	Mar. 30–Apr. 4, 2009	Knoxville, USA	
22th	2011	Annecy, France	

The PHOBOS experiment measures total charged particle multiplicity and particle correlations. The PHENIX experiment is designed to measure electrons, muons, hadrons and photons. The STAR experiment concentrates on measurements of hadron production over a large solid angle.

The future of the field is connected with the construction of the Large Hadron Collider (LHC) at CERN (Pb on Pb reactions at $\sqrt{s_{\rm NN}} = 5.5$ TeV). Nevertheless, the performance of new experiments at lower energies is also very important, since this allows us to study the energy dependence of many characteristics of the particle production. Within the more recent project at the SPS, in the years 1999–2003, the NA49 Collaboration recorded Pb+Pb collisions at 20, 30, 40 and 80 GeV per nucleon. More information on the development of the experimental situation in the last thirty years may be found in the series of the *Quark Matter Proceedings* whose list is given in Table 1.2 [1].

[1]It is not quite clear what the first conference in this series was. An alternative for the first position listed in Table 1.2 is the *First Workshop on Ultra-Relativistic Nuclear Collisions*, organized at the Lawrence Berkeley Laboratory in 1979, LBL report 8957.

1.2 Theoretical methods

In the ultra-relativistic heavy-ion collisions very large numbers of particles are produced (we deal with so called large particle *multiplicities*). For example, in the central Au+Au collisions at RHIC, at the highest beam energy $\sqrt{s_{\rm NN}} = 200$ GeV, the total charged particle multiplicity is about 5300 [23]. Hence, the number of produced particles exceeds the number of initial nucleons by a factor of 10. In this situation, different theoretical methods are used, which are suitable for description of large macroscopic systems, e.g., thermodynamics, hydrodynamics, kinetic (transport) theory, field theory at finite temperature and density, non-equilibrium field theory, Monte-Carlo simulations.

Many estimates of the effects in high-energy nuclear collisions are done on the basis of purely *thermodynamic* or *statistical* considerations. However, the hadronic systems produced in the collisions are not static. The need for the dynamical description involves rich applications of *relativistic hydrodynamics.* Furthermore, since the matter produced in the collisions lives only for a short while, it is natural to expect that its spacetime evolution proceeds far away from equilibrium. Consequently, there exists a growing interest in applying and developing *transport theories* which are suitable for the description of non-equilibrium processes. In the similar spirit, one tries to describe the heavy-ion reactions with the help of *microscopic Monte-Carlo simulations* which usually represent an extrapolation of low energy models of hadron-hadron collisions. Last but not least, the physics of the ultra-relativistic heavy-ion collisions triggered fast development of the *quantum theory of fields in and out of equilibrium.* Using the methods of field theory one can study the in-medium properties of particles. Moreover, this approach allows also for the formulation of the kinetic equations satisfied by the particle distribution functions.

1.3 Quantum chromodynamics

Generally speaking, during high-energy nuclear collisions a many-body system of *strongly* interacting particles is produced. The fundamental theory of the strong interactions is *Quantum Chromodynamics* (QCD), the theory of quarks and gluons which are confined in hadrons, i.e., baryons and mesons.

From the historical perspective we may say that the development of QCD started with the 1963 proposal of Gell-Mann [24] and Zweig suggesting that the structure of hadrons could be explained by the existence of smaller particles inside hadrons (at that time u, d and s quarks). In 1964 Greenberg [25] and in 1965 Han with Nambu [26] proposed that quarks possessed an additional degree of freedom, that was later called the *color charge.* Han and Nambu noted that quarks might interact via exchanges of an octet of vector gauge bosons (later *gluons*). Feynman and Bjorken argued that high-energy experiments should reveal the existence of partons, i.e., particles that are parts of hadrons. Those suggestions were spectacularly veri-

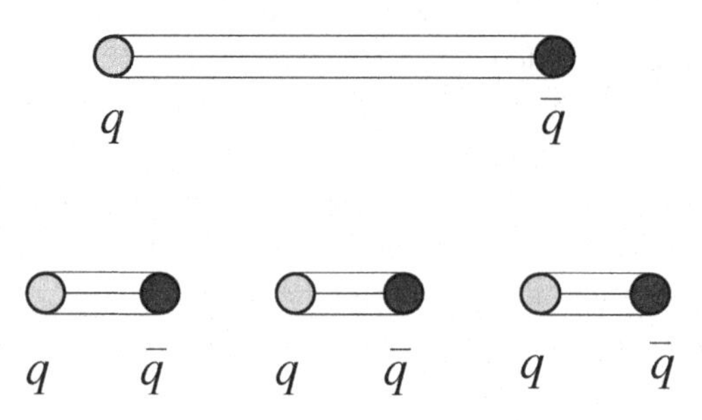

Fig. 1.1 Schematic view of the confinement mechanism. The separating $q\bar{q}$ pair stretches the color string until the increasing potential energy is sufficient to create another $q\bar{q}$ pair. More pairs may be produced in this way, which leads to the formation of final state hadrons.

fied in the *deep inelastic scattering* of electrons on protons, the experiments carried out at the Stanford Linear Accelerator Center (SLAC) in 1969. The partons were identified with quarks. The discovery of asymptotic freedom in the strong interactions by Gross, Politzer and Wilczek [27–29] allowed for making precise predictions of the results of many high-energy experiments in the framework of the perturbative quantum field theory [30] — the asymptotic freedom is the property that the interaction between particles becomes weaker at shorter distances.

Probably the most striking feature of QCD is the color confinement [31–33], which is the other side of the asymptotic freedom. This is the phenomenon that color charged particles (such as quarks and gluons) cannot be isolated as separate objects. In other words, quarks and gluons cannot be directly observed. The physical concept of confinement may be illustrated by a string which is spanned between the quarks when we try to separate them, see Fig. 1.1. If the quarks are pulled apart too far, large energy is deposited in the string which breaks into smaller pieces. As a result the quarks form new hadrons produced from the pieces of the initial string.

Such a qualitative picture of confinement is supported strongly by the numerical calculations. On the other hand, at the moment there is no analytic approach or approximation that describes the behavior of QCD at large distances. Lacking this, we often feel that we do not understand fully the mechanism of confinement and its proof is missing. This situation is reflected by the fact that the Clay Mathematics Institute of Cambridge includes the confinement problem as one of the seven Millennium Problems and offers a prize of one million dollars for the proof [34].

In some sense, the *nuclear* force between baryons and mesons can be viewed as a residual force acting between quarks and gluons, in the analogous way as the *chemical* (van der Waals) force is the residual electromagnetic interaction. Since QCD is a complicated non-linear theory, the *complete* description of the relativistic heavy-ion collision based exclusively on first principles is impossible in practice. Almost in all cases we have to use models, although QCD may be successfully applied to describe subprocesses of complicated collisions or to deliver an input for modeling. In particular, the lattice simulations of QCD give us information about the equation of state of strongly interacting hot matter, which may be then used as an input for the hydrodynamical codes. In addition, theoretical calculations based on QCD inspire new measurements.

More information about QCD (although still very much restricted) will be given in Chap. 5. We refer also to the general textbooks discussing QCD, for example, Refs. [35, 36].

1.4 Quark-gluon plasma

The main challenge of the ultra-relativistic heavy-ion collisions is the observation of the two phase transitions predicted by QCD, i.e., the *deconfinement* and *chiral phase transitions.* As we have mentioned above, at Earth conditions (i.e., at low energy densities) quarks and gluons are confined in hadrons. However, with increasing

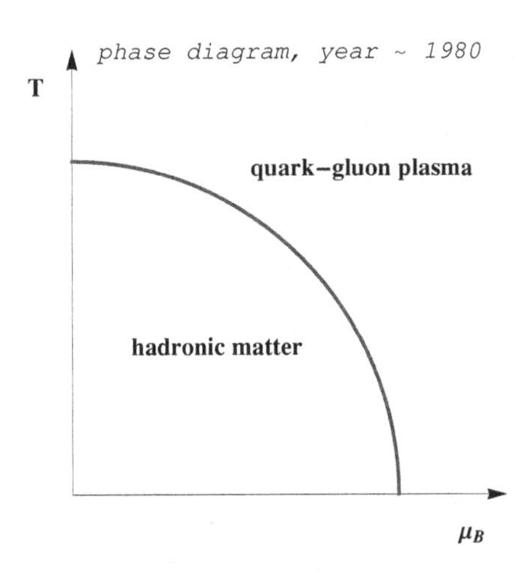

Fig. 1.2 The first phase diagram of strongly interacting matter was introduced in the paper by Cabibbo and Parisi in 1975 [37]. It looked similar to the plot shown here. The line distinguishes two regions in the two-dimensional space of temperature, T, and baryon chemical potential, μ_B. For smaller values of T and/or μ_B (the points below the curve) the matter is made out of hadrons, whereas at sufficiently high values of T or μ_B (the points above the curve) the matter is made up of deconfined quarks and gluons.

temperature (heating) and/or increasing baryon density (compression), a phase transition may occur to the state where the ordinary hadrons do not exist anymore, and where quarks and gluons become the correct degrees of freedom. In 1975, soon after the discovery of asymptotic freedom in the strong interactions, Collins and Perry argued that "superdense matter (found in neutron-star cores, exploding black holes, and the early big-bang universe) consists of quarks rather than of hadrons" [38]. In the same year, Cabibbo and Parisi identified the limiting Hagedorn temperature with the temperature of the phase transition from hadronic to quark matter [37]. They also sketched the first phase diagram of strongly interacting matter, see Figs. 1.2 and 1.3. The collective phenomena in gauge theories were studied then by Kislinger and Morley [39, 40]. Freedman and McLerran computed the three-loop contributions to the thermodynamic potential [41–43]. Calculations at finite temperature were performed by Shuryak [44, 45], who in 1978 introduced the name *quark-gluon plasma* (QGP), and by Kapusta [46]. The first quantitative considerations concerning the possibility of the formation of hot quark matter in relativistic heavy-ion collision were given by Chin [47] [2].

The present experimental evidence indicates that in the ultra-relativistic heavy-ion collisions an extended and very dense system of strongly interacting matter is indeed formed. It differs in many aspects from the systems formed in elementary hadron-hadron reactions. On Feb. 10, 2000, CERN announced officially [48] that "a compelling evidence now exists for the formation of a new state of matter at energy densities about 20 times larger than in the center of atomic nuclei and temperatures about 100000 times higher than in the center of the sun". This announcement followed the analysis of many experimental data collected during 15 years of heavy-ion experiments at the SPS.

The first RHIC data confirmed the overall picture that emerged from the studies at lower energies. However, several new features of the collision process were observed, e.g., higher particle multiplicities, increased production of antiparticles, strong collective phenomena, and lower baryon number density in the central rapidity region. The data collected in the next runs, especially in the deuteron-gold collisions which were used as a reference measurement, brought the evidence for strong quenching of very energetic particles traversing the medium created in the central Au+Au collisions. The streams of such particles, called jets, appear in elementary collisions where they form two, flying back-to-back, groups of hadrons. The RHIC data indicate that such back-to-back correlations between very energetic hadrons are lost in central Au+Au collisions. This effect may be understood easily by the assumption that one of the jets is absorbed by the medium because it has a longer distance to traverse in the medium.

The notorious question is asked if the quark-gluon plasma has been indeed dis-

[2] Probably, for the first time the idea of using the ultra-relativistic heavy-ion collisions to produce and study new forms of matter was introduced during the *Workshop on BeV Collisions of Heavy Ions: How and Why*, which was held in Bear Mountain, New York, Nov. 29 - Dec. 1, 1974, BNL-AUI report, 1975.

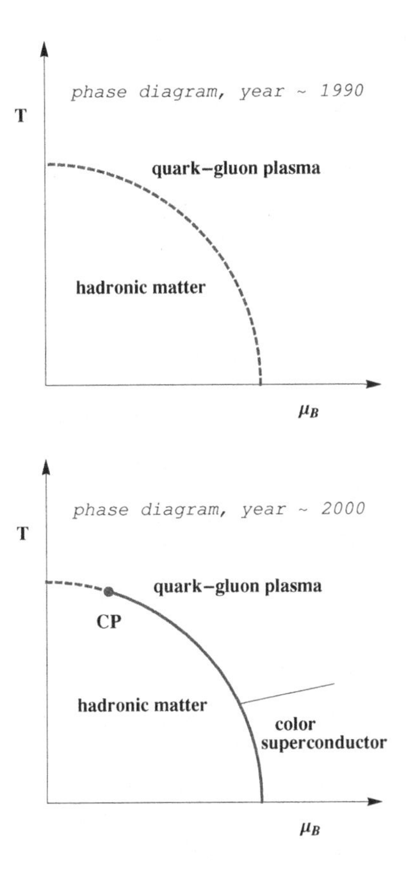

Fig. 1.3 Time evolution of our views on the phase diagram of strongly interacting matter. Around 1990 most physicists believed that there was a crossover between the hadronic matter and the quark-gluon plasma, i.e., a sudden change in the energy density, but not real phase transition (dashed line in the upper part). By the year 2000 this opinion had changed. At present, we expect that there is a line of the first order phase transition ending in a critical point (solid thick line in the lower part). In addition, at very high baryon density (large baryon chemical potential) there is a color superconductivity region, for a review see [49]. The crossover transition takes place at low baryon density and high temperature (dashed line in the lower part).

covered in the relativistic heavy-ion collisions. If we have in mind the asymptotic state, where due to the asymptotic freedom the plasma is treated as an ideal gas of quarks and gluons, then the question seems to be quite difficult or even impossible to answer. Although different "plasma signatures" have been proposed, we have no definite proof that such an asymptotic state has been reached. On the other hand, the phenomena such as the absorption of jets or simple estimates of the energy density accessible at the center of relativistic heavy-ion collisions indicate that the produced matter cannot consist of hadrons — the hadronic sizes are simply too big to allow for the treatment of hadrons in the initial stages of the collisions as well isolated degrees of freedom. Consequently, the matter produced in the ultra-relativistic heavy-ion collision is definitely a system of interacting quarks and gluons. It is much denser than that produced in more elementary hadronic or proton-nucleus collisions. Moreover, the system created in heavy-ion collisions exhibits high level of thermalization and shows strong collective behavior. Accepting these facts, it is natural to admit that the produced matter is an interacting quark-gluon plasma [50–53]. This point of view has been adopted in this book. The remaining problem is, however, to establish more precisely the physical properties of the plasma. In particular, it is important to conclude how strongly interacting system it is [54–56].

Clearly, further systematic studies are necessary now to extract more detailed information about the dense medium found in the heavy-ion experiments. Certainly, only now this field of physics has come into its mature age. Lead on lead collisions at the LHC, offering the initial energy density 50 to 100 times larger than that of normal nuclear matter, will be the next source of very intriguing data.

1.5 Chiral symmetry

There exist six different types (flavors) of quarks: *up, down, strange, charmed, bottom* and *top*. The bottom and the top quark are sometimes called the beauty and the true one. Let us now restrict our considerations to the subsector of QCD describing only up and down quarks. At normal Earth conditions, the most common hadrons, i.e., protons, neutrons and pions are made of these two flavors. The masses of up and down quarks are very small so they are usually ignored in most of the practical calculations, see Table 1.3.

In the limit of vanishing masses the left- and right-handed quarks become decoupled from each other and QCD becomes invariant under their interchange. One of the consequences of this fact is that there are left- and right-handed quark currents which are separately conserved, instead of only the vector current which is conserved in the massive case. Symmetry between the left- and right-handed quarks implies also that each state of the theory should have a degenerate partner of the opposite parity. On the other hand, we know that hadrons have well defined parity, and no such parity partners are observed! The paradox is resolved by the phenomenon of the *spontaneous breakdown of chiral symmetry*: the chiral symmetry of the interaction is broken by the true ground state of the theory, see Fig. 1.4.

This mechanism was recognized first by Nambu, already in the pre-QCD times [57–59]. There are many examples of such situations in other fields of physics. For instance, the exact translational symmetry is broken by the ground states of solids which are periodic crystals. According to the famous Goldstone theorem [60, 61], spontaneous breaking of any continuous symmetry is connected with the existence of *soft modes*.

Table 1.3 Mass, electric charge, the third component of the isospin, strangeness (S), charm (C), bottom (B), and top (T) of quarks. All quarks have spin $\frac{1}{2}\hbar$, baryon number $\frac{1}{3}$, and lepton number 0. The values of the masses are taken from Ref. [62]. We note that these are *current masses* that appear in the perturbative QCD calculations. They differ from the *constituent masses* that are effective masses of the strongly interacting quarks.

flavour	up (u)	down (d)	strange (s)	charmed (c)	bottom (b)	top (t)
m	1–5 MeV	3–9 MeV	75–170 MeV	1.15–1.35 GeV	4.0–4.4 GeV	174.3 ± 5.1 GeV
Q	$\frac{2}{3}e$	$-\frac{1}{3}e$	$-\frac{1}{3}e$	$\frac{2}{3}e$	$-\frac{1}{3}e$	$\frac{2}{3}e$
I_3	$\frac{1}{2}$	$-\frac{1}{2}$	0	0	0	0
S	0	0	-1	0	0	0
C	0	0	0	+1	0	0
B	0	0	0	0	-1	0
T	0	0	0	0	0	+1

In QCD such soft modes correspond to pseudoscalar pions. Their existence solves the paradox — infinitely soft pseudoscalar modes can be added to any state, changing the parity without any change in energy. In reality pions have a small mass and are sometimes called pseudo-Goldstone bosons. This is due to the fact that the masses of up and down quarks are not exactly zero and the chiral symmetry is approximate.

In a very hot and dense hadronic medium, ordinary hadrons lose their identities and the quark-gluon plasma is produced. In this case the ground state of the strong interactions is significantly modified and the chiral symmetry is expected to be restored; in the natural way the left- and right-handed (practically massless) quark excitations in the plasma are the chiral partners to each other.

Observation of the signals of the chiral phase transition is an exciting perspective of the experiments with heavy ions. It is possible that the two phase transitions, i.e., deconfinement and chiral restoration, do not happen simultaneously. One can imagine that with the increasing temperature we first have deconfinement (but still the quarks have non-zero effective masses) and later the chiral phase transition to massless quarks. In 1983 the lattice simulations of both SU(2) and SU(3) gauge theories indicated, however, that the two phase transitions took place at almost the same temperature [63]. The coincidence of the two temperatures has been confirmed by more recent calculations, for example, see [64]. The issue whether this

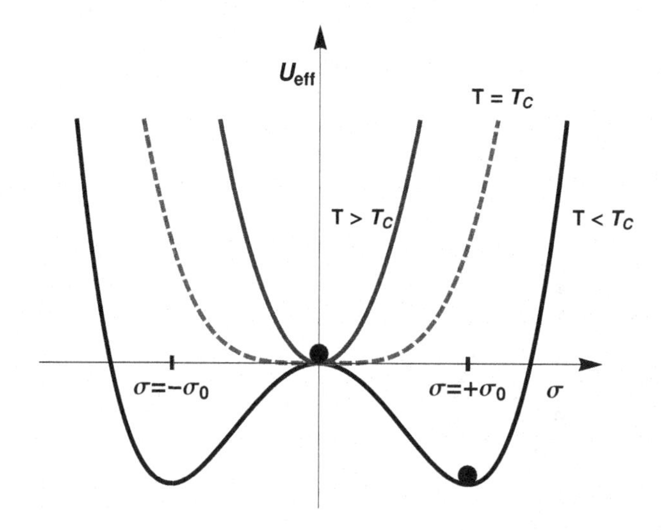

Fig. 1.4 The mechanism of the spontaneous symmetry breaking may be illustrated by a symmetric, temperature dependent potential $U_{\rm eff}(\sigma)$. For $T > T_{\rm c}$ the potential has one (global) minimum and the ground state of the field σ corresponds to the case $\sigma = 0$. If the temperature becomes smaller than $T_{\rm c}$, the potential develops two different local minima. Although the potential is still a symmetric function of σ, the ground state of the field is randomly changed to either $+\sigma_0$ or $-\sigma_0$ (to $+\sigma_0$ in this plot).

coincidence is exact or approximate is under discussion now, but in any case the two temperatures seem to be very close to each other. This is why most physicists identify the two phase transitions. A theoretical possibility still exists, however, that the two phase transitions differ significantly at a finite baryon chemical potential.

1.6 Hot and dense nuclear matter

The study of high-energy nuclear reactions gives us important information about properties of hot and dense hadronic matter — by this we mean here the matter present at the late stages of the collisions where hadrons may be regarded as the correct degrees of freedom. Heavy-ion collisions are the only way to compress and heat up nuclear matter in laboratory conditions. Information extracted from data can be useful for the construction of adequate models of neutron stars and supernova explosions. Already at energies of the order of a few GeV per nucleon one encounters many interesting and well established phenomena like, e.g., collective flows or subthreshold production of particles. Central collisions of symmetric heavy ions at 1 GeV per nucleon (so called *relativistic regime*) are likely to yield about 3 times normal nuclear matter density. The particles inside such a system do not propagate completely freely: their Compton wavelength may be comparable with their mean free path. In this situation, we expect that some of the particle properties (e.g., hadron masses, widths or coupling constants) can be changed. These *in-medium modifications* can lead to the experimentally observed phenomena. For example, the change of the ρ meson mass and/or width in dense matter can influence the measured dilepton spectrum. Nowadays, one attempts to connect in-medium modifications of hadron properties with the *partial* restoration of chiral symmetry.

1.7 Units and notation

In this book we use the natural system of units where the velocity of light in vacuum, c, the Planck constant h divided by 2π, $\hbar$, and the Boltzmann constant, k_B, are all equal to unity, $c = \hbar = k_B = 1$. The exception are the sections devoted to the kinetic theory where the convention $h = 1$ is sometimes more useful. With the choice $c = \hbar = k_B = 1$ it becomes unnecessary to write $c, \hbar$, and k_B in the equations, thus, we save space and trouble. The dimensional analysis may be always used to unambiguously reinsert those constants into various expressions. It is suggested to the readers who are not familiar with the natural system of units that they first work out Ex. 7.1. For electromagnetic quantities we have adopted the Heaviside-Lorentz system of units where ϵ_0 is set equal to unity and the factors 4π are absent in the Maxwell equations.

The field of the relativistic heavy-ion collisions includes many subfields with commonly accepted notation. In the book discussing phenomena belonging to such different subfields it becomes difficult to use simple and unequivocal notation. We have tried to avoid the situations where the same symbols are used for different physical quantities but sometimes such situations are inevitable. We hope that it does not lead to much confusion, since the proper meaning follows usually from the physical context. The most common symbols used in the book are presented in Tables 1.4–1.7. We note that spatial three-vectors and color two-vectors (in the space of color isotopic charge and color hypercharge) are indicated by letters in boldface.

In the text we often refer to Lorentz transformations. By this we mean the Lorentz boosts, mostly in the direction of the beam axis, or the proper orthochronous Lorentz transformations. Again the correct meaning follows from the context.

Table 1.4 Symbols of the physical quantities used in the book (part 1).

spacetime variables	
$g_{\mu\nu} = \mathrm{diag}(1,-1,-1,-1)$	metric tensor
$x^\mu = (x^0, x^1, x^2, x^3) = (t, x, y, z)$	spacetime coordinates
$r = \sqrt{x^2+y^2}$	distance from the collision axis
$\phi = \arctan(y/x)$	azimuthal angle
$\theta = \arctan(r/z)$	polar angle
$\tau = \sqrt{t^2-z^2}$	(longitudinal) proper time
$\eta_\parallel = \frac{1}{2}\ln\frac{t+z}{t-z}$	spacetime rapidity
kinematical variables describing single particles	
$m \quad (m_N, m_\pi)$	mass (nucleon mass, pion mass)
$p^\mu = (p^0, p^1, p^2, p^3) = (E, p_x, p_y, p_z)$	four-momentum
$p^0 = p_0 = E$	energy
$E_p = \sqrt{m^2+p^2}$	mass-shell energy
$p^3 = -p_3 = p_z = p_\parallel$	longitudinal momentum of a particle
$p_\perp = \sqrt{p_x^2+p_y^2}$	transverse momentum of a particle
$\phi_p = \arctan(p_y/p_x)$	momentum azimuthal angle
$m_\perp = \sqrt{m^2+p_\perp^2}$	transverse mass
$\mathrm{y} = \frac{1}{2}\ln\frac{E+p_\parallel}{E-p_\parallel}$	rapidity
$\eta = \frac{1}{2}\ln\frac{\lvert\mathbf{p}\rvert+p_\parallel}{\lvert\mathbf{p}\rvert-p_\parallel}$	pseudorapidity
variables characterizing nucleus-nucleus, nucleon-nucleus, and nucleon-nucleon collisions	
A, B	atomic mass numbers (also labels characterizing nuclei)
ρ_A	Woods-Saxon function
$N_{\mathrm{part}}, N_{\mathrm{spec}}$	number of participants, number of spectators
n, w	number of binary collisions, number of wounded nucleons
$\mathbf{b}, b$	impact vector, impact parameter
c	centrality
$s = (p_1+p_2)^2$, $t = (p_1-p_1')^2$, $u = (p_1-p_2')^2$	Mandelstam variables
$f(s,t)$	scattering amplitude
θ	scattering angle
$\sigma_{\mathrm{tot}}, \sigma_{\mathrm{in}}, \sigma_{\mathrm{el}}$	nucleon-nucleon total, inelastic, elastic cross section
σ_{incl}	inclusive cross section
$\delta(s,\mathbf{b})$	phase shift
$t(\mathbf{b}), T_A(\mathbf{b}), T_{AB}(\mathbf{b})$	thickness functions
R_{AB}	nuclear modification factor
$\mathcal{P}_1, \mathcal{P}_2$	invariant one- and two-particle inclusive distributions

Table 1.5 Symbols of the physical quantities used in the book (part 2).

Symbol	Meaning
kinetic-theory variables	
$f(x,p)$ $\tilde{f} = (2\pi\hbar)^3 f = h^3 f$	phase-space distribution function
ϵ	parameter of the equilibrium distribution functions $\epsilon = +1$ for bosons and $\epsilon = -1$ for fermions
$\bar{f}(x,p) = 1 + h^3 \epsilon f(x,p)$	statistical correction factors in collision terms
$\Phi[f(x,p)]$	functional used to define entropy current
$W_{\alpha,\beta}(x,p)$	Wigner function
$S(x,p)$	emission (source) function
$d\Gamma_{\rm cl} = \frac{dV d^3p}{(2\pi\hbar)^3}$	classical phase-space element
$d\Gamma_{\rm inv} = dV\, dt\, \frac{d^3p}{E_p}$	Lorentz invariant phase-space element
$N^\mu = (n^0, \mathbf{n})$	particle number current
$j^\mu = (\rho, \mathbf{j})$	electric current
$T^{\mu\nu}$	energy-momentum tensor
S^μ, $\Delta\overline{\Gamma}$	entropy current, statistical weight
$\Delta\Gamma$	differential transition rate
$\Delta\sigma$	differential cross section
$\Phi_{\rm i}$, $F_{\rm i}$, $v_{\rm M}$	flux, invariant flux, Møller velocity
C, C_{kl}	collision terms
$W(p, p_1 \vert p', p'_1)$, W_{kl}, $W_{kl\vert ij}$	transition rates
QED and QCD variables	
A^μ, A^μ_a	four-potential, color four-potential
$F^{\mu\nu}$, $F^{\mu\nu}_a$	field tensor, color field tensor
$\mathcal{E}$	longitudinal electric field F^{30}
$\boldsymbol{\mathcal{E}} = (\mathcal{E}^3, \mathcal{E}^8)$	"neutral" components of longitudinal chromoelectric field (F_3^{30}, F_8^{30})
$\boldsymbol{\epsilon}_i$, $\boldsymbol{\eta}_{ij}$	color charges of quarks $(i = 1,2,3)$ and "charged" gluons $(i,j = 1,2,3; i \neq j)$
λ^a	Gell-Mann matrices
g, $\alpha_s = g^2/(4\pi)$	strong coupling constants
B	bag constant

Table 1.6 Symbols of the physical quantities used in the book (part 3).

thermodynamic variables	
N	multiplicity, baryon number
n	multiplicity, particle density, baryon number density
$T \quad (T_{\rm i}, T_{\rm f})$	temperature (initial temperature, final temperature)
$\mu \; (\mu_B)$	chemical potential (baryon chemical potential)
$P, \quad V$	pressure, volume
$E, \quad \varepsilon = E/V$	energy, energy density
$W = E + PV$	enthalpy
$w = W/N$	enthalpy per baryon
$\tilde{w} = W/V$	enthalpy density
S	entropy
$s = S/N$	entropy per baryon
$\sigma = S/V$	volume entropy density
$c_{\rm s}$	sound velocity
$\lambda = c_{\rm s}^2$	sound velocity squared
$T_{\rm H}$	Hagedorn limiting temperature
$\rho(m)$	hadron mass spectrum
g_π	degrees of freedom of the pion gas
$g_{\rm qgp}$	degrees of freedom of the weakly-interacting quark-gluon plasma
	thermodynamic variables characterizing relativistic gas of
$n_{\rm cl}, P_{\rm cl}, ...$	classical massive particles (cl),
$n_{\rm b}, P_{\rm b}, ...$	massless bosons (b), and
$n_{\rm f}, P_{\rm f}, ...$	massless fermions (f)
hydrodynamic variables	
$u^\mu = (u^0, \mathbf{u}) = \gamma(1, \mathbf{v})$	fluid four-velocity
$\mathbf{v} = (v_x, v_y, v_z),$	fluid three-velocity
$\gamma = (1 - v^2)^{-1/2}$	gamma Lorentz factor
$v_\perp = \sqrt{v_x^2 + v_y^2}$	transverse flow
$\vartheta_\perp = \frac{1}{2} \ln \frac{1+v_\perp}{1-v_\perp}$	transverse fluid rapidity
$\vartheta = \frac{1}{2} \ln \frac{1+v}{1-v}$	fluid rapidity (note that this definition does not require that $\mathbf{v}$ has only the longitudinal component)
$\frac{d}{d\tau} = u^\mu \partial_\mu$	total time derivative
$c_{\rm s}$	sound velocity
$\Phi(T)$	potential used in the Baym formalism
$\Phi_{\rm L}(t, z)$	potential used in the Landau formalism
$\chi(T, \vartheta)$	Khalatnikov potential
$Y = \ln(T/T_{\rm i})$	logarithm of temperature

Table 1.7 Symbols of the physical quantities used in the book (part 4).

terminology for interferometry	
$C(\mathbf{p}_1, \mathbf{p}_2)$	two-particle correlation function
$\rho(\mathbf{p}_1, \mathbf{p}_2)$	density matrix
$\mathbf{k} = \frac{1}{2}(\mathbf{p}_1 + \mathbf{p}_2)$	average three-momentum
$\mathbf{q} = \mathbf{p}_1 - \mathbf{p}_2, \quad q = q_1 - q_2$	relative three- and four-momentum
$\boldsymbol{\beta}$	velocity of a pair
$\mathbf{q}^*, \quad \mathbf{r}^*$	relative three-momentum and relative distance in the pair rest frame
i = out, side, long	
q_i	ith component of $\mathbf{q}$ in the out-side-long frame
$\tilde{q}_i$	relative four-momentum in the out-side-long frame for the case where only q_i is different from zero
R_i	HBT radii
special functions	
$\theta(x)$	Heaviside function (unit step function)
$\delta(x)$	Dirac delta function
$I_n(x)$	modified Bessel functions of the first kind
$K_n(x)$	modified Bessel functions of the second kind
$\Gamma(x)$	Euler gamma function
$\zeta(x)$	Riemann zeta function
$\mathrm{Li}_n(x)$	polylogarithm function
B_n^*	Bernoulli numbers

Bibliography to Chapter 1

[1] R. Stock, "Relativistic nucleus nucleus collisions: From the BEVALAC to RHIC," *J. Phys.* **G30** (2004) S633–S648.

[2] G. Baym, "RHIC: From dreams to beams in two decades," *Nucl. Phys.* **A698** (2002) XXIII–XXXII.

[3] H. Satz, "Statistical mechanics of quarks and hadrons, Proc. of the Int. Symp., Univ. Bielefeld, Germany, Aug. 24–31, 1980," (North-Holland Publ., Amsterdam, 1981).

[4] M. Jacob and H. Satz, "Quark matter formation and heavy ion collisions, Proc. of the Bielefeld Workshop, Bielefeld, Germany, May 10–14, 1982," (World Scientific, Singapore, 1982).

[5] T. W. Ludlam and H. E. Wegner, "Quark Matter '83 Proceedings, 3rd International Conference on Ultra-Relativistic Nucleus-Nucleus Collisions, Upton, USA, September 26–30, 1983," *Nucl. Phys.* **A418** (1984) 1c–678c.

[6] K. Kajantie, "Quark Matter '84 Proceedings, 4th International Conference on Ultra-Relativistic Nucleus-Nucleus Collisions, Helsinki, Finland, June 17–21, 1984," (Lecture Notes in Physics, 221, Berlin, Germany, Springer, 1985) 305 p.

[7] L. S. Schroeder and M. Gyulassy, "Quark Matter '86 Proceedings, 5th International Conference on Ultra-Relativistic Nucleus-Nucleus Collisions, Pacific Grove, USA, April 13–17, 1986," *Nucl. Phys.* **A461** (1987).

[8] H. Satz, H. J. Specht, and R. Stock, "Quark Matter '87 Proceedings, 6th International Conference on Ultra-Relativistic Nucleus-Nucleus Collisions, Nordkirchen, Germany, August 24–28, 1987," *Z. Phys.* **C38** (1988) 1–370.

[9] G. A. Baym, P. Braun-Munzinger, and S. Nagamiya, "Quark Matter '88 Proceedings, 7th International Conference on Ultra-Relativistic Nucleus-Nucleus Collisions, Lenox, USA, September 26–30, 1988," *Nucl. Phys.* **A498** (1989) 1c–628c.

[10] J. P. Blaizot, C. Gerschel, B. Pire, and A. Romana, "Quark Matter '90 Proceedings, 8th International Conference on Ultra-Relativistic Nucleus-Nucleus Collisions, Menton, France, May 7–11, 1990," *Nucl. Phys.* **A525** (1991).

[11] T. C. Awes, F. E. Obenshain, F. Plasil, M. R. Strayer, and C. Y. Wong, "Quark Matter '91 Proceedings, 9th International Conference on Ultra-Relativistic Nucleus-Nucleus Collisions, Gatlinburg, USA, November 11–15, 1991," *Nucl. Phys.* **A544** (1992).

[12] E. Stenlund, H. A. Gustafsson, A. Oskarsson, and I. Otterlund, "Quark Matter '93 Proceedings, 10th International Conference on Ultrarelativistic Nucleus-Nucleus Collisions, Borlaenge, Sweden, June 20–24, 1993," *Nucl. Phys.* **A566** (1994).

[13] A. M. Poskanzer, J. W. Harris, and L. S. Schroeder, "Quark Matter '95 Proceedings, 11th International Conference on Ultra-Relativistic Nucleus-Nucleus Collisions, Monterey, USA, January 9–13, 1995," *Nucl. Phys.* **A590** (1995).

[14] P. Braun-Munzinger, H. J. Specht, R. Stock, and H. Stoecker, "Quark Matter '96 Proceedings, 12th International Conference on Ultra-Relativistic Nucleus-Nucleus Collisions, Heidelberg, Germany, May 20–24, 1996," *Nucl. Phys.* **A610** (1996).

[15] T. Hatsuda, Y. Miake, K. Yagi, and S. Nagamiya, "Quark Matter '97 Proceedings, 13th International Conference on Ultra-Relativistic Nucleus-Nucleus Collisions, Tsukuba, Japan, December 1–5, 1997," *Nucl. Phys.* **A638** (1998).

[16] L. Riccati, M. Masera, and E. Vercellin, "Quark Matter '99 Proceedings, 14th International Conference on Ultra-Relativistic Nucleus-Nucleus Collisions, Torino, Italy, May 10–15, 1999," *Nucl. Phys.* **A661** (1999) 1–765.

[17] T. J. Hallman, D. E. Kharzeev, J. T. Mitchell, and T. S. Ullrich, "Quark Matter '01

Proceedings, 15th International Conference on Ultra-Relativistic Nucleus-Nucleus Collisions, Stony Brook, USA, January 15–20, 2001," *Nucl. Phys.* **A698** (2002).
[18] H. Gutbrod, J. Aichelin, and K. Werner, "Quark Matter '02 Proceedings, 16th International Conference on Ultra-Relativistic Nucleus-Nucleus Collisions, Nantes, France, July 18–24, 2002," *Nucl. Phys.* **A715** (2003).
[19] H. G. Ritter and X. N. Wang, "Quark Matter '04 Proceedings, 17th International Conference on Ultra-Relativistic Nucleus-Nucleus Collisions, Oakland, USA, January 11–17, 2004," *J. Phys.* **G30** (2004).
[20] T. Csorgo, P. Levai, G. David, and G. Papp, "Quark Matter '05 Proceedings, 18th International Conference on Ultra-Relativistic Nucleus-Nucleus Collisions, Budapest, Hungary, August 4–9, 2005," *Nucl. Phys.* **A774** (2006).
[21] Y.-G. Ma *et al.*, "Quark Matter '06 Proceedings, 19th International Conference on Ultra-Relativistic Nucleus-Nucleus Collisions, Shanghai, P.R. China, November 14–20, 2006," *J. Phys.* **G34** (2007).
[22] J. Alam *et al.*, "Quark Matter '08 Proceedings, 20th International Conference on Ultra-Relativistic Nucleus-Nucleus Collisions, Jaipur, India, February 4–10, 2008," *J. Phys.* **G35** (2008).
[23] **PHOBOS** Collaboration, B. B. Back *et al.*, "Centrality and energy dependence of charged-particle multiplicities in heavy ion collisions in the context of elementary reactions," *Phys. Rev.* **C74** (2006) 021902.
[24] M. Gell-Mann, "A schematic model of baryons and mesons," *Phys. Lett.* **8** (1964) 214–215.
[25] O. W. Greenberg, "Spin and unitary spin independence in a paraquark model of baryons and mesons," *Phys. Rev. Lett.* **13** (1964) 598–602.
[26] M. Y. Han and Y. Nambu, "Three-triplet model with double SU(3) symmetry," *Phys. Rev.* **139** (1965) B1006–B1010.
[27] D. J. Gross and F. Wilczek, "Ultraviolet behavior of non-abelian gauge theories," *Phys. Rev. Lett.* **30** (1973) 1343–1346.
[28] D. J. Gross and F. Wilczek, "Asymptotically free gauge theories. 1," *Phys. Rev.* **D8** (1973) 3633–3652.
[29] H. D. Politzer, "Reliable perturbative results for strong interactions?," *Phys. Rev. Lett.* **30** (1973) 1346–1349.
[30] **CTEQ** Collaboration, R. Brock *et al.*, "Handbook of perturbative QCD: Version 1.0," *Rev. Mod. Phys.* **67** (1995) 157–248.
[31] K. G. Wilson, "Confinement of quarks," *Phys. Rev.* **D10** (1974) 2445–2459.
[32] G. 't Hooft, "On the phase transition towards permanent quark confinement," *Nucl. Phys.* **B138** (1978) 1.
[33] G. 't Hooft, "Topology of the gauge condition and new confinement phases in non-abelian gauge theories," *Nucl. Phys.* **B190** (1981) 455.
[34] Millennium Problems, `http://www.claymath.org/millennium/`.
[35] F. J. Ynduráin, "Quantum Chromodynamics," (Springer Verlag, New York, 1983).
[36] K. Huang, "Quarks, Leptons, and Gauge Fields," (World Scientific, Singapore, 1982).
[37] N. Cabibbo and G. Parisi, "Exponential hadronic spectrum and quark liberation," *Phys. Lett.* **B59** (1975) 67.
[38] J. C. Collins and M. J. Perry, "Superdense matter: neutrons or asymptotically free quarks?," *Phys. Rev. Lett.* **34** (1975) 1353.
[39] M. B. Kislinger and P. D. Morley, "Collective phenomena in gauge theories. 1. The plasmon effect for Yang-Mills fields," *Phys. Rev.* **D13** (1976) 2765.
[40] M. B. Kislinger and P. D. Morley, "Collective phenomena in gauge theories. 2.

Renormalization in finite temperature field theory," *Phys. Rev.* **D13** (1976) 2771.
[41] B. A. Freedman and L. D. McLerran, "Fermions and gauge vector mesons at finite temperature and density. 1. Formal techniques," *Phys. Rev.* **D16** (1977) 1130.
[42] B. A. Freedman and L. D. McLerran, "Fermions and gauge vector mesons at finite temperature and density. 2. The ground state energy of a relativistic electron gas," *Phys. Rev.* **D16** (1977) 1147.
[43] B. A. Freedman and L. D. McLerran, "Fermions and gauge vector mesons at finite temperature and density. 3. The ground state energy of a relativistic quark gas," *Phys. Rev.* **D16** (1977) 1169.
[44] E. V. Shuryak, "Quark-gluon plasma and hadronic production of leptons, photons and pions," *Phys. Lett.* **B78** (1978) 150.
[45] E. V. Shuryak, "Quantum chromodynamics and the theory of superdense matter," *Phys. Rept.* **61** (1980) 71–158.
[46] J. I. Kapusta, "Quantum chromodynamics at high temperature," *Nucl. Phys.* **B148** (1979) 461–498.
[47] S. A. Chin, "Transition to hot quark matter in relativistic heavy ion collision," *Phys. Lett.* **B78** (1978) 552–555.
[48] CERN Press Release Feb. 10, 2000, `http://press.web.cern.ch/press/PressReleases/Releases2000/PR01.00EQuarkGluonMatter.html`.
[49] M. G. Alford, A. Schmitt, K. Rajagopal, and T. Schafer, "Color superconductivity in dense quark matter," *Rev. Mod. Phys.* **80** (2008) 1455–1515.
[50] **BRAHMS** Collaboration, I. Arsene *et al.*, "Quark gluon plasma and color glass condensate at RHIC? The perspective from the BRAHMS experiment," *Nucl. Phys.* **A757** (2005) 1–27.
[51] **PHENIX** Collaboration, K. Adcox *et al.*, "Formation of dense partonic matter in relativistic nucleus nucleus collisions at RHIC: Experimental evaluation by the PHENIX collaboration," *Nucl. Phys.* **A757** (2005) 184–283.
[52] **PHOBOS** Collaboration, B. B. Back *et al.*, "The PHOBOS perspective on discoveries at RHIC," *Nucl. Phys.* **A757** (2005) 28–101.
[53] **STAR** Collaboration, J. Adams *et al.*, "Experimental and theoretical challenges in the search for the quark gluon plasma: The STAR collaboration's critical assessment of the evidence from RHIC collisions," *Nucl. Phys.* **A757** (2005) 102–183.
[54] M. Gyulassy and L. McLerran, "New forms of QCD matter discovered at RHIC," *Nucl. Phys.* **A750** (2005) 30–63.
[55] E. V. Shuryak, "What RHIC experiments and theory tell us about properties of quark-gluon plasma?," *Nucl. Phys.* **A750** (2005) 64–83.
[56] J.-P. Blaizot, "Theoretical overview: Towards understanding the quark- gluon plasma," *J. Phys.* **G34** (2007) S243–252.
[57] Y. Nambu, "Axial vector current conservation in weak interactions," *Phys. Rev. Lett.* **4** (1960) 380–382.
[58] Y. Nambu and G. Jona-Lasinio, "Dynamical model of elementary particles based on an analogy with superconductivity. I," *Phys. Rev.* **122** (1961) 345–358.
[59] Y. Nambu and G. Jona-Lasinio, "Dynamical model of elementary particles based on an analogy with superconductivity. II," *Phys. Rev.* **124** (1961) 246–254.
[60] J. Goldstone, "Field theories with superconductor solutions," *Nuovo Cim.* **19** (1961) 154–164.
[61] J. Goldstone, A. Salam, and S. Weinberg, "Broken symmetries," *Phys. Rev.* **127** (1962) 965–970.
[62] **Particle Data Group** Collaboration, C. Caso *et al.*, "Review of particle physics,"

Eur. Phys. J. **C3** (1998) 1–794.
[63] J. B. Kogut *et al.*, "Deconfinement and chiral symmetry restoration at finite temperatures in SU(2) and SU(3) gauge theories," *Phys. Rev. Lett.* **50** (1983) 393.
[64] **RBC** Collaboration, F. Karsch, "Equation of state and more from lattice regularized QCD," *J. Phys.* **G35** (2008) 104096.

Chapter 2

Basic Dictionary

In this Chapter we introduce the basic terminology used to describe the geometry and kinematics of heavy-ion collisions. We define commonly used concepts such as: participants, spectators, transverse mass, rapidity, pseudorapidity, wounded nucleons. We present also different methods of the determination of the reaction plane and discuss the phenomena of collective flow, stopping, and transparency.

2.1 Participants, spectators, and impact parameter

In the ultra-relativistic heavy-ion collisions, the energy per nucleon in the center-of-mass frame is much larger than the nucleon mass. At such very high energies, simple geometric concepts are often used. For example, one separates so called *participants* from *spectators*, see Fig. 2.1. If we assume that all nucleons propagate along parallel, straight line trajectories, then the nucleons which do not meet any other nucleons on their way are called spectators (there can be target and projectile spectators in a collision). Other nucleons which interact with each other are called participants. The participants which suffered at least one inelastic collision are called the *wounded nucleons*. Since the inelastic processes dominate at very high energies, very often the terms "participants" and "wounded nucleons" are regarded as synonyms. The precise definition of the wounded nucleons, in a reference to the Glauber model of multiple scattering processes, will be given later in Sec. 3.5. We note that at the considered very high beam energies the binding energy of the nucleons in nuclei as well as the energies of the excited nuclear states may be ignored, hence, only the spatial distribution of the nucleons in the nuclei and the value of the nucleon-nucleon cross section at a given beam energy have relevance for the outcome of the collision.

A two-dimensional vector connecting centers of the colliding nuclei in the plane transverse to the nucleon trajectories is called the *impact vector*, and its length is the *impact parameter*. In particle as well as in nuclear physics it is practical to introduce a coordinate system, where the spatial z-axis is parallel to the beam of the accelerator, and where the impact vector $\mathbf{b}$ points in x-direction. The two axes,

x and z, span the *reaction plane* of a given collision.

The very important class of *central collisions* corresponds to the zero impact parameter [1]. The measurements averaged over different impact parameters are called the *minimum-bias* data. The value of the impact parameter determines the number of the participants, $N_{\rm part}$, as well as the number of the spectators, $N_{\rm spec}$. An estimate of $N_{\rm part}$ allows us to compare proton-nucleus (pA) and nucleus-nucleus (AA) results to pp data by means of a simple rescaling (there are obviously two participating nucleons in a pp collision). This facilitates also comparisons of different heavy-ion collisions. Experimentally, the value of $N_{\rm spec}$ may be inferred from the measurement of the energy deposited in the zero-degree calorimeter (ZDC). At RHIC, each experiment is equipped with a pair of such calorimeters which are placed close to the beam line but far away from the center of the interaction region (at the distance of approximately 18 m). Being placed behind the dipole magnets which sweep away charged particles, ZDC's measure the energy of the spectator neutrons only. Coincidences between the ZDC counters serve also as the trigger for the collision events.

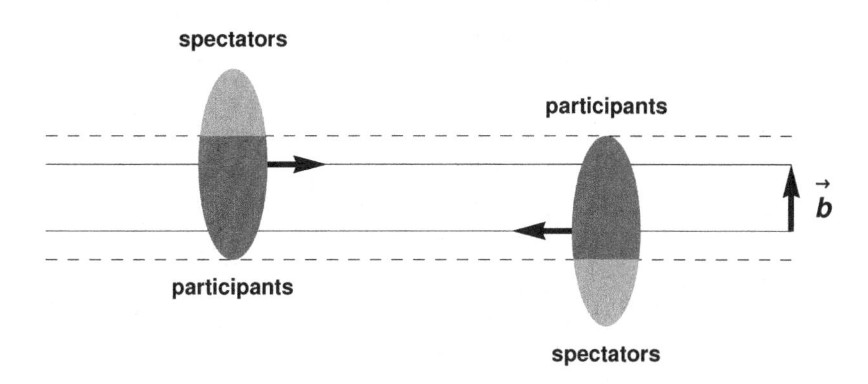

Fig. 2.1 Participants (darker regions) and spectators (lighter regions) in a nuclear collision. The impact vector is denoted by $\mathbf{b}$. The two colliding nuclei are Lorentz contracted. For the relativistic energy $\sqrt{s_{\rm NN}} = 5$ GeV the Lorentz gamma factor is 2.7 (this corresponds roughly to the case shown in the figure). For the LHC energy $\sqrt{s_{\rm NN}} = 5.5$ TeV the gamma factor will reach 2700 (!). At such large energies, the velocities of the nuclei are practically equal to the speed of light.

[1] In practice one considers a group of events which are characterized by the smallest values of the impact parameter. The concept of centrality is discussed more thoroughly in Sec. 2.3.

2.2 Kinematical variables

2.2.1 *Transverse mass*

The component of a three-vector $\mathbf{A}$ parallel to z-axis is usually denoted by $\mathbf{A}_{\parallel}$, and the transverse component is $\mathbf{A}_{\perp} = \mathbf{A} - \mathbf{A}_{\parallel}$. The transverse mass of a particle is defined as

$$m_{\perp} = \sqrt{m^2 + \mathbf{p}_{\perp}^2}, \tag{2.1}$$

where m and $\mathbf{p}$ are the particle's mass and three-momentum, respectively [2]. Clearly, the transverse mass is invariant under Lorentz boosts along the beam axis. The measured $m_{\perp}$-distribution of the produced particles is typically of the exponential form (for not too large transverse momenta, $p_{\perp} < 1$–2 GeV)

$$\frac{dN}{2\pi\, m_{\perp}\, dm_{\perp}} = A \exp\left(-m_{\perp}/\lambda\right). \tag{2.2}$$

The two parameters A and λ are obtained from the fits to the experimental data [3]. A typical value of λ for pions measured at the SPS and RHIC is about 200 MeV. Similarity of the distribution (2.2) to the Boltzmann thermal factor is the reason why λ is very often described as an effective temperature of the spectrum. It is also called, more appropriately, the *inverse slope parameter.*

The form of Eq. (2.2) reflects the experimental fact that most of the produced particles have small transverse momenta: for $\lambda = 200$ MeV the average transverse momentum is 400 MeV and about 70% of the produced pions have transverse momenta smaller than 500 MeV. An intriguing similarity of the experimental spectra to the Boltzmann distribution was first realized by Hagedorn. We discuss his ideas in more detail in Sec. 6.2.

2.2.2 *Rapidity and pseudorapidity*

Since we deal with relativistic energies, it is useful to use the *rapidity* instead of the standard velocity. It is defined by the equation

$$\mathrm{y} = \frac{1}{2} \ln \frac{(E + p_{\parallel})}{(E - p_{\parallel})} = \operatorname{arctanh}\left(\frac{p_{\parallel}}{E}\right) = \operatorname{arctanh}\left(\mathrm{v}_{\parallel}\right). \tag{2.3}$$

Here E is the energy of a particle, $E = \sqrt{m^2 + \mathbf{p}^2}$, and $\mathrm{v}_{\parallel} = p_{\parallel}/E$ is the longitudinal component of the velocity. Rapidity is additive under Lorentz boosts along the

[2] The "transverse" quantities are sometimes denoted by the subscript T, e.g., m_T or p_T. The "longitudinal" quantities are then denoted by the subscript L, e.g., p_L.

[3] The $m_{\perp}$-dependent inverse slope parameter λ is obtained from the measured spectrum with the help of the formula

$$\lambda = -\left[\frac{d}{dm_{\perp}} \ln\left(\frac{dN}{m_{\perp} dm_{\perp}}\right)\right]^{-1}.$$

The observed weak dependence of λ on $m_{\perp}$ for $p_{\perp} < 1$–2 GeV speaks for usefulness of the formula (2.2).

z-axis (checking of this and other properties of the rapidity variable is left as Ex. 7.6). Using the rapidity and the transverse mass, we can calculate the energy and the longitudinal momentum of a particle from the equations

$$E = p^0 = m_\perp \cosh \mathrm{y} \tag{2.4}$$

and

$$p_{\|} = m_\perp \sinh \mathrm{y}. \tag{2.5}$$

In the similar way one defines the pseudorapidity variable η, namely

$$\eta = \frac{1}{2} \ln \frac{(|\mathbf{p}| + p_{\|})}{(|\mathbf{p}| - p_{\|})} = \ln \left(\cot \frac{\theta}{2} \right) = - \ln \left(\tan \frac{\theta}{2} \right), \tag{2.6}$$

where θ is the scattering angle. In analogy to Eqs. (2.4) and (2.5) we have

$$|\mathbf{p}| = p_\perp \cosh \eta \tag{2.7}$$

and

$$p_{\|} = p_\perp \sinh \eta. \tag{2.8}$$

After simple manipulations we also find

$$\sin \theta = \frac{1}{\cosh \eta}. \tag{2.9}$$

In the limit of small hadron masses, $m \longrightarrow 0$, the rapidity and the pseudorapidity become equal. For finite masses the relations between the rapidity and the pseudorapidity are more complicated

$$\mathrm{y} = \frac{1}{2} \ln \left[\frac{\sqrt{p_\perp^2 \cosh^2 \eta + m^2} + p_\perp \sinh \eta}{\sqrt{p_\perp^2 \cosh^2 \eta + m^2} - p_\perp \sinh \eta} \right], \tag{2.10}$$

$$\eta = \frac{1}{2} \ln \left[\frac{\sqrt{m_\perp^2 \cosh^2 \mathrm{y} - m^2} + m_\perp \sinh \mathrm{y}}{\sqrt{m_\perp^2 \cosh^2 \mathrm{y} - m^2} - m_\perp \sinh \mathrm{y}} \right]. \tag{2.11}$$

Equations (2.10) or (2.11) can be used to find a connection between the rapidity distribution of particles and the pseudorapidity distribution

$$\frac{dN}{d\eta\, d^2p_\perp} = \sqrt{1 - \frac{m^2}{m_\perp^2 \cosh^2 \mathrm{y}}} \frac{dN}{d\mathrm{y}\, d^2p_\perp} = \frac{|\mathbf{p}|}{E} \frac{dN}{d\mathrm{y}\, d^2p_\perp}. \tag{2.12}$$

In the center-of-mass frame, the region of the phase-space where $\mathrm{y} \approx \eta \approx 0$ is called the *central rapidity region* or the *midrapidity region*. On the other hand, the regions corresponding to the initial rapidities of the projectile and target ($\mathrm{y} \approx \mathrm{y}_P$, $\mathrm{y} \approx \mathrm{y}_T$) are called the projectile and target *fragmentation regions*, respectively. Particle production in the central region is of special interest, since the particles with zero rapidity are either the new particles created during the collision process or they are particles already present in the beams but undergoing several rescattering processes which substantially change their initial purely longitudinal momenta. In the central region, the relation between the pseudorapidity distribution and the rapidity distribution, Eq. (2.12), involves only the simple factor

$$\left.\frac{dN}{d\eta\, d^2p_\perp}\right|_{\eta=0} = \frac{p_\perp}{m_\perp} \left.\frac{dN}{d\mathrm{y}\, d^2p_\perp}\right|_{\mathrm{y}=0}. \tag{2.13}$$

The factor $p_\perp/m_\perp$ is nothing else but the transverse velocity of a particle. It is always smaller than one and makes the pseudorapidity distribution smaller than the corresponding rapidity distribution (at $\mathrm{y} \approx 0$). This effect is very often responsible for flattening of the pseudorapidity distributions — the experimental Gaussian in y, after multiplication by $p_\perp/m_\perp$, becomes a flat pseudorapidity distribution [1–3].

2.2.3 *Light-cone variable x*

For *inclusive experiments* [4] $a + b \to c + X$ we define the light-cone variable x as

$$x = \frac{E^c + p_\parallel^c}{E^a + p_\parallel^a}. \tag{2.14}$$

We find easily the following relation between x and y

$$x = \frac{m_\perp^c}{m^a} \exp\left(\mathrm{y} - \mathrm{y}^a\right), \tag{2.15}$$

or conversely

$$\mathrm{y} = \mathrm{y}^a + \ln x - \ln \frac{m_\perp^c}{m^a}. \tag{2.16}$$

Equation (2.15) shows that x is invariant under Lorentz boosts along the beam axis.

One may check, for the explicit calculation see for example [5], that for $x \gg 0$ and high center-of-mass energies, the light-cone variable coincides with the Feynman scaling variable x_F

$$x_F = \frac{p_\parallel^c}{p_{\mathrm{max}}^c}. \tag{2.17}$$

We note that $p_\parallel^c$ and p_{max}^c in the definition (2.17) are measured in the center-of-mass frame.

2.2.4 *Experimental rapidity and transverse-momentum distributions*

In this Section we give first examples of the data describing hadron production in ultra-relativistic heavy-ion collisions. The upper part of Fig. 2.2 shows the rapidity

[4] In the inclusive experiments of the type discussed here X is "anything", i.e., an arbitrary number of particles produced in addition to the particle c. The concept of exclusive and inclusive experiments was introduced by Feynman in [4].

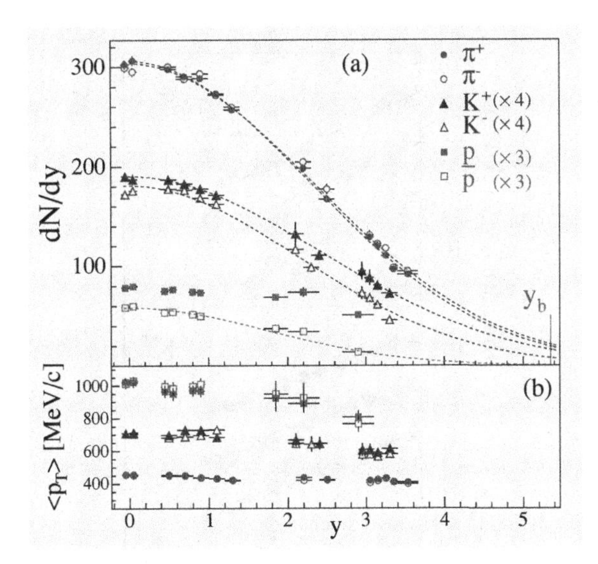

Fig. 2.2 The measurement of the BRAHMS Collaboration at BNL (Au+Au collisions at $\sqrt{s_{\rm NN}} = 200$ GeV, the most central events). Rapidity distributions **(a)** and average transverse momenta **(b)** of charged pions, charged kaons, protons and antiprotons. Reprinted figure with permission from [6]. Copyright (2009) by the American Physical Society.

distribution of pions, kaons, protons, and antiprotons measured by the BRAHMS Collaboration at BNL in the most central Au+Au collisions at $\sqrt{s_{\rm NN}} = 200$ GeV [6]. The lower part of Fig. 2.2 shows the corresponding average transverse momenta of the hadrons. One of the goals of the BRAHMS experiment was to study a broad rapidity range, not covered by other experiments [5]. The rapidity distributions measured by BRAHMS are well described by rather broad Gaussians centered at $y = 0$. One can see that the numbers of positive and negative pions are almost equal in the whole range of the rapidity. On the other hand, the numbers of positive kaons and protons are significantly larger than the numbers of negative kaons and antiprotons, respectively. These differences grow with the rapidity of hadrons. The average transverse momenta of hadrons at $y = 0$ are: 450 MeV for pions, 700 MeV for kaons, and 1 GeV for protons. For $y > 1$ the average momenta slowly decrease. We may conclude that the results delivered by BRAHMS show small variations of the physical quantities within one unit of rapidity around $y = 0$, which may be interpreted as the onset of the boost-invariance in this region, see Sec. 2.7. For $y > 1$, the physical observables clearly change with respect to their values at $y = 0$.

In Fig. 2.3 we show the transverse-mass spectra of pions and kaons measured at midrapidity in the central Pb+Pb collisions at the energy $E_{\rm lab} = 40$ A GeV (triangles), 80 A GeV (squares), and 158 A GeV (circles). This measurement was

[5] Large values of pseudorapidity are also available for PHOBOS, however without possibility of particle identification.

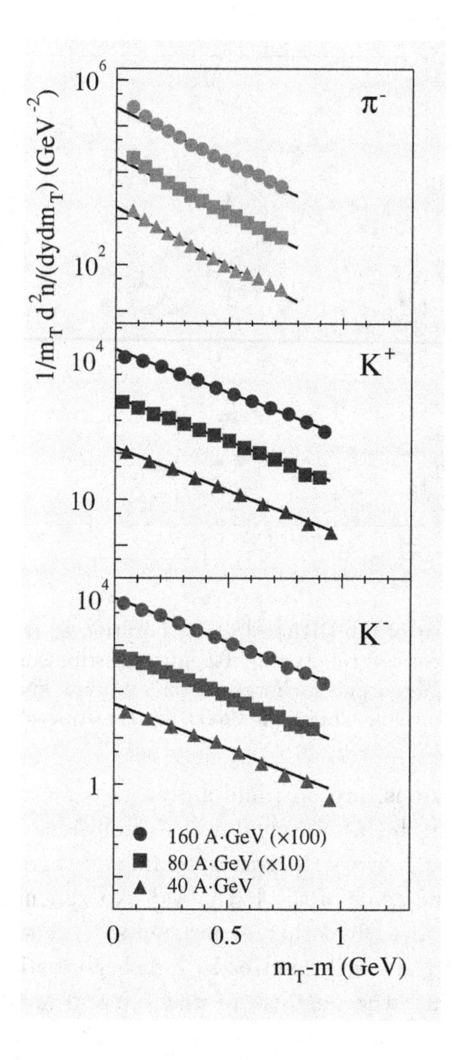

Fig. 2.3 The measurements done by the NA49 Collaboration at CERN. The transverse-mass spectra of π^-, K^+, and K^- at midrapidity ($|y| < 0.1$ for kaons and $0 < y < 0.2$ for pions) in the central Pb+Pb collisions at the energy $E_{\rm lab} = 40$ A GeV (triangles), 80 A GeV (squares), and 158 A GeV (circles). The lines are the exponential fits to the spectra in the interval 0.2 GeV $< m_T - m <$ 0.7 GeV. The values for 80 A GeV and 158 A GeV are rescaled by the factors of 10 and 100, respectively. Reprinted figure with permission from [7]. Copyright (2009) by the American Physical Society.

done by the NA49 Collaboration at CERN [7]. We observe a characteristic fall-off of the spectra, which is very well described by the exponential formula (2.2). The fitted values of the inverse slope of pions are: $\lambda = 169$ MeV, 179 MeV, and 180 MeV for the beam energies: 40 A GeV, 80 A GeV, and 158 A GeV, respectively. Figure 2.4 shows the rapidity distributions measured for the same colliding systems [7].

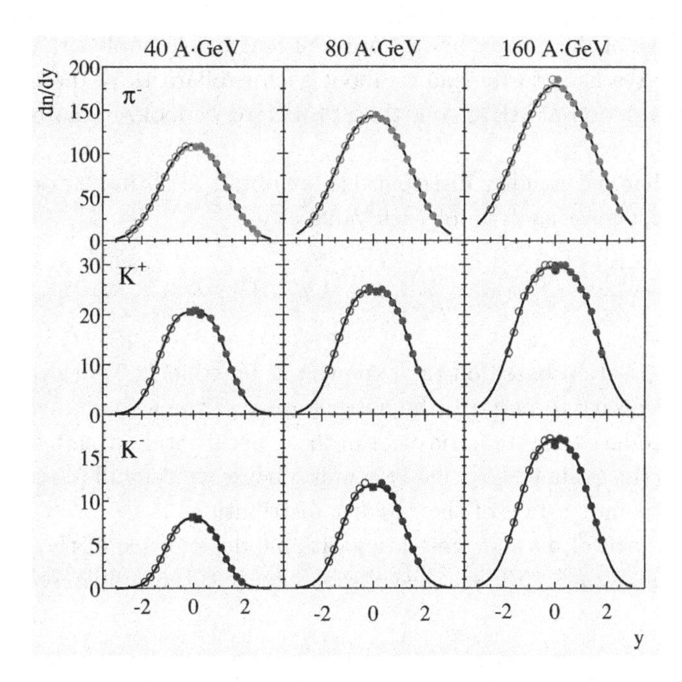

Fig. 2.4 The measurements done by the NA49 Collaboration at CERN. The rapidity distributions of π^-, K^+, and K^- in the central Pb+Pb collisions at the energy $E_{\rm lab} = 40$ A GeV, 80 A GeV, and 158 A GeV. The closed symbols indicate the measured points, whereas the open points are reflection of the measured points with respect to the axis y = 0. Reprinted figure with permission from [7]. Copyright (2009) by the American Physical Society.

We observe that most of particles are produced at midrapidity, and the range of the rapidity distribution is determined by the rapidity of the beam

$$\mathrm{y}_{\rm lab}^{\rm beam} = \mathrm{arccosh}\left(\frac{E_{\rm lab}}{A\, m_{\rm N}}\right). \tag{2.18}$$

Equation (2.18) yields (with $m_{\rm N}$ denoting the nucleon mass) $\mathrm{y}_{\rm lab}^{\rm beam} \approx 5.8$ for $E_{\rm lab} = 158$ A GeV, and $\mathrm{y}_{\rm lab}^{\rm beam} \approx 4.4$ for $E_{\rm lab} = 40$ A GeV. The corresponding range in the RHIC experiments at the maximum energy of 200 GeV is $2 \times \mathrm{arccosh}(100\,\mathrm{GeV}/m_{\rm N}) = 2 \times 5.36 = 10.7$ [6].

Although the center-of-mass energy available at RHIC is one order of magnitude larger than the maximum SPS energy, the rapidity distributions measured by BRAHMS are rather smooth extrapolation of the results obtained at CERN. In particular, the pion density at midrapidity increased only by 50% (from about 200 to about 300 for negative or positive pions). If the particles were produced uniformly in the firecylinder characterized by the transverse radius R, this would mean

[6] Note that RHIC is a collider and the value 10.7 corresponds to the measurement done in a hypothetical reference frame connected with one of the nuclei.

that the change of the energy by one order of magnitude implies the increase of R by 20% only. We have to remember about such similarities in the global features of hadron production at RHIC and the SPS before we look, in more detail, at the differences.

We note that the rapidity distributions are obtained by integration of the measured transverse-mass spectra for each value of y

$$\frac{dN}{dy} = \int d^2p_\perp \frac{dN}{dy\, d^2p_\perp} = \int_m^\infty m_\perp\, dm_\perp \frac{dN}{dy\, m_\perp\, dm_\perp}. \tag{2.19}$$

Here an extrapolation based on the exponential function (2.2) or other reasonable function is necessary to perform the integration over the whole range of $m_\perp$. This procedure introduces a systematic error in the experimental estimates of the hadron yields (also in the estimates of the mean multiplicities in the full phase space, which are obtained by integration of the rapidity distributions).

The experimental data suggest a logarithmic dependence of the particle pseudorapidity density, $dN_{\rm ch}/d\eta$, on the collision energy (for example, see [8, 9])

$$\left.\frac{dN_{\rm ch}}{d\eta}\right|_{\eta=0} \propto \ln\sqrt{s_{\rm NN}}. \tag{2.20}$$

Since the width of the rapidity distribution also increases with the energy as $\ln\sqrt{s_{\rm NN}}$ [7], we expect a logarithmic-squared dependence of the total multiplicity

$$N_{\rm ch} \propto \ln^2\sqrt{s_{\rm NN}}. \tag{2.21}$$

2.3 Centrality

So far we have interpreted the most central collisions as those corresponding to the smallest values of the impact parameter. In this Section we are going to introduce the quantitative measure of the centrality and relate it directly to the impact parameter. In the experiments with ultra-relativistic heavy ions, the centrality c is defined as the percentile of events with the largest multiplicity (as registered in detectors), or with the largest number of participants (as determined from ZDC's). We denote this number generically as n. Results of different measurements are then presented for various centrality classes. For example, in Fig. 2.5 the pseudorapidity distributions of hadrons measured by the PHOBOS Collaboration [10] were arranged into 6 different centrality classes. The centrality class 0–6% corresponds to the most central collisions characterized by the smallest values of the impact parameter and the largest values of the participating nucleons, $N_{\rm part} \sim 340$. In such collisions also

[7] This is so because the beam rapidity $y_{\rm lab}^{\rm beam}$ at high energies is proportional to $\ln\sqrt{s_{\rm NN}}$.

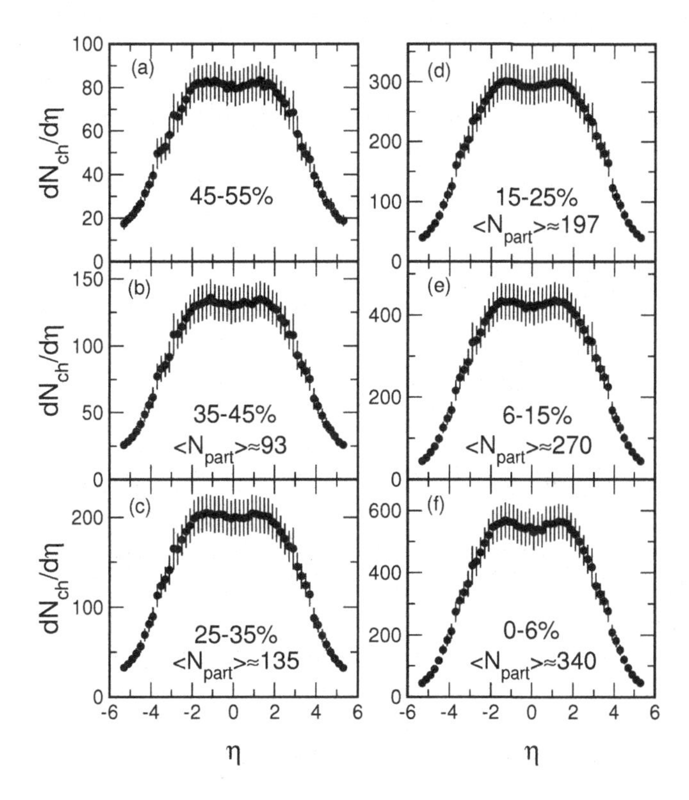

Fig. 2.5 Pseudorapidity distributions of the charged particles measured by the PHOBOS Collaboration in Au+Au collisions at $\sqrt{s_{\rm NN}} = 130$ GeV. The measurements were done for six different centrality classes. Reprinted figure with permission from [10]. Copyright (2009) by the American Physical Society.

the largest multiplicities of the produced hadrons are obtained. With the increasing centrality, the number of the participants is reduced, and already for the centrality class 34–45%, the mean number of the participants drops below 100. *Note that the large centrality corresponds to the large value of the impact parameter and, hence, to a peripheral rather than to a central collision!*

An interesting feature of the pseudorapidity distributions is the flat plateau with a small dip in the central region around the value $\eta \approx 0$. This shape is different from the shape of the rapidity distribution shown in Fig. 2.2. The reason for this difference is the Jacobian appearing in Eq. (2.12), which makes the pseudorapidity distribution at $\eta = \mathrm{y} = 0$ always smaller than the rapidity distribution [1–3].

From the experimental point of view the centrality is a convenient criterion allowing to divide the data. On the other hand, theoreticians need to assign an impact parameter, b, to a given centrality. The impact parameter is in a sense more basic, since it determines the initial geometry of the collision and appears across the formalism. Theoretical calculations in heavy-ion physics input b in order to obtain predictions. Having done the calculation, the question arises as to which centrality class the model results should be compared. For that purpose one typically applies the Glauber model in order to compute the numbers of wounded nucleons or binary collisions at a given b, which are subsequently related to multiplicities or number of participants. Since these are measured in the experiment, one is able to identify b with c.

In fact, as long as we are not interested in the effects of fluctuations such an effort is not necessary, since under general assumptions which hold very well in relativistic heavy-ion collisions, we have the relation

$$c(N) \simeq \frac{\pi b^2(N)}{\sigma_{\rm in}^{AB}}, \qquad \text{for } b < \bar{R}, \tag{2.22}$$

where $\sigma_{\rm in}^{AB}$ is the total inelastic nucleus-nucleus cross section, and $\bar{R}$ is of the order of the sum of the radii of the colliding nuclei (for identical nuclei $\bar{R} = 2R$ and $c \approx b^2/(4R^2)$). The centrality $c(N)$ is the centrality of events with the multiplicity higher than N, while $b(N)$ is the value of the impact parameter for which the average multiplicity $\bar{n}(b)$ is equal to N.

Note that the centrality c in Eq. (2.22) corresponds to the *centrality class* denoted commonly as 0–c (for example 0–10%), and the value of the impact parameter b is the *maximum* value in this class. To find a "typical" value of b in the class one may use Eq. (2.22) with c divided by 2. For the centrality classes of the form $c = c_1 - c_2$, for example $c =$ 20–40%, one may find the typical value of b by using Eq. (2.22) with c replaced by $(c_1 + c_2)/2$.

Equation (2.22) holds to a high accuracy for all but most peripheral collisions. It is geometric in nature, and does not involve explicitly the variable n needed to categorize the data (multiplicities, number of participants, number of binary collisions, etc.). At first glance, this fact may seem a bit surprising.

2.3.1 *Competition of archers...*

To illustrate the geometric nature of Eq. (2.22) we shall consider now, as an example, a competition of archers [11]. Each of them is shooting only once at a target of radius $\bar{R}$. They are rather poor, such that they shoot randomly. They are paid accordingly to their aim: more central, higher reward. We are not allowed to watch the competition, hence do not know which spot on the target has been hit, but later we review the reward records. Suppose a large number of archers scored (here we take only 10 in order to write down the results explicitly, see Fig. 2.6), and are ranked according to their prizes which are: 100\$, 50\$, 50\$, 25\$, 25\$, 25\$, 10\$, 10\$, 10\$, 10\$. The archer that got the highest prize (100\$ in this case) had to hit the bull's eye. Since he represents 10% of all archers, and they were shooting randomly, we can immediately determine (neglecting the statistical error) the radius b of the bull's eye, since 10% is the ratio of the area of the bull's eye to the total area of the target: $10\% = \pi b^2/(\pi R^2)$. Therefore $b = R\sqrt{10\%}$. Now, imagine another

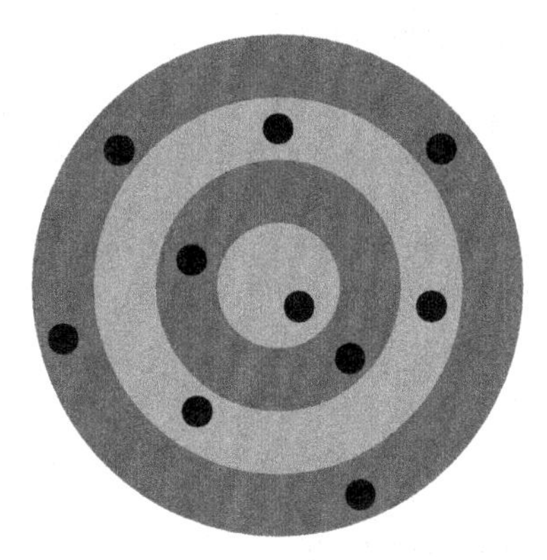

Fig. 2.6 Archers shoot randomly at the target. The black dots describe their results. Knowing the rewards given to the archers we are able to conclude about the size of the bull's-eye of the target. Similarly, in the heavy-ion experiment we can make an estimate of the size of the overlapping region of two nuclei, if the number of the produced particles (a reward in this case) is a monotonic function of this size.

competition is held, with all rules the same but the prizes differently assigned to the rings of the target. Suppose the ten archers got 500\$, 100\$, 100\$, 50\$, 50\$, 50\$, 10\$, 10\$, 10\$, 10\$. Again, we can determine that the 10% of the highest rewards correspond to hitting the central spot, and can determine its radius b exactly as before. Note that in the determination of b we are not using the actual values of the rewards at all — the function used can be any monotonic function of the centrality. The rewards are only used to *categorize* the data. Once this is done, we can identify the c "most central" archers and determine b according to Eq. (2.22), irrespectively of the function used for categorizing. Our example can be translated into heavy-ion collisions in the following way: archery competition — heavy-ion experiment, archer that scored — event, rewards in competition I — number of participants, rewards in competition II — multiplicity of produced particles, percentile of highest-scoring archers — centrality, radii of rings on the target — impact parameters.

The above example shows the essence of the argument. There are, however, two additional features which need to be considered. First, a collision at a particular impact parameter b produces values of n which are statistically distributed around some mean value $\bar{n}(b)$ with a distribution width $\Delta n(b)$. Nevertheless, Eq. (2.22), formally valid at $\Delta n(b) \ll \bar{n}(b)$, is accurate even for realistically large $\Delta n(b)$ [11], such as obtained from statistical models of particle production. Second, there are boundary effects near $b \sim \bar{R}$ — at lower values of b the inelastic cross section is the cross section for colliding *black disks*, whereas at the boundary the target gradually becomes transparent.

2.3.2 ... *and heavy-ion experiment*

We now proceed with a formal derivation. Let $P(n)$ denote the probability of obtaining value n for the categorizing function (multiplicity of produced particles, number of participants, number of binary collisions, etc.). For simplicity of the language we call it the *multiplicity*, having in mind it could be any of these quantities. The centrality c is defined as the cumulant of $P(n)$, namely

$$c(N) = \sum_{n=N}^{\infty} P(n). \tag{2.23}$$

Thus $c(N)$ is the probability of obtaining an event with multiplicity larger or equal to N. A particular value of multiplicity n may be collected from collisions with various impact parameters b', thus we can write

$$c(N) = \sum_{n=N}^{\infty} \int_0^{\infty} \frac{2\pi b' db'}{\sigma_{\text{in}}^{AB}} \rho(b') P(n|b'), \tag{2.24}$$

where $2\pi b'db'$ is the area of the ring between impact parameters b' and $b' + db'$, the quantity $\rho(b')$ is the probability of an event (inelastic collision) at impact parameter b', and $P(n|b')$ is the conditional probability of producing multiplicity n provided the impact parameter is b'. The function $\rho(b')$ is unity for b' below $\bar{R}$, and drops smoothly to zero at b' around $\bar{R}$, reflecting the washed-out shape of the nuclear density functions at the edges. The interpretation of Eq. (2.24) is clear: the probabilities for hitting the ring between b' and $b' + db'$, the probability for an event to occur at b', and the probability to produce multiplicity n (provided the event occurred at b') are multiplied, as requested by the classical nature of the problem. Since we have

$$\sum_{n=1}^{\infty} P(n|b') = 1, \tag{2.25}$$

and, by definition,

$$\int_0^{\infty} 2\pi b' db' \rho(b') = \sigma_{\text{in}}^{AB}, \tag{2.26}$$

we verify the proper normalization in Eq. (2.24), namely $c(1) = 1$. Furthermore, for heavy nuclei we may use the continuity limit,

$$\sum_{n=N}^{\infty} \to \int_N^{\infty} dn = \int_0^{\infty} dn\, \theta(n - N). \tag{2.27}$$

The function $P(n|b')$ is not known. Yet, by the statistical nature of the particle production, and by experience of various models, we expect that for large values of n it is narrowly peaked around an average value $\bar{n}(b')$. Thus we take the limit of an infinitely-narrow distribution, $P(n|b') = \delta(n - \bar{n}(b'))$. In this case

$$\begin{aligned} c(N) &= \int_0^{\infty} dn\, \theta(n - N) \int_0^{\infty} \frac{2\pi b' db'}{\sigma_{\text{in}}^{AB}} \rho(b') \delta(n - \bar{n}(b')) \\ &= \int_0^{\infty} \frac{2\pi b' db'}{\sigma_{\text{in}}^{AB}} \rho(b') \theta(\bar{n}(b') - N). \end{aligned} \tag{2.28}$$

Since $\bar{n}(b')$ is a monotonically *decreasing* function of b', we have $\theta(\bar{n}(b') - N) = \theta(b(N) - b')$, where $b(N)$ is the solution of the equation $\bar{n}(b) = N$. Therefore

$$\begin{aligned} c(N) &= \int_0^\infty \frac{2\pi b' db'}{\sigma_{\text{in}}^{AB}} \rho(b') \theta(b(N) - b') \\ &= \int_0^{b(N)} \frac{2\pi b' db'}{\sigma_{\text{in}}^{AB}} \rho(b') = \frac{\sigma_{\text{in}}^{AB}(b(N))}{\sigma_{\text{in}}^{AB}}, \end{aligned} \tag{2.29}$$

where $\sigma_{\text{in}}^{AB}(b(N))$ is the inelastic cross section accumulated from $b' \leq b(N)$. Equation (2.29) is a generalization of formula (2.22). In Ref. [12] it has been quoted in the context of the Glauber model. We notice that although c and b depend implicitly on N, their relation does not explicitly involve N.

As mentioned above our discussion has been restricted here to the case where the effects of fluctuations may be neglected. Recently, the effects of different definitions of the centrality bins on correlation and fluctuation observables in heavy-ion collision have been presented in Ref. [13]

2.4 Reaction plane

In this Section we present various experimental methods used to determine the reaction plane. Its simple geometrical concept was introduced earlier in Sec. 2.1.

2.4.1 *Sphericity method*

At low and intermediate energies the traditional method used to determine the reaction plane of a collision is based on the analysis of the kinetic-energy tensor

$$S^{ij} = \sum_{n=1}^{N} w_n p_{(n)}^i p_{(n)}^j. \tag{2.30}$$

Here $p_{(n)}^i$ is the ith component of the three-momentum $\mathbf{p}_{(n)}$ of the nth particle registered in an event consisting of N particles $(i, j = 1, 2, 3)$ [14–16]. All the measurements are done in the center-of-mass frame of the colliding nuclei. The quantities w_n are weights. In the literature one finds different forms of w_n, e.g., $w_n = 1/p_{(n)}$ and $w_n = 1/p_{(n)}^2$ in [14] or $w_n = 1/(2m_n)$ in [15, 16]. Different choices of the weights serve to optimize the determination of the reaction plane.

The tensor S^{ij} is symmetric by construction and has six independent variables. They may be identified with three eigenvalues and three Euler angles that are obtained from the diagonalization procedure. These parameters define the shape and orientation of an ellipsoid. The reaction plane is determined by the eigenvector defining the largest axis of the ellipsoid e_z and the beam direction, see Fig. 2.7.

At ultra-relativistic energies the method described above does not work — the longitudinal momenta are much larger than the transverse momenta and the eigenvector defining the largest axis coincides with the beam direction. In this case, one

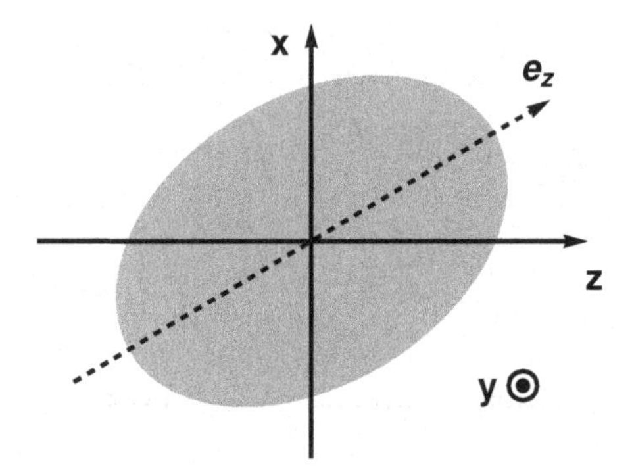

Fig. 2.7 The graphical representation of the sphericity tensor S^{ij} as an ellipsoid projected on the reaction plane. The reaction plane is determined by the beam axis z and the vector defining the largest axis of the ellipsoid e_z.

may consider the two-dimensional sphericity tensor, as proposed by Ollitrault in Refs. [17,18],

$$S_\perp^{ij} = \sum_{n=1}^{N} w_n p^i_{\perp(n)} p^j_{\perp(n)}. \tag{2.31}$$

Here the sum runs again over all particles registered in one event and $p^i_{\perp(n)}$ is the ith component of the transverse momentum vector $\mathbf{p}_{\perp(n)}$ $(i, j = 1, 2)$. Here also different choices for the weights w_n are used. In particular, $w_n \sim 1/p^2_{\perp(n)}$ in [18,19].

The tensor $S_\perp^{ij}$ has three independent components and is fully determined by its two eigenvalues s_1 and s_2 (one usually assumes that $s_1 \geq s_2$) and the angle $\Psi_{\rm RP}$ between the laboratory $x_{\rm lab}$-axis and the eigenvector associated with the eigenvalue s_1 $(-\pi/2 \leq \Psi_{\rm RP} \leq \pi/2)$, see Fig. 2.8. Instead of the eigenvalues s_1 and s_2 we may also choose the variables

$$\bar{s} = \frac{1}{2}\left(s_1 + s_2\right), \quad \Delta s = \frac{s_1 - s_2}{s_1 + s_2}. \tag{2.32}$$

Then, the two-dimensional sphericity tensor $S_\perp^{ij}$ may be written in the matrix form as

$$S_\perp = \bar{s} \begin{pmatrix} 1 + \Delta s \cos 2\Psi_{\rm RP} & \Delta s \sin 2\Psi_{\rm RP} \\ \Delta s \sin 2\Psi_{\rm RP} & 1 - \Delta s \cos 2\Psi_{\rm RP} \end{pmatrix}. \tag{2.33}$$

At ultra-relativistic energies, the physical system created initially in non-central collisions has smaller size in the direction of the impact parameter than in the perpendicular direction (when projected on the transverse plane). This effect causes that the matter expands preferentially in the direction of the impact parameter, see Sec. 2.5.2. Since the angle $\Psi_{\rm RP}$ defines the direction of the maximum kinetic-energy flux we identify this direction with the orientation of the reaction plane.

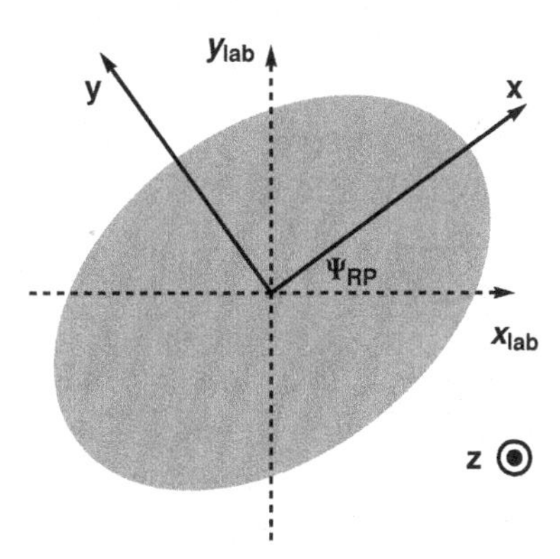

Fig. 2.8 The graphical representation of the sphericity tensor $S_{\perp}^{ij}$ as an ellipse. The angle $\Psi_{\rm RP}$ determines the direction of the x-axis with respect to the laboratory $x_{\rm lab}$-axis.

2.4.2 *Transverse-momentum method*

The transverse-momentum method was introduced by Danielewicz and Odyniec [20]. In this approach the reaction plane is defined by the direction of the beam and the vector $\mathbf{Q}$ defined by the equation

$$\mathbf{Q} = \sum_{n=1}^{N} w_n \mathbf{p}_{\perp(n)}. \tag{2.34}$$

The standard choice for the weights w_n is

$$w_n = \begin{cases} +1 & \text{if} \quad \mathrm{y} > \mathrm{y_{cms}} + \delta \\ 0 & \text{if} \quad \mathrm{y_{cms}} - \delta \le \mathrm{y} \le \mathrm{y_{cms}} + \delta \\ -1 & \text{if} \quad \mathrm{y} < \mathrm{y_{cms}} - \delta. \end{cases} \tag{2.35}$$

Here $\mathrm{y_{cms}}$ is the center-of-mass rapidity of the colliding system and the parameter δ is introduced to exclude the particles from the midrapidity region, which add random fluctuations to the calculations [21]. The positive and negative weights are necessary to obtain a non-zero signal. Due to the overall momentum conservation, the contribution of the particles with $\mathrm{y} > \mathrm{y_{cms}} + \delta$ is compensated by the contribution of the particles with $\mathrm{y} < \mathrm{y_{cms}} - \delta$. The quantity $\mathbf{Q}$ may be interpreted as the transverse-momentum transfer between the target and projectile regions.

2.4.3 *Fourier analysis method*

The sphericity and transverse-momentum methods yield comparable results, as shown for example in [16]. These two methods may be regarded as the two special cases of a more general method which is based on the Fourier expansion of the momentum distributions in the azimuthal angle. In the Fourier analysis method one considers the two-dimensional vectors $\mathbf{Q}_k = (Q_{k,x}, Q_{k,y})$ defined by the equations

$$Q_{k,x} = \sum_{n=1}^{N} w_n \cos(k\phi_p^{(n)}) \equiv Q_k \cos(k\Psi_k),$$
$$Q_{k,y} = \sum_{n=1}^{N} w_n \sin(k\phi_p^{(n)}) \equiv Q_k \sin(k\Psi_k). \tag{2.36}$$

Here $\phi_p^{(n)}$ is the azimuthal angle of the transverse momentum of the nth particle and the sum runs over N particles in an event. The weight w_n is usually taken to be the transverse energy or the transverse momentum. For odd harmonics the weights in the projectile rapidity region are positive, while in the target rapidity region they are taken with a negative sign. The orientation of the reaction plane may be calculated from the equation

$$\Psi_{\mathrm{RP}} = \Psi_1 = \arctan\left(\frac{Q_{1,y}}{Q_{1,x}}\right). \tag{2.37}$$

At ultra-relativistic energies, the practical application of Eq. (2.37) may be difficult and one uses the expression related to the second harmonics

$$\Psi_{\mathrm{RP}} = \Psi_2 = \frac{1}{2}\arctan\left(\frac{Q_{2,y}}{Q_{2,x}}\right). \tag{2.38}$$

This approach is equivalent to the use of the transverse sphericity tensor, see Eq. (2.31).

It must be emphasized that in our discussion of the methods of determination the reaction plane we have not included the effects of fluctuations. In particular, one distinguishes between the reaction plane and the *participant plane*. The participant plane is connected with the initial distribution of the participants realized in an event. Even for the fixed impact parameter such distributions may vary due to the statistical fluctuations. The experimental methods pin down the position of the participant plane rather than the position of the reaction plane. On the other hand, the reaction plane is the basic concept entering theoretical modeling where the position of the impact vector specifies the initial condition. In this situation it is clear that more precise comparing of the data with the model calculations requires that the fluctuations are taken into account.

2.5 Collective flows

At present the extraction of the reaction (participant) plane is one aspect of the very advanced *flow analysis* of the collisions. In this type of the investigations one represents the momentum distribution of the produced particles in the form

$$\frac{dN}{dy d^2p_\perp} = \frac{dN}{2\pi p_\perp dp_\perp dy}\left[1 + \sum_{k=1}^{\infty} 2v_k \cos\left(k(\phi_p - \Psi_{\mathrm{RP}})\right)\right], \tag{2.39}$$

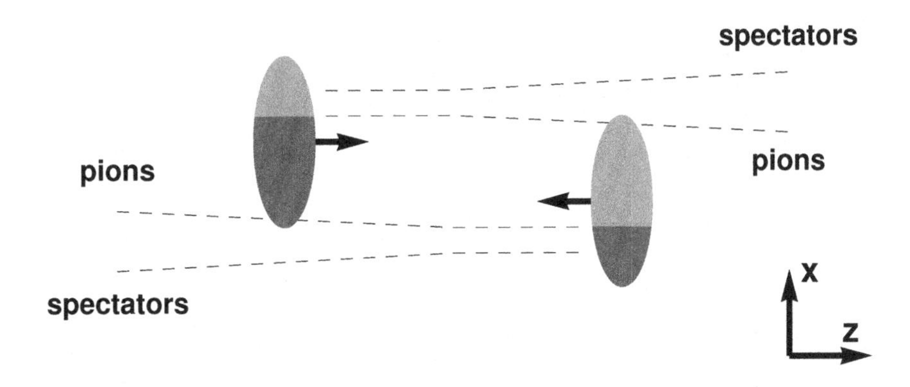

Fig. 2.9 Schematic view of the directed flow observed at relativistic energies. For positive and large rapidities ($y \sim y_P$) the spectators are deflected towards positive values of x. For positive and small rapidities ($y \geq 0$) the produced particles have negative v_1, hence they are deflected towards negative values of x.

where $\Psi_{\rm RP}$ is the angle defining the reaction plane. Clearly, the averaging of Eq. (2.39) over the azimuthal angle gives the transverse-momentum distribution Eq. (2.2). The coefficients v_k characterize the momentum anisotropy. The coefficient v_1 is called the *directed flow*, whereas the coefficient v_2 is called the *elliptic flow*. In general, the coefficients v_k are functions of rapidity and transverse momentum, $v_k = v_k(y, p_\perp)$, and in this form often called the kth harmonic *differential flow*. The *integrated flow* describes the values of v_k averaged over certain ranges of transverse momentum and rapidity. The use of the names *directed flow* or *elliptic flow* is related to the fact that the coefficients v_1 and v_2 are quantitative measures of the phenomena understood as the collective, hydrodynamic-like expansion of matter produced in heavy-ion collisions.

2.5.1 *Directed flow*

At *low* energies, the directed flow is manifested by the reflection of incoming matter by the first produced regions of highly compressed nuclear matter. The nucleons moving with positive rapidities are deflected towards positive x values, while those moving with negative rapidities are deflected towards negative x values. The magnitude of the deflection probes the compressibility of the nuclear matter. It also probes the system at early time because the deflection takes place during the passing time of the colliding nuclei. At the SPS energies the situation is already quite complex [24]. At positive rapidities the proton directed flow is positive, while the pion directed flow is negative, see Fig. 2.9. This suggests a different origin of v_1 of protons and pions. At the RHIC energies the directed flow of charged particles is negative, whereas the v_1 of the spectator neutrons is positive. This trend in the data suggests different behavior of the matter created in the central region and in

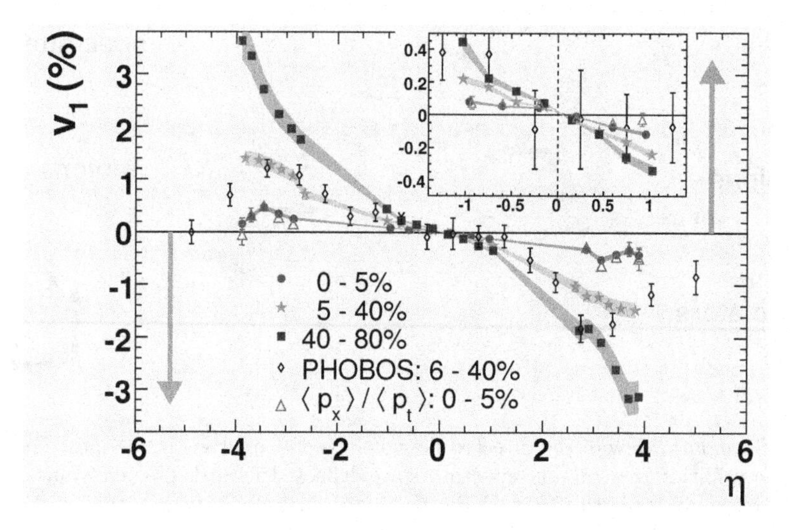

Fig. 2.10 The directed flow v_1 of charged particles as measured by STAR [22] for three centralities in Au+Au collisions at $\sqrt{s_{NN}} = 200$ GeV. The arrows indicate v_1 for spectator neutrons, and their positions on the pseudorapidity axis correspond to the beam rapidity. The smaller window shows the midrapidity region in more detail. The figure includes also the PHOBOS results [23]. Reprinted figure with permission from [22]. Copyright (2009) by the American Physical Society.

the target/projectile fragmentation regions. By the way, at midrapidity the directed flow measured at RHIC is very small, of the order of 1%, see Fig. 2.10. We note, that due to the symmetry reasons, $v_1(-\mathrm{y}) = -v_1(\mathrm{y})$, the directed flow is exactly zero at $\mathrm{y} = 0$.

2.5.2 *Elliptic flow*

The elliptic flow, whose quantitative measure is v_2, characterizes the azimuthal asymmetry of the momentum distributions. At lower energies v_2 is negative, indicating that more momentum is transferred in the direction perpendicular to the reaction plane (out-of-plane flow). At higher energies the elliptic flow becomes positive, hence the excess of the momentum is observed in the reaction plane (in-plane flow). Such energy dependence is interpreted as the effect of spectators. At lower energies the spectators block the expansion of matter in the reaction plane and the matter is squeezed out of the reaction plane. At higher energies the spectators move sufficiently fast, leaving free space for the in-plane expansion of matter.

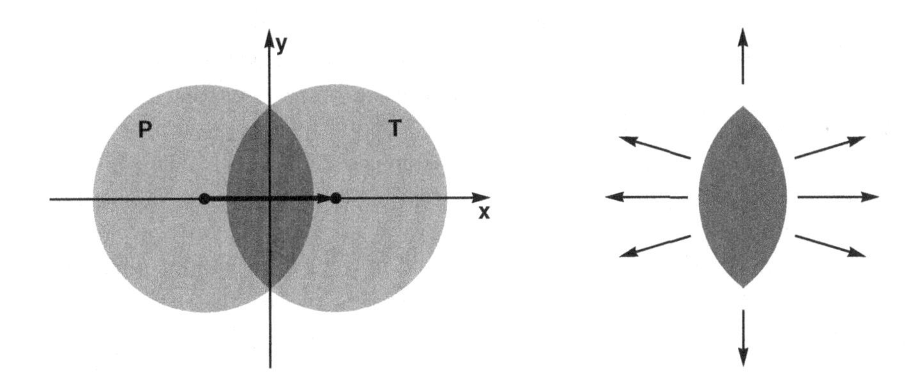

Fig. 2.11 In non-central collisions the region of the particle production has an almond shape in the transverse plane. Due to the interaction of the produced particles the spatial asymmetry leads to the azimuthal asymmetry of the momentum distributions. At ultra relativistic energies, the expansion is stronger in the reaction plane — the produced matter is not blocked by spectators.

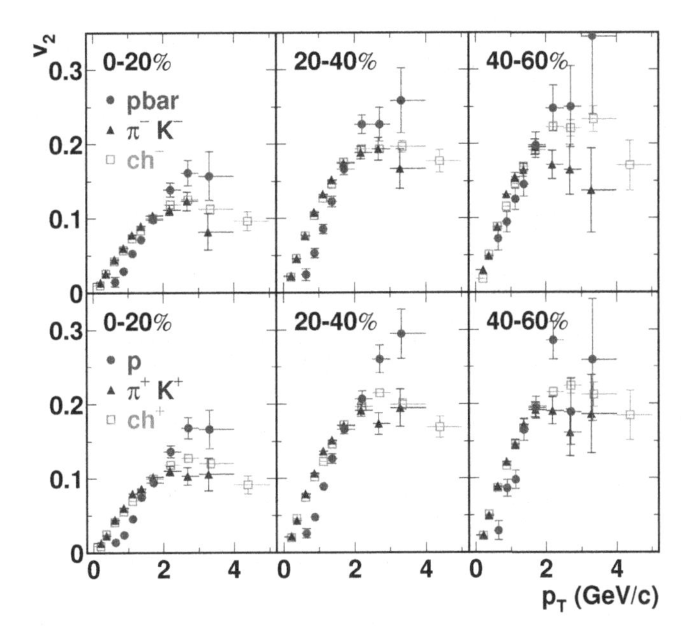

Fig. 2.12 Transverse-momentum dependence of the elliptic flow coefficient $v_2(\mathrm{y}=0)$ as measured by PHENIX [25] at $\sqrt{s_{\mathrm{NN}}} = 200$ GeV for combined π^- and K^- (top) or π^+ and K^+(bottom), and compared with $\bar{p}$ (top) and p (bottom). The results for inclusive negative (top) and positive (bottom) charged particle distributions are plotted as open squares. From left to right, the three different centrality selections are shown: 0–20%, 20–40%, and 40–60%. Reprinted figure with permission from [25]. Copyright (2009) by the American Physical Society.

The first measurements of the elliptic flow at RHIC [26] showed that v_2 is quite large and approaches the predictions of perfect hydrodynamics. Generally speaking, the origin of the non-zero elliptic flow is the interaction between particles produced in the initially asymmetric region of space — an almond-like shape formed by the two overlapping nuclei in the non-central collisions, see Fig. 2.11. The interaction between the particles in such an asymmetric region leads to the momentum anisotropy shown in Fig. 2.12.

This mechanism may be most easily interpreted in the hydrodynamic approach, where the largest pressure gradient acts in the reaction plane. In particular, the approaches based on the perfect hydrodynamics yield large elliptic flow, since perfect hydrodynamics corresponds formally to the limit where the cross sections are infinite. Thus, the agreement of the observed v_2 with the perfect hydrodynamics was interpreted as a signature of the very fast thermalization of the system and was considered as one of the most important discoveries at RHIC, which led to the concept of the strongly interacting quark-gluon plasma (sQGP).

2.6 Stopping and transparency

The relativistic heavy-ion collisions can proceed in two different ways. In the collisions with large *stopping power* the baryons from the colliding nuclei are stopped in the middle of the reaction zone, and a dense baryon-rich matter is produced at midrapidity. On the other hand, in the *transparent* collisions (negligible stopping) the initial baryons are not slowed down, and the two baryon-rich regions are

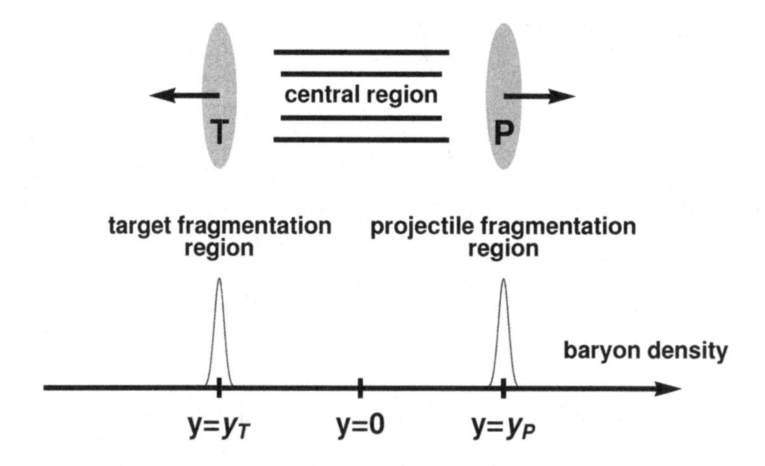

Fig. 2.13 In the transparent collisions the two colliding nuclei are not slowed down and the two baryon-rich regions are well separated from each other by the central region, where a baryon-free quark-gluon plasma is expected to occur. Here y_T is the target rapidity, and y_P is the projectile rapidity.

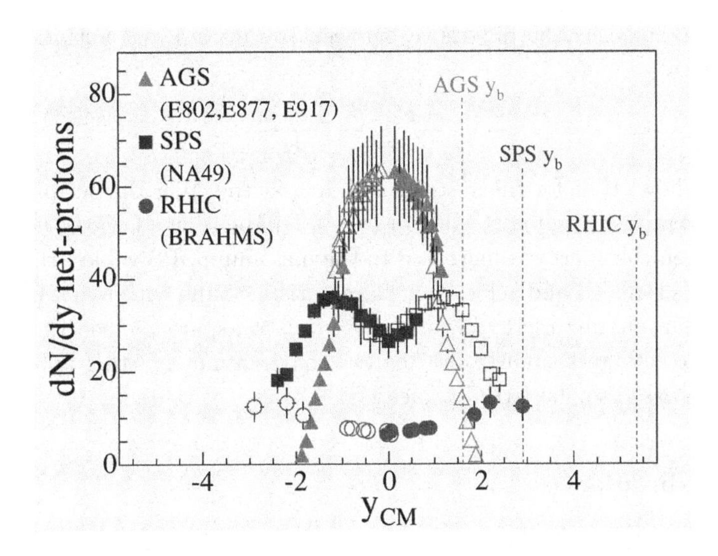

Fig. 2.14 The net-proton distributions measured in different experiments. Reprinted figure with permission from [27]. Copyright (2009) by the American Physical Society.

separated from each other (see Fig. 2.13).

In the transparent collisions, the matter produced in the central region has very small net baryon number and practically zero baryon chemical potential. In this case, the possibility of creation of the quark-gluon plasma is of special interest, since theoretical calculations for such a state of matter are especially advanced (most of the simulations of QCD on a lattice have been performed in the case of vanishing baryon chemical potential). In the collisions with large stopping power we might have the opportunity to produce the quark-gluon plasma at large values of the chemical potential and, in this way, explore the other, less known parts of the phase diagram of the strongly interacting matter.

There is a long expectation that at *extreme* large energies all collisions become transparent and the central region becomes invariant under Lorentz boosts, see our discussion in Sec. 2.7. As we shall see in the next Chapters, this symmetry is used in many models of heavy-ion collisions. It substantially simplifies theoretical calculations as it reduces the number of independent variables and dimensions.

At the AGS experiments with Au beams (Au+Au at 11.4 GeV per nucleon) the baryon distribution is approximately Gaussian with a peak in the central region and the width significantly narrower than that observed in the case of smaller Si+Al systems. This observation indicates the strong baryon stopping. On the other hand, the experiments at the SPS and RHIC show partial transparency.

In order to illustrate this behavior, in Fig. 2.14 we show the compilation done by the BRAHMS Collaboration [27], which shows the data on the net-proton distributions measured in differenet heavy-ion collisions. The net-proton rapidity distribution is defined as the difference between the proton and antiproton rapidity distributions,

$$\frac{dN_{\text{protons}}}{dy} - \frac{dN_{\text{antiprotons}}}{dy}. \tag{2.40}$$

Figure 2.14 shows that for the systems colliding at the AGS the net-proton density in the central rapidity region is quite large, ~ 60, indicating a large baryon stopping. However, when the energy is increased to the maximum SPS values, the net-proton distribution is reduced and achieves a characteristic shape with two separated maxima. This is a characteristic feature of partial transparency, see Fig. 2.13. The net-proton density gets smaller and the two maxima move farther apart, when the energy is increased to the RHIC values.

2.7 Boost-invariance

The concept of the boost-invariant character of the particle production may be traced back to the seminal paper by Feynman [4]. Feynman envisions the production of hadrons at very high energies as the radiation of some sort of the field. At high energy, due to the Lorentz transformation the radiated field is contracted to the narrow range in the z direction. This means that the Fourier transform leads to a uniform distribution of the field energy in the longitudinal momentum variable $p_\parallel$. If we assume that the field energy is distributed among various kinds of particles in fixed ratios, one may conclude that the mean number of particles of any kind, multiplied by their energy, is also distributed uniformly. We may then write

$$dN = F(p_\perp) \frac{dp_\parallel d^2 p_\perp}{E_p}. \tag{2.41}$$

Equation (2.41) is regarded by Feynman as the limiting case of the more general expression,

$$dN = f(p_\perp, x_F) \frac{dp_\parallel d^2 p_\perp}{E_p}, \tag{2.42}$$

where

$$x_F = \frac{2p_\parallel}{\sqrt{s}}, \tag{2.43}$$

with s being the Mandelstam variable and $p_\parallel$ measured in the center-of-mass frame, compare Eq. (2.17). In the limit $x_F \ll 1$ we have $f(p_\perp, x_F) \to F(p_\perp)$ and Eq. (2.42) is reduced to Eq. (2.41).

Equation (2.42) represents the *Feynman scaling.* We shall concentrate now, however, in more detail on Eq. (2.41). Introducing the rapidity variable we obtain the momentum distribution

$$\frac{dN}{dy d^2 p_\perp} = F(p_\perp), \tag{2.44}$$

which is independent of rapidity (for $p_{\|} \ll \sqrt{s}$).

Generally speaking, the boost-invariance is the symmetry of the physical systems with respect to Lorentz boosts along the beam axis. It imposes special constraints on the form of the physical quantities. For example, the thermodynamic functions used in the relativistic hydrodynamics, such as the temperature, pressure, or the energy density, are Lorentz scalars. The boost-invariance in this cases means that they may depend only on the transverse coordinates and the longitudinal proper time $\tau = \sqrt{t^2 - z^2}$. Similarly, the rapidity distribution dN/dy is boost-invariant if it is independent of rapidity.

From the formal point of view, a scalar field $\psi(x)$ has the following transformation rule

$$\psi(x) \to \psi'(x'), \qquad \psi'(x') = \psi(x), \tag{2.45}$$

where x, x' are spacetime coordinates connected by the Lorentz transformation L, namely $x' = Lx$. The scalar field is invariant under Lorentz boosts along the z axis if the transformed field in the new spacetime point x' coincides with the original field at that point,

$$\psi'(x') = \psi(x'). \tag{2.46}$$

Combining Eqs. (2.45) and (2.46), one obtains the constraint

$$\psi(x') = \psi(x), \tag{2.47}$$

which means that ψ may depend only on the transverse variables x, y and the longitudinal proper time $\tau = \sqrt{t^2 - z^2}$, as we have stated above.

It is also interesting to analyze the boost-invariant four-vector field. The general transformation rule in this case is

$$u^\mu(x) \to u'^\mu(x'), \qquad u'^\mu(x') = L^\mu{}_\nu u^\nu(x). \tag{2.48}$$

The boost-invariance demands again that the transformed "new" field u'^μ in the "new" spacetime point x' coincides with the original "old" field at this point (we illustrate this concept in Fig. 2.15, considering for simplicity rotations rather than Lorentz boosts)

$$u'^\mu(x') = u^\mu(x'). \tag{2.49}$$

Equations (2.48) and (2.49) lead to the condition

$$u^\mu(x') = L^\mu{}_\nu u^\nu(x), \tag{2.50}$$

which states that the transformed boost-invariant field may be obtained by the simple substitution of the argument, $x \to x'$.

As an example we may consider the four-vector field which describes the hydrodynamic flow of matter produced in heavy-ion collisions. With the condition that

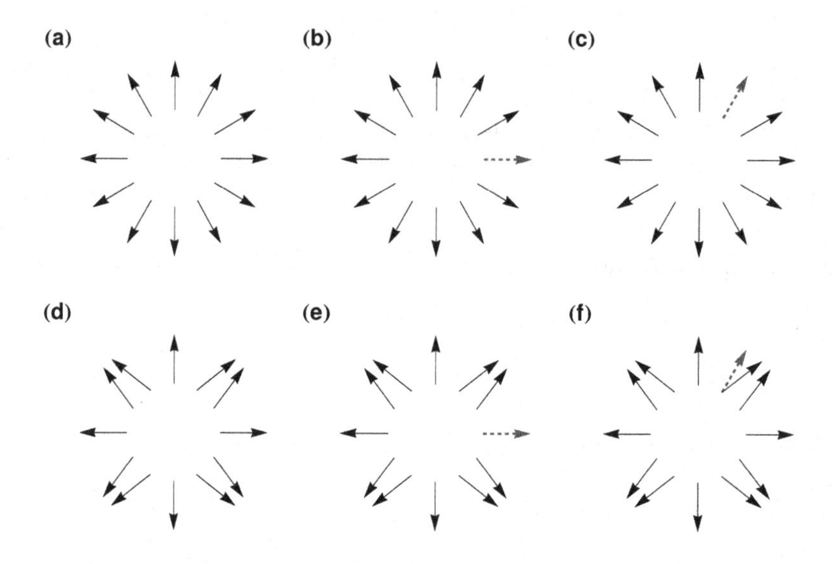

Fig. 2.15 Schematic illustration of the rotational invariance of the vector field. If the vector field is rotationally invariant, as in panel **(a)**, the rotation of a vector yields the "new" vector placed in the "new" place that coincides with the "old" vector in the same "new" place. For example, the dashed arrow in **(b)**, after the rotation by $\pi/3$, coincides with the dashed vector in **(c)**. This property does not hold for the non-symmetric field in **(d)**. The rotation of the dashed vector in **(e)** by $\pi/3$ yields the the dashed vector in **(f)**, that does not agree with the old vector at that place.

the flow is zero for $z = 0$ one may check that the boost-invariant form of such a flow is

$$u^\mu = \gamma(1, v_x, v_y, v_z) = \bar{\gamma}(\tau, x, y)\,\frac{t}{\tau}\left(1, \frac{\tau}{t}\bar{v}_x(\tau, x, y), \frac{\tau}{t}\bar{v}_y(\tau, x, y), \frac{z}{t}\right), \tag{2.51}$$

where, to fulfill the normalization condition for the four-velocity, $u^\mu u_\mu = 1$, one assumes

$$\bar{\gamma} = \frac{1}{\sqrt{1 - \bar{v}_x^2 - \bar{v}_y^2}}. \tag{2.52}$$

As may be inferred from Eq. (2.51), the functions $\bar{v}_x$, $\bar{v}_y$, and $\bar{\gamma}$ are the transverse components of the fluid velocity and the corresponding Lorentz gamma factor, all determined in the plane $z = 0$ (where also $\tau = t$). The longitudinal flow has the scaling form

$$v_z = \frac{z}{t}. \tag{2.53}$$

It is interesting to note that the requirement of the boost-invariance leads to a more general form of the velocity, namely, $v_z = (Az - Bt)/(At - Bz)$, where A and B are constants. Only if we add the additional constraint that $v_z = 0$ at $z = 0$, as mentioned above, the boost-invariant form is reduced to Eq. (2.53).

We have spent some time discussing the boost invariance because this symmetry is very frequently used in the models describing the evolution of matter formed in relativistic heavy-ion collisions. It reduces the number of independent variables and dimensions, facilitating theoretical calculations. The popularity of using this symmetry has its origin in the famous Bjorken paper [28], where he implemented this symmetry into hydrodynamic equations and made estimates of the initial energy density accessible in the collisions. Our discussion of the experimental data indicated that the boost-invariance may be regarded as the good approximation only for the central region of the most energetic heavy-ion collisions. Similarly the Feynman arguments lead to the boost-invariant spectrum for $p_\parallel \ll \sqrt{s}$ (in the center-of-mass frame).

Bibliography to Chapter 2

[1] D. E. Lyon, C. Risk, and D. M. Tow, "Angular distributions from multiparticle production models," *Phys. Rev.* **D3** (1971) 104–108.

[2] P. Carruthers and M. Duong-Van, "New scaling law based on the hydrodynamical model of particle production," *Phys. Lett.* **B41** (1972) 597–601.

[3] P. Carruthers and M. Doung-van, "Rapidity and angular distributions of charged secondaries according to the hydrodynamical model of particle production," *Phys. Rev.* **D8** (1973) 859–874.

[4] R. P. Feynman, "Very high-energy collisions of hadrons," *Phys. Rev. Lett.* **23** (1969) 1415–1417.

[5] C. Y. Wong, "Introduction to high-energy heavy ion collisions," (Singapore, Singapore: World Scientific, 1994) 516 p.

[6] **BRAHMS** Collaboration, I. G. Bearden *et al.*, "Charged meson rapidity distributions in central Au+Au collisions at $\sqrt{s_{\rm NN}} = 200$ GeV," *Phys. Rev. Lett.* **94** (2005) 162301.

[7] **NA49** Collaboration, S. V. Afanasiev *et al.*, "Energy dependence of pion and kaon production in central Pb + Pb collisions," *Phys. Rev.* **C66** (2002) 054902.

[8] A. Budzanowski, "By small steps towards "The beginning" what have we learned from first results of the PHOBOS detector at RHIC?," *Acta Phys. Pol.* **B33** (2002) 33.

[9] **PHOBOS** Collaboration, B. B. Back *et al.*, "Centrality and energy dependence of charged-particle multiplicities in heavy ion collisions in the context of elementary reactions," *Phys. Rev.* **C74** (2006) 021902.

[10] **PHOBOS** Collaboration, B. B. Back *et al.*, "Charged-particle pseudorapidity density distributions from Au+Au collisions at $\sqrt{s_{\rm NN}} = 130$ GeV," *Phys. Rev. Lett.* **87** (2001) 102303.

[11] W. Broniowski and W. Florkowski, "Geometric relation between centrality and the impact parameter in relativistic heavy ion collisions," *Phys. Rev.* **C65** (2002) 024905.

[12] R. Vogt, "Relation of hard and total cross sections to centrality," *Heavy Ion Phys.* **9** (1999) 339–348.
[13] V. P. Konchakovski, M. Hauer, G. Torrieri, M. I. Gorenstein, and E. L. Bratkovskaya, "Forward-backward correlations in nucleus-nucleus collisions: baseline contributions from geometrical fluctuations," *Phys. Rev.* **C79** (2009) 034910.
[14] J. Cugnon, J. Knoll, C. Riedel, and Y. Yariv, "Event by event emission pattern analysis of the intranuclear cascade," *Phys. Lett.* **B109** (1982) 167–170.
[15] M. Gyulassy, K. A. Frankel, and H. Stoecker, "Do nuclei flow at high-energies?," *Phys. Lett.* **B110** (1982) 185–188.
[16] H. H. Gutbrod *et al.*, "Squeeze-out of nuclear matter as a function of projectile energy and mass," *Phys. Rev.* **C42** (1990) 640–651.
[17] J.-Y. Ollitrault, "Anisotropy as a signature of transverse collective flow," *Phys. Rev.* **D46** (1992) 229–245.
[18] J.-Y. Ollitrault, "Determination of the reaction plane in ultrarelativistic nuclear collisions," *Phys. Rev.* **D48** (1993) 1132–1139.
[19] S. Voloshin and Y. Zhang, "Flow study in relativistic nuclear collisions by Fourier expansion of Azimuthal particle distributions," *Z. Phys.* **C70** (1996) 665–672.
[20] P. Danielewicz and G. Odyniec, "Transverse Momentum Analysis of Collective Motion in Relativistic Nuclear Collisions," *Phys. Lett.* **B157** (1985) 146–150.
[21] H. H. Gutbrod *et al.*, "A new component of the collective flow in relativistic heavy-ion collisions," *Phys. Lett.* **B216** (1989) 267–271.
[22] **STAR** Collaboration, B. I. Abelev *et al.*, "System-size independence of directed flow at the Relativistic Heavy-Ion Collider," *Phys. Rev. Lett.* **101** (2008) 252301.
[23] **PHOBOS** Collaboration, B. B. Back *et al.*, "Energy dependence of directed flow over a wide range of pseudorapidity in Au+Au collisions at RHIC," *Phys. Rev. Lett.* **97** (2006) 012301.
[24] H. Stoecker, "Collective flow signals the quark-gluon plasma," *Nucl. Phys.* **A750** (2005) 121–147.
[25] **PHENIX** Collaboration, S. S. Adler *et al.*, "Elliptic flow of identified hadrons in Au+Au collisions at $\sqrt{s_{NN}} = 200$ GeV," *Phys. Rev. Lett.* **91** (2003) 182301.
[26] **STAR** Collaboration, C. Adler *et al.*, "Elliptic flow from two- and four-particle correlations in Au+Au collisions at $\sqrt{s_{NN}} = 130$ GeV," *Phys. Rev.* **C66** (2002) 034904.
[27] **BRAHMS** Collaboration, I. G. Bearden *et al.*, "Nuclear stopping in Au+Au collisions at $\sqrt{s_{NN}} = 200$ GeV," *Phys. Rev. Lett.* **93** (2004) 102301.
[28] J. D. Bjorken, "Highly relativistic nucleus-nucleus collisions: The central rapidity region," *Phys. Rev.* **D27** (1983) 140–151.

Chapter 3

Glauber Model

In realistic situations the separation between spectators and participants is not so sharp as in the simple geometric picture of Fig. 2.1. A more elaborate estimate of the number of participating nucleons can be done within the Glauber model [1–4] which treats a nucleus-nucleus collision as a multiple nucleon-nucleon collision process. In the Glauber model, the nucleon distributions in nuclei are random and given by the nuclear density profiles, whereas the elementary nucleon-nucleon collision is characterized by the total inelastic cross section $\sigma_{\rm in}$. Initially, the Glauber model was applied only to elastic collisions. In this case a nucleon does not change its properties in the individual collisions, so all nucleon interactions can be well described by the same cross section. Applying the Glauber model to inelastic collisions, we assume that after a single inelastic collision an excited nucleon-like object is created that interacts basically with the same inelastic cross section with other nucleons.

3.1 Eikonal approximation

In this Section we introduce the eikonal [1] approximation. It is used below as one of the elements of the Glauber multiple scattering model, see also [6]. The eikonal approximation is the classical approximation to the angular momentum. It may be applied to the standard expansion of the elastic scattering amplitude into the angular-momentum eigenstates defined by the orbital number l,

$$f(s,t) = \frac{1}{2ip} \sum_l (2l+1) \left[e^{2i\delta_l} - 1 \right] P_l(\cos\theta). \tag{3.1}$$

Here s and t are the Mandelstam variables [7], i.e., the center-of-mass energy squared and the invariant momentum transfer squared,

$$s = (p_1 + p_2)^2, \qquad t = (p_1 - p_1')^2. \tag{3.2}$$

In Eq. (3.2) p_1 is the four-momentum of the projectile, p_2 is the four-momentum of the target, and p_1' is the four-momentum of the scattered projectile particle. The

[1] The word *eikonal* is of Greek origin. It is related to the words *icon* and *image* [5].

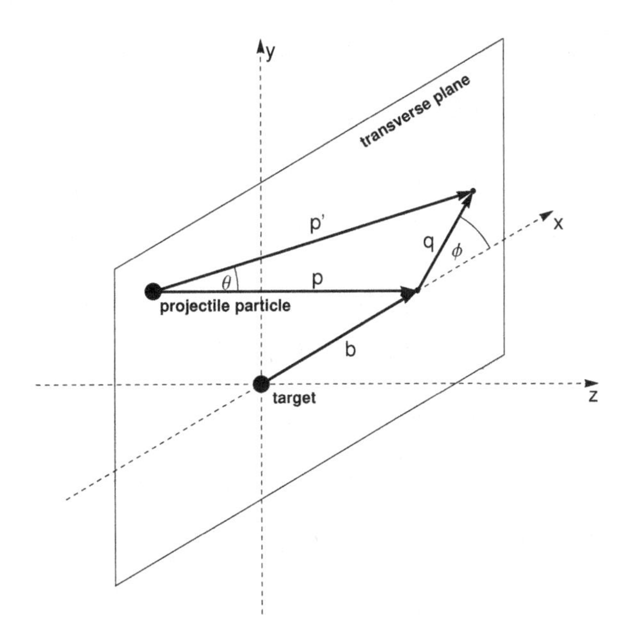

Fig. 3.1 Geometry of the high-energy elastic scattering process. The projectile particle moves along the z-axis. The momentum transfer three-vector $\mathbf{q}$ lies in the transverse plane.

variables s and t may be expressed by the three-momentum of the projectile p and the scattering angle θ, both determined in the center-of-mass frame. If the particles have the mass m, the appropriate relations have the form

$$s = 4(m^2 + p^2), \qquad \frac{2t}{s - 4m^2} = \cos\theta - 1. \tag{3.3}$$

In Eq. (3.1) P_l is the Legendre polynomial of the lth order, depending on the cosine of the scattering angle θ, and δ_l is the phase shift. We note that the phase shifts contain the whole information about the scattering process and in general may contain a non-zero imaginary part [2].

High-energy elastic scattering processes are far from being spherically symmetric, hence the large values of l dominate in Eq. (3.1) and we may write

$$p\,b = l + \frac{1}{2}, \tag{3.4}$$

where b is the impact parameter. For large l and small scattering angles θ the Legendre polynomial $P_l(\cos\theta)$ may be approximated by the formula

$$P_l(\cos\theta) = \int\limits_0^{2\pi} \frac{d\phi}{2\pi} e^{i(2l+1)\sin(\theta/2)\cos(\phi)}. \tag{3.5}$$

[2] The popular way of writing the amplitude is then $f = 1/(2ip)\sum_l (2l+1)\left[\eta_l e^{2i\delta_l} - 1\right] P_l$, where η_l is the *inelasticity coefficient*.

At high energy, the momentum transfer vector $\mathbf{q} = \mathbf{p}' - \mathbf{p}$ lies in the transverse plane, see Fig. 3.1, and we may rewrite the argument of the exponential function in (3.5) as a scalar product of $\mathbf{q}$ and $\mathbf{b}$,

$$(2l+1)\sin\left(\frac{\theta}{2}\right)\cos(\phi) = 2p\sin\left(\frac{\theta}{2}\right)\frac{l+1/2}{p}\cos(\phi) = \mathbf{q}\cdot\mathbf{b}. \tag{3.6}$$

In this way we find the convenient expression

$$P_l(\cos\theta) = \int_0^{2\pi}\frac{d\phi}{2\pi}e^{i\mathbf{q}\cdot\mathbf{b}}. \tag{3.7}$$

After replacing l by b we may treat b as the continuous variable (with $db = dl/p$ and $d^2b = b\,db\,d\phi$). In this approximation, the scattering amplitude has the form

$$f(s,t) = \frac{ip}{2\pi}\int d^2b\, e^{i\mathbf{q}\cdot\mathbf{b}}\left[1 - e^{i\chi(s,\mathbf{b})}\right], \tag{3.8}$$

where the phase shift of the projectile is defined as

$$\chi(s,\mathbf{b}) = 2\,\delta(s,\mathbf{b}). \tag{3.9}$$

The total cross section may be obtained from the forward scattering amplitude with the help of the optical theorem

$$\sigma_{\text{tot}} = \frac{4\pi}{p}\text{Im}\, f(s,t=0) = 2\int d^2b\left[1 - \text{Re}\, e^{i\chi(s,\mathbf{b})}\right]. \tag{3.10}$$

On the other hand, the elastic cross section is obtained by squaring the amplitude and integrating over the solid angle. Since the scattering is concentrated in the forward direction, the integration over the solid angle may be replaced by the integral over the space orthogonal to the momentum vector $\mathbf{p}$, see Fig. 3.1,

$$d\Omega = \frac{d^2q}{p^2}. \tag{3.11}$$

Using this property we obtain

$$\begin{aligned}\sigma_{\text{el}} &= \int\frac{d^2q}{4\pi^2}\int d^2b\int d^2b' e^{i\mathbf{q}\cdot\mathbf{b}}\left[1 - e^{i\chi(s,\mathbf{b})}\right]e^{-i\mathbf{q}\cdot\mathbf{b}'}\left[1 - e^{i\chi(s,\mathbf{b}')}\right]^* \\ &= \int d^2b\left|1 - e^{i\chi(s,\mathbf{b})}\right|^2.\end{aligned} \tag{3.12}$$

Finally, the inelastic cross section is

$$\sigma_{\text{in}} = \sigma_{\text{tot}} - \sigma_{\text{el}} = \int d^2b\left(1 - \left|e^{i\chi(s,\mathbf{b})}\right|^2\right). \tag{3.13}$$

We note that σ_{in} is different from zero only if $\chi(s,\mathbf{b})$ has a non zero imaginary part.

3.2 Nucleon-nucleon collisions

The main feature of the nucleon-nucleon (pp) interaction is that the total cross section for this process is about 40 mb, and this value is approximately constant in the energy range: 3 GeV $< \sqrt{s} <$ 100 GeV. The inelastic cross section in this range gives the main contribution to the total cross section. A certain subclass of the inelastic processes is the *diffractive dissociation* process. In this process a nucleon is only slightly excited and a small number of particles is produced, which is in contrast to the typical *nondiffractive inelastic* events. The diffractive processes represent about 10% of all inelastic collisions.

3.2.1 *Energy dependence of particle production*

In the following we shall discuss the inelastic processes having in mind only the non-diffractive processes. In such a typical inelastic nucleon-nucleon collision a certain number of charged particles is produced. The average charged particle multiplicity may be described by the phenomenological formula [8]

$$\overline{N}_{\rm NN} = 0.88 + 0.44 \ln \frac{s}{s_0} + 0.118 \left(\ln \frac{s}{s_0} \right)^2, \tag{3.14}$$

where $s_0 = 1$ GeV. Another phenomenological formula may be used to describe the average charged particle multiplicity at midrapidity

$$\left. \frac{d\overline{N}_{\rm NN}}{d\eta} \right|_{\eta=0} = 2.5 - 0.25 \ln \frac{s}{s_0} + 0.023 \left(\ln \frac{s}{s_0} \right)^2. \tag{3.15}$$

Equation (3.15) is a parametrization of the $p\bar{p}$ data obtained by the UA5 and the CDF group in the range 50 GeV $< \sqrt{s} <$ 2000 GeV [9].

3.2.2 *Thickness function*

Let us consider a nucleon-nucleon collision at a given energy $\sqrt{s}$ and at an impact parameter b, see Fig. 3.2 (a). According to our discussion presented in Sec. 3.1, Eq. (3.13), we may introduce the probability of having a nucleon-nucleon inelastic collision

$$p\,(\mathbf{b}) = \left(1 - \left| e^{i\chi(\mathbf{b})} \right|^2 \right) \equiv t\,(\mathbf{b})\, \sigma_{\rm in}. \tag{3.16}$$

The function $t\,(\mathbf{b})$, defined by Eq. (3.16), is called the nucleon-nucleon *thickness function.* The integral of $p\,(\mathbf{b})$ over the whole range of the impact parameter should be normalized to $\sigma_{\rm in}$. Thus, the thickness function is normalized to unity

$$\int d^2b\, t\,(\mathbf{b}) = 1. \tag{3.17}$$

For collisions with unpolarized beams $t\,(\mathbf{b})$ depends only on the magnitude of $\mathbf{b}$.

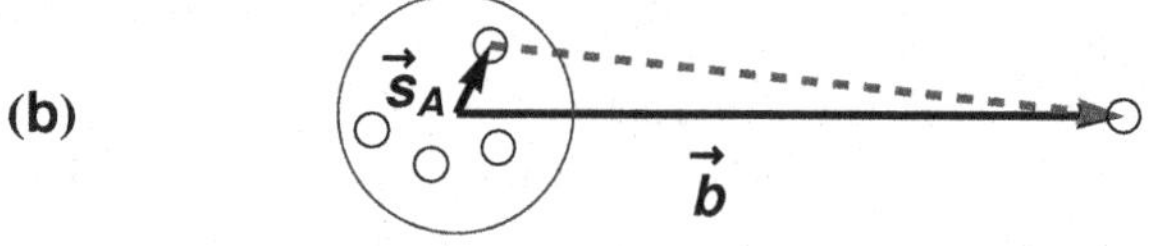

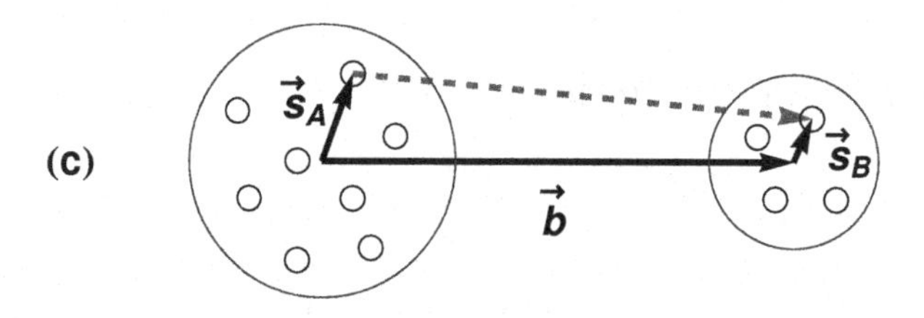

Fig. 3.2 A nucleon-nucleon collision **(a)**, a nucleon-nucleus collision **(b)**, and a nucleus-nucleus collision **(c)** seen in the plane transverse to the collision axis z. The impact vector is denoted in each case by $\mathbf{b}$. The positions of nucleons in the nuclei A and B are denoted by the vectors $\mathbf{s}_A$ and $\mathbf{s}_B$, respectively.

3.3 Nucleon-nucleus collisions

3.3.1 *Nuclear density profiles*

We consider next the nucleon-nucleus collision in the Glauber framework. The probability of finding a nucleon in the nucleus with the atomic mass number A is the usual baryon density divided by the number of baryons in the nucleus. For large nuclei, one commonly uses the Woods-Saxon function [10]

$$\rho_A(r) = \frac{\rho_0}{A\left(1+\exp\left[\frac{r-r_0}{a}\right]\right)}, \tag{3.18}$$

with the parameters [3]:

$$r_0 = (1.12A^{1/3} - 0.86A^{-1/3})\,\mathrm{fm}, \tag{3.19}$$

$$a = 0.54\,\mathrm{fm}, \tag{3.20}$$

and

$$\rho_0 = 0.17\,\mathrm{fm}^{-3}. \tag{3.21}$$

[3] Our definition of $\rho_A(r)$ includes A in the denominator, because we want to interpret $\rho_A(r)$ as the probability distribution.

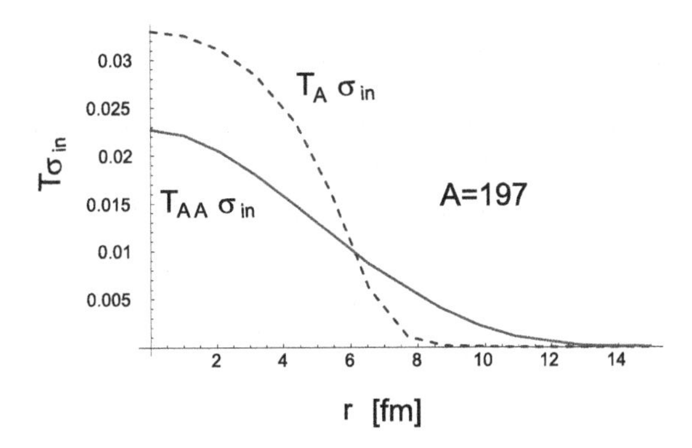

Fig. 3.3 Thickness functions $T_A(r)$ and $T_{AA}(r)$ for the gold nucleus ($A = 197$), multiplied by the nucleon-nucleon inelastic cross section $\sigma_{\rm in}$= 30 mb. The baryon distribution in the nucleus is described by the Woods-Saxon function (3.18) with the parameters given by Eqs. (3.19)–(3.21). The nucleon-nucleon thickness function is approximated by the Dirac delta function.

The parameter ρ_0 is the nuclear saturation density. In simple estimates one may also use a sharp-cutoff distribution

$$\rho_A(r) = \frac{3}{4\pi R^3}\,\theta\,(R-r)\,, \tag{3.22}$$

where $\theta(x)$ is the step function and the radius $R \approx 1.12A^{1/3}$ fm is determined by counting the number of nucleons in the nucleus,

$$\frac{4}{3}\pi R^3 \rho_0 = A. \tag{3.23}$$

The nucleon-nucleus thickness function for the nucleus A is obtained from a geometric consideration depicted in Fig. 3.2 (b) and the assumption that the nucleon positions in the nucleus A are not changed during the collision process. In this way we obtain

$$T_A\,(\mathbf{b}) = \int dz_A \int d^2 s_A\, \rho_A(\mathbf{s}_A, z_A)\, t(\mathbf{s}_A - \mathbf{b}). \tag{3.24}$$

Here the transverse coordinates are denoted by the vector $\mathbf{s}_A$, and we use notation

$$\rho_A(\mathbf{s}_A, z_A) = \rho_A\left(\sqrt{\mathbf{s}_A^2 + z_A^2}\right). \tag{3.25}$$

Equation (3.17) implies the normalization condition

$$\int d^2 b\, T_A\,(\mathbf{b}) = 1. \tag{3.26}$$

3.3.2 *Independent collisions*

The quantity $T_A(\mathbf{b})\,\sigma_{\rm in}$ is the probability that a single nucleon-nucleon collision takes place in a nucleon-nucleus collision at the impact parameter $\mathbf{b}$. Treating all possible nucleon-nucleon collisions in the nucleon-nucleus collision as completely independent and characterized by the same cross section, we easily find the probability of having n such collisions. The latter is expressed by the binomial distribution

$$P(n;A;\mathbf{b}) = \binom{A}{n}\left[1 - T_A(\mathbf{b})\,\sigma_{\rm in}\right]^{A-n}\left[T_A(\mathbf{b})\,\sigma_{\rm in}\right]^n. \qquad (3.27)$$

The average number of binary nucleon-nucleon collisions may be calculated from Eq. (3.27) which gives

$$\overline{n}(A;\mathbf{b}) = \sum_{n=1}^{A} nP(n;A;\mathbf{b}) = A\,T_A(\mathbf{b})\;\sigma_{\rm in}. \qquad (3.28)$$

Similarly, we find

$$\overline{n^2}(A;\mathbf{b}) = \sum_{n=1}^{A} n^2 P(n;A;\mathbf{b}), \qquad (3.29)$$

which gives us the variance

$$\Delta n^2(A;\mathbf{b}) = \overline{n^2}(A;\mathbf{b}) - \overline{n}^2(A;\mathbf{b}) = A\,T_A(\mathbf{b})\;\sigma_{\rm in}\left[1 - T_A(\mathbf{b})\,\sigma_{\rm in}\right]. \qquad (3.30)$$

Figure 3.3 shows that the condition $T_A(\mathbf{b})\,\sigma_{\rm in} \ll 1$ is quite well satisfied for the realistic nuclear density profiles. In this situation, for $n \ll A$ Eq. (3.27) may be approximated by the Poisson distribution

$$P(n;A;\mathbf{b}) = \frac{\left[\overline{n}(A;\mathbf{b})\right]^n}{n!}\exp\left[-\overline{n}(A;\mathbf{b})\right]. \qquad (3.31)$$

Since the scale at which the nucleon-nucleon thickness function varies is typically smaller than the scale at which the nuclear density changes, we may often replace $t(\mathbf{s}_A - \mathbf{b})$ in Eq. (3.24) by the delta function $\delta^{(2)}(\mathbf{s}_A - \mathbf{b})$. In this approximation $T_A(\mathbf{b})$ is the nuclear density projected onto the transverse plane

$$T_A(\mathbf{b}) = \int dz_A\,\rho_A(\mathbf{b}, z_A), \qquad (3.32)$$

and the average number of the collisions is

$$\overline{n}(A;\mathbf{b}) = A\,\sigma_{\rm in}\int dz_A\,\rho_A(\mathbf{b}, z_A). \qquad (3.33)$$

3.4 Nucleus-nucleus collisions

Finally, we define the thickness function for the nucleus-nucleus collision. A geometric consideration shown in Fig. 3.2 (c) leads to the formula

$$T_{AB}(\mathbf{b}) = \int dz_A \int d^2s_A\, \rho_A(\mathbf{s}_A, z_A) \int dz_B \int d^2s_B\, \rho_B(\mathbf{s}_B, z_B)\, t(\mathbf{b} + \mathbf{s}_B - \mathbf{s}_A), \tag{3.34}$$

with the corresponding normalization condition

$$\int d^2b\, T_{AB}(\mathbf{b}) = 1. \tag{3.35}$$

The quantity $T_{AB}(\mathbf{b})\,\sigma_{\rm in}$ is the *averaged* probability that a nucleon-nucleon collision takes place in a nucleus-nucleus collision characterized by the impact parameter $\mathbf{b}$. In the limit $t(\mathbf{b}) \to \delta^{(2)}(\mathbf{b})$ we may write

$$\begin{aligned} T_{AB}(\mathbf{b}) &= \int d^2s_A \int dz_A\, \rho_A(\mathbf{s}_A, z_A) \int dz_B\, \rho_B(\mathbf{s}_A - \mathbf{b}, z_B) \\ &= \int d^2s_A\, T_A(\mathbf{s}_A)\, T_B(\mathbf{s}_A - \mathbf{b}), \end{aligned} \tag{3.36}$$

or in a more symmetric form

$$T_{AB}(\mathbf{b}) = \int d^2s\, T_A\left(\mathbf{s} + \frac{1}{2}\mathbf{b}\right) T_B\left(\mathbf{s} - \frac{1}{2}\mathbf{b}\right). \tag{3.37}$$

The nucleus-nucleus thickness function $T_{AB}(\mathbf{b})$ can be used to calculate the probability of having n inelastic binary nucleon-nucleon collisions in a nucleus-nucleus collision at the impact parameter $\mathbf{b}$. Similarly to the nucleon-nucleus case, see Eq. (3.27), we obtain

$$P(n; AB; \mathbf{b}) = \binom{AB}{n} \left[1 - T_{AB}(\mathbf{b})\,\sigma_{\rm in}\right]^{AB-n} \left[T_{AB}(\mathbf{b})\,\sigma_{\rm in}\right]^n. \tag{3.38}$$

The results for the average number of the collisions $\overline{n}(AB; \mathbf{b})$ and the dispersion $\Delta n^2(AB; \mathbf{b})$ are of the same form as Eqs. (3.28) and (3.30), for example

$$\overline{n}(AB; \mathbf{b}) = AB\, T_{AB}(\mathbf{b})\, \sigma_{\rm in}. \tag{3.39}$$

For $n \ll AB$ one may also approximate Eq. (3.38) by the Poisson distribution.

3.4.1 *Total inelastic cross section*

The total probability of an inelastic nuclear collision is the sum over n from $n = 1$ to $n = AB$

$$P_{\rm in}(AB; \mathbf{b}) = \sum_{n=1}^{AB} P(n; AB; \mathbf{b}) = 1 - \left[1 - T_{AB}(\mathbf{b})\,\sigma_{\rm in}\right]^{AB}. \tag{3.40}$$

From Eq. (3.40), by integrating over the impact parameter space, one may obtain the total inelastic cross section for the collision of the two nuclei A and B (see also [11])

$$\sigma_{\rm in}^{AB} = \int d^2b \left(1 - \left[1 - T_{AB}\left(\mathbf{b}\right)\sigma_{\rm in}\right]^{AB}\right). \tag{3.41}$$

Using the thickness function for the Au+Au collisions we find $\sigma_{\rm in}^{\rm AuAu} = 6.8$ b for $\sigma_{\rm in} = 30$ mb and $\sigma_{\rm in}^{\rm AuAu} = 7.0$ b for $\sigma_{\rm in} = 40$ mb. We note that those cross sections are larger than the geometric cross section $\sigma_{\rm geo}^{\rm AuAu} = 4\pi R^2 \approx 5\pi A^{2/3} = 5.3$ b. This is due to the tails of the Woods-Saxon distribution (3.18), which make possible that a nucleon-nucleon collision occurs in the nuclear collision at the impact parameter b larger than $2R$.

In order to derive Eq. (3.41) we used, from the very beginning, the averaged probability for nucleon-nucleon collisions given by the thickness function (3.34). In more realistic calculations we proceed in a different way. We first consider the case where the positions of nucleons in the target and projectile nucleus are fixed, and the averaging is done later. The probability of an inelastic collision for a fixed nucleon configuration is given by the more accurate formula

$$1 - \prod_{j=1}^{A}\prod_{i=1}^{B}\left[1 - t\left(\mathbf{b} + \mathbf{s}_i^B - \mathbf{s}_j^A\right)\sigma_{\rm in}\right]. \tag{3.42}$$

The probability of an inelastic *nuclear collision* at the impact parameter $\mathbf{b}$ is then obtained by averaging of Eq. (3.42) over the nucleon positions. If the positions are uncorrelated we may use the formula

$$\begin{aligned} P_{\rm in}\left(AB;\mathbf{b}\right) = &\int d^2s_1^A T_A(\mathbf{s}_1^A)\cdots d^2s_A^A T_A(\mathbf{s}_A^A) \int d^2s_1^B T_B(\mathbf{s}_1^B)\cdots d^2s_B^B T_B(\mathbf{s}_B^B) \\ &\times \left\{1 - \prod_{j=1}^{A}\prod_{i=1}^{B}\left[1 - t\left(\mathbf{b} + \mathbf{s}_i^B - \mathbf{s}_j^A\right)\sigma_{\rm in}\right]\right\}. \end{aligned} \tag{3.43}$$

The integration of Eq. (3.43) over the impact parameter gives the inelastic nuclear cross section $\sigma_{\rm in}^{AB}$. One can notice that Eqs. (3.40) and (3.43) differ from each other. The more accurate formula Eq. (3.43) is much more complicated to handle and cannot be simply reduced to Eq. (3.40). Only in the case of nucleon-nucleus collisions the two methods are equivalent. Since there is no good analytic method to evaluate Eq. (3.43) for large values of A and B, one is most often satisfied with Eqs. (3.40) and (3.41) only. These equations are called the optical limit of the Glauber model [4].

[4] In this approach, the center-of-mass correlations, i.e., the conditions $\mathbf{s}_1^A + \cdots + \mathbf{s}_A^A = 0$ and $\mathbf{s}_1^B + \cdots + \mathbf{s}_B^B = 0$ may be also taken into account in the analytic way, for example, see [12].

The precise calculations of the cross sections, based on Eq. (3.43), are of course possible with the help of the Monte Carlo simulations. Different Monte Carlo realizations of the Glauber model are publicly available at the moment. One of them is the program GLISSANDO [13]. With $\sigma_{\rm in} = 42$ mb GLISSANDO gives the total inelastic cross section $\sigma_{\rm in}^{\rm AuAu} = 6.4$ b. This value is used in the hydrodynamic calculations presented in Sec. 22.1.1.

3.4.2 *Exclusion of elastic processes*

Our expressions for the probabilities $P\,(n; AB; \mathbf{b})$ include the elastic process corresponding to the case $n = 0$. If we are interested only in the inelastic nucleus-nucleus collisions, we should renormalize our results by the factor

$$\frac{\sum_{n=0}^{AB} P\,(n, AB; \mathbf{b})}{\sum_{n=1}^{AB} P\,(n, AB; \mathbf{b})} = \left(1 - [1 - T_{AB}\,(\mathbf{b})\,\sigma_{\rm in}]^{AB}\right)^{-1}. \tag{3.44}$$

Hence, the average number of binary nucleon-nucleon collisions in an *inelastic* nucleus-nucleus collision at the impact parameter $\mathbf{b}$ is

$$\overline{n}_{\rm in}\,(AB; \mathbf{b}) = \frac{A\,B\,T_{AB}\,(\mathbf{b})\,\sigma_{\rm in}}{\left(1 - [1 - T_{AB}\,(\mathbf{b})\,\sigma_{\rm in}]^{AB}\right)}. \tag{3.45}$$

Further averaging over the impact parameters yields

$$\begin{aligned} \overline{n}_{\rm in}^{AB} &= \frac{\int d^2b\;\overline{n}_{\rm in}\,(AB; \mathbf{b})\,(d\sigma_{\rm in}^{AB}/d^2b)}{\int d^2b\;(d\sigma_{\rm in}^{AB}/d^2b)} \\ &= \frac{\int d^2b\;\overline{n}\,(AB; \mathbf{b})}{\int d^2b\,\left(1 - [1 - T_{AB}\,(\mathbf{b})\,\sigma_{\rm in}]^{AB}\right)} = A\,B\,\frac{\sigma_{\rm in}}{\sigma_{\rm in}^{AB}}. \end{aligned} \tag{3.46}$$

3.5 Wounded nucleons

The Glauber model can be used also to calculate the number of the participants. To be more precise we distinguish between the *participants which may interact elastically* and the *participants which interact only inelastically*. Following Ref. [14], the latter are called the *wounded nucleons* [5]. The probability that the nucleon i

[5] Very often the difference between the participants and the wounded nucleons is neglected, and these two terms are treated equivalently. Since the elastic part of the nucleon-nucleon cross section is much smaller than the inelastic part, this approach is admitable as long as we do not analyze details of the collision process. It is also worth to mention that the experimental groups use very often their own definitions of the number of participants, influenced by the technical possibilities of determining $N_{\rm part}$ or by the specific Monte-Carlo implementations of the Glauber model.

from B is wounded, i.e., it collides inelastically with anyone of the nucleons in the nucleus A is [14]

$$p\left(\mathbf{s}_i^B; A; \mathbf{s}_1^A, ..., \mathbf{s}_A^A; \mathbf{b}\right) \equiv p_A\left(\mathbf{s}_i^B\right) = 1 - \prod_{j=1}^{A}\left[1 - t\left(\mathbf{b} + \mathbf{s}_i^B - \mathbf{s}_j^A\right)\sigma_{\text{in}}\right]. \quad (3.47)$$

Now the impact parameter $\mathbf{b}$ is the distance between the trajectories of the centers of the nuclei as defined in Sec. 2.1, see Figs. 2.1 and 3.2 (c). After integrating over different configurations of the nucleons in B we obtain the probability that w_B nucleons in B suffered at least one collision

$$\begin{aligned} P\left(w_B; B; A; \mathbf{s}_1^A, ..., \mathbf{s}_A^A; \mathbf{b}\right) &= \binom{B}{w_B}\left[1 - \overline{p}\left(A; \mathbf{s}_1^A, ..., \mathbf{s}_A^A; \mathbf{b}\right)\right]^{B-w_B} \\ &\times \left[\overline{p}\left(A; \mathbf{s}_1^A, ..., \mathbf{s}_A^A; \mathbf{b}\right)\right]^{w_B}, \end{aligned} \quad (3.48)$$

where

$$\begin{aligned} \overline{p}\left(A; \mathbf{s}_1^A, ..., \mathbf{s}_A^A; \mathbf{b}\right) &= \int d^2s^B\, T_B\left(\mathbf{s}^B\right)\, p\left(\mathbf{s}^B; A; \mathbf{s}_1^A, ..., \mathbf{s}_A^A; \mathbf{b}\right) \quad (3.49) \\ &= \int d^2s^B\, T_B\left(\mathbf{b} - \mathbf{s}^B\right)\left(1 - \prod_{j=1}^{A}\left[1 - t\left(\mathbf{s}^B - \mathbf{s}_j^A\right)\sigma_{\text{in}}\right]\right). \end{aligned}$$

The average number of the wounded nucleons in B (at fixed impact parameter $\mathbf{b}$) is

$$\begin{aligned} \overline{w}_B\left(B; A; \mathbf{s}_1^A, ..., \mathbf{s}_A^A; \mathbf{b}\right) &= \sum_{w_B=1}^{B} w_B\, P\left(w_B; B; A; \mathbf{s}_1^A, ..., \mathbf{s}_A^A; \mathbf{b}\right) \quad (3.50) \\ &= \sum_{w_B=1}^{B} w_B \binom{B}{w_B}(1 - \overline{p})^{B-w_B}\, \overline{p}^{w_B} = B\overline{p}\left(A; \mathbf{s}_1^A, ..., \mathbf{s}_A^A; \mathbf{b}\right). \end{aligned}$$

Averaging over the configurations of the nucleons in the nucleus A gives

$$\begin{aligned} \overline{w}_B(B; A; \mathbf{b}) &= B\int d^2s_1^A T_A\left(\mathbf{s}_1^A\right)\cdots\int d^2s_A^A T_A\left(\mathbf{s}_A^A\right) \quad (3.51) \\ &\times \int d^2s^B\, T_B\left(\mathbf{b} - \mathbf{s}^B\right)\left(1 - \prod_{j=1}^{A}\left[1 - t\left(\mathbf{s}^B - \mathbf{s}_j^A\right)\sigma_{\text{in}}\right]\right) \\ &= B\int d^2s^B\, T_B\left(\mathbf{b} - \mathbf{s}^B\right)\left(1 - \left[1 - \int d^2s^A T_A\left(\mathbf{s}^A\right) t\left(\mathbf{s}^B - \mathbf{s}^A\right)\sigma_{\text{in}}\right]^A\right). \end{aligned}$$

This result can be simplified if we assume that the nucleon-nucleon thickness function can be approximated by the Dirac delta function, $t(\mathbf{b}) \longrightarrow \delta^{(2)}(\mathbf{b})$. In this case

$$\overline{w}_B(B;A;\mathbf{b}) = B\int d^2s^B\, T_B\left(\mathbf{b}-\mathbf{s}^B\right)\left(1-\left[1-T_A\left(\mathbf{s}^B\right)\sigma_{\rm in}\right]^A\right). \tag{3.52}$$

Since the number of wounded nucleons in the collision of A and B is the sum of the wounded nucleons in the nucleus A and B, we obtain

$$\begin{aligned}\overline{w}(A;B;\mathbf{b}) = {} & A\int d^2s\, T_A(\mathbf{b}-\mathbf{s})\left(1-\left[1-\sigma_{\rm in}T_B(\mathbf{s})\right]^B\right) \\ & + B\int d^2s\, T_B(\mathbf{b}-\mathbf{s})\left(1-\left[1-\sigma_{\rm in}T_A(\mathbf{s})\right]^A\right).\end{aligned} \tag{3.53}$$

Finally, averaging over different impact parameters we find

$$\overline{w}_{AB} = \frac{\int d^2b\, \overline{w}(A;B;\mathbf{b})\left(d\sigma_{\rm in}^{AB}/d^2b\right)}{\int d^2b\, \left(d\sigma_{\rm in}^{AB}/d^2b\right)}. \tag{3.54}$$

If we want to exclude the elastic collisions, in analogy to Eq. (3.46) we may use the formula

$$\overline{w}_{\rm in}^{AB} = \frac{\int d^2b\, \overline{w}(A;B;\mathbf{b})}{\int d^2b\, \left(d\sigma_{\rm in}^{AB}/d^2b\right)}. \tag{3.55}$$

Note that for $A = B = 1$ Eq. (3.55) gives $\overline{w}_{\rm in}^{AB} = 2$. Of course, this is an expected result for the nucleon-nucleon collisions.

It is important to realize that the number of the binary nucleon-nucleon collisions is very much different from the number of the wounded nucleons (participants). In Table 3.1 we show the results of the calculations done for Au+Au collisions at different values of the impact parameter b. We used two values of the nucleon-nucleon cross section: $\sigma_{\rm in} = 30$ mb, and $\sigma_{\rm in} = 40$ mb. We observe that the number of the binary collisions, $\overline{n}(b)$, is usually much larger than the number of the wounded nucleons, $\overline{w}(b)$. Moreover, $\overline{n}(b)$ depends more strongly on $\sigma_{\rm in}$ than $\overline{w}(b)$ does.

3.6 Soft and hard processes

It is an experimental fact that pions (the most abundant particles produced in a nucleon-nucleon as well as in a nucleus-nucleus collision) have on average small transverse momenta, $p_\perp \sim 400$ MeV (see our discussion in Sec. 2.2.1 and Fig. 2.3). The processes leading to the production of such low-energetic pions are called *soft processes.* On the other hand, the pions with large transverse momenta, $p_\perp > 1$–2 GeV, are produced by *hard processes.* The soft processes cannot be described directly by perturbative QCD. In this case the strong coupling constant is

Table 3.1 The numbers of binary collisions, $\overline{n}(b)$, and the numbers of wounded nucleons, $\overline{w}(b)$, for Au+Au collisions ($A = 197$) at different values of the impact parameter b. The results are presented for two different values of the nucleon-nucleon inelastic cross section: $\sigma_{\rm in} = 30$ mb (the second and the third column), and $\sigma_{\rm in} = 40$ mb (the fifth and the sixth column). The fourth and seventh columns give geometric estimates of the centrality class of the collisions with the impact parameters *smaller* than b (the fourth column is for $\sigma_{\rm in}^{\rm AuAu} = 6.8$ b, whereas the sixth column is for $\sigma_{\rm in}^{\rm AuAu} = 7.0$ b).

b [fm]	$\overline{n}(b)$	$\overline{w}(b)$	c	$\overline{n}(b)$	$\overline{w}(b)$	c
0	881	370	0.00	1174	378	0.00
1	859	363	0.00	1146	371	0.00
2	801	344	0.02	1068	354	0.02
3	717	315	0.04	957	326	0.04
4	617	280	0.07	823	291	0.07
5	587	241	0.12	783	251	0.11
6	397	200	0.17	530	211	0.16
7	298	160	0.23	397	170	0.22
8	209	122	0.29	279	131	0.29
9	136	88	0.37	182	95	0.36
10	82	58	0.46	109	64	0.45

large and the nonperturbative effects, which are very difficult to deal with, are important. Contrary, the hard processes involve large momentum transfers connected with a small value of the strong coupling constant. Hence, they can be described successfully by the methods of perturbative QCD.

In the previous Section we have calculated the number of the wounded nucleons, $\overline{w}_{AB}$, and the number of binary nucleon-nucleon collisions, $\overline{n}_{AB}$, in a collision of the two nuclei A and B. An interesting question is whether the knowledge of $\overline{w}_{AB}$ and $\overline{n}_{AB}$ may be used to make an estimate of the multiplicity of the particles produced in a nuclear collision, provided the information about the multiplicity of the particles produced in a more elementary nucleon-nucleon collision (at the same energy) is available. For hard processes it is natural to assume that the number of the produced particles scales with the number of binary collisions. In this case the scattering processes are well localized and the interference effects between different collisions may be neglected. For soft processes the appropriate scaling is more difficult to find. In fact, it is a postulate of the wounded-nucleon model proposed in Ref. [14] that the multiplicity of soft particles scales with the number of the wounded nucleons. We discuss this model in more detail below.

Table 3.2 Estimates of the charged particle multiplicities obtained from the wounded nucleon model, $\frac{1}{2}\,\overline{w}_{AA}\,\overline{N}_{\rm NN}$, compared with the measured multiplicities, $\overline{N}_{AA}$, for different reactions studied by the NA49 [16] and PHOBOS Collaborations [17]. The last column shows the ratio of the measured multiplicity and the model prediction.

Expt.	$E_{\rm lab}/A$ [GeV]	$\sqrt{s_{\rm NN}}$ [GeV]	$\overline{N}_{AA}$	$\overline{w}_{AA}$	$\frac{1}{2}\,\overline{w}_{AA}\,\overline{N}_{\rm NN}$	r
NA49	40	8.8	693	349	875	0.79
NA49	80	12.3	1029	349	1059	0.97
NA49	158	17.3	1413	362	1307	1.08
PHOBOS	(9000)	130.0	4200	355	2902	1.45

3.7 Wounded-nucleon model

Białas, Bleszyński and Czyż argued [14] that the average multiplicity in a collision of two nuclei with the mass numbers A and B is

$$\overline{N}_{AB} = \frac{1}{2}\,\overline{w}_{AB}\,\overline{N}_{\rm NN}, \tag{3.56}$$

where $\overline{N}_{\rm NN}$ is the average multiplicity in proton-proton (nucleon-nucleon) collisions, and $\overline{w}_{AB}$ is the average number of the wounded nucleons (calculated in the Glauber framework). The energy dependence of $\overline{N}_{\rm NN}$ is described by Eq. (3.14). The motivation for the use of Eq. (3.56) came from the interpretation of the nucleon-nucleus interactions. The formula (3.56) with an additional expression for the dispersion of multiplicity distributions form the main ingredients of the *wounded-nucleon model* of the nucleus-nucleus collisions [14].

The question may be asked about the physical picture leading to the main assumption of the model. The simplified answer is that the particles are produced by the decays of nucleons which are excited by soft collisions. Multiple soft collisions change only the excited states of the nucleons, but they do not contribute to the particle creation process. Each excited nucleon produces particles at the moment when it leaves the interaction region, so the multiplicity is proportional to the number of the excited nucleons (which are nothing other but the wounded nucleons in this case). More detailed analysis of the foundations of the wounded nucleon model refers to the issues of the formation time of hadrons. For a recent discussion see [15].

To see how Eq. (3.56) works in practice we may analyze the *most central* collisions at the SPS and RHIC. The NA49 Collaboration measured the total multiplicities of charged pions and kaons at different beam energies: $E_{\rm lab} = 40$, 80, and 158 A GeV [16]. In each case the number of the wounded nucleons was also calculated with the help of the Monte-Carlo simulations. Those results are shown in Table 3.2 (the fourth and the fifth column) and used to obtain the wounded-nucleon-model

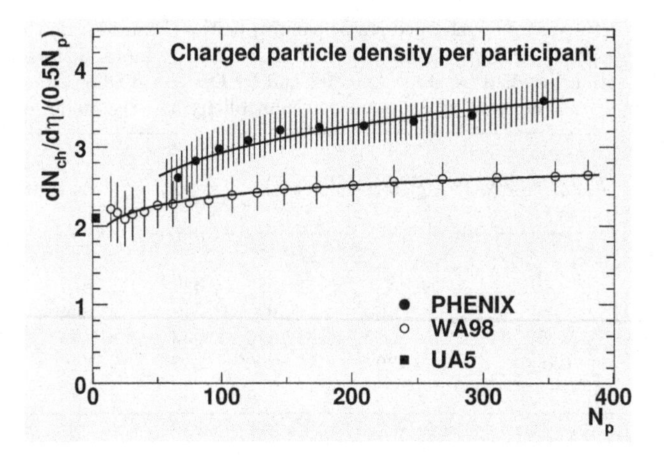

Fig. 3.4 The charged particle pseudorapidity density as a function of the number of the participants. The measurement of the PHENIX group at RHIC, $\sqrt{s_{\rm NN}} = 130$ GeV, is compared to the measurement done by the WA98 group at the SPS, $\sqrt{s_{\rm NN}} = 17.3$ GeV. Reprinted from [18] with permission from Elsevier.

estimate (the sixth column). The last column gives the ratio of the experimentally measured multiplicity to the model prediction. Table 3.2 contains also the result of the PHOBOS Collaboration [17] for the most central Au+Au collisions at $\sqrt{s_{\rm NN}} = 130$ GeV. We observe that the ratio r is close to unity, which indicates the general agreement of the wounded-nucleon model with the data. Nevertheless, at smaller energies the wounded-nucleon model overpredicts the measured multiplicity, whereas at higher energies (especially at RHIC) the predicted multiplicity is too small. The increase of the ratio r with the energy may indicate that the role of the hard processes becomes more and more important as we pass from low to high energies. Naturally, a contribution to the particle production coming from the hard processes scales with the number of the binary collisions, which is much larger than the number of the wounded nucleons (see Table 3.1).

The concept of the wounded-nucleon model may be applied also to the collisions at fixed beam energy. In this case we analyze the collisions with different impact parameters and observe if the multiplicity of the produced particles scales with the number of the wounded nucleons. In other words, we check the condition

$$\frac{\overline{N}_{AB}(\mathbf{b})}{\overline{w}(A;B;\mathbf{b})} = \text{const.}, \tag{3.57}$$

or, restricting ourselves to the central region in rapidity, we may also verify the formula

$$\frac{1}{\overline{w}(A;B;\mathbf{b})} \left. \frac{d\overline{N}_{AB}(\mathbf{b})}{d\eta} \right|_{\eta=0} = \text{const.} \tag{3.58}$$

The results of such investigations performed at RHIC by the PHENIX Collaboration [18] are shown in Fig. 3.4 (Au+Au collisions at $\sqrt{s_{\rm NN}} = 130$ GeV). We

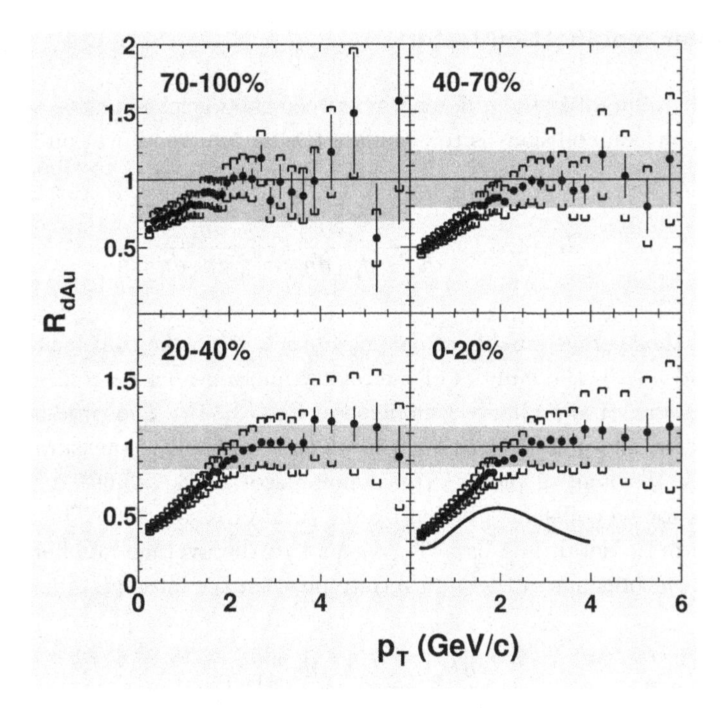

Fig. 3.5 The nuclear modification factor R_{dAu} as measured by the PHOBOS Collaboration at BNL. Reprinted figure with permission from [20]. Copyright (2009) by the American Physical Society.

can see that the number of the produced charged particles N_{ch} grows slowly with the increasing number of the participants, N_{part}. This effect is a deviation from the wounded-model conjecture (3.58). Similarly to our previous discussion of the energy dependence, the excess of the produced particles in this case may be also connected with an additional hard contribution proportional to the number of the binary collisions. To quantify this contribution one usually fits the experimental results to the formula

$$\frac{d\overline{N}_{AB}}{d\eta}(\mathbf{b})\bigg|_{\eta=0} = \frac{\alpha}{2}\,\overline{w}_{AB}(\mathbf{b}) + \beta\,\overline{n}_{AB}(\mathbf{b})\,. \tag{3.59}$$

The relative magnitude of the coefficient α and β characterizes the role of the hard processes. The PHENIX measurement gives $\beta/\alpha \approx 0.15$. Similar values were also obtained by PHOBOS [19].

3.8 Nuclear modification factor

A simple way to quantify the differences between the nucleus-nucleus collisions and the nucleon-nucleon collisions is to calculate the nuclear modification factor,

$$R_{AB}(p_\perp) = \frac{1}{\overline{n}_{AB}} \frac{d^2\overline{N}_{AB}}{dp_\perp d\eta} / \frac{1}{\sigma_{\rm tot}^{pp}} \frac{d\sigma_{\rm incl}^{pp}}{dp_\perp d\eta}. \tag{3.60}$$

Here $\overline{N}_{AB}$ is the average number of particles produced in the collisions of the nuclei A and B, and $\overline{n}_{AB}$ is the number of the binary nucleon-nucleon collisions obtained in the framework of the Glauber model, see Eq. (3.39). The produced particles are typically the charged hadrons or, what is more difficult to measure, the neutral hadrons, e.g., the neutral pions. The denominator of Eq. (3.60) is the inclusive cross section for pp collisions divided by the total cross section. This quantity, as explained in more detail in Chap. 30, is equal to the average number of particles produced in pp collisions in the appropriate phase-space interval,

$$\frac{dN_{pp}}{dp_\perp d\eta} = \frac{1}{\sigma_{\rm tot}^{pp}} \frac{d\sigma_{\rm incl}^{pp}}{dp_\perp d\eta}. \tag{3.61}$$

If the collisions of the nuclei A and B are simple superpositions of the elementary pp collisions, the scaling with the number of binary collisions should hold, and the nuclear modification factor is expected to be equal to 1.

In Fig. 3.5 we present the results obtained by the PHOBOS Collaboration [20]. The nuclear modification factor $R_{\rm dAu}$ (A = deuteron, B = gold) is shown as a function of the transverse momentum of charged hadrons for four different centrality classes. One can see that within the experimental errors $R_{\rm dAu}$ levels off at large momenta, $p_\perp \sim 2$ GeV, and approaches unity. This type of behavior indicates that the scaling with the number of binary collisions is indeed appropriate at large momenta.

In Fig. 3.6 we show the results delivered by the BRAHMS Collaboration [21]. In this case the three panels on the left-hand-side describe the data at midrapidity, $\eta = 0$, whereas the three panels on the right-hand-side describe the data at $\eta = 2.2$. The two upper panels show the nuclear modification factor $R_{\rm AuAu}$ for central collisions (centrality class 0–10%), the two panels in the center show $R_{\rm AuAu}$ for peripheral collisions (centrality class 40–60%), and the two lower panels show the ratio of the central data to the peripheral data, $R_{\rm AuAu}(0\text{–}10\%)/R_{\rm AuAu}(40\text{–}60\%)$. At first we may notice a similarity between the peripheral Au+Au collisions and d+Au collisions shown in Fig. 3.5. In both cases the nuclear modification factor levels off at $p_\perp \sim 2$ GeV. This is what one may expect, since the peripheral Au+Au collisions involve a relatively small number of the participants, and they may be

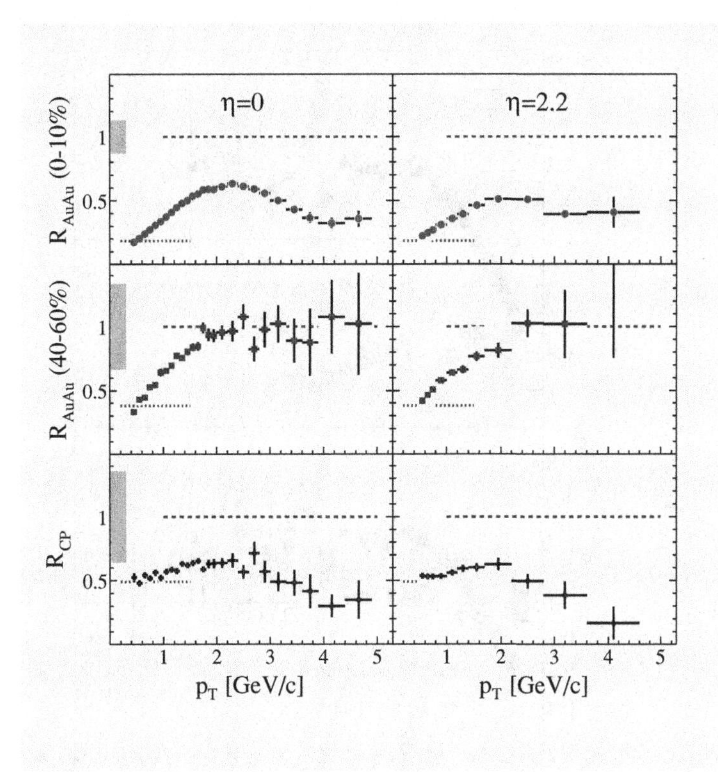

Fig. 3.6 The nuclear modification factors R_{AuAu} for central and peripheral collisions (the upper and central two panels), and their ratio (the lower two panels). The measurement of the BRAHMS Collaboration at BNL. Reprinted figure with permission from [21]. Copyright (2009) by the American Physical Society.

equivalent to asymmetric collisions where a smaller nucleus hits a larger one. Interestingly, the results at $\eta = 0$ and $\eta = 2.2$ are very similar, indicating that the same production mechanism is valid at different rapidities. On the other hand, the nuclear modification factor measured in the central Au+Au collisions shows a clear deviation from the scaling with the number of the binary collisions. For the central Au+Au collisions the measured values of R_{AuAu} are always below 1 and differ significantly from those measured in peripheral collisions.

Such a behavior of the nuclear modification factor R_{AuAu} in the central Au+Au collisions was independently and consistently found by the four RHIC experiments [20–23]. For example, in the upper panel of Fig. 3.7 one can see the result of the PHENIX Collaboration [22] confirming that R_{AuAu} for central Au+Au collisions is much smaller than R_{dAu} for the minimum bias d+Au events and shows no leveling off at unity.

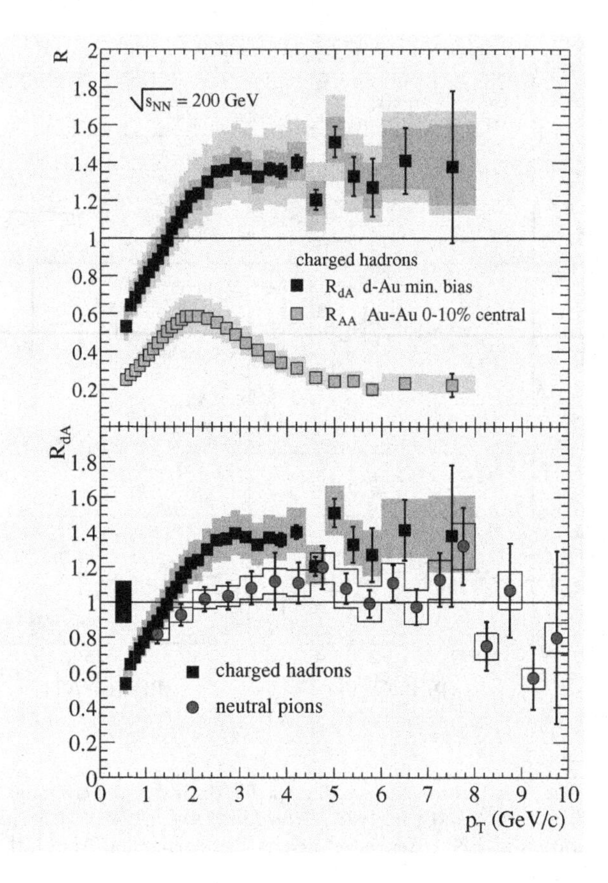

Fig. 3.7 The nuclear modification factors R_{dAu} and R_{AuAu} measured by the PHENIX Collaboration at BNL. Reprinted figure with permission from [22]. Copyright (2009) by the American Physical Society.

Bibliography to Chapter 3

[1] R. J. Glauber, "High Energy Collision Theory," in "Lectures in Theoretical Physics", W. E. Brittin and L. G. Dunham eds. (Interscience Publishers, New York, 1959) Vol. 1, p. 315.

[2] R. J. Glauber, in "High Energy Physics and Nuclear Structure", G. Alexander ed. (North Holland Publishing Co., Amsterdam, 1967) p. 311.

[3] R. J. Glauber, in "High Energy Physics and Nuclear Structure", S. Devons ed. (Plenum Press, New York, 1970) p. 207.

[4] R. J. Glauber, "Quantum Optics and Heavy Ion Physics," *Nucl. Phys.* **A774** (2006) 3–13.

[5] K. V. Shajesh, "Eikonal approximation," report prepared at the University of Oklahoma, http://www.nhn.ou.edu/ shajesh/eikonal/sp.pdf.

[6] D. E. Kharzeev and J. Raufeisen, "High energy nuclear interactions and QCD: An introduction," `nucl-th/0206073`.

[7] S. Mandelstam, "Determination of the pion-nucleon scattering amplitude from

dispersion relations and unitarity. general theory," *Phys. Rev.* **112** (1958) 1344–1360.

[8] **Aachen-CERN-Heidelberg-Munich** Collaboration, W. Thome *et al.*, "Charged particle multiplicity distributions in pp collisions at ISR energies," *Nucl. Phys.* **B129** (1977) 365.

[9] **CDF** Collaboration, F. Abe *et al.*, "Pseudorapidity distributions of charged particles produced in $\bar{p}p$ interactions at $\sqrt{s} = 630$ GeV and 1800 GeV," *Phys. Rev.* **D41** (1990) 2330.

[10] R. D. Woods and D. S. Saxon, "Diffuse surface optical model for nucleon-nuclei scattering," *Phys. Rev.* **95** (1954) 577–578.

[11] W. Czyz and L. C. Maximon, "High-energy, small angle elastic scattering of strongly interacting composite particles," *Annals Phys.* **52** (1969) 59–121.

[12] A. Bialas, W. Czyz, and L. Lesniak, "Additive quark model of multiparticle production and nucleus-nucleus collisions at high energies," *Phys. Rev.* **D25** (1982) 2328.

[13] W. Broniowski, M. Rybczynski, and P. Bozek, "GLISSANDO: GLauber Initial-State Simulation AND mOre.," *Comput. Phys. Commun.* **180** (2009) 69–83.

[14] A. Bialas, M. Bleszynski, and W. Czyz, "Multiplicity distributions in nucleus-nucleus collisions at high energies," *Nucl. Phys.* **B111** (1976) 461.

[15] A. Bialas, "Wounded nucleons, wounded quarks: An update," *J. Phys.* **G35** (2008) 044053.

[16] **NA49** Collaboration, S. V. Afanasiev *et al.*, "Energy dependence of pion and kaon production in central Pb + Pb collisions," *Phys. Rev.* **C66** (2002) 054902.

[17] **PHOBOS** Collaboration, B. B. Back *et al.*, "Charged-particle pseudorapidity density distributions from Au+Au collisions at $\sqrt{s_{\rm NN}} = 130$ GeV," *Phys. Rev. Lett.* **87** (2001) 102303.

[18] **PHENIX** Collaboration, A. Milov, "Charged particle multiplicity and transverse energy in Au Au collisions at $\sqrt{s_{\rm NN}} = 130$ GeV," *Nucl. Phys.* **A698** (2002) 171–176.

[19] **PHOBOS** Collaboration, B. B. Back *et al.*, "Collision geometry scaling of Au+Au pseudorapidity density from $\sqrt{s_{\rm NN}} = 19.6$ GeV to 200 GeV," *Phys. Rev.* **C70** (2004) 021902.

[20] **PHOBOS** Collaboration, B. B. Back *et al.*, "Centrality dependence of charged hadron transverse momentum spectra in d+Au collisions at $\sqrt{s_{\rm NN}} = 200$ GeV," *Phys. Rev. Lett.* **91** (2003) 072302.

[21] **BRAHMS** Collaboration, I. Arsene *et al.*, "Transverse momentum spectra in Au+Au and d+Au collisions at $\sqrt{s_{\rm NN}} = 200$ GeV and the pseudorapidity dependence of high p_T suppression," *Phys. Rev. Lett.* **91** (2003) 072305.

[22] **PHENIX** Collaboration, S. S. Adler *et al.*, "Absence of suppression in particle production at large transverse momentum in $\sqrt{s_{\rm NN}} = 200$ GeV d+Au collisions," *Phys. Rev. Lett.* **91** (2003) 072303.

[23] **STAR** Collaboration, J. Adams *et al.*, "Evidence from d+Au measurements for final-state suppression of high p_T hadrons in Au+Au collisions at RHIC," *Phys. Rev. Lett.* **91** (2003) 072304.

Chapter 4

Spacetime Picture of Ultra-Relativistic Heavy-Ion Collisions

In this Chapter we discuss the spacetime evolution of matter created in relativistic heavy-ion collisions. Our presentation follows the time development of the system. We start with the introduction of various mechanisms of particle production and end up with the discussion of the methods used to determine the spacetime dimensions of the system at *freeze-out*, i.e., at the stage where the final hadrons are emitted.

The *string models*, the *parton cascade models*, and the theory of *color glass condensate* are briefly presented in Sec. 4.1. In Sec. 4.2 the intriguing question of the presumable very fast thermalization/equilibration of the produced quark-gluon plasma is reviewed. If the state of the local thermodynamic equilibrium is indeed reached and maintained, the subsequent evolution of the system may be described in terms of the relativistic perfect-fluid hydrodynamics, whose main features are summarized in Sec. 4.3 [1]. The *hydrodynamic expansion* causes that the system becomes more and more dilute. The *phase transition* from the quark-gluon plasma to the hadronic gas takes place. Here the phenomenological approaches get in close contact with the fundamental theory, since the hydrodynamic description may use the form of the equation of state that is obtained directly from the lattice simulations of QCD. In particular, the information about the phase transition may be incorporated into the hydrodynamic description.

Further expansion triggers a transition from the strongly interacting hadronic gas to the weakly interacting system of hadrons which move freely to detectors. Such decoupling of hadrons is called the *freeze-out*. The freeze-out is generally a complicated process that may proceed through several stages — the most important ones, known as the *thermal/kinetic freeze-out* and the *chemical freeze-out*, are discussed in Secs. 4.4 and 4.5, respectively. The spacetime dimensions of the system at the thermal freeze-out may be inferred from the *interferometry* measurements which are shortly summarized in Sec. 4.6.

The relativistic character of the particle production, namely, the time dilation effect, is reflected in the spacetime picture where the fast particles are produced

[1] In the natural way, the chronological development of the ideas was quite different. At first the perfect-fluid hydrodynamics was applied, and only later its successes in reproducing the RHIC data with early starting times suggested early thermalization.

later, while the slow ones are produced earlier (in the center-of-mass frame). This means that the evolution of matter in the central region is "measured" by the longitudinal proper time $\tau = \sqrt{t^2 - z^2}$ rather than by the ordinary laboratory time t (see our discussion of the boost-invariance in Sec. 2.7). This characteristic feature is illustrated in Fig. 4.1, where a sequence of different stages in the evolution of matter is shown.

4.1 Particle production processes

The result of the multiple nucleon-nucleon collisions discussed in the previous Chapter is that the two colliding nuclei evolve rapidly into an extended, hot and dense system of quarks and gluons. There exist several frameworks to describe this transition, for example: QCD string breaking, QCD parton cascades, or color glass condensate evolving into glasma and later into the quark-gluon plasma. In all cases, the process of the particle production may be characterized by the decoherence time $\tau_{\rm dec}$ which is required to form the incoherent distribution of quarks and gluons from the highly coherent nuclear wave functions. Theoretical interpretations of the data measured at RHIC suggest that this time is very short, $\tau_{\rm dec} \ll 1$ fm. This is so, because the decoherence time $\tau_{\rm dec}$ should be smaller than the equilibration time $\tau_{\rm therm}$, and the latter has been found to be a fraction of a fermi at RHIC.

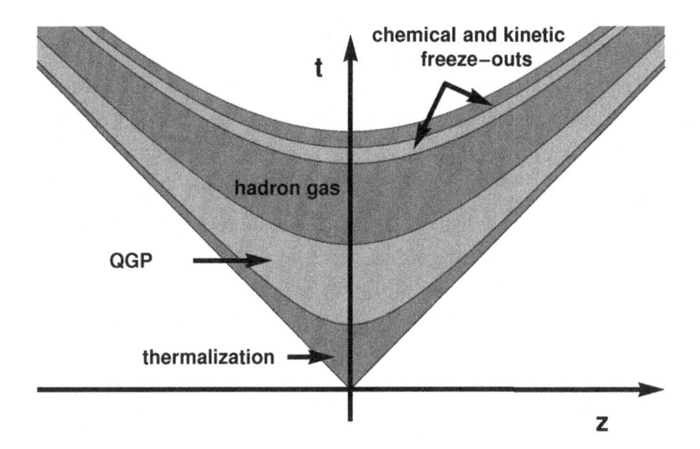

Fig. 4.1 The spacetime diagram of ultra-relativistic nuclear collisions. In the center-of-mass frame, partons moving fast hadronize later than those moving slowly. Consequently, at very high energies the evolution of the system at midrapidity is governed by the longitudinal proper time $\tau = \sqrt{t^2 - z^2}$, rather than by the ordinary time t. Note, that this picture breaks in the fragmentation regions (i.e., at large values of $|\eta|$) where physical processes have different character.

4.1.1 *String decays*

In the string picture, the nuclei pass through each other and the collisions of the nucleons lead to the formation of color strings. In analogy to a quark-antiquark string depicted in Fig. 1.1, the strings formed in nucleon-nucleon collisions may be imagined as quark - diquark pairs connected by the color field. Such systems may be naively treated as the excited nucleons. In the next step, the strings decay/fragment forming quarks and gluons or directly hadrons. The hadrons (sometimes clusters of hadrons) are modeled as smaller pieces of the original string. Fragmentation of strings into hadrons is described in the framework of the Monte-Carlo simulations which originate from the Lund Model [1, 2], its Jetset version [3, 4], and the Dual Parton Model [5–8].

Nowadays, the most popular models which incorporate string dynamics are HIJING (heavy-ion jet interaction generator) [9–13] and RQMD (relativistic quantum molecular dynamics) [14–19]. Besides the string decays these two models include other important physical effects. For example, following the Fritiof [20, 21] and especially the Pythia version of the Lund Model [22, 23], HIJING includes the production of minijets. The minijets are pairs of partons which go in the opposite directions like standard jets but have transverse momentum which might be regarded as a lower limit for hard processes, $p_\perp \geq 2$ GeV. This means that the production rate of minijets may be still obtained from the perturbative QCD. At present, HIJING is the only event generator for nuclear collisions, which incorporates such perturbative-QCD effects.

A different physical picture is offered by RQMD. In this case, a diquark from the initially spanned quark-diquark string may collide with nucleons. Such collisions lead to deacceleration of the diquark and, consequently, to large stopping power and relatively large baryon/energy density in the central region. Moreover, the RQMD model combines the string dynamics with the classical propagation of hadrons and formation of resonances. The interactions between the particles drive the system towards the local equilibrium and develop the collective flow (already in the pre-equilibrium stage). Another characteristic feature of RQMD is that overlapping color strings may fuse and form so called *color ropes* [24, 25].

We note that many other models for the relativistic heavy-ion collisions have been also developed, for example: ATTILA [26], VENUS [27, 28], HIJET [29, 30], or MCMHA [31]. Unfortunately, all these models, including HIJING and RQMD, are not able to describe successfully the data collected at the RHIC energies.

Production of quarks and gluons by the decays of strings may be also incorporated into kinetic equations. This case, being a simple model of production of the quark-gluon plasma from color fields, will be discussed in more detail in Chap. 11.

4.1.2 *Parton cascade model*

The parton cascade model [32–36] is based solely on the perturbative QCD. The colliding nuclei are treated as clouds of quarks and gluons which penetrate through each other. Multiple hard scatterings between partons as well as the gluon radiation produce large energy and entropy density.

The initial state is viewed as an ensemble of quarks and gluons determined by the quark and gluon distribution functions $q_f\left(x, Q^2\right)$ and $g(x, Q^2)$. The Bjorken x variable is defined as the ratio of the longitudinal momentum of the constituent of a hadron to the hadron longitudinal momentum in the reference frame where the hadron has very large energy. The quantity Q^2 is the parton virtuality. The fact that partons are confined in hadrons and cannot exist as free particles implies that they propagate off-shell and Q^2 is an additional independent variable. A limited information about the transverse momenta is additionally used to model the phase-space distributions of partons before the interaction. The subsequent time evolution of the parton phase-space distributions is governed by a relativistic Boltzmann equation with a collision term that contains dominant perturbative QCD interactions. Recently, the parton cascade models have been intensively studied in the context of the early thermalization problem. We shall come back to this issue in Sec. 4.2.

It is important to emphasize that both the string approach and the parton cascade model encounter conceptual problems and limitations. The string picture becomes invalid at very high energies, when the strings overlap and cannot be treated as independent objects. On the other hand, the parton approach is invalid at lower energies, where parton scatterings involve momentum transfers which are too small to be described by perturbation theory.

4.1.3 *Color glass condensate*

The color glass condensate [37–40] is nowadays regarded as the universal form of matter that controls the high-energy scattering of strongly interacting matter. The high-energy limit is understood in this case as the limit where the energy of collisions goes to infinity but the typical momentum transfer is finite. This is not the short distance perturbative limit where both the momentum transfer and the energy go to infinity. The high-energy limit includes non-perturbative phenomena, although it is also a weak coupling limit, $\alpha_s \ll 1$.

The idea of the color glass condensate was motivated by the HERA data indicating that the gluon density is rising rapidly as a function of decreasing x. This implies that the transverse density of gluons also increases. However, since the total cross sections rise slowly at high energies, the gluons must "fit" inside the size of the hadron, and the density of gluons should become limited or *saturated* [41–44].

The emerging physical picture in the transverse plane is that a hadron may

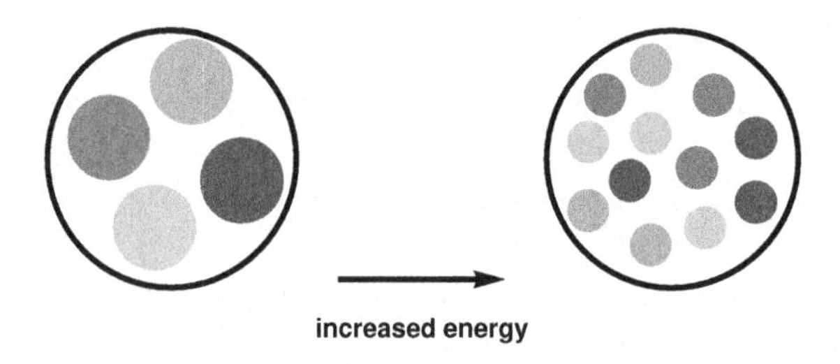

Fig. 4.2 With the increased energy, the colliding hadron looks like a denser system of smaller gluons.

be viewed as a tightly packed system of gluons that are larger than a certain size scale. With the increased energy more gluons are added but their size is smaller, see Fig. 4.2. In the momentum space there is a critical momentum Q_s characterizing the filling: the gluons with the momentum smaller than Q_s are tightly packed, while the gluons with the larger momenta may still fill the empty space. The saturation scale increases as the energy increases, hence the total number of gluons may increase without the bound.

Since the low x gluons are closely packed together, their interaction strength becomes weak. Such a weakly coupled gluon system is called a color glass condensate. The word "color" in the name appears because the gluons are colored. The word "glass" describes the property of the low x gluons, whose evolution is slow compared to other time scales present in the problem. The word "condensate" reflects the large gluon occupation numbers.

The gluons at small x are generated by gluons at larger values of x. In fact, the fast gluons may be treated as the sources of the slow gluons described in terms of the classical color fields. The fast gluons are Lorentz contracted and redistributed on the two very thin sheets representing the two colliding hadrons/nuclei. The sheets are perpendicular to the beam axis. The fast gluons produce the color electric and magnetic fields which also exist only in the sheets and are mutually orthogonal [2], see Fig. 4.3. Because of the uncertainty principle, it may seem that the gluons associated with those fields have large x, not small. This paradox is resolved by the fact that the small x gluons are described by the Fourier transform of the vector potential that exists also in the region outside the sheets. In this way the gluons with small x have indeed the large longitudinal extent.

Immediately after the collision, i.e., just after the passage of the two gluonic sheets through each other, the longitudinal electric and magnetic fields are produced. This form of matter is called the *glasma* [45]. The physical situation resembles the case of the string models, see Fig. 4.4, however, in the string models only the longitudinal color electric field is present while the glasma consists of both

[2]We should be familiar with this picture, see Sec. 2.7 where Feynman's general ideas about hadronic collisions were introduced.

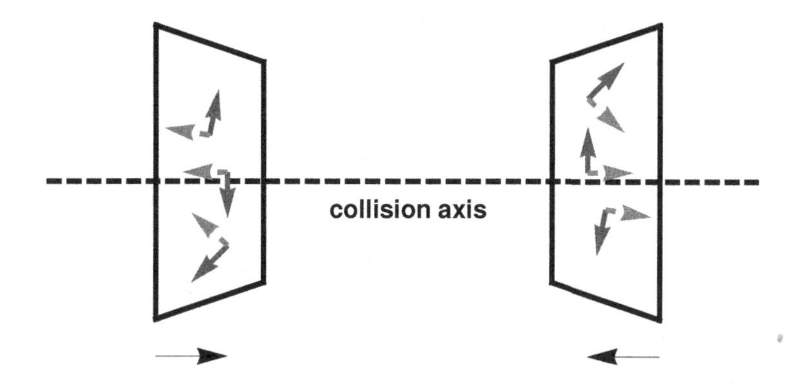

Fig. 4.3 The color electric and magnetic fields describing low x gluons of the color glass condensate (CGC) befor the collision exist only in the sheets and are mutually orthogonal.

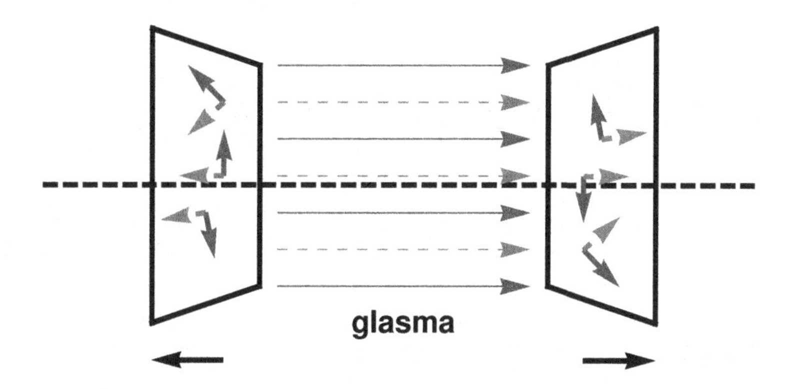

Fig. 4.4 After the collision, in addition to the transverse CGC fields on the sheets there are longitudinal color electric and magnetic fields forming *glasma*.

the electric and magnetic fields. The glasma fields decay due to the classical rearrangement of the fields into radiation of gluons with $p_\perp \sim Q_s$. Also the decays due to the quantum pair creation are possible. In this way the quark-gluon plasma is produced.

4.2 Thermalization

The experimental data obtained in the RHIC experiments favor a very short thermalization/equilibration time, $\tau_{\rm therm} < 1$ fm. The support for this idea comes mainly from the observation of the large elliptic flow, an effect explained by the hydrodynamic expansion with an early starting time $\tau_{\rm i}$ [3], see our discussion in Sec. 2.5.2. Within the concept of a strongly coupled quark-gluon plasma such a short time appears in the natural way, as the strongly interacting system equilibrates very fast *per se*. It is still, however, an open problem if the plasma in the early stage is indeed strongly coupled. In the last decade many different approaches were used to address the problem of very early thermalization and very often such attempts considered the plasma as a weakly interacting system. Below, following Ref. [46], we briefly review several approaches discussing this intriguing issue.

i) The equilibration problem was studied within the parton cascade model. In its early form, the model included only binary collisions [47]. Further important developments by Geiger and Muller took into account the gluon radiation in the initial and final states [32–34, 36]. More recently, the advanced numerical codes have been developed, which emphasize the role of the multi-particle processes [48–50]. In Refs. [48, 50] the production and absorption $2 \leftrightarrow 3$ processes are taken into account, whereas in Ref. [49] the three-particle collisions $3 \leftrightarrow 3$ are studied. Within both approaches the equilibration is claimed to be significantly speeded-up when compared to the equilibration driven by the binary collisions, however the determined thermalization times remain larger than 1 fm for RHIC conditions.

ii) In the papers mentioned above, one usually assumed that the initial partons are produced by hard or semi-hard interactions of partons in the incident nuclei. In the "bottom-up" thermalization scenario [51], the initial state is described by the QCD saturation mechanism that is incorporated in the framework of the color glass condensate. Thus, the initial state is dominated by the small x gluons of transverse momentum of order Q_s, the saturation scale. Those gluons are freed from the incoming nuclei after a time $1/Q_s$. Weak coupling techniques are used, because Q_s is expected to be much larger than $\Lambda_{\rm QCD}$ at sufficiently high collision energies (one expects $Q_s \sim 1$ GeV for RHIC, and $Q_s \sim 2$–3 GeV for the LHC [51]). The calculations performed within the "bottom-up" thermalization scenario [51], where the binary and $2 \leftrightarrow 3$ processes are taken into account, give an equilibration time of at least 2.6 fm/c [52], see also [53].

iii) The most common concept is, of course, that the equilibration of the system is an effect of parton rescattering. An interesting phenomenon occurs, however, that the equilibration is speeded-up by instabilities generated in an anisotropic quark-gluon plasma [54, 55]. This is so because the growth of the unstable modes is associated with the isotropization of the momentum distributions, which helps to achieve the full equilibration. Moreover, the instabilities are much "faster" than the collisions in the weak coupling regime [54–57].

iv) In several papers the authors argued that the very process of particle production leads to the equilibrium state without any secondary interactions. For

[3] The equilibration time $\tau_{\rm therm}$ denotes the moment when the state of the local thermal equilibrium is reached. In this book we usually identify it with the time when the hydrodynamic description may be initialized, $\tau_{\rm therm} \sim \tau_{\rm i}$.

example, Refs. [58, 59] refer to the Schwinger mechanism of the particle production in the strong chromoelectric field. However, this approach explains the equilibration of the transverse momentum only. The approach of Refs. [60, 61], where the longitudinal momentum is also thermal, includes the Hawking-Unruh effect: an observer moving with an acceleration a experiences the influence of a thermal bath with an effective temperature $a/2\pi$, similar to the one present in the vicinity of a black hole horizon.

v) An interesting physical scenario was also formulated, where the thermalization is not an effect of collisions but a consequence of the chaotic dynamics of the non-Abelian classical color fields, coupled or not to the classical colored particles [62–64].

We summarize different approaches to the early-thermalization problem with the conclusion that newly developed simulations based on perturbative QCD as well as new non-perturbative frameworks may explain the thermalization times of about 1 fm. Still, the problem exists to explain much shorter thermalization times used in the hydrodynamic codes that successfully describe the data.

4.3 Hydrodynamic expansion

The production and rescattering of partons is typically described in the framework of the relativistic kinetic theory — the only approach which is capable of dealing with non-equilibrium phenomena. This is the reason why the field of the ultra-relativistic heavy-ion collisions triggered broad interest in the development of the kinetic theory [4]. RQMD and the parton cascade model are good examples of the progress done in this field.

If the thermalization rate is sufficiently fast, a locally thermalized quark-gluon plasma is created. In this case, the subsequent evolution of the system may be described by the equations of the relativistic perfect-fluid hydrodynamics [5]. The use of the relativistic hydrodynamics simplifies very much our description of the collision process (provided the thermalization indeed takes place at a certain stage). The hydrodynamic equations describe local conservation laws of energy, momentum, baryon number, strangeness, etc., and require knowledge of the equation of state of the matter. The important point here is that the equation of state may be taken in the most sophisticated form as that delivered by the lattice simulations of QCD. In this way the hydrodynamic calculations form a straightforward link between the QCD first-principle calculations and the dynamic properties of the expanding

[4]The basic information about the relativistic kinetic theory can be found in Chaps. 8 and 9. An example of the application of the kinetic theory to describe formation of the quark-gluon plasma in heavy-ion collisions is presented in Chap. 11.

[5]The general introduction to relativistic hydrodynamics of a perfect fluid is given in Chaps. 13 and 14. The use of the hydrodynamic equations to describe the evolution of matter in the ultra-relativistic heavy-ion collisions is presented in Chaps. 20–22.

fireball formed and experimentally studied in ultra-relativistic heavy-ion collisions.

The lattice QCD results show a very rapid change of the thermodynamic properties of the hadronic matter, an effect indicating the presence of the phase transition from the hadron gas at low temperatures ($T \ll T_c \sim 170$ MeV) to the quark-gluon plasma at high temperatures ($T \gg T_c$). The order of the phase transition is studied currently by several groups. The accumulated evidence coming from the lattice QCD simulations suggests that the phase transition at $\mu_B = 0$ is a smooth crossover — we discuss this issue in more detail in Sec. 5.4.

If the phase transition were of the first order, a mixed phase consisting of the plasma and the hadron gas would be created. In such a case one may consider three steps in the evolution of matter as it passes the phase transition. At first, to a good approximation the matter is an adiabatically expanding quark-gluon plasma, later the matter expands as a mixed phase of the plasma and the hadron gas, finally the matter expands as the hadron gas. Of course, the plasma formed during a collision should reorganize itself into hadrons again, since finally only hadrons will reach our detectors.

4.4 Thermal freeze-out

The *thermal* or *kinetic freeze-out* is the stage in the evolution of matter when the hadrons practically stop to interact. In other words, the thermal freeze-out is a transition from a strongly coupled system (very likely evolving from one local equilibrium state to another) to a weakly coupled one (consisting of essentially free streaming particles). It is triggered by the expansion of matter, which causes a rapid growth of the mean free path, $\lambda_{\rm mfp}$, of particles. The thermal freeze-out happens when the timescale connected with the collisions, $\tau_{\rm coll} \sim \lambda_{\rm mfp}$, becomes larger than the expansion timescale, $\tau_{\rm exp}$. In this case the particles depart from each other so fast that the collision processes become ineffective. We may formulate this condition as the inequality

$$\tau_{\rm coll} \geq \tau_{\rm exp}, \tag{4.1}$$

where the magnitude of the collision time is determined by the product of the average cross section and the particle density,

$$\tau_{\rm coll} \sim \frac{1}{\sigma n}, \tag{4.2}$$

whereas the magnitude of the expansion time is characterized by the divergence of the four-velocity field, u^μ, describing the hydrodynamic flow of matter [65],

$$\tau_{\rm exp} \sim \frac{1}{\partial_\mu u^\mu}. \tag{4.3}$$

Very often a simplified criterion is assumed which says that the thermal freeze-out happens at the time [6] when the mean free path of hadrons is of the same order as the size of the system.

In general, the particles with different cross sections (as, e.g., strange and non-strange particles) may have different freeze-out points, i.e., they may decouple from the rest of matter at different times. Similarly, different types of processes may be turned off at different times, as is discussed in more detail in the next Section. With these remarks in mind it is clear that freeze-out is a complicated dynamic process [66]. However, if the expansion of the matter is fast, the thermal freeze-out process is also fast and deserves its name; the momentum distributions of particles are frozen during a sudden decoupling process and do not change in time anymore when the particles move freely to detectors. In this case the measurement of the transverse-momentum spectra reveals information about the state of matter just before the thermal freeze-out.

In fact, the experimental data show that the hadron spectra in the soft region, $p_\perp \leq 2$ GeV, have thermal character, although the original thermal distributions are modified by the collective transverse flow and decays of resonances, see Fig. 4.5. This observation strongly supports the usefulness of the concept of the thermal freeze-out. In particular, the model calculations allow us to determine the temperature and strength of the transverse flow at the decoupling stage [7].

4.5 Chemical freeze-out

The thermal freeze-out described above may take place after the *chemical freeze-out.* The concept of the latter is based on the observation that at low colliding energies the inelastic cross sections are typically smaller than the elastic cross sections. Therefore, as the hadronic system cools down the inelastic collisions between its constituents are very likely to cease before the elastic collisions. The moment when the inelastic collisions stop is defined as the chemical freeze-out. Naturally, the temperatures corresponding to the two freeze-outs satisfy the condition

$$T_{\rm chem} > T_{\rm therm}, \tag{4.4}$$

where $T_{\rm chem}$ is the temperature of the chemical freeze-out (inferred from the studies of the ratios of hadron multiplicities, see remarks below) and $T_{\rm therm}$ is the temperature of the thermal freeze-out (inferred from the studies of the transverse-momentum spectra).

As the system evolves from the chemical to thermal freeze-out, the dominant processes are elastic collisions (such as, for example, $\pi+\pi \to \rho \to \pi+\pi$ or $\pi+N \to \Delta \to \pi + N$) and strong decays of heavier resonances which populate the yields of

[6] Strictly speaking, one should talk about the collection of spacetime points that form a three-dimensional freeze-out hypersurface in the four-dimensional Minkowski space.

[7] Popular hydro-inspired parameterizations used to determine the thermal properties of the system at the thermal freeze-out will be discussed in Chap. 25.

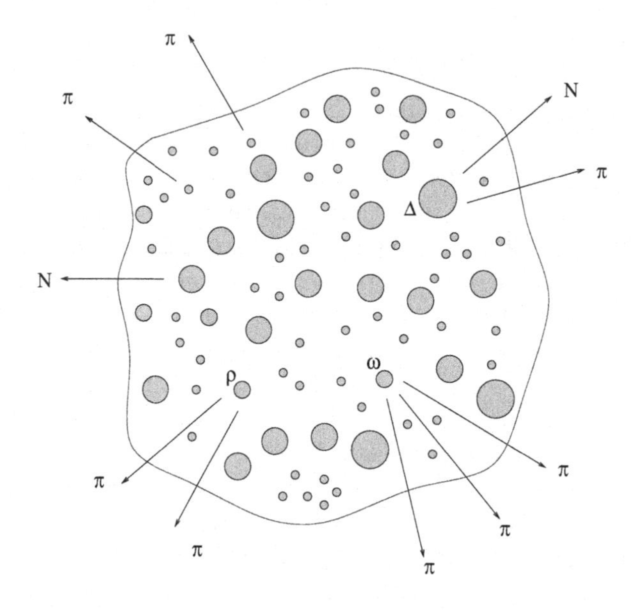

Fig. 4.5 A schematic physical picture adopted in the thermal models of particle production. At a certain stage of the evolution of the system, a gas of stable hadrons and resonances is formed. The final (measured) multiplicities of hadrons consist of primary particles, present in the hot fireball, and of secondary particles coming from the decays of resonances. Reprinted from [67] with permission from Acta Phys. Pol. **B**.

stable hadrons. Accepting this scenario and assuming that hadrons form an ideal expanding gas, we may describe the measured ratios of hadron abundances (frozen at the chemical freeze-out) by a few thermodynamic parameters characterizing the matter at the chemical freeze-out [68]. In fact, such approach (called commonly the thermal approach or the statistical approach) turned out to be very successful in describing the data at the AGS [69, 70], SPS [70–74], and RHIC energies [75, 76].

An interesting fact was observed by Cleymans and Redlich — the chemical freeze-out parameters (temperature, chemical potentials and fugacities) at the CERN/SPS, BNL/AGS and GSI/SIS correspond to a unique value (1 GeV) of the energy per hadron [77, 78]. Another intriguing theoretical observation was made by Becattini et al. — the statistical models are able to reproduce the particle ratios in e^+e^- annihilation processes [79, 80] and $pp\,(p\bar{p})$ collisions [81]. Moreover, at very high energies the temperature obtained from the model calculations, $T_{\rm chem} \sim$ 170 MeV, turned out to be the same for both elementary and nuclear collisions, although the final-state hadronic interactions (possibly leading to thermalization) are completely absent in the e^+e^- case. This coincidence may indicate that chemical equilibrium is pre-established by the hadronization process [58, 59, 79–83] and explain partially the early thermalization problem discussed in Sec. 4.2.

4.5.1 *Little Bang*

It is interesting to note that there are analogies between the physics of the ultra-relativistic heavy-ion collisions and the physics of the Early Universe [84]. We know that the observed microwave radiation has a Bose-Einstein spectrum, whose temperature is redshifted by the cosmological expansion from the original 3000 K to the observed 2.7 K. *Nota bene*, this is the finest example of the Bose-Einstein distribution known in Nature — deviations from the ideal curve found by the COBE experiment are observed at the level smaller than 10^{-5}. In the ultra-relativistic heavy-ion collisions the measured hadron transverse-momentum spectra are blueshifted due to the strong transverse flow which shares similarities with the Hubble flow [85]. Hence, in both the Big Bang and heavy-ion collisions, the observed momentum spectra contain thermal physics modified by collective dynamics. Another analogy is that the microwave background radiation decoupled long after the nucleosynthesis took place. This resembles the situation with the two freeze-outs discussed above.

The analogies outlined above are used sometimes to justify the name Little Bang for the process of the production of hot and dense matter in the ultra-relativistic heavy-ion collisions. This name is attractive but when we use it we should keep in mind that the two processes have also different aspects. For example, due to the gravitational attractive forces the expansion of the Universe is slower [86].

4.6 Hanbury Brown – Twiss interferometry

The radar technology developed during the Second World War initiated the field of radio astronomy. In the 1950s Hanbury Brown and Twiss showed that it was possible to determine the angular sizes of astronomical radio sources and stars from the correlations of signal intensities, rather than amplitudes. A similar idea was later used in the particle and nuclear physics, making possible to gain information about the spacetime geometry of the collision region [87–90] (for review papers see [91–96]). In spite of the differences in the astronomical and high-energy measurements, the studies of the emission regions formed in hadronic/nuclear interactions have become known as the HBT interferometry (more recently, the name *femtoscopy* is used).

To some extent, the results of the HBT analysis are model dependent. The physical significance of the measured quantities, such as the system size or lifetime, relies on the theoretical input used to interpret the data. This point will be clarified later in Chap. 17, where the most popular approaches of modeling the HBT correlations are presented. At the moment, we may only emphasize that the HBT interferometry measures the range of the correlations rather than the actual size of the system. In any case, the comparison of the HBT data with model predictions restricts the class of the realistic models. In this Section we introduce the basic concepts and terminology used in the HBT technique. This will allow us to present and discuss the most recent HBT data coming from RHIC and the SPS.

The fundamental object in the HBT interferometry is the two-particle correlation function $C(\mathbf{p}_1, \mathbf{p}_2)$, measured for pairs of identical particles such as $\pi^+\pi^+$, $\pi^-\pi^-$, or K^+K^+. In general, it is defined by the expression [97]

$$C(\mathbf{p}_1, \mathbf{p}_2) = \frac{\mathcal{P}_2(\mathbf{p}_1, \mathbf{p}_2)}{\mathcal{P}_1(\mathbf{p}_1)\mathcal{P}_1(\mathbf{p}_2)}, \tag{4.5}$$

where $\mathcal{P}_1(\mathbf{p})$ is the invariant inclusive one-particle distribution function in the space of rapidity and transverse-momentum,

$$\mathcal{P}_1(\mathbf{p}) = E_p \frac{dN}{d^3p} = \frac{dN}{dy d^2p_\perp}, \tag{4.6}$$

and $\mathcal{P}_2(\mathbf{p}_1, \mathbf{p}_2)$ is the analogous two-particle distribution

$$\mathcal{P}_2(\mathbf{p}_1, \mathbf{p}_2) = E_{p_1} E_{p_2} \frac{dN}{d^3p_1 d^3p_2} = \frac{dN}{dy_1 d^2p_{1\perp} dy_2 d^2p_{2\perp}}. \tag{4.7}$$

Equations (4.6) and (4.7) imply that the correlation function (4.5) transforms like a Lorentz scalar.

The distributions of pairs, $\mathcal{P}_2(\mathbf{p}_1, \mathbf{p}_2)$, and single particles, $\mathcal{P}_1(\mathbf{p})$, are obtained by measuring the particles in each event and then by calculating the average over an ensemble of events. This procedure means that the correlation function (4.5) may be expressed by the inclusive cross sections

$$C(\mathbf{p}_1, \mathbf{p}_2) = \frac{1}{\sigma_{\rm tot}} \frac{d\sigma^{\rm incl}}{d^3p_1 d^3p_2} / \left(\frac{1}{\sigma_{\rm tot}} \frac{d\sigma^{\rm incl}}{d^3p_1} \frac{1}{\sigma_{\rm tot}} \frac{d\sigma^{\rm incl}}{d^3p_2} \right). \tag{4.8}$$

For a general discussion of the inclusive measurements we refer to Sec. 30.5.

In Eq. (4.5) we may change variables and use the average momentum

$$\mathbf{k} = \frac{1}{2} \left(\mathbf{p}_1 + \mathbf{p}_2 \right), \tag{4.9}$$

and the difference of the two momenta

$$\mathbf{q} = \mathbf{p}_1 - \mathbf{p}_2. \tag{4.10}$$

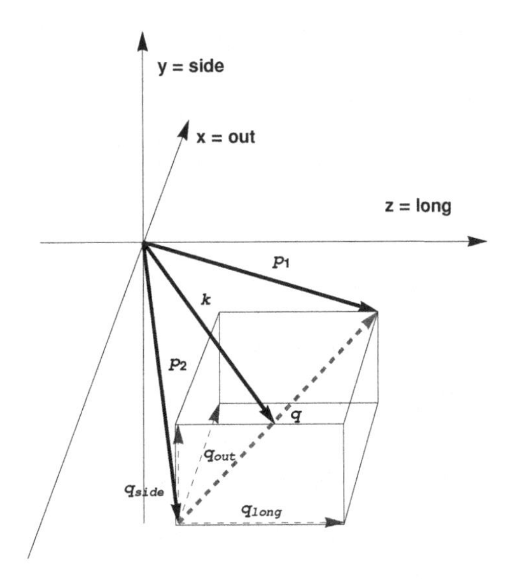

Fig. 4.6 The out-side-long coordinate system used in the standard HBT analysis of the correlation functions. The vector $\mathbf{k}$ lies in the $x-z$ plane. By making the Lorentz boost along the collision axis we may also set $\mathbf{k}_{\parallel} = 0$. In this way we change to the special frame that is called the longitudinally comoving system (LCMS).

In this way, we introduce the correlation function depending on $\mathbf{k}$ and $\mathbf{q}$ [8]

$$C(\mathbf{k}+\mathbf{q}/2, \mathbf{k}-\mathbf{q}/2) \to C(\mathbf{k},\mathbf{q}). \tag{4.11}$$

As $\mathbf{q}$ becomes very large (for a fixed value of $\mathbf{k}$), the correlation between particles is lost and the function $C(\mathbf{k},\mathbf{q})$ approaches unity. On the other hand, at small momentum differences the correlation function should tend to 2, due to the Bose-Einstein interference of non-interacting particles emitted from a perfectly chaotic source, see Sec. 17.1.3. The characteristic falloff of the correlation function with increasing values of $|\mathbf{q}|$ gives hints about the size of the emitting source at the time when the observed particles no longer interact with other particles, i.e., at the thermal freeze-out. This is why we discuss the HBT interferometry in the end of the Chapter describing the spacetime evolution of matter.

Clearly, the direction of the vector $\mathbf{q}$ is important in the analysis of the correlation functions, since the system may have different characteristics in different directions. Following Bertsch and Pratt, in most of the HBT measurements one uses special coordinates. They are determined in a Cartesian frame with the z axis along the collision axis (the so called longitudinal or *long* axis), and with two other axes defined separately for each pair: the x axis is parallel to $\mathbf{k}_\perp$ (the so called *out* axis), whereas the y axis is perpendicular to both the longitudinal and out axes (the so called *side* axis), see Fig. 4.6.

[8] For simplicity of notation we keep the same letter to denote the correlation function depending on $\mathbf{k}$ and $\mathbf{q}$.

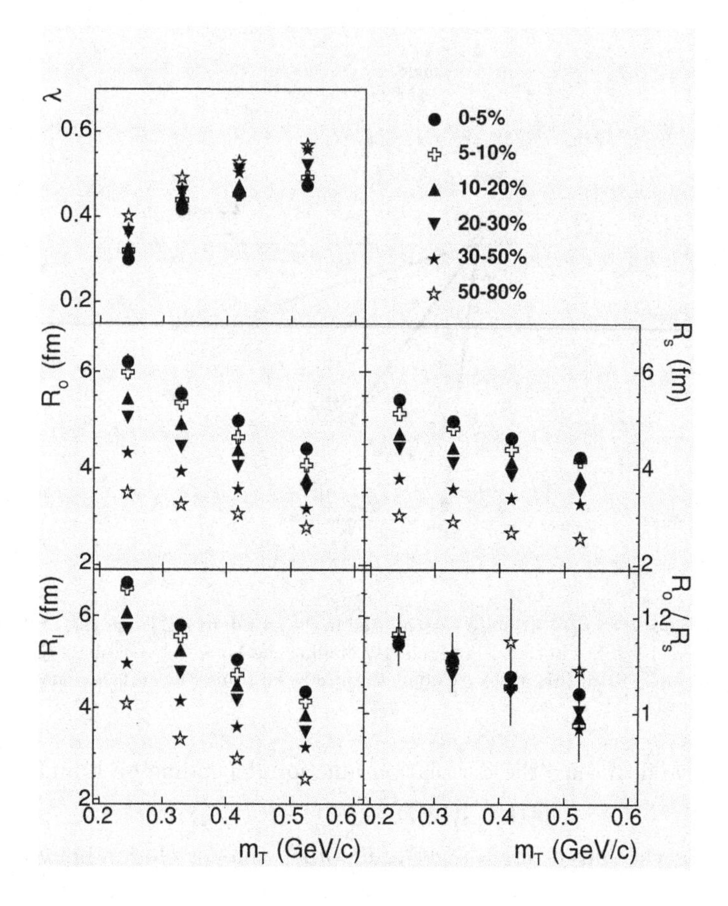

Fig. 4.7 Pion HBT radii vs. $m_T = \sqrt{k_T^2 + m_\pi^2}$ measured by the STAR Collaboration at midrapidity in six different centrality windows. Reprinted figure with permission from [98]. Copyright (2009) by the American Physical Society.

For boost-invariant and cylindrically symmetric systems the correlation function may depend only on the value of $k_\perp$ and three values of the relative momentum, $\mathbf{q} = (q_{\rm out}, q_{\rm side}, q_{\rm long})$. In this case (with the Coulomb interaction removed) we may parameterize the correlation function in the following way,

$$C(k_\perp, \mathbf{q}) = 1 + \lambda \exp\left[-R_{\rm long}^2(k_\perp) q_{\rm long}^2 - R_{\rm out}^2(k_\perp) q_{\rm out}^2 - R_{\rm side}^2(k_\perp) q_{\rm side}^2\right], \quad (4.12)$$

where the quantities $R_{\rm out}$, $R_{\rm side}$ and $R_{\rm long}$ are called commonly the HBT radii, while λ is the intercept parameter.

For systems which are cylindrically asymmetric one should use a more general parameterizations than that given by Eq. (4.12). In this case we talk about azimuthally sensitive HBT interferometry (azHBT) [99, 100]. The latter brings in

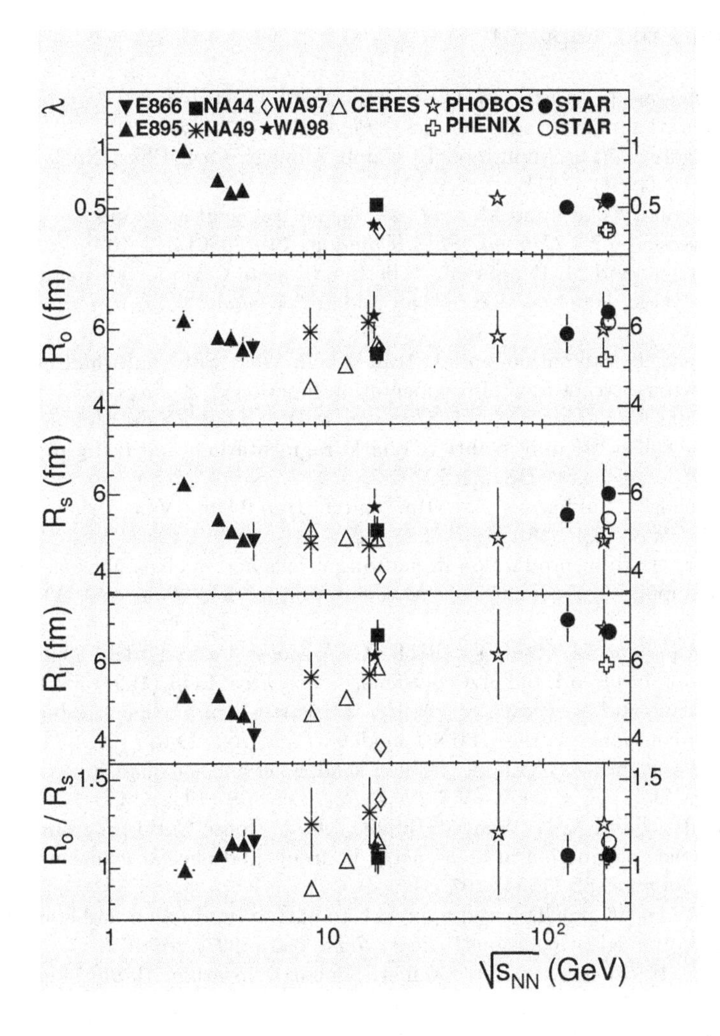

Fig. 4.8 Energy dependence of the pion HBT parameters for central Au+Au, Pb+Pb, and Pb+Au collisions at midrapidity and $k_\perp = 0.2$ GeV. The figure shows the compilations of the results obtained by NA44, NA49, CERES, PHENIX, PHOBOS and STAR Collaborations. Reprinted figure with permission from [98]. Copyright (2009) by the American Physical Society.

information about the dependence of shape and flow on the azimuthal angle ϕ. This information is complementary to the data on the transverse-momentum elliptic flow coefficient v_2.

The examples of the experimental data showing the pion HBT radii measured at midrapidity (where $\mathbf{k}_{\parallel} \approx 0$) and averaged over the azimuthal angle are shown in Figs. 4.7 and 4.8. Two remarkable features of these data are that the ratio $R_{\rm out}/R_{\rm side}$ is close to unity and the radii exhibit weak energy dependence. We shall come back to the discussion of the issues related to the HBT measurements in Chaps. 17 and 22.

Bibliography to Chapter 4

[1] B. Andersson, G. Gustafson, G. Ingelman, and T. Sjostrand, "Parton fragmentation and string dynamics," *Phys. Rept.* **97** (1983) 31.

[2] B. Andersson, "The Lund model," Camb. Monogr. Part. Phys. Nucl. Phys. Cosmol. **7** (1997) 1–471.

[3] T. Sjostrand, "The Lund Monte Carlo for jet fragmentation and e^+e^- physics: Jetset version 6.2," *Comput. Phys. Commun.* **39** (1986) 347–407.

[4] T. Sjostrand and M. Bengtsson, "The Lund Monte Carlo for jet fragmentation and e^+e^- physics: Jetset version 6.3: an update," *Comput. Phys. Commun.* **43** (1987) 367.

[5] A. Capella, U. Sukhatme, and J. Tran Thanh Van, "Soft multi-hadron production from partonic structure and fragmentation functions," *Z. Phys.* **C3** (1979) 329–337.

[6] A. Capella, U. Sukhatme, C.-I. Tan, and J. Tran Thanh Van, "Jets in small p_T hadronic collisions, universality of quark fragmentation, and rising rapidity plateaus," *Phys. Lett.* **B81** (1979) 68.

[7] A. Capella, U. Sukhatme, C.-I. Tan, and J. Tran Thanh Van, "Dual parton model," *Phys. Rept.* **236** (1994) 225–329.

[8] J. Ranft, "Hadron production in hadron-nucleus and nucleus-nucleus collisions in the dual monte carlo multichain fragmentation model," *Phys. Rev.* **D37** (1988) 1842.

[9] X.-N. Wang and M. Gyulassy, "HIJING: A Monte Carlo model for multiple jet production in pp, pA and AA collisions," *Phys. Rev.* **D44** (1991) 3501–3516.

[10] X.-N. Wang and M. Gyulassy, "A Systematic study of particle production in $p + p(\bar{p})$ collisions via the HIJING model," *Phys. Rev.* **D45** (1992) 844–856.

[11] X.-N. Wang and M. Gyulassy, "Gluon shadowing and jet quenching in $A + A$ collisions at $\sqrt{s_{\rm NN}} = 200$ GeV," *Phys. Rev. Lett.* **68** (1992) 1480–1483.

[12] M. Gyulassy and X.-N. Wang, "HIJING 1.0: A Monte Carlo program for parton and particle production in high-energy hadronic and nuclear collisions," *Comput. Phys. Commun.* **83** (1994) 307.

[13] X.-N. Wang, "A pQCD-based approach to parton production and equilibration in high-energy nuclear collisions," *Phys. Rept.* **280** (1997) 287–371.

[14] H. Sorge, H. Stocker, and W. Greiner, "Poincare invariant Hamiltonian dynamics: Modelling multi-hadronic interactions in a phase space approach," *Annals Phys.* **192** (1989) 266–306.

[15] H. Sorge, H. Stocker, and W. Greiner, "Relativistic quantum molecular dynamics approach to nuclear collisions at ultrarelativistic energies," *Nucl. Phys.* **A498** (1989) 567c–576c.

[16] C. Hartnack *et al.*, "Quantum Molecular Dynamics: A microscopic model from UNILAC to CERN energies," *Nucl. Phys.* **A495** (1989) 303c–320c.

[17] J. Aichelin, "Quantum molecular dynamics: A Dynamical microscopic n body approach to investigate fragment formation and the nuclear equation of state in heavy ion collisions," *Phys. Rept.* **202** (1991) 233–360.

[18] S. A. Bass *et al.*, "Microscopic models for ultrarelativistic heavy ion collisions," *Prog. Part. Nucl. Phys.* **41** (1998) 255–369.

[19] M. Bleicher *et al.*, "Relativistic hadron hadron collisions in the ultra- relativistic quantum molecular dynamics model," *J. Phys.* **G25** (1999) 1859–1896.

[20] B. Andersson, G. Gustafson, and B. Nilsson-Almqvist, "A model for low p_T hadronic reactions, with generalizations to hadron-nucleus and nucleus-nucleus collisions," *Nucl. Phys.* **B281** (1987) 289.

[21] B. Nilsson-Almqvist and E. Stenlund, "Interactions between hadrons and nuclei: The Lund Monte Carlo, FRITIOF version 1.6," *Comput. Phys. Commun.* **43** (1987) 387.
[22] H.-U. Bengtsson and T. Sjostrand, "The Lund Monte Carlo for hadronic processes: PYTHIA version 4.8," *Comput. Phys. Commun.* **46** (1987) 43.
[23] T. Sjostrand, "High-energy physics event generation with PYTHIA 5.7 and JETSET 7.4," *Comput. Phys. Commun.* **82** (1994) 74–90.
[24] H. Sorge, M. Berenguer, H. Stocker, and W. Greiner, "Color rope formation and strange baryon production in ultrarelativistic heavy ion collisions," *Phys. Lett.* **B289** (1992) 6–11.
[25] H. Sorge, "Flavor Production in Pb (160 A GeV) on Pb Collisions: Effect of Color Ropes and Hadronic Rescattering," *Phys. Rev.* **C52** (1995) 3291–3314.
[26] M. Gyulassy, "ATTILA (all type target independent LUND algorithm) for nuclear collisions at approximately 200 A GeV," invited talk at the Int. Conference on Intermediate Energy Nuclear Physics, Balatonfured, Hungary, June 8-13, 1987.
[27] K. Werner, "Analysis of energy flow in ^{16}O nucleus collisions at 60 GeV and 200 GeV by the multistring model VENUS," *Z. Phys.* **C42** (1989) 85.
[28] K. Werner, "Strings, pomerons, and the VENUS model of hadronic interactions at ultrarelativistic energies," *Phys. Rept.* **232** (1993) 87–299.
[29] T. W. Ludlam, A. Pfoh, and A. Shor, "HIJET. A Monte Carlo event generator for p-nucleus and nucleus-nucleus collisions," Brookhaven Nat. Lab. Upton – BNL-37196 (REC.FEB.86) 9p.
[30] A. Shor and R. S. Longacre, "Effects of secondary interactions in proton-nucleus and nucleu-nucleus collisions using the HIJET event generator," *Phys. Lett.* **B218** (1989) 100.
[31] Y. Iga, R. Hamatsu, S. Yamazaki, and H. Sumiyoshi, "Monte Carlo event generator MCMHA for high-energy hadron-nucleus collisions and intranuclear cascade interactions," *Z. Phys.* **C38** (1988) 557–564.
[32] K. Geiger and B. Muller, "Dynamics of parton cascades in highly relativistic nuclear collisions," *Nucl. Phys.* **B369** (1992) 600–654.
[33] K. Geiger, "Thermalization in ultrarelativistic nuclear collisions. 1. Parton kinetics and quark gluon plasma formation," *Phys. Rev.* **D46** (1992) 4965–4985.
[34] K. Geiger, "Thermalization in ultrarelativistic nuclear collisions. 2. Entropy production and energy densities at RHIC and LHC," *Phys. Rev.* **D46** (1992) 4986–5005.
[35] K. Geiger, "Particle production in high-energy nuclear collisions: A Parton cascade cluster hadronization model," *Phys. Rev.* **D47** (1993) 133–159.
[36] K. Geiger, "Space-time description of ultrarelativistic nuclear collisions in the QCD parton picture," *Phys. Rept.* **258** (1995) 237–376.
[37] L. D. McLerran and R. Venugopalan, "Computing quark and gluon distribution functions for very large nuclei," *Phys. Rev.* **D49** (1994) 2233–2241.
[38] L. D. McLerran and R. Venugopalan, "Gluon distribution functions for very large nuclei at small transverse momentum," *Phys. Rev.* **D49** (1994) 3352–3355.
[39] L. D. McLerran and R. Venugopalan, "Green's functions in the color field of a large nucleus," *Phys. Rev.* **D50** (1994) 2225–2233.
[40] E. Iancu and R. Venugopalan, "The color glass condensate and high energy scattering in QCD," `hep-ph/0303204`.
[41] L. V. Gribov, E. M. Levin, and M. G. Ryskin, "Semihard processes in QCD," *Phys. Rept.* **100** (1983) 1–150.
[42] A. H. Mueller and J.-W. Qiu, "Gluon recombination and shadowing at small

values of x," *Nucl. Phys.* **B268** (1986) 427.
[43] K. J. Golec-Biernat and M. Wusthoff, "Saturation effects in deep inelastic scattering at low Q^2 and its implications on diffraction," *Phys. Rev.* **D59** (1999) 014017.
[44] K. J. Golec-Biernat and M. Wusthoff, "Saturation in diffractive deep inelastic scattering," *Phys. Rev.* **D60** (1999) 114023.
[45] T. Lappi and L. McLerran, "Some features of the glasma," *Nucl. Phys.* **A772** (2006) 200–212.
[46] S. Mrowczynski, "Instabilities driven equilibration of the quark-gluon plasma," *Acta Phys. Polon.* **B37** (2006) 427–454.
[47] D. H. Boal, "Thermalization in ultrarelativistic heavy ion collisions: Parton cascade approach," *Phys. Rev.* **C33** (1986) 2206–2208.
[48] Z. Xu and C. Greiner, "Thermalization of gluons in ultrarelativistic heavy ion collisions by including three-body interactions in a parton cascade," *Phys. Rev.* **C71** (2005) 064901.
[49] X.-M. Xu, Y. Sun, A.-Q. Chen, and L. Zheng, "Triple-gluon scatterings and early thermalization," *Nucl. Phys.* **A744** (2004) 347–377.
[50] Z. Xu and C. Greiner, "The role of the gluonic $gg \leftrightarrow ggg$ interactions in early thermalization in ultrarelativistic heavy-ion collisions," *Eur. Phys. J.* **C49** (2007) 187–191.
[51] R. Baier, A. H. Mueller, D. Schiff, and D. T. Son, "Bottom-up thermalization in heavy ion collisions," *Phys. Lett.* **B502** (2001) 51–58.
[52] R. Baier, A. H. Mueller, D. Schiff, and D. T. Son, "Does parton saturation at high density explain hadron multiplicities at RHIC?," *Phys. Lett.* **B539** (2002) 46–52.
[53] A. El, Z. Xu, and C. Greiner, "Thermalization of a color glass condensate and review of the 'Bottom-Up' scenario," *Nucl. Phys.* **A806** (2008) 287–304.
[54] S. Mrowczynski, "Color collective effects at the early stage of ultrarelativistic heavy ion collisions," *Phys. Rev.* **C49** (1994) 2191–2197.
[55] P. Arnold, J. Lenaghan, G. D. Moore, and L. G. Yaffe, "Apparent thermalization due to plasma instabilities in quark gluon plasma," *Phys. Rev. Lett.* **94** (2005) 072302.
[56] J. Randrup and S. Mrowczynski, "Chromodynamic Weibel instabilities in relativistic nuclear collisions," *Phys. Rev.* **C68** (2003) 034909.
[57] A. Rebhan, P. Romatschke, and M. Strickland, "Hard-loop dynamics of non-Abelian plasma instabilities," *Phys. Rev. Lett.* **94** (2005) 102303.
[58] A. Bialas, "Fluctuations of string tension and transverse mass distribution," *Phys. Lett.* **B466** (1999) 301–304.
[59] W. Florkowski, "Schwinger tunneling and thermal character of hadron spectra," *Acta Phys. Polon.* **B35** (2004) 799–808.
[60] D. Kharzeev and K. Tuchin, "From color glass condensate to quark gluon plasma through the event horizon," *Nucl. Phys.* **A753** (2005) 316–334.
[61] P. Castorina, D. Kharzeev, and H. Satz, "Thermal hadronization and Hawking-Unruh radiation in QCD," *Eur. Phys. J.* **C52** (2007) 187–201.
[62] T. S. Biro, C. Gong, B. Muller, and A. Trayanov, "Hamiltonian dynamics of Yang-Mills fields on a lattice," *Int. J. Mod. Phys.* **C5** (1994) 113–149.
[63] S. Sengupta, P. K. Kaw, and J. C. Parikh, "Relativistic particle simulation of a colored parton plasma," *Phys. Lett.* **B446** (1999) 104–110.
[64] V. M. Bannur, "Statistical mechanics of Yang-Mills classical mechanics," *Phys. Rev.* **C72** (2005) 024904.
[65] C. M. Hung and E. V. Shuryak, "Equation of state, radial flow and freeze-out in

high energy heavy-ion collisions," *Phys. Rev.* **C57** (1998) 1891–1906.
[66] L. V. Bravina, I. N. Mishustin, J. P. Bondorf, A. Faessler, and E. E. Zabrodin, "Microscopic study of freeze-out in relativistic heavy ion collisions at SPS energies," *Phys. Rev.* **C60** (1999) 044905.
[67] W. Broniowski, A. Baran, and W. Florkowski, "Thermal approach to rhic," *Acta Phys. Polon.* **B33** (2002) 4235–4258.
[68] J. Cleymans and H. Satz, "Thermal hadron production in high-energy heavy ion collisions," *Z. Phys.* **C57** (1993) 135–148.
[69] P. Braun-Munzinger, J. Stachel, J. P. Wessels, and N. Xu, "Thermal equilibration and expansion in nucleus-nucleus collisions at the AGS," *Phys. Lett.* **B344** (1995) 43–48.
[70] F. Becattini, J. Cleymans, A. Keranen, E. Suhonen, and K. Redlich, "Features of particle multiplicities and strangeness production in central heavy ion collisions between 1.7-A- GeV/c and 158-A-GeV/c," *Phys. Rev.* **C64** (2001) 024901.
[71] F. Becattini, M. Gazdzicki, A. Keranen, J. Manninen, and R. Stock, "Study of chemical equilibrium in nucleus nucleus collisions at AGS and SPS energies," *Phys. Rev.* **C69** (2004) 024905.
[72] J. Sollfrank, M. Gazdzicki, U. W. Heinz, and J. Rafelski, "Chemical freezeout conditions in central S-S collisions at 200-A/GeV," *Z. Phys.* **C61** (1994) 659–666.
[73] P. Braun-Munzinger, J. Stachel, J. P. Wessels, and N. Xu, "Thermal and hadrochemical equilibration in nucleus-nucleus collisions at the SPS," *Phys. Lett.* **B365** (1996) 1–6.
[74] P. Braun-Munzinger, I. Heppe, and J. Stachel, "Chemical equilibration in Pb + Pb collisions at the SPS," *Phys. Lett.* **B465** (1999) 15–20.
[75] P. Braun-Munzinger, D. Magestro, K. Redlich, and J. Stachel, "Hadron production in Au Au collisions at RHIC," *Phys. Lett.* **B518** (2001) 41–46.
[76] W. Florkowski, W. Broniowski, and M. Michalec, "Thermal analysis of particle ratios and p_T spectra at RHIC," *Acta Phys. Polon.* **B33** (2002) 761–769.
[77] J. Cleymans and K. Redlich, "Unified description of freeze-out parameters in relativistic heavy ion collisions," *Phys. Rev. Lett.* **81** (1998) 5284–5286.
[78] J. Cleymans and K. Redlich, "Chemical and thermal freeze-out parameters from 1-A-GeV to 200-A-GeV," *Phys. Rev.* **C60** (1999) 054908.
[79] F. Becattini, "A thermodynamical approach to hadron production in e^+e^- collisions," *Z. Phys.* **C69** (1996) 485–492.
[80] F. Becattini, P. Castorina, J. Manninen, and H. Satz, "The thermal production of strange and non-strange hadrons in e^+e^- collisions," *Eur. Phys. J.* **C56** (2008) 493–510.
[81] F. Becattini and U. W. Heinz, "Thermal hadron production in pp and $p\bar{p}$ collisions," *Z. Phys.* **C76** (1997) 269–286.
[82] F. Becattini, "Thermal hadron production in high energy collisions," *J. Phys.* **G23** (1997) 1933–1940.
[83] U. W. Heinz, "Hadronic observables: Theoretical highlights," *Nucl. Phys.* **A638** (1998) 357c–364c.
[84] U. W. Heinz, "The little bang: Searching for quark-gluon matter in relativistic heavy-ion collisions," *Nucl. Phys.* **A685** (2001) 414–431.
[85] M. Chojnacki, W. Florkowski, and T. Csorgo, "On the formation of Hubble flow in Little Bangs," *Phys. Rev.* **C71** (2005) 044902.
[86] J. Letessier and J. Rafelski, "Hadrons and quark - gluon plasma," Camb. Monogr. Part. Phys. Nucl. Phys. Cosmol. **18** (2002) 1–397.
[87] G. Goldhaber, S. Goldhaber, W.-Y. Lee, and A. Pais, "Influence of Bose-Einstein

statistics on the antiproton proton annihilation process," *Phys. Rev.* **120** (1960) 300–312.
[88] G. I. Kopylov and M. I. Podgoretsky, "Correlations of identical particles emitted by highly excited nuclei," *Sov. J. Nucl. Phys.* **15** (1972) 219–223.
[89] G. I. Kopylov and M. I. Podgoretsky, "Multiple production and interference of particles emitted by moving sources," *Sov. J. Nucl. Phys.* **18** (1974) 336–341.
[90] G. I. Kopylov, "Like particle correlations as a tool to study the multiple production mechanism," *Phys. Lett.* **B50** (1974) 472–474.
[91] G. Baym, "The physics of Hanbury Brown-Twiss intensity interferometry: From stars to nuclear collisions," *Acta Phys. Polon.* **B29** (1998) 1839–1884.
[92] U. A. Wiedemann and U. W. Heinz, "Particle interferometry for relativistic heavy-ion collisions," *Phys. Rept.* **319** (1999) 145–230.
[93] U. W. Heinz and B. V. Jacak, "Two-particle correlations in relativistic heavy-ion collisions," *Ann. Rev. Nucl. Part. Sci.* **49** (1999) 529–579.
[94] R. M. Weiner, "Boson interferometry in high energy physics," *Phys. Rept.* **327** (2000) 249–346.
[95] T. Csorgo, "Particle interferometry from 40 MeV to 40 TeV," *Heavy Ion Phys.* **15** (2002) 1–80.
[96] B. Tomasik and U. A. Wiedemann, "Central and non-central HBT from AGS to RHIC," `hep-ph/0210250`.
[97] S. Pratt, "Coherence and coulomb effects on pion interferometry," *Phys. Rev.* **D33** (1986) 72–79.
[98] **STAR** Collaboration, J. Adams *et al.*, "Pion interferometry in Au+Au collisions at $\sqrt{s_{\rm NN}} = 200$ GeV," *Phys. Rev.* **C71** (2005) 044906.
[99] S. V. Voloshin, LBNL Annual Report R20, 1998, http://ie.lbl.gov/nsd1999/rnc/RNC.htm.
[100] **E895** Collaboration, M. A. Lisa *et al.*, "Azimuthal dependence of pion interferometry at the AGS," *Phys. Lett.* **B496** (2000) 1–8.

Chapter 5

More about Quarks and Gluons

This Chapter introduces more information about the formal aspects of *quantum chromodynamics* (QCD) . After short remarks of Sec. 5.1, where we present the historical development of the concept of a *color charge* as the source of strong interactions, the QCD Lagrangian and the phenomenon of *asymptotic freedom* is introduced in Sec. 5.2. The idea of the asymptotic freedom is then used in Sec. 5.3 as the motivation to consider in more detail the equation of state of the *weakly-interacting quark-gluon plasma* (wQGP) — a practically non-interacting gas of quarks and gluons [1].

In realistic situations one has to consider the case where the strong coupling constant remains large and the only way to establish the thermodynamic properties of the strongly interacting quark-gluon system is via the *lattice simulations of QCD* that are characterized in Sec. 5.4. Such simulations are based on first principles and deliver the most advanced information about the phase diagram of the strongly-interacting matter. They predict the existence of a phase transition from the hadron gas to the quark-gluon plasma, however, the order of the phase transition depends on the number of quark flavors, the quark masses, and the value of the baryon chemical potential μ_B. For three quark flavors with realistic masses and $\mu_B = 0$, the *crossover* phase transition rather than a genuine phase transition is found. This means that the values of the thermodynamic parameters change very abruptly in the narrow range of the temperature, however, no real discontinuities in the behavior of thermodynamic quantities are observed.

The Chapter is closed with Sec. 5.5 describing a special "rearrangement" of the standard eight gluonic fields. Two of them are treated as the neutral fields,

[1] In the introductory sections we have argued that the successes of the perfect-fluid hydrodynamics applied to ultra-relativistic heavy-ion collisions indicate that the plasma studied at RHIC behaves like a strongly-interacting system. However, there are still approaches where the plasma is treated as a weakly-interacting system, see our discussion of the early thermalization problem in Sec. 4.2. Thus, the strength of the interactions in the plasma created in the laboratory conditions is the question under debate. In any case, the asymptotic limit where the plasma is treated as an ideal gas rather than as a perfect fluid is a convenient reference point for many calculations. It is also conceivable that the chances for the creation of the weakly-interacting plasma in the laboratory grow with the increasing beam energies.

while the remaining six ones are combined together to form the three charged gluon fields. The quarks and charged gluons couple to the neutral gluons in the very similar way as the electrically charged particles couple to the electromagnetic field (of course, in addition there exist interactions between charged quarks and gluons). Such a choice of the basis in the color space emphasizes similarities to *quantum electrodynamics*. It is used in Chap. 11 where different processes are treated in the *abelian approximation*.

5.1 Color

In the early days of the quark model of hadrons it was realized that there were difficulties connected with the Pauli principle for quarks. There exist baryons such as, e.g., Δ^{++}, Δ^- or Ω^-, which have spin 3/2 and consist of three quarks of the same flavor: uuu, ddd and sss, respectively. In this case, all quarks have the same spin orientation and identical coordinate wave functions. In order to avoid the conflict with the Pauli principle, it was assumed that there exists a quantum number called *color*. The three different states of color are usually called: *red*, *green* and *blue*. The concept of color was introduced by Greenberg in 1964 and, independently, by Han and Nambu in 1965 [1, 2]. The further direct experimental evidence for color came from the measurements of the e^+e^- annihilations and the $\pi^0 \to \gamma\gamma$ decays.

Color is the source of the force which holds quarks together in hadrons. In some respect, it resembles ordinary electric charge being the source of electromagnetic interactions described by *Quantum Electrodynamics* (QED). For example, positive and negative color charges of the same kind attract each other forming mesons. On the other hand, three different colors can also attract each other. In this case baryons are formed. This phenomenon has of course no analog in QED. Both mesons and baryons are color-neutral objects as a whole, similarly as atoms consisting of negatively charged electrons and a positively charged nucleus are electrically neutral. Colloquially, we describe baryons and mesons as *white* objects. This terminology is, of course, based on the analogy with the way all real colors are made up of three primary colors. As each fundamental force in Nature, the color force is mediated by the exchange of some sort of quanta. In the case of QCD, they are called *gluons* (since they glue quarks together). There are eight types of gluons, which is a consequence of the fact that QCD is a *gauge theory* based on the SU(3) color group.

5.2 Gauge symmetry

The QCD Lagrangian has the form

$$\mathcal{L} = -\frac{1}{4}\sum_a F^a_{\mu\nu}F^{\mu\nu}_a + \sum_f^{N_f} \overline{\Psi}_f \left(i\gamma^\mu\partial_\mu - g\gamma^\mu \sum_a A^a_\mu \frac{\lambda^a}{2} - m_f \right) \Psi_f, \tag{5.1}$$

where Ψ_f are the quark fields ($f = u, d, s, c, b, t$ is the quark flavor index) and A_μ^a are the gluon fields ($a = 1, ..., 8$ is the color index). In Eq. (5.1), the quantity g is the strong coupling constant, γ^μ are the Dirac matrices, and λ^a are the Gell-Mann matrices:

$$\lambda^1 = \begin{pmatrix} 0 & 1 & 0 \\ 1 & 0 & 0 \\ 0 & 0 & 0 \end{pmatrix}, \quad \lambda^2 = \begin{pmatrix} 0 & -i & 0 \\ i & 0 & 0 \\ 0 & 0 & 0 \end{pmatrix}, \quad \lambda^3 = \begin{pmatrix} 1 & 0 & 0 \\ 0 & -1 & 0 \\ 0 & 0 & 0 \end{pmatrix},$$

$$\lambda^4 = \begin{pmatrix} 0 & 0 & 1 \\ 0 & 0 & 0 \\ 1 & 0 & 0 \end{pmatrix}, \quad \lambda^5 = \begin{pmatrix} 0 & 0 & -i \\ 0 & 0 & 0 \\ i & 0 & 0 \end{pmatrix},$$

$$\lambda^6 = \begin{pmatrix} 0 & 0 & 0 \\ 0 & 0 & 1 \\ 0 & 1 & 0 \end{pmatrix}, \quad \lambda^7 = \begin{pmatrix} 0 & 0 & 0 \\ 0 & 0 & -i \\ 0 & i & 0 \end{pmatrix},$$

$$\sqrt{3}\,\lambda^8 = \begin{pmatrix} 1 & 0 & 0 \\ 0 & 1 & 0 \\ 0 & 0 & -2 \end{pmatrix}. \tag{5.2}$$

The Gell-Mann matrices (divided by a factor of two) are the generators of the SU(3) group. They satisfy the commutation relations

$$\left[\frac{\lambda_a}{2}, \frac{\lambda_b}{2}\right] = i f_{abc} \frac{\lambda_c}{2}, \tag{5.3}$$

which define the Lie algebra of the SU(3) group (from now on we use the convention that the symbols of sums over repeated color indices may be omitted). The coefficients f_{abc} are called the structure constants of the Lie algebra. They are totally antisymmetric symbols with respect to the permutation of the indices $a\,b\,c$. Different from zero coefficients of the SU(3) group have the following values:

$$f^{123} = 1, \quad f^{147} = \frac{1}{2}, \quad f^{156} = -\frac{1}{2}, \quad f^{246} = \frac{1}{2}, \quad f^{257} = \frac{1}{2},$$

$$f^{345} = \frac{1}{2}, \quad f^{367} = -\frac{1}{2}, \quad f^{458} = \frac{\sqrt{3}}{2}, \quad f^{678} = \frac{\sqrt{3}}{2}. \tag{5.4}$$

The non-linear gluon field strength in Eq. (5.1) is defined by the expression

$$F_a^{\mu\nu} = \partial^\mu A_a^\nu - \partial^\nu A_a^\mu - g\, f_{abc} A_\mu^b A_\nu^c. \tag{5.5}$$

For fixed flavor, the quark wave function has three components (each of them is a Dirac bispinor)

$$\Psi = \begin{pmatrix} \psi_{\text{red}} \\ \psi_{\text{green}} \\ \psi_{\text{blue}} \end{pmatrix} \tag{5.6}$$

and the local SU(3) gauge transformations have the form

$$\Psi \rightarrow \Psi' = U(x)\Psi, \qquad U(x) \in \mathrm{SU}(3). \tag{5.7}$$

It is important to remember that for local transformations the matrices U depend on the spacetime position x. Lagrangian (5.1) remains invariant under transformations (5.7) if, at the same time, the gluon fields change according to the rule

$$A^\mu = A^\mu_a \frac{\lambda_a}{2} \rightarrow A'^{\,\mu} = U(x) A^\mu U^{-1}(x) - \frac{i}{g} U(x) \partial^\mu U^{-1}(x). \tag{5.8}$$

Derivation of this fact is the subject of Ex. 7.12.

QCD predicts weakening of the strong interaction at short distances or high momenta (the phenomenon called asymptotic freedom) and yields the running coupling constant

$$g^2(Q^2) = \frac{16\pi^2}{(11 - \frac{2}{3}N_f)\ln(Q^2/\Lambda^2_{\mathrm{QCD}})} \stackrel{Q^2 \rightarrow \infty}{\longrightarrow} 0, \tag{5.9}$$

with the QCD scale parameter Λ_{QCD} of about 200 MeV [2]. The vanishing of the QCD coupling constant at short distances implies that the interactions of quarks and gluons are negligible in the limit of very high temperature [3–5], which supports the concept of creation of the practically non-interacting quark-gluon plasma at extreme, perhaps "cosmological" temperatures [6] [3].

5.3 Equation of state of weakly-interacting quark-gluon plasma

Suppose that the interactions in the quark-gluon plasma may be indeed neglected. At the temperature T and quark baryon chemical potential μ, the energy densities and pressures of gluons, massless quarks and antiquarks are given by the following expressions [4]

$$\varepsilon_g = 16 \frac{\pi^2}{30} T^4, \qquad P_g = \frac{1}{3}\varepsilon_g, \tag{5.10}$$

$$\varepsilon_q + \varepsilon_{\bar{q}} = 6N_f \left(\frac{7\pi^2}{120} T^4 + \frac{1}{4}\mu^2 T^2 + \frac{1}{8\pi^2}\mu^4 \right),$$

$$P_q + P_{\bar{q}} = \frac{1}{3}\left(\varepsilon_q + \varepsilon_{\bar{q}}\right). \tag{5.11}$$

We note that the quark baryon chemical potential μ is one third of the baryon chemical potential μ_B,

$$\mu = \frac{1}{3}\mu_B. \tag{5.12}$$

The factor N_f may be taken here as the number of flavors which are effectively massless at a given temperature (for $T < 200$ MeV one may take $N_f = 2$, and for the temperature range 200 MeV $< T < 1$ GeV, where the strange quark can be considered light, $N_f = 3$).

[2]The generally adopted convention is to define the value of $g(Q^2)$ at the mass of the Z^0 boson. With $\alpha_s(m^0_Z) = g^2(m^0_Z)/4\pi = 0.118$ one finds $\Lambda_{\mathrm{QCD}} = 217$ MeV.

[3]The situation at high baryon density and low temperature is different because the Pauli principle forbids most scattering processes among quarks, except for those near the Fermi surface [7].

[4]The necessary integrals are elaborated in Secs. 8.7.2 and 8.7.3.

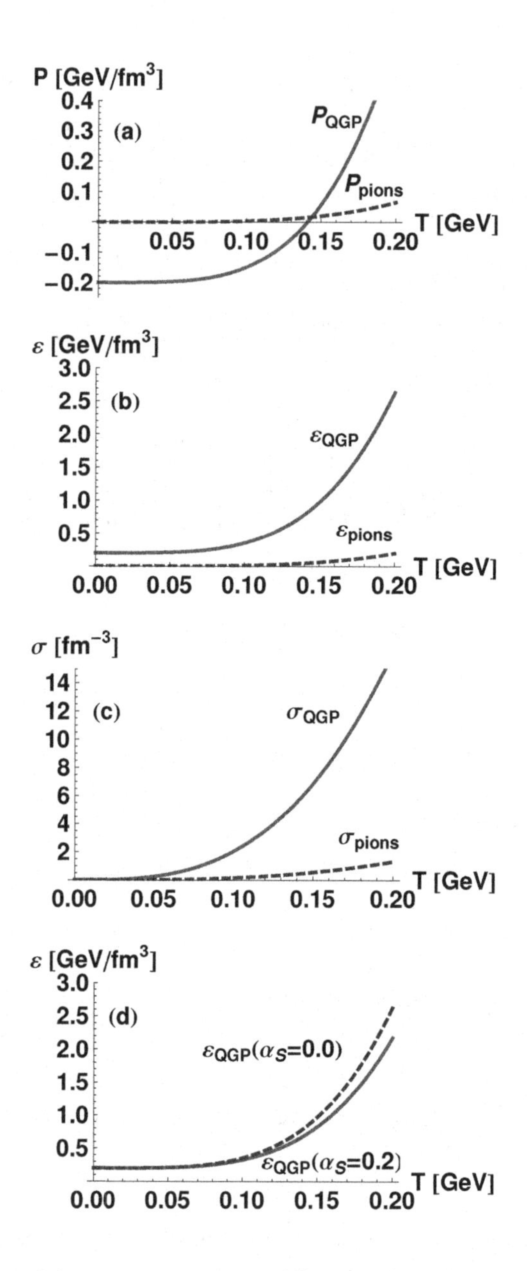

Fig. 5.1 **(a)** Pressure, **(b)** energy density, and **(c)** entropy density of the massless pion gas (dashed lines) and of the weakly interacting quark-gluon plasma (solid lines) as calculated from Eqs. (5.10)–(5.15) with $N_f = 2$, $\mu = 0$, and $B^{1/4} = 200$ MeV. These parameters give the critical temperature $T_c \approx 144$ MeV and the corresponding critical energy density $\varepsilon_c \approx 0.85$ GeV/fm^3. The part **(d)** displays the effect of the finite coupling constant, $\alpha_s = g^2/4\pi$, on the energy density of the plasma (in the leading order of the perturbative expansion given by Eq. (5.16)).

Before Eqs. (5.10) and (5.11) can be used to obtain other thermodynamic quantities of interest, one important effect has to be taken into account. Free quarks and gluons can only propagate if the true QCD vacuum state is destroyed and replaced by the perturbative vacuum. Such a modification of the ground state costs certain amount of energy per unit volume, quantitatively expressed by the MIT-bag constant $B \simeq (150\text{MeV} - 200\text{MeV})^4$. With this remark in mind, the equation of state of the weakly-interacting quark-gluon plasma can be written in the form

$$\varepsilon_{\text{qgp}} = \varepsilon_g(T) + \varepsilon_q(T,\mu) + \varepsilon_{\overline{q}}(T,\mu) + B \tag{5.13}$$

or

$$P_{\text{qgp}} = P_g(T) + P_q(T,\mu) + P_{\overline{q}}(T,\mu) - B, \tag{5.14}$$

where the partial pressure of gluons and quarks can be calculated directly from Eqs. (5.10) and (5.11). The negative sign of B in (5.14) is a consequence of the thermodynamic relations (see Eq. (5.19) below). It can be understood as a signal of the instability of the perturbative vacuum. The latter collapses unless it is supported by the high enough pressure of partons. More information about the physical significance of the bag constant may be found in the original paper about the Bag model of hadrons [8].

A lower limit of stability of the plasma is obtained by condition $P_{\text{qgp}} = 0$. More realistically, the plasma phase becomes unstable against formation of a gas of hadrons when its pressure is equal to that of a hadron gas at the same temperature and chemical potential. In order to gain simple quantitative results, we can restrict ourselves to the baryon-free case ($\mu = 0$) and assume that hadronic matter is represented by a gas of massless, noninteracting pions characterized by the following equation of state

$$\varepsilon_\pi = 3\,\frac{\pi^2}{30}\,T^4, \qquad P_\pi = \frac{1}{3}\varepsilon_\pi. \tag{5.15}$$

Thermodynamical stability requires that the phase with the larger pressure dominates, and phase equilibrium is achieved when $P_{\text{qgp}}(T_c) = P_\pi(T_c)$, with T_c being the critical temperature (the Maxwell construction). For $N_f = 2$ this simple model predicts a first-order phase transition at $T_c \approx 0.72\, B^{\frac{1}{4}}$ ($T_c \approx 144$ MeV for $B^{\frac{1}{4}} = 200$ MeV, $\varepsilon_c \approx 0.85$ GeV/fm^3, see Fig. 5.1).

Certainly, the model of the phase transition described above is oversimplified. One way of its improvement is to consider more hadrons (in addition to pions) and to include the interactions between quarks and gluons. In the first-order perturbation theory ($N_f = 2$, $\alpha_s = g^2/4\pi$), the modification of the plasma equation of state is [9]

$$\begin{aligned}\varepsilon_{\text{qgp}} = &\left(1 - \frac{15}{4\pi}\alpha_s\right)\frac{8\pi^2}{15}T^4 + \left(1 - \frac{50}{21\pi}\alpha_s\right)\frac{7\pi^2}{10}T^4 \\ &+ \left(1 - \frac{2}{\pi}\alpha_s\right)\frac{3}{\pi^2}\mu^2\left(\pi^2 T^2 + \frac{\mu^2}{2}\right) + B.\end{aligned} \tag{5.16}$$

One can see that such perturbative interactions lead to a reduction of the energy density and pressure. More on this subject may be found in the review article [10].

5.4 Lattice QCD

The fundamental predictions concerning the QCD phase transitions can be obtained from the numerical studies of QCD on a discretized spacetime lattice [11, 12]. In this approach one calculates the partition function by the Monte-Carlo methods. We recall that the grand canonical partition function is defined by the relation

$$Z = \mathrm{Tr}\left[\exp\left(-\frac{\hat{H} - \mu_B \hat{N}_B}{T}\right)\right] \equiv \exp\left(-\frac{\Omega}{T}\right), \tag{5.17}$$

where $\hat{H}$ is the Hamiltonian, $\hat{N}_B$ is the baryon number operator, T is the temperature and μ_B is the baryon chemical potential (as usual we use the natural system of units where $\hbar = c = k_B = 1$). From this expression all other standard thermodynamic quantities (pressure, baryon number, entropy and energy) can be determined from the relations:

$$P = -\left(\frac{\partial \Omega}{\partial V}\right)_{T,\mu_B}, \quad N_B = -\left(\frac{\partial \Omega}{\partial \mu_B}\right)_{T,V}, \quad S = -\left(\frac{\partial \Omega}{\partial T}\right)_{V,\mu_B} \tag{5.18}$$

and

$$E = TS - PV + \mu_B N_B. \tag{5.19}$$

The partition function (5.17) for $\mu_B = 0$ is equivalently given by the path integral

$$Z = \int dA\, d\Psi\, d\bar{\Psi} \exp\left(-\int_V d^3x \int_0^{1/T} d\tau\, \mathcal{L}\left(A, \Psi, \bar{\Psi}\right)\right). \tag{5.20}$$

Here the spatial integration is done over the volume V, while the time coordinate x^0 is "Wick rotated" to purely imaginary values, $\tau = ix^0$, and the integration range of τ is determined by the temperature of the system. The bosonic gluon (fermionic quark) fields have to be periodic (antiperiodic) functions of τ in the interval $0 \le \tau \le \beta = 1/T$.

In principle, the technique used in the lattice approach should accurately describe the quark-gluon plasma as well as the hadronic phase but, in practice, its accuracy especially at low temperature is severely limited by the finite size effects and other technical difficulties. In particular, the calculations involving fermions on the lattice are much more time consuming, hence the numerical results are less statistically meaningful in this case.

We shall not enter a deeper discussion of the lattice QCD methods because for our purposes it is important to know merely the main results of such studies. Such results may be directly applied in different phenomenological approaches. Thus, in the next two sections we shall review the main facts delivered by the lattice QCD simulations about the equation of state of strongly interacting matter. The reader interested in more formal aspects of the lattice QCD may consult, for example, Ref. [13].

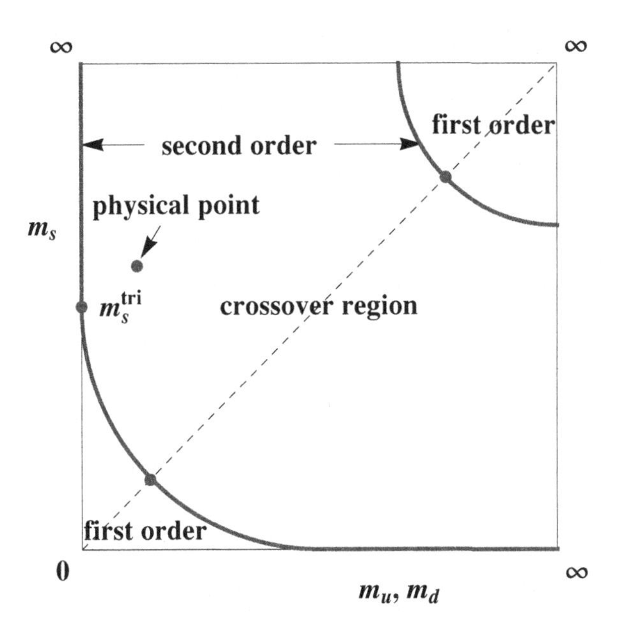

Fig. 5.2 Dependence of the critical behavior in QCD on the values of the quark masses is represented by the diagram popularly known as the "Columbia plot" [14], see also [15]. This diagram summarizes our expectations on the nature of the transition that have been derived from studies of simpler systems with the same global symmetries as QCD.

5.4.1 *Order of phase transition at* $\mu_B = 0$

The results of the lattice QCD simulations concerning the order of the phase transition depend strongly on the number of quark flavors and on the quark masses. For vanishing baryon chemical potential, $\mu_B = 0$, those results are summarized in Fig. 5.2. In the limit $m_u, m_d, m_s \to \infty$ one recovers the pure SU(3) gauge theory with a deconfinement phase transition that is first order. In the limit $m_u, m_d, m_s \to 0$ one deals with a chiral phase transition that is again first order. For $m_u = m_d = 0$ and m_s larger than the tricritical point $m_s^{\rm tri}$ the transition is of second order. The "physical point" corresponds to small m_u and m_d, while $m_s > m_s^{\rm tri}$ [16]. According to our current best knowledge, this point corresponds to the crossover transition, where a sudden but continuous jump in the energy density is observed. This jump in the energy density appears together with a more gradual increase of pressure.

As an example, in Fig. 5.3 we show the results of the lattice simulations of QCD at the finite temperature obtained in Ref. [17]. They were obtained for physical masses of the light quarks and the strange quark. Those results have been recently

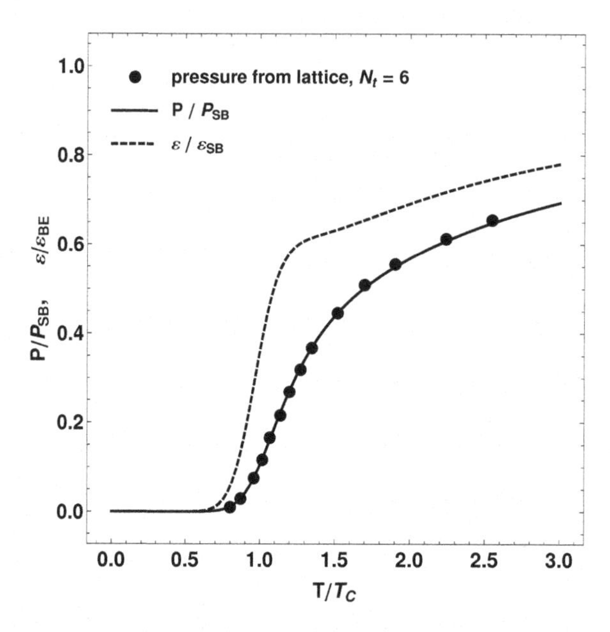

Fig. 5.3 Thermodynamic variables obtained from the lattice simulations of QCD: The dots represent the temperature dependence of pressure calculated in [17]. The quantity N_t is the number of the time points on the grid used in the calculations, and P_{SB} is the Stefan-Boltzmann limit for pressure — the result obtained for the weakly-interacting plasma, see Eq. (8.86) which for $N_f = 3$ gives $P/T^4 = 5.2$. The solid line is the fit to the lattice results [17] based on Eq. (5.21). The dashed line shows the corresponding temperature dependence of the energy density (scaled by the corresponding Stefan-Boltzmann limit for the energy density, $\varepsilon_{PB} = 3P_{SB}$). The value of the critical temperature $T_c = 173$ MeV.

parameterized in Ref. [18] with the help of the compact expression

$$P = cT^4 f\left(T_c/T\right), \quad f(x) = \frac{1 + e^{-\frac{a}{b}}}{1 + e^{\frac{x-a}{b}}} e^{-\lambda x}, \tag{5.21}$$

where the dimensionless fit parameters equal: $a = 0.91$, $b = 0.11$, $c = 5.2$ and $\lambda = 1.08$. The fit results are represented by the solid line in Fig. 5.3.

Indeed, with realistic parameters, as those used in [17], a smooth crossover is found. This means that no genuine phase transition is observed (no real discontinuity in thermodynamic variables exists) but a rapid change over a small temperature range is found. Moreover, the pressure attains the black-body limit at larger temperatures slower than the energy density. This indicates that non perturbative effects are present in the plasma. Those non perturbative effects are not completely understood. Their existence indicates that the treatment of the plasma as a massless gas of non interacting partons is only a crude approximation.

Since we deal with the crossover rather than with the genuine phase transition, the strict meaning of the critical temperature is lost. However, its concept is still used to denote the position of the temperature range where thermodynamic variables exhibit strong variations. In the lattice calculations discussed here $T_c = 173$ MeV.

5.4.2 *Critical point*

Lattice simulations of QCD at finite chemical potential are much more difficult than those performed at zero value of μ_B. The reason for this difficulty is the appearance of the complex fermionic determinant in the path integral approach. This leads to considerable cancellations between different field configurations and unstable numerical results. Several methods have been proposed to deal with this problem. For example, it has been suggested to expand the fermionic determinant into a Taylor series and to calculate the expansion coefficient at $\mu_B = 0$ [19]. Another way is to introduce the purely imaginary chemical potential $\mu_B = -i|\mu_B|$ [20,21]. In this case the fermionic determinant is real, however, at the end of the calculations one has to perform the analytic continuation back to the real values of μ_B. Yet another procedure is called the reweighting method [22, 23]. In this case the fermionic determinant at non-zero μ_B is treated as an observable rather than as a part of the integration measure.

The recent results [24] indicate that the crossover transition extends to the critical point $(T_{\rm crit}, \mu^B_{\rm crit})$, see Fig. 1.3 where the critical point is denoted as CP. At this point the transition is of the second order. For smaller temperatures and larger chemical potentials, we deal with the first order transition. From the model calculations, we expect that the line of the first-order transition ends at the μ_B-axis.

The experimental studies of the QCD phase diagram over an extended region of high baryon density will be the subject of the next heavy-ion experiments: FAIR at GSI Darmstadt [25] and SHINE/NA61 at CERN [26]. The main aim of these experiments is the location of the critical point and searches for the first-order phase transition in QCD at finite baryon density. Of course, the experimental studies of new regions in the QCD phase diagram requires that new theoretical tools, for example, special hydrodynamic codes [27], are developed in parallel.

5.5 Color isotopic charge and color hypercharge

In this Section we come back to the formal aspects of QCD discussed at the beginning of this Chapter in Sec. 5.2. We shall introduce a special basis for the gluon fields, which makes a relation between QCD and QED especially transparent. In this basis the two real gluon fields behave like the neutral fields, whereas the remaining six real gluon fields are grouped into three complex fields describing the charged gluon fields. Both quarks and charged gluons carry two types of the color charge: the *color isotopic charge* and *color hypercharge*. These two charges describe their coupling to the two neutral gluon fields in the similar way as the electric charge describes the coupling of the charged electron to the neutral electromagnetic field. The use of such basis turned out to be very convenient in the construction of the kinetic equations for the quark-gluon plasma, see Chap. 11.

5.5.1 *Quarks*

Out of eight generators of the SU(3) group, there are only two which commute with each other, namely

$$T^3 \equiv \frac{\lambda_3}{2} = \frac{1}{2}\begin{pmatrix} 1 & 0 & 0 \\ 0 & -1 & 0 \\ 0 & 0 & 0 \end{pmatrix}, \qquad T^8 \equiv \frac{\lambda_8}{2} = \frac{1}{2\sqrt{3}}\begin{pmatrix} 1 & 0 & 0 \\ 0 & 1 & 0 \\ 0 & 0 & -2 \end{pmatrix}. \tag{5.22}$$

These two generators may be interpreted as the operators of the color charges carried by the quarks. Following Ref. [28] we shall call them the color isotopic charge and the color hypercharge. The values of the quarks charges are obtained as the eigenvalues $\epsilon_i^{(3)}$ and $\epsilon_i^{(8)}$ in the equations

$$T^3\,\Psi_i = \epsilon_i^{(3)}\,\Psi_i, \qquad T^8\,\Psi_i = \epsilon_i^{(8)}\,\Psi_i, \tag{5.23}$$

where the field Ψ_i has only its ith component different from zero:

$$\Psi_1 = \begin{pmatrix} \Psi_{\rm red} \\ 0 \\ 0 \end{pmatrix}, \quad \Psi_2 = \begin{pmatrix} 0 \\ \Psi_{\rm green} \\ 0 \end{pmatrix}, \quad \Psi_3 = \begin{pmatrix} 0 \\ 0 \\ \Psi_{\rm blue} \end{pmatrix}. \tag{5.24}$$

The solution of Eqs. (5.23) is

$$\epsilon_1 = \frac{1}{2}\left(1, \sqrt{\frac{1}{3}}\right), \quad \epsilon_2 = \frac{1}{2}\left(-1, \sqrt{\frac{1}{3}}\right), \quad \epsilon_3 = \left(0, -\sqrt{\frac{1}{3}}\right), \tag{5.25}$$

where we introduced the notation based on the two-dimensional vectors in the space of the color isotopic charge and color hypercharge

$$\boldsymbol{\epsilon}_i = \left(\epsilon_i^{(3)}, \quad \epsilon_i^{(8)}\right). \tag{5.26}$$

We note that the vectors $\boldsymbol{\epsilon}_i$ may be treated as the mathematical representation of three different (red, green, and blue) color charges. Each charge $\boldsymbol{\epsilon}_i$ carried by a quark has its opposite partner $-\boldsymbol{\epsilon}_i$ carried by an antiquark, so a quark-antiquark pair is neutral or white. Moreover, the sum of the three charges (5.25) is also zero, $\boldsymbol{\epsilon}_1 + \boldsymbol{\epsilon}_2 + \boldsymbol{\epsilon}_3 = 0$, hence three quarks or three antiquarks may also form a neutral/white object.

5.5.2 *Gluons*

The color isotopic charge and hypercharge are also carried by the gluons. However, in this case we have to consider the charge operators in a different space. The gluon fields, represented by the objects

$$A^\mu \equiv \frac{1}{2} A_a^\mu \lambda_a, \tag{5.27}$$

do not form the set of base states for the fundamental representation of SU(3) as quarks but form the basis for the adjoint representation [5]. The action of the appropriate charge operators on the gluon fields is defined by the calculation of the commutators of the generators T^3 and T^8 with those fields [29],

$$\mathrm{Ad}\, T^3 = [T^3, \cdot], \qquad \mathrm{Ad}\, T^8 = [T^8, \cdot]. \tag{5.28}$$

One immediately finds that the generators $\tau_3 = T^3$ and $\tau_8 = T^8$ are the eigenstates of the new operators $\mathrm{Ad}\, T^3$ and $\mathrm{Ad}\, T^8$ with zero eigenvalues. Other non-trivial eigenstates (τ_{12}, τ_{21}, τ_{13}, τ_{31}, τ_{23}, and τ_{32}) are defined by the relations:

$$\tau_{12} = \frac{1}{\sqrt{2}}(T_1 + iT_2) = \frac{1}{\sqrt{2}} \begin{pmatrix} 0\,1\,0 \\ 0\,0\,0 \\ 0\,0\,0 \end{pmatrix}, \quad \tau_{21} = \frac{1}{\sqrt{2}}(T_1 - iT_2) = \tau_{12}^\dagger, \tag{5.29}$$

$$\tau_{13} = \frac{1}{\sqrt{2}}(T_4 + iT_5) = \frac{1}{\sqrt{2}} \begin{pmatrix} 0\,0\,1 \\ 0\,0\,0 \\ 0\,0\,0 \end{pmatrix}, \quad \tau_{31} = \frac{1}{\sqrt{2}}(T_4 - iT_5) = \tau_{13}^\dagger, \tag{5.30}$$

$$\tau_{23} = \frac{1}{\sqrt{2}}(T_6 + iT_7) = \frac{1}{\sqrt{2}} \begin{pmatrix} 0\,0\,0 \\ 0\,0\,1 \\ 0\,0\,0 \end{pmatrix}, \quad \tau_{32} = \frac{1}{\sqrt{2}}(T_6 - iT_7) = \tau_{23}^\dagger. \tag{5.31}$$

Indeed, the straightforward calculation shows ($i, j = 1, 2, 3; i \neq j$; no summation over repeated indices)

$$\left[T^3, \tau_{ij}\right] = \eta_{ij}^{(3)} \tau_{ij}, \qquad \left[T^8, \tau_{ij}\right] = \eta_{ij}^{(8)} \tau_{ij}, \tag{5.32}$$

where the charges $\boldsymbol{\eta}_{ij}$ have the form

$$\boldsymbol{\eta}_{ij} = \boldsymbol{\epsilon}_i - \boldsymbol{\epsilon}_j. \tag{5.33}$$

Equation (5.33) reflects a quark-antiquark structure of a gluon. The three special cases are

$$\boldsymbol{\eta}_{12} = (1, 0), \qquad \boldsymbol{\eta}_{13} = \left(\frac{1}{2}, \frac{1}{\sqrt{3}}\right), \qquad \boldsymbol{\eta}_{23} = \left(-\frac{1}{2}, \frac{1}{\sqrt{3}}\right). \tag{5.34}$$

[5] In the discussion of the color charges of gluons we follow the lectures on the Standard Model by Krzysztof Golec-Biernat, http://phoebe.ifj.edu.pl/ golec/czastki.pdf (in Polish).

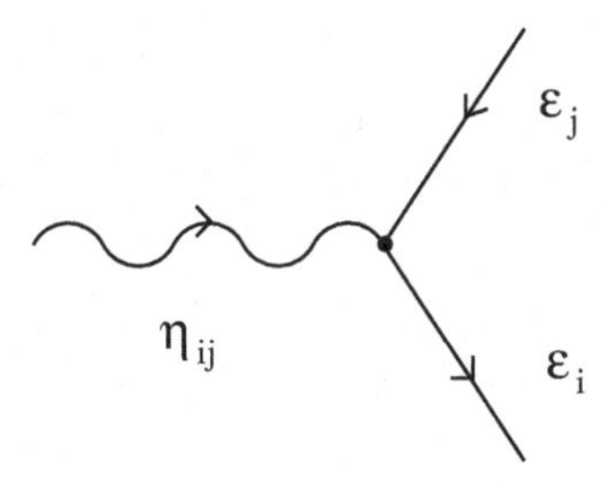

Fig. 5.4 Absorption of the charged gluon by a quark. The initial color charge ϵ_j is changed to ϵ_i.

The remaining three charges are obtained from the rule

$$\eta_{ji} = -\eta_{ij}, \tag{5.35}$$

which follows immediately from Eq. (5.33).

The gluon field (5.27) rewritten in the basis of the τ generators (for simplicity of notation with skipped Lorentz indices) has the form

$$\begin{aligned} A &= \frac{1}{2} A_a \lambda_a \\ &= A^3 \tau_3 + A^8 \tau_8 + \left[\frac{A^1 - iA^2}{\sqrt{2}} \tau_{12} + \frac{A^4 - iA^5}{\sqrt{2}} \tau_{13} + \frac{A^6 - iA^7}{\sqrt{2}} \tau_{23} + \text{h.c.} \right] \end{aligned} \tag{5.36}$$

or

$$\begin{aligned} A = G^3 \tau_3 + G^8 \tau_8 + G^{12} \tau_{12} + G^{13} \tau_{13} \\ + G^{23} \tau_{23} + G^{21} \tau_{21} + G^{31} \tau_{31} + G^{32} \tau_{32}. \end{aligned} \tag{5.37}$$

Here we made the following identifications:

$$G_3 = A_3, \qquad G_8 = A_8, \tag{5.38}$$

$$G_{12} = \frac{1}{\sqrt{2}} (A_1 - iA_2), \; G_{13} = \frac{1}{\sqrt{2}} (A_4 - iA_5), \; G_{23} = \frac{1}{\sqrt{2}} (A_6 - iA_7), \tag{5.39}$$

$$G_{21} = G_{12}^\dagger, \qquad G_{31} = G_{13}^\dagger, \qquad G_{32} = G_{23}^\dagger. \tag{5.40}$$

Clearly, the fields G^3 and G^8 are neutral since we have

$$\begin{aligned} \left[T^3, G^3\right] = 0, \qquad \left[T^8, G^3\right] = 0, \\ \left[T^3, G^8\right] = 0, \qquad \left[T^8, G^8\right] = 0. \end{aligned} \tag{5.41}$$

On the other hand, the fields G^{ij} carry the color charges $\boldsymbol{\eta}_{ij}$ (here again $i, j = 1, 2, 3$; $i \neq j$; no summation over repeated indices),

$$\left[T^3, G^{ij}\right] = \eta_{ij}^{(3)} G^{ij}, \qquad \left[T^8, G^{ij}\right] = \eta_{ij}^{(8)} G^{ij}. \tag{5.42}$$

In the new basis, the quark-gluon interaction term in the Lagrangian (5.1) takes the form (neglecting the flavor indices)

$$\begin{aligned}\mathcal{L}_{\text{int}} = & -g\overline{\Psi}\left(\tau_3\gamma_\mu G_3^\mu + \tau_8\gamma_\mu G_8^\mu\right)\Psi \\ & -g\left[\overline{\Psi}\left(\tau_{21}\gamma_\mu G_{21}^\mu + \tau_{31}\gamma_\mu G_{31}^\mu + \tau_{32}\gamma_\mu G_{32}^\mu\right)\Psi + \text{h.c.}\right].\end{aligned} \tag{5.43}$$

Equation (5.43) shows that the color charge of a quark is not changed during the interaction with the fields G^3 and G^8. Hence, these two types of gluons may be treated similarly as photons in QED. On the other hand, interaction of the quarks with the gluon fields G^{ij} changes the color state of quarks: absorption of the gluon G^{ij} by a quark with the charge $\boldsymbol{\epsilon}_j$ produces a quark with the charge $\boldsymbol{\epsilon}_i$, see Fig. 5.4.

In the end of our discussion we have to emphasize that the division of the gluon field into the charged and neutral part is gauge dependent, i.e., it is not invariant under transformations (5.7) and (5.8). Nevertheless, after fixing the gauge, the method described above may be applied and it turns out to be very useful for the physical interpretation of many processes.

Bibliography to Chapter 5

[1] O. W. Greenberg, "Spin and unitary spin independence in a paraquark model of baryons and mesons," *Phys. Rev. Lett.* **13** (1964) 598–602.

[2] M. Y. Han and Y. Nambu, "Three-triplet model with double SU(3) symmetry," *Phys. Rev.* **139** (1965) B1006–B1010.

[3] D. J. Gross and F. Wilczek, "Ultraviolet behavior of non-abelian gauge theories," *Phys. Rev. Lett.* **30** (1973) 1343–1346.

[4] D. J. Gross and F. Wilczek, "Asymptotically free gauge theories. 1," *Phys. Rev.* **D8** (1973) 3633–3652.

[5] H. D. Politzer, "Reliable perturbative results for strong interactions?," *Phys. Rev. Lett.* **30** (1973) 1346–1349.

[6] J. C. Collins and M. J. Perry, "Superdense matter: neutrons or asymptotically free quarks?," *Phys. Rev. Lett.* **34** (1975) 1353.

[7] M. G. Alford, A. Schmitt, K. Rajagopal, and T. Schafer, "Color superconductivity in dense quark matter," *Rev. Mod. Phys.* **80** (2008) 1455–1515.

[8] A. Chodos, R. L. Jaffe, K. Johnson, C. B. Thorn, and V. F. Weisskopf, "A new extended model of hadrons," *Phys. Rev.* **D9** (1974) 3471–3495.

[9] S. A. Chin, "Transition to hot quark matter in relativistic heavy-ion collision," *Phys. Lett.* **B78** (1978) 552–555.

[10] D. H. Rischke, "The quark-gluon plasma in equilibrium," *Prog. Part. Nucl. Phys.* **52** (2004) 197–296.

[11] K. G. Wilson, "Confinement of quarks," *Phys. Rev.* **D10** (1974) 2445–2459.

[12] M. Creutz, "Monte Carlo study of quantized SU(2) gauge theory," *Phys. Rev.* **D21** (1980) 2308–2315.
[13] C. Y. Wong, "Introduction to high-energy heavy ion collisions," (Singapore, World Scientific, 1994) 516 p.
[14] F. R. Brown *et al.*, "On the existence of a phase transition for QCD with three light quarks," *Phys. Rev. Lett.* **65** (1990) 2491–2494.
[15] E. Laermann and O. Philipsen, "Status of lattice QCD at finite temperature," *Ann. Rev. Nucl. Part. Sci.* **53** (2003) 163–198.
[16] H. Satz, "The States of Matter in QCD," `arXiv:0903.2778`.
[17] Y. Aoki, Z. Fodor, S. D. Katz, and K. K. Szabo, "The equation of state in lattice QCD: With physical quark masses towards the continuum limit," *JHEP* **01** (2006) 089.
[18] T. S. Biro and J. Zimanyi, "Entropy of expanding QCD matter," *Phys. Lett.* **B650** (2007) 193–196.
[19] C. R. Allton *et al.*, "The QCD thermal phase transition in the presence of a small chemical potential," *Phys. Rev.* **D66** (2002) 074507.
[20] P. de Forcrand and O. Philipsen, "The QCD phase diagram for small densities from imaginary chemical potential," *Nucl. Phys.* **B642** (2002) 290–306.
[21] M. D'Elia and M.-P. Lombardo, "Finite density QCD via imaginary chemical potential," *Phys. Rev.* **D67** (2003) 014505.
[22] Z. Fodor and S. D. Katz, "Lattice determination of the critical point of QCD at finite T and mu," *JHEP* **03** (2002) 014.
[23] Z. Fodor and S. D. Katz, "A new method to study lattice QCD at finite temperature and chemical potential," *Phys. Lett.* **B534** (2002) 87–92.
[24] Z. Fodor and S. D. Katz, "Critical point of QCD at finite T and mu, lattice results for physical quark masses," *JHEP* **04** (2004) 050.
[25] **CBM** Collaboration, S. Chattopadhyay, "Physics at high baryon density at FAIR," *J. Phys.* **G35** (2008) 104027.
[26] **NA61/SHINE** Collaboration, M. Gazdzicki, "Ion Program of Na61/Shine at the CERN SPS," `arXiv:0812.4415`.
[27] K. Paech, H. Stoecker, and A. Dumitru, "Hydrodynamics near a chiral critical point," *Phys. Rev.* **C68** (2003) 044907.
[28] K. Huang, "Quarks, leptons, and gauge fields," (World Scientific, Singapore, 1982).
[29] W. Greiner and B. Muller, "Theoretical physics. Vol. 2: Quantum mechanics. Symmetries," (Springer, Berlin, 1989) 368 p.

Chapter 6

More about Hadrons

In the following Chapter we provide more information about hadrons and their thermodynamic properties. A short characteristics of the hadronic resonance states is presented in Sec. 6.1. Then, in Sec. 6.2 we briefly review the *statistical bootstrap model*. Nowadays, this model is of the historical importance mainly, however, several significant concepts were conceived in its framework. They are still quite interesting and remain lively discussed. For example, the statistical bootstrap model leads to the concept of the *limiting hadronic temperature*, called also the *Hagedorn temperature*. It also introduces the concept that hadrons form clusters which in turn form heavier clusters etc.

At present, the Hagedorn temperature is commonly understood as the temperature of the phase transition from the hadron gas to the quark-gluon plasma. However, there exist new interesting developments concerning this idea and one of them, concerning the differences in the mass spectrum of baryons and mesons, is discussed in Sec. 6.3.

In the late stages of the spacetime evolution, the matter created in the relativistic heavy-ion collisions is typically treated as an *expanding hadron gas*. Therefore, to correctly describe the late stages it is very important to know the thermodynamic properties of the system of hadrons. The theoretical description of the hadron gas relies on the *Dashen-Ma-Bernstein formalism* which may be regarded as the relativistic generalization of the *virial expansion*. These formal frameworks are discussed in Secs. 6.4 and 6.5.

Finally, in Sec. 6.6 we describe the construction of the *realistic equation of state of the strongly interacting matter* that agrees with the hadron-gas model at low temperatures and coincides with the lattice simulations of QCD at high temperatures. This equation of state is valid for the case of zero baryon chemical potential and has been successfully used in the hydrodynamic calculations describing the evolution of matter created in heavy-ion collisions at the RHIC energies.

6.1 Resonances

In the 1950s the known hadrons included: nucleons (protons and neutrons), pions, hyperons (Λ, Σ, Ξ), and kaons. Meanwhile, the particle tables have become much more longer and contain hundreds of hadronic species. For example, the statistical hadronization models SHARE [1,2] and THERMINATOR [3], based on the Particle Data Group publication [4], include 371 well established light-flavor states, i.e., the hadrons composed of up, down, and strange quarks. The hadron list comprises 256 baryons and 115 mesons (counted with their isospin degeneracies but without the spin degrees of freedom).

Most hadrons decay by strong interactions, so they do not live long enough to be detected directly. They are identified through their decay products. The mass of a decaying particle is equal to the total energy of the products measured in the rest frame. According to the uncertainty principle connecting energy and time, the short lifetime of a particle leads to the substantial uncertainty of its energy/mass (of the order $\hbar/\Delta t$). For example, the Δ-resonance is formed and decays in πN scattering ($\pi\ N \rightarrow\ \Delta \rightarrow \pi\ N$). The decay of the Δ means that the normalization of its wave function decreases exponentially

$$|\psi\,(t)|^2 = |\psi\,(0)|^2\, e^{-t/\tau}, \tag{6.1}$$

where

$$\tau = \frac{1}{\Gamma} \tag{6.2}$$

is called the lifetime of the state. Thus, the time dependence of the wave function is of the form

$$\psi\,(t) \sim e^{-iMt} e^{-\Gamma t/2}, \tag{6.3}$$

where M is the mass of the state (the energy in the rest frame). As a function of the center-of-mass energy E of the πN system, the state is described by the Fourier transform

$$\psi(E) = \int \psi\,(t)\, e^{iEt} dt \sim \frac{1}{E - M + \frac{i\Gamma}{2}}. \tag{6.4}$$

The corresponding πN reaction rate is, see Ex. 7.14,

$$|\psi(E)|^2 \sim \frac{1}{(E-M)^2 + \left(\frac{\Gamma}{2}\right)^2}. \tag{6.5}$$

This function has a peak at M with a width determined by Γ. Equation (6.5) is called the Breit-Wigner resonance formula, and M and Γ are known as the mass and width of the unstable hadron called a *resonance*.

6.2 Statistical bootstrap model

The statistical bootstrap model (SBM) [5–10] is a model of strong interactions based on the observation that hadrons form bound and resonance states. This leads to the concept of a possibly unlimited sequence of heavier and heavier bound and resonance states, each being a possible constituent of a still heavier resonance. The number of such states in the mass interval $(m, m + dm)$ is denoted by $\rho(m)dm$, and we call the function $\rho(m)$ the SBM mass spectrum. The requirement that resonances are formed from other resonances in the self-consistent manner leads to the *bootstrap condition* or *bootstrap equation* for the mass spectrum $\rho(m)$. The solution of the bootstrap equation shows that the mass spectrum for large masses m has to grow exponentially, as found by Hagedorn already in 1965 [5]. As a consequence, any thermodynamics employing this mass spectrum has a singular temperature $T_{\rm H}$ generated by the asymptotics $\rho(m) \sim \exp(m/T_{\rm H})$. At present $T_{\rm H}$ (*the Hagedorn temperature*) is interpreted as the temperature where the phase transition from the hadron gas to the quark-gluon plasma occurs [11].

We can illustrate the ideas described above with the help of a simple toy model in which clusters are composed of resonances with vanishing kinetic energy. In this case we write

$$\rho(m) = \delta(m - m_0) + \sum_{n=2}^{\infty} \frac{1}{n!} \int \delta(m - \sum_{i=1}^{n} m_i) \prod_{i=1}^{n} \rho(m_i)\, dm_i. \tag{6.6}$$

The Laplace transform of Eq. (6.6) is

$$\int \rho(m) \exp[-\beta m]\, dm = \exp[-\beta m_0] + \sum_{n=2}^{\infty} \frac{1}{n!} \prod_{i=1}^{n} \int \rho(m_i)\, \exp[-\beta m_i] dm_i. \tag{6.7}$$

Now we define

$$z(\beta) = \exp[-\beta m_0] \tag{6.8}$$

and

$$G(z) = \int \exp[-\beta m] \rho(m)\, dm. \tag{6.9}$$

Thus, the bootstrap equation (6.6) takes the form

$$z = 2G(z) - \exp[G(z)] + 1. \tag{6.10}$$

Differentiation of Eq. (6.10) gives

$$1 = 2G'(z) - \exp[G(z)]G'(z) = 2G'(z) - (2G(z) + 1 - z)\, G'(z), \tag{6.11}$$

or

$$1 = G'(z)\, (1 + z - 2G(z))\,. \tag{6.12}$$

Most easily this equation can be solved by the substitution of a series in the form

$$G(z) = \sum_n c_n\, z^n. \tag{6.13}$$

One can check that the coefficients c_n satisfy the following recurrent relation

$$c_{n+1} = \frac{1}{n+1} \left(-n \, c_n + 2 \sum_{m=1}^{n} m \, c_m \, c_{n+1-m} \right). \tag{6.14}$$

A straightforward numerical study of the radius of convergence of the series (6.13) with the coefficients (6.14) shows that $|z|$ must be smaller than $z_{\rm H} \approx 0.386$. In connection with Eqs. (6.7) and (6.8), the finite radius of convergence means that there is a limiting temperature for the system of hadrons

$$\frac{1}{T_{\rm H}} = -\left(\frac{1}{m_0}\right) \ln\left(z_{\rm H}\right) \approx \frac{0.95}{m_0}. \tag{6.15}$$

Putting $m_0 = m_\pi$ we find the value

$$T_{\rm H} = 145 \text{ MeV}, \tag{6.16}$$

which can be interpreted as the temperature of the phase transition [11]. We note that in the analytic treatment of Eq. (6.10) we first construct the function $z(G)$, see [12]. The parabola-like maximum of $z(G)$ implies a branch cut singularity of $G(z)$ starting at $z_{\rm H} = \ln 4 - 1 \approx 0.386$. This singularity leads, through the inverse Laplace transform, to the exponential mass spectrum $\rho(m) \sim \exp(m/T_{\rm H})$.

Coming back to Eqs. (6.8) and (6.13) we find

$$G(z) = \sum_n c_n \, \exp\left[-\beta n m_0\right] \tag{6.17}$$

and

$$\rho\,(m) = \sum_n c_n \, \delta\left(n m_0 - m\right). \tag{6.18}$$

Hence, in the toy model discussed here the masses of the clusters are multiples of the pion mass. Certainly, this is not a realistic case. To deal with this and other problems, more elaborate versions of the bootstrap model were developed and analyzed. In particular, such modifications included: a larger number of "input" particles, the full relativistic phase space, symmetry constraints, baryon number (strangeness) conservation, and finite particle volumes.

6.3 Hagedorn temperatures for mesons and baryons

Nowadays, SBM as one of many pre-QCD approaches does not play an important role. At the same time when SBM was being developed, the quarks and gluons were discovered and accepted as the building blocks of matter. Nevertheless, the SBM prediction of the exponential grow of the hadron mass spectrum still attracts a lot of attention. Speaking more quantitatively, SBM predicts the asymptotic behavior of the density of hadron states in a form

$$\rho\,(m) \approx c m^a \exp\left[m/T_{\rm H}\right], \tag{6.19}$$

where a is a negative power ($a \leq -5/2$ [5,6], or $a \leq -3$ [7]). The parameters of the asymptotic spectrum should be determined from the fits to the experimental mass spectra. Hagedorn and Ranft [6] fit the function

$$\rho_{\text{HR}}(m) = \frac{c}{(m^2 + m_0^2)^{5/4}} \exp(m/T_{\text{H}}), \tag{6.20}$$

with m_0=0.5 GeV, and got T_{H}= 160 MeV. A similar value for T_{H} was obtained by Frautschi.

Starting from the 1960s, it has been believed that there is one universal scale in the asymptotic spectrum of hadrons, and its value is about 160 MeV. However, the old fits had a rather poor spectrum to their disposal, sufficiently dense only in the range of masses up to 1 GeV. Meanwhile, the particle tables became more complete [4]. Quite recently, a new analysis [13, 14] of the hadron states [4] showed that the Hagedorn temperatures of mesons and baryons are *significantly different.* Both the meson and baryon spectra increase exponentially up to 1.8 GeV, however the baryon spectrum grows much more rapidly. Application of the form (6.20) with m_0 = 500 MeV gives: T_{meson} = 197 MeV and T_{baryon} = 141 MeV, see Fig. 6.1 where the cumulants of the meson and baryon experimental mass spectra and the fits based on Eq. (6.20) are shown. Note, that the cumulants give the number of

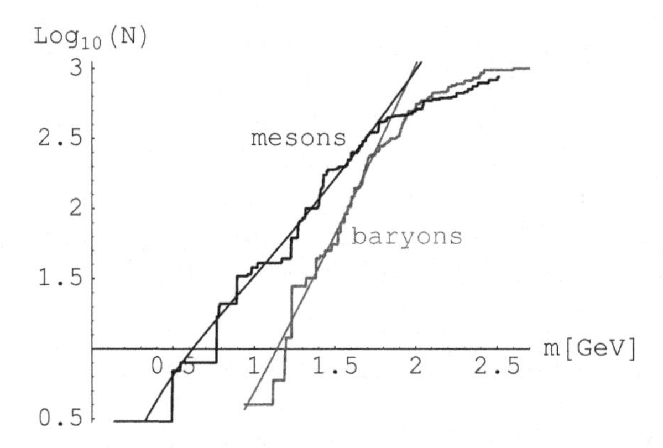

Fig. 6.1 Cumulants of the meson and baryon experimental mass spectra and the fits based on Eq. (6.20), plotted as functions of mass. Reprinted from [13] with permission from Elsevier.

states with the mass smaller than m, namely

$$N(m) = \int_0^m dm' \, \rho(m'). \tag{6.21}$$

A challenge remains to explain the different behavior of mesonic and baryonic spectra. First of all, the models with a finite number of degrees of freedom are not capable of reproducing the exponentially rising spectrum. On the other hand, the models which do reproduce the exponential grow, very often lead to the same rate of the growth for mesons and baryons. In particular, this is the case of SBM where baryons are formed by attaching mesons to the "input" baryon [8] and, as a result, the baryon spectrum grows in the same way as the meson spectrum. An interesting possibility of the explanation of the effect is offered by the Dual String Models. For more discussion on this point we refer the reader to Refs. [13–15].

The flattening of the meson and baryon mass distributions above $m = 2$ GeV reflects our poor knowledge of the resonance states in this region. In fact, such states may exist and their inclusion may improve the exponential form of the mass distributions. For example, it has been argued [16] that extra Hagedorn states may contribute to fast chemical equilibration times of known baryons and kaons in heavy-ion collisions studied at RHIC.

6.4 Dashen-Ma-Bernstein formalism

The formal treatment of the resonance states in a hadronic gas in thermal equilibrium was elaborated by Dashen, Ma, and Bernstein [17], and by Dashen and Rajaraman [18, 19]. More recently, this issue was studied and further developed by Weinhold, Friman, and Nörenberg [20–24] (see also [25–27]). The fundamental formula of this approach is the equation defining the number of resonances per unit volume and per unit invariant mass, M, produced in the two-body channel of particles 1 and 2

$$\frac{dn}{dM} = g \int \frac{d^3p}{(2\pi)^3} \frac{1}{\pi} \frac{d\delta_{12}(M)}{dM} \frac{1}{\exp\left(\frac{\sqrt{M^2+\mathbf{p}^2}}{T}\right) \pm 1}. \tag{6.22}$$

Here $\delta_{12}(M)$ is the phase shift for the scattering of particles 1 and 2, g is a spin-isospin factor, T is the temperature, and the $\pm$ sign reflects the statistics of the resonance ($+1$ for fermions and -1 for bosons).

The authors of Ref. [20] pointed out that in many works the spectral function of the resonance is used as the weight in Eq. (6.22) instead of the derivative of the phase shift. Such a procedure is acceptable for narrow resonances, since in this case both the spectral function and the derivative of the phase shift are very sharply peaked at the resonance position $m_{\rm R}$, yielding

$$\frac{d\delta_{12}(M)}{dM} \simeq \pi \delta(M - m_{\rm R}). \tag{6.23}$$

In this situation one obtains the narrow-resonance limit

$$n^{(\text{narrow})} = g \int \frac{d^3p}{(2\pi)^3} \frac{1}{\exp\left(\frac{\sqrt{m_{\mathrm{R}}^2+\mathbf{p}^2}}{T}\right) \pm 1}. \tag{6.24}$$

It is important to emphasize that for wide resonances, or when the effects of tails are investigated, the difference between the correct formula (6.22) and the approximate formula (6.24) is important, not only conceptually but also numerically.

6.5 Virial expansion — non-relativistic treatment

Since the crucial role of the phase shifts in the theoretical treatment of interacting gases is not always properly emphasized, in this Section we give a simple example showing how the phase shifts allow us to determine most of the thermodynamic properties of the non-relativistic and non-ideal gases. The physical intuition we gain from this example may be directly applied to relativistic systems, such as the hadron gas formed in relativistic heavy-ion collisions. We note that from the formal point of view, we analyze the first term in the virial expansion, i.e., the expansion of pressure in powers of $1/V$. Our discussion is concentrated on the non-relativistic gas of identical boson particles which undergo resonant scattering on each other. We show that in some cases the interacting system behaves effectively as a non-interacting gas which includes besides the input particles also resonant states as separate degrees of freedom. Our discussion follows the treatment of this problem by Landau and Lifshitz [28], however, several aspects are discussed in greater detail.

We start our discussion with the definition of the thermodynamic potential Ω which accounts only for one- and two-particle states

$$\Omega = -T \ln\left\{1 + \sum_n e^{(\mu - E_{1,n})/T} + \sum_n e^{(2\mu - E_{2,n})/T}\right\}. \tag{6.25}$$

The first sum is over one-particle energy levels, denoted as $E_{1,n}$, while the second sum is over two-particle levels, denoted as $E_{2,n}$. In the low-density (Boltzmann) limit the chemical potential μ is negative and its magnitude is large, hence the logarithm in Eq. (6.25) may be approximated by the three terms

$$\Omega = -T \left\{\sum_n e^{(\mu - E_{1,n})/T} + \sum_n e^{(2\mu - E_{2,n})/T} \right.$$
$$\left. -\frac{1}{2}\sum_n e^{(\mu - E_{1,n})/T} \sum_m e^{(\mu - E_{1,m})/T}\right\}. \tag{6.26}$$

In the classical approximation, the summation over the states may be replaced by the phase-space integrals. In this way we obtain

$$\Omega = -T \left\{\int d\Gamma_{\mathrm{cl},1}\, e^{(\mu - E(p_1,q_1))/T} + \frac{1}{2}\int d\Gamma_{\mathrm{cl},1} d\Gamma_{\mathrm{cl},2}\, e^{(2\mu - E(p_1,q_1;p_2,q_2))/T} \right.$$
$$\left. - \frac{1}{2}\int d\Gamma_{\mathrm{cl},1}\, e^{(\mu - E(p_1,q_1))/T} \int d\Gamma_{\mathrm{cl},2}\, e^{(\mu - E(p_2,q_2))/T}\right\}, \tag{6.27}$$

where we use the notation

$$d\Gamma_{\rm cl,1} = \frac{d^3q_1 d^3p_1}{(2\pi\hbar)^3}, \qquad d\Gamma_{\rm cl,2} = \frac{d^3q_2 d^3p_2}{(2\pi\hbar)^3}, \tag{6.28}$$

and the factor 1/2 in the second term in Eq. (6.27) accounts for the proper counting of identical particles. We further assume that the one-particle energies include only the kinetic parts, $T(p_1) = p_1^2/(2m)$ and $T(p_2) = p_2^2/(2m)$, while the two-particle energies include also the interaction term $U(q_2 - q_1)$ depending on the difference of the position coordinates. In this case we may write

$$\Omega = -T \left\{ \frac{V e^{\mu/T}}{(2\pi\hbar)^3} \int d^3p_1 \, e^{-T(p_1)/T} \right.$$
$$\left. + \frac{e^{2\mu/T}}{2} \int d\Gamma_{\rm cl,1} d\Gamma_{\rm cl,2} \, e^{-(T(p_1)+T(p_2))/T} \left(e^{-U(q_1-q_2)/T} - 1 \right) \right\}. \tag{6.29}$$

In the next step we introduce the center-of-mass and relative (internal) coordinates defined by the equations

$$\mathbf{P} = \mathbf{p}_1 + \mathbf{p}_2, \qquad \mathbf{p} = \frac{1}{2}(\mathbf{p}_1 - \mathbf{p}_2), \tag{6.30}$$

$$\mathbf{Q} = \frac{1}{2}(\mathbf{q}_1 + \mathbf{q}_2), \qquad \mathbf{q} = (\mathbf{q}_1 - \mathbf{q}_2). \tag{6.31}$$

The thermodynamic potential may be now rewritten in the following form

$$\Omega = -T \left\{ \frac{V e^{\mu/T}}{(2\pi\hbar)^3} \int d^3p_1 \, e^{-T(p_1)/T} \right.$$
$$\left. + \frac{e^{2\mu/T}}{2} \int d\Gamma_{\rm cl} \, e^{-\frac{P^2}{4mT}} \int d\gamma_{\rm cl} \, e^{-\frac{p^2}{mT}} \left(e^{-U(q)/T} - 1 \right) \right\}, \tag{6.32}$$

where we introduced the integration measures

$$d\Gamma_{\rm cl} = \frac{d^3Q d^3P}{(2\pi\hbar)^3}, \qquad d\gamma_{\rm cl} = \frac{d^3q d^3p}{(2\pi\hbar)^3}. \tag{6.33}$$

One observes that the second term in Eq. (6.32) becomes a product of two partition functions, the first factor describes free particles with the mass $2m$, and the second factor describes the particles with the reduced mass $m/2$, which interact via the central potential U. We define now the interacting partition function

$$Z_2 = \frac{1}{2} \int d\gamma_{\rm cl} \, e^{-\frac{p^2}{mT}} \left(e^{-U(q)/T} - 1 \right)$$
$$= \frac{1}{2} \int d\gamma_{\rm cl} \, e^{-\frac{p^2}{mT} - \frac{U(q)}{T}} - \frac{1}{2} \int d\gamma_{\rm cl} \, e^{-\frac{p^2}{mT}} \tag{6.34}$$

and come back to the quantum description. In the quantum approach the states in the central potential are labeled by the numbers n, l and m. Since the energy levels are degenerate in m, it is enough to use n and l, and introduce the degeneracy factor $2l + 1$. Hence, the partition function (6.34) may be expressed as a difference of the two terms,

$$Z_2 = \sum_{n,l} (2l+1)\, e^{-\frac{E(n,l)}{T}} - \sum_{n,l} (2l+1)\, e^{-\frac{E_0(n,l)}{T}}, \tag{6.35}$$

where the energies are given by the magnitude of the wave vectors

$$E(n,l) = \frac{\hbar^2 k^2(n,l)}{m}, \qquad E_0(n,l) = \frac{\hbar^2 k_0^2(n,l)}{m}. \tag{6.36}$$

The factor 1/2 (appearing in front of the two sums in Eq. (6.34)) has been omitted since in the quantum case the symmetry between particles should be taken into account by the inclusion of the proper values of l (in our case we deal with bosons and we should include only even values of the orbital momentum). The subscript 0 denotes that the energy values E_0 and the momenta k_0 are calculated for vanishing value of the potential U. From the point of the mathematical manipulations it is convenient to close the system in a large spherical box of radius R which may be taken to be arbitrarily large. In the end of the calculations we may check that our final result is independent of the specific choice of R. The asymptotic form of the solutions of the Schrödinger equation with a given orbital momentum l has the form

$$\psi = \frac{\text{const.}}{r} \sin\left(kr - \frac{1}{2} l\pi + \delta_l(k)\right), \tag{6.37}$$

where δ_l is the phase shift. By demanding that the wave function vanishes at the edge of the box (for $r = R$) we find the condition

$$kR - \frac{1}{2} l\pi + \delta_l(k) = n\pi. \tag{6.38}$$

This is exactly the condition leading to the discrete energy values appearing in Eq. (6.36). For non-interacting particles $\delta_l = 0$ so that for fixed l we obtain

$$\frac{R}{\pi} dk_0 = dn. \tag{6.39}$$

On the other hand, for interacting particles we have

$$\left(\frac{R}{\pi} + \frac{d\delta_l}{dk}\right) dk = dn. \tag{6.40}$$

Changing from summation to integration in Eq. (6.35) and denoting k_0 in the second term simply by k we obtain the formula

$$\begin{aligned} Z_2 = &\sum_l \int_0^\infty dk \left(\frac{R}{\pi} + \frac{d\delta_l}{dk}\right) (2l+1)\, e^{-\frac{\hbar^2 k^2}{mT}} \\ &-\sum_l \int_0^\infty dk \frac{R}{\pi}\, (2l+1)\, e^{-\frac{\hbar^2 k}{mT}}, \end{aligned} \tag{6.41}$$

which leads to the concise result

$$Z_2 = \sum_l (2l+1) \int_0^\infty dk \, \frac{d\delta_l}{\pi \, dk} \, e^{-\frac{\hbar^2 k^2}{mT}} \tag{6.42}$$

where the dependence on the sphere radius R has been canceled [1]. Combining Eqs. (6.32) and (6.42) we find

$$\begin{aligned} \Omega = -T \Bigg\{ & \frac{V e^{\mu/T}}{(2\pi\hbar)^3} \int d^3 p_1 \, e^{-\frac{p_1^2}{2mT}} \\ & + \frac{V e^{2\mu/T}}{(2\pi\hbar)^3} \sum_l (2l+1) \int d^3 P \, e^{-\frac{P^2}{4mT}} \int_0^\infty dk \, \frac{d\delta_l}{\pi \, dk} \, e^{-\frac{\hbar^2 k^2}{mT}} \Bigg\}. \end{aligned} \tag{6.43}$$

The number of particles is obtained by differentiation of the potential Ω with respect to the chemical potential and changing the sign, in this way we find

$$\begin{aligned} \frac{N}{V} = & \frac{e^{\mu/T}}{(2\pi\hbar)^3} \int d^3 p_1 \, e^{-\frac{p_1^2}{2mT}} \\ & + 2 \frac{e^{2\mu/T}}{(2\pi\hbar)^3} \sum_l (2l+1) \int d^3 P \, e^{-\frac{P^2}{4mT}} \int_0^\infty dp \, \frac{d\delta_l}{\pi \, dp} \, e^{-\frac{p^2}{mT}}. \end{aligned} \tag{6.44}$$

We observe that the number density of particles has been changed from its noninteracting density given by the first term in Eq. (6.44). The second term in (6.44) gives a correction from the scattering states. Its interpretation is most clearly seen in the case where the phase shift increases suddenly by π, in this case the derivative of the phase shift becomes the delta function,

$$\delta_l(p) = \pi \theta(p - p_{R,l}), \qquad \frac{d\delta_l}{\pi \, dp} = \delta(p - p_{R,l}), \tag{6.45}$$

and the last integration in Eq. (6.44) can be done explicitly

$$\begin{aligned} \frac{N}{V} = & \frac{e^{(\mu+m)/T}}{(2\pi\hbar)^3} \int d^3 p_1 \, e^{-\left(m + \frac{p_1^2}{2m}\right)\frac{1}{T}} \\ & + 2 \frac{e^{2(\mu+m)/T}}{(2\pi\hbar)^3} \sum_l (2l+1) \int d^3 P \, e^{-\left(2m + \frac{P^2}{4m}\right)\frac{1}{T}} \, e^{-\frac{p_{R,l}^2}{mT}}. \end{aligned} \tag{6.46}$$

Here we have added and subtracted the rest mass energy in the arguments of the exponentials trying to recover a form of the kinetic energy, which is the non-relativistic expansion of the relativistic formula $\sqrt{m^2 + p^2} \approx m + p^2/(2m)$. In the

[1] We note that for non-interacting system the summation range over l is restricted by $l_{\max} = 2kR/\pi$. For the interacting system the sum over l is naturally cut by the dynamic properties of the two-particle system.

non-relativistic regime we have $E_{R,l} = p_{R,l}^2/m \ll m$, hence the equivalent form of Eq. (6.46) is (with a rescaled chemical potential $\mu + m \to \mu$)

$$\begin{aligned}\frac{N}{V} &= \frac{e^{\mu/T}}{(2\pi\hbar)^3}\int d^3p_1 \exp\left[-\left(m+\frac{p_1^2}{2m}\right)\frac{1}{T}\right] \\ &+\frac{2\,e^{2\mu/T}}{(2\pi\hbar)^3}\sum_l (2l+1) \\ &\times \int d^3P \exp\left[-\left((2m+E_{R,l})+\frac{P^2}{2(2m+E_{R,l})}\right)\frac{1}{T}\right].\end{aligned} \tag{6.47}$$

Equation (6.47) has a very attractive physical interpretation. The particle number density is modified by the particles which form two-particle clusters or resonances (hence the factor 2 in front of the second term). The mass of the clusters is $2m + E_{R,l}$. This is the sum of the masses of the input particles that has been modified by the interaction term. This is exactly what we expect in the non-relativistic approximation. In addition we see that the clusters behave as particles with the spin l.

The non-relativistic case discussed above illustrates how resonances appear in the treatment of the interacting gas. Generalized to the relativistic systems, this example shows how the thermodynamic system of interacting pions may be approximated by the non-interacting gas of pions and all pion resonances. It is worth emphasizing that in the case where the phase shifts have no sudden jump, the integration over the whole energy range should be done with the weight given by the derivative of the phase shift in a given channel. Such expressions are the correct thermodynamic formulas for inclusion of the resonances, which should be applied especially in the cases where the resonances are broad.

6.6 EOS of strongly interacting matter

One of the ultimate aims of the physics of heavy-ion collisions is to extract the information about the equation of state of strongly interacting matter [30]. Since thermodynamic quantities characterizing hot and dense matter formed in the collisions are not accessible in direct measurements, the standard procedure is to use different forms of the equations of state in model calculations and to check which form leads to the best description of the data. In this way we select the equations of state that may be regarded as the realistic ones.

The successes of the thermal models indicate that for low temperatures, $T < T_c$, the realistic equation of state of strongly interacting matter is reduced to the equation of state of the hadron gas. On the other hand, at high temperatures, $T > T_c$, it should agree with the results of the lattice simulations of QCD. In the case of

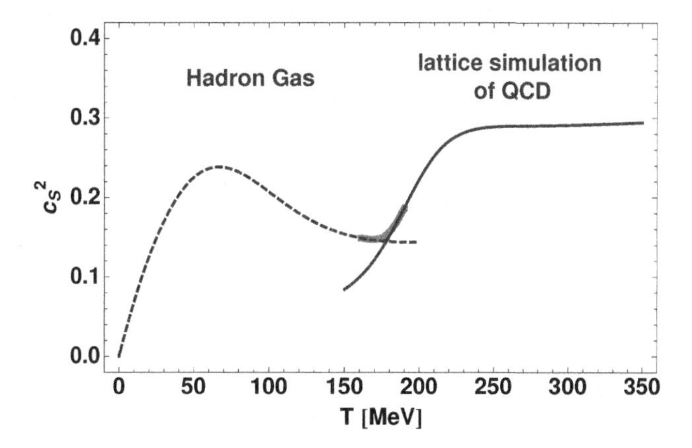

Fig. 6.2 Temperature dependence of the square of the sound velocity at zero baryon density, $c_s^2(T) = \partial P/\partial \varepsilon$. The result of the lattice simulations of QCD [29] is represented by the solid line, whereas the result obtained with the ideal hadron-gas model is represented by the dashed line. A piece of the thick solid line shows the simplest interpolation between those two calculations. In this case the critical temperature T_c equals 170 MeV. It is defined as the place where the sudden change of the thermodynamic variables occurs, see Fig. 6.3.

the vanishing baryon chemical potential, where the lattice calculations deliver the most reliable results, such an equation of state was worked out in Ref. [31]. It was successfully used in many hydrodynamic calculations.

The starting point for the construction of the equation state in [31] is the analysis of the temperature dependence of the sound velocity. For zero baryon chemical potential the sound velocity is defined by the formula

$$c_s^2 = \frac{\partial P}{\partial \varepsilon} = \frac{\sigma}{T}\frac{\partial T}{\partial \sigma}. \tag{6.48}$$

In Fig. 6.2 we show the sound velocity calculated in the hadron-gas model (dashed line) and the sound velocity obtained from the lattice simulations of QCD (solid line). The complete sound-velocity function is obtained in Ref. [31] by assuming that those two results should be connected with the simple interpolating function in the neighborhood of the critical temperature (the piece of the thick solid line).

Thermodynamic systems at $\mu_B = 0$ are characterized by only one independent thermodynamic variable, for example, the temperature T. This means that the knowledge of the sound velocity as a function of T allows us to calculate other thermodynamic variables. As the special case we may consider the entropy density where

$$\sigma(T_2) = \sigma(T_1) \exp\left[\int_{T_1}^{T_2} \frac{dT'}{T' c_s^2(T')}\right]. \tag{6.49}$$

Here T_1 and T_2 are two values of the temperature. It is convenient to choose T_1 to be sufficiently low, such that $\sigma(T_1)$ corresponds to the entropy of the pion gas and

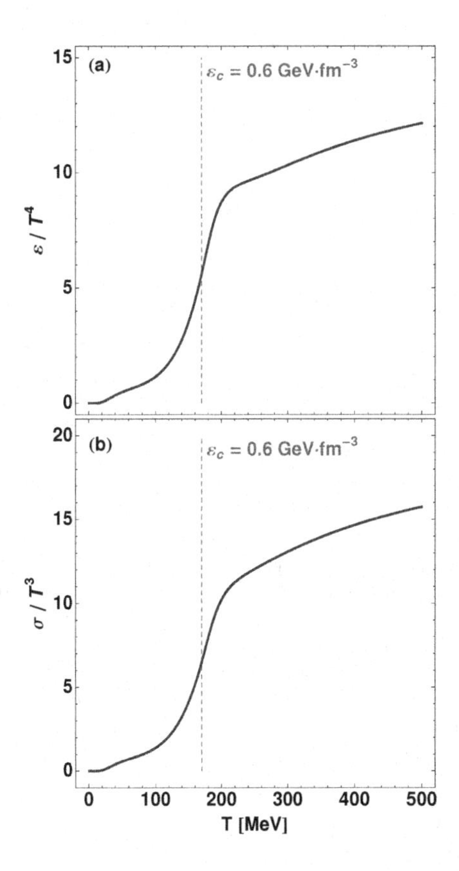

Fig. 6.3 The energy and entropy densities, scaled by T^4 and T^3, respectively, shown as functions of the temperature. One observes a sudden but smooth change of ε/T^4 and σ/T^3 at $T \sim T_c$. The vertical line indicates the critical energy density corresponding to $T_c = 170$ MeV. With the considered equation of state one finds $\varepsilon_c = 0.6\,\mathrm{GeV/fm}^3$. The presented thermodynamic functions follow directly from the temperature-dependent sound velocity shown in Fig. 6.2.

may be easily calculated (at low temperatures the pions give the main contributions to the thermodynamic properties of the hadron gas).

The thermodynamic variables obtained from the sound velocity depicted in Fig. 6.2 are shown in Figs. 6.3 and 6.4. In Fig. 6.3 we plot the energy and entropy densities scaled by the appropriate powers of T. We observe sudden changes in the behavior of the quantities ε/T^4 and σ/T^3 in the vicinity of the critical temperature $T \sim T_c$. This behavior reflects the presence of the crossover phase transition.

Interestingly, the value of the critical energy density, $\varepsilon_c = 0.6\,\mathrm{GeV/fm}^3$, is similar to that obtained in the naive model of the phase transition discussed in Sec. 5.3,

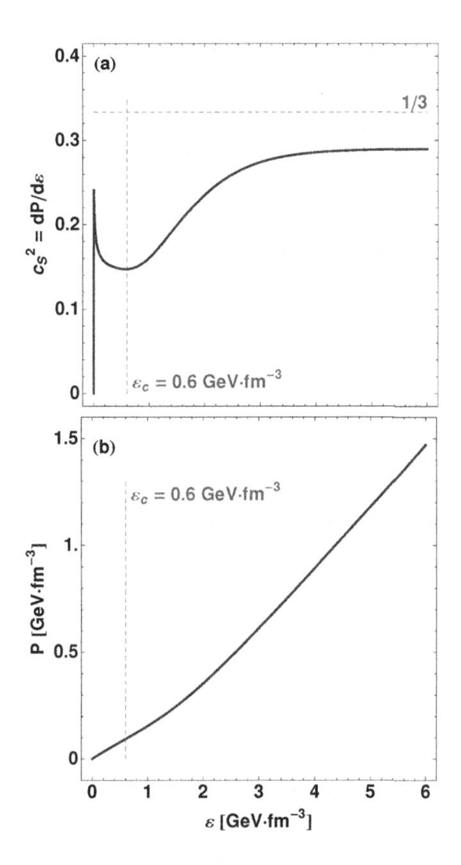

Fig. 6.4 The sound velocity and pressure shown as functions of the energy density. Notation the same as in Fig. 6.3.

see also Ex. 7.2. We note, however, that the behavior of the sound velocity shown in Fig. 6.2 is quite different from the behavior expected during the first-order phase transition. In the latter case, the changes of the pressure at the critical temperature are zero. This implies that the sound velocity drops to zero at $T = T_c$.

In Fig. 6.4 we plot the sound velocity and pressure as the functions of the energy density. One observes that the values of the sound velocity remain well below the ideal-gas limit for massless particles, $c_s^2 = 1/3$, even at very high energy densities. This behavior suggests that in the considered energy range the particle interactions in the plasma are not negligible.

It is worth to emphasize that the equation of state introduced above is limited to the case $\mu_B = 0$. We have discussed it in more detail, since it is used in the hydrodynamic calculations presented in the next Chapters. The inclusion of the non zero baryon number density leads to rich structures in the phase diagram, for example, see the review article [32].

Bibliography to Chapter 6

[1] G. Torrieri *et al.*, "SHARE: Statistical hadronization with resonances," *Comput. Phys. Commun.* **167** (2005) 229–251.
[2] G. Torrieri, S. Jeon, J. Letessier, and J. Rafelski, "SHAREv2: Fluctuations and a comprehensive treatment of decay feed-down," *Comput. Phys. Commun.* **175** (2006) 635–649.
[3] A. Kisiel, T. Taluc, W. Broniowski, and W. Florkowski, "THERMINATOR: Thermal heavy-ion generator," *Comput. Phys. Commun.* **174** (2006) 669–687.
[4] **Particle Data Group** Collaboration, C. Caso *et al.*, "Review of particle physics," *Eur. Phys. J.* **C3** (1998) 1–794.
[5] R. Hagedorn, "Statistical thermodynamics of strong interactions at high- energies," *Nuovo Cim. Suppl.* **3** (1965) 147–186.
[6] R. Hagedorn and J. Ranft, "Statistical thermodynamics of strong interactions at high-energies. 2. Momentum spectra of particles produced in *pp* collisions," *Nuovo Cim. Suppl.* **6** (1968) 169–354.
[7] S. Frautschi, "Statistical bootstrap model of hadrons," *Phys. Rev.* **D3** (1971) 2821–2834.
[8] W. Nahm, "Analytical solution of the statistical bootstrap model," *Nucl. Phys.* **B45** (1972) 525–553.
[9] J. Yellin, "An explicit solution of the statistical bootstrap," *Nucl. Phys.* **B52** (1973) 583–594.
[10] R. Hagedorn, "The long way to the statistical bootstrap model," invited talk at NATO Advanced Study Workshop on Hot Hadronic Matter: Theory and Experiment, Divonne-les-Bains, France, June 27 - July 1, 1994.
[11] N. Cabibbo and G. Parisi, "Exponential hadronic spectrum and quark liberation," *Phys. Lett.* **B59** (1975) 67.
[12] J. Letessier and J. Rafelski, "Hadrons and quark - gluon plasma," Camb. Monogr. Part. Phys. Nucl. Phys. Cosmol. **18** (2002) 1–397.
[13] W. Broniowski and W. Florkowski, "Different Hagedorn temperatures for mesons and baryons," *Phys. Lett.* **B490** (2000) 223–227.
[14] W. Broniowski, "Distinct Hagedorn temperatures from particle spectra: A higher one for mesons, a lower one for baryons," `hep-ph/0008112`.
[15] W. Broniowski, W. Florkowski, and L. Y. Glozman, "Update of the Hagedorn mass spectrum," *Phys. Rev.* **D70** (2004) 117503.
[16] J. Noronha-Hostler, C. Greiner, and I. A. Shovkovy, "Fast equilibration of hadrons in an expanding fireball," *Phys. Rev. Lett.* **100** (2008) 252301.
[17] R. Dashen, S.-K. Ma, and H. J. Bernstein, "S matrix formulation of statistical mechanics," *Phys. Rev.* **187** (1969) 345–370.
[18] R. F. Dashen and R. Rajaraman, "Narrow resonances in statistical mechanics," *Phys. Rev.* **D10** (1974) 694.
[19] R. F. Dashen and R. Rajaraman, "Effective elementarity of resonances and bound states in statistical mechanics," *Phys. Rev.* **D10** (1974) 708.
[20] W. Weinhold, "Zur thermodynamik des pion-nukleon-systems," GSI report 96-1 (1995) 67.

[21] W. Weinhold, "Zur thermodynamik des pion-nukleon-systems," Diplomarbeit, TH Darmstadt, 1995.

[22] W. Weinhold, B. L. Friman, and W. Noerenberg, "Thermodynamics of an interacting πN system," *Acta Phys. Polon.* **B27** (1996) 3249–3253.

[23] W. Weinhold, B. Friman, and W. Norenberg, "Thermodynamics of Delta resonances," *Phys. Lett.* **B433** (1998) 236–242.

[24] W. Weinhold, "Thermodynamik mit resonanzzustaenden," PhD Thesis, TU Darmstadt, 1998.

[25] K. G. Denisenko and S. Mrowczynski, "Deltas in hadron gas," *Phys. Rev.* **C35** (1987) 1932.

[26] A. B. Larionov, W. Cassing, M. Effenberger, and U. Mosel, "(p, π^{+-}) correlations in central heavy-ion collisions at 1 A GeV to 2 A GeV," *Eur. Phys. J.* **A7** (2000) 507–518.

[27] J. R. Pelaez, "The SU(2) and SU(3) chiral phase transitions within chiral perturbation theory," *Phys. Rev.* **D66** (2002) 096007.

[28] L. D. Landau and E. M. Lifshitz, "Statistical physics, part 1," (Butterworth-Heinemann, Oxford, 2001).

[29] Y. Aoki, Z. Fodor, S. D. Katz, and K. K. Szabo, "The equation of state in lattice QCD: With physical quark masses towards the continuum limit," *JHEP* **01** (2006) 089.

[30] H. Stoecker and W. Greiner, "High-energy heavy-ion collisions: Probing the equation of state of highly excited hadronic matter," *Phys. Rept.* **137** (1986) 277–392.

[31] M. Chojnacki and W. Florkowski, "Temperature dependence of sound velocity and hydrodynamics of ultra-relativistic heavy-ion collisions," *Acta Phys. Polon.* **B38** (2007) 3249–3262.

[32] D. H. Rischke, "The quark-gluon plasma in equilibrium," *Prog. Part. Nucl. Phys.* **52** (2004) 197–296.

Chapter 7

Exercises to PART I

Exercise 7.1. *Natural system of units.*
Given the Planck constant $\hbar = 6.58 \cdot 10^{-22}$ MeV s, the speed of light in vacuum $c = 2.998 \cdot 10^{8}$ m/s, and the Boltzmann constant $k_{\rm B} = 1.38 \cdot 10^{-16}$ erg/K (1 eV = 1.6 $\cdot 10^{-12}$ erg). **i)** What length is equivalent to 1 GeV in the natural system of units where $\hbar = c = k_{\rm B} = 1$. Express your result in fm. **ii)** What temperature (in K) is equivalent to 100 MeV? **iii)** Change the standard unit used for the cross sections, 1 milibarn = 1 mb = 10^{-31} m^2, to fm^2.

Answers:

$$1\,\text{GeV} = 5.07\,\text{fm}^{-1}\,, \tag{7.1}$$

$$100\,\text{MeV} = 116 \cdot 10^{10}\,\text{K}\,, \tag{7.2}$$

$$10\,\text{mb} = 1\,\text{fm}^2\,. \tag{7.3}$$

Exercise 7.2. *Energy density of normal nuclear matter.*
Calculate the energy density of normal nuclear matter.

Answer: The easiest way is to use the value of the nuclear saturation density $\rho_0 = 0.17\,\text{fm}^{-3}$ and to multiply it by the nucleon mass $m_{\rm N} = 940$ MeV. This gives

$$\varepsilon_0 \approx 0.16\,\text{GeV}/\text{fm}^3\,. \tag{7.4}$$

Find other ways of making this estimate.

Exercise 7.3. *Kinetic energy of a truck.*
The weight of a truck is 10 tons and it is moving at a speed of 100 km/h. Calculate its kinetic energy in eV.

Exercise 7.4. *Participants and spectators.*
i) Calculate the number of the participant nucleons in a central collision of the two nuclei characterized by the atomic numbers A and B (the impact parameter is exactly zero, $b = 0$). Assume that the nuclei have sharp surfaces and their density distribution is uniform and equal to the saturation density $\rho_0 = 0.17/\text{fm}^3$. **ii)** Find the numerical value of $N_{\rm part}$ for the central S+Au reaction.

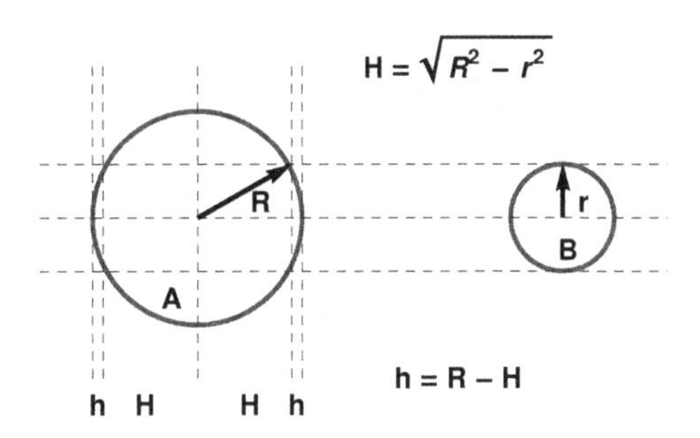

Fig. 7.1 Geometry of A+B collision, explanation of the notation used in Ex. 7.4.

Answers:
We assume $B \leq A$ and use the following notation: $R = 1.12A^{1/3}$, $r = 1.12B^{1/3}$, $H = \sqrt{R^2 - r^2}$, and $h = R - H$. Simple geometric considerations, see Fig. 7.1, lead to the formula

$$N_{\rm part} = B + \left[2\pi r^2 H + \frac{2}{3}\pi h^2 (3R - h) \right] \rho_0. \tag{7.5}$$

For $A = 197$ and $B = 32$ one finds $N_{\rm part} = 113$.

Exercise 7.5. *Transverse-momentum spectra.*
i) For the exponential distribution function given by Eq. (2.2) show that the average transverse mass, $\langle m_\perp \rangle$, and the average transverse momentum, $\langle p_\perp \rangle$, are given by the expressions,

$$\langle m_\perp \rangle = \frac{2\lambda^2 + 2\lambda m + m^2}{\lambda + m}, \tag{7.6}$$

$$\langle p_\perp \rangle = \frac{m^2 K_2(m/\lambda)}{\lambda + m} \, e^{m/\lambda}, \tag{7.7}$$

where m is the particle's mass. *Hint:* In the calculation of the average transverse momentum use the definition of the modified Bessel function of the second kind, Eq. (31.6) in Sec. 31.2. **ii)** In the limit $m/\lambda \ll 1$ show that

$$\langle m_\perp \rangle \approx 2\lambda + \frac{m^2}{\lambda},$$

$$\langle p_\perp \rangle \approx 2\lambda + \frac{m^2}{2\lambda},$$

which leads to $\langle m_\perp \rangle - \langle p_\perp \rangle \approx m^2/(2\lambda)$. *Hint:* In Eq. (7.7) use the asymptotic expansion (31.10).

Exercise 7.6. *Properties of rapidity.*
i) Calculate the rapidities of the projectile nuclei in the SPS fixed-target experiments with the beam energy of 60 and 200 GeV per nucleon. **ii)** Prove that rapidities are additive under Lorentz boosts along the beam axis. **iii)** Show that for a high-energy particle one can measure independently its rapidity and longitudinal position.

Exercise 7.7. *Properties of pseudorapidity.*
Using elementary trigonometric identities derive Eqs. (2.6) and (2.9).

Exercise 7.8. *Mandelstam variables.*
i) Besides the two Mandelstam variables s and t, defined by Eq. (3.2), we may introduce the third variable $u = (p_1 - p_2')^2$, where p_2' is the four-momentum of the target particle after the collision. Show that the following relation holds:

$$s + t + u = m_1^2 + m_2^2 + m_1'^2 + m_2'^2. \tag{7.8}$$

Here we allow for the general situation where the masses of the particles before and after the collision are different, $m_1 \neq m_1'$, $m_2 \neq m_2'$. **ii)** Show that Eq. (3.3) is valid for the elastic collisions of identical particles when $m_1 = m_1'$, $m_2 = m_2'$, and $m_1 = m_2 = m$.

Exercise 7.9. *Boost-invariance.*
Show that the longitudinal velocity of the form

$$v_z = \frac{Az - Bt}{At - Bz}, \tag{7.9}$$

where A and B are constants, is boost-invariant. *Hint:* Calculate the corresponding four-vector $u^\mu = \gamma(1, v_z)$. Apply the longitudinal Lorentz boost and check that u^μ is boost-invariant .

Exercise 7.10. *Thickness functions.*
i) Show that the nucleon-nucleus thickness function for the sharp-cutoff baryon distribution and $t(\mathbf{b}) = \delta^{(2)}(\mathbf{b})$ is given by the formula

$$T_A\left(b\right) = \frac{3\sqrt{R^2 - b^2}}{2\pi R^3}\theta\left(R - b\right). \tag{7.10}$$

ii) Calculate the nucleon-nucleus thickness function for the Gaussian baryon distribution

$$\rho_A(\mathbf{s}_A, z_A) = \frac{1}{(2\pi)^{3/2}\sigma_A^3}\exp\left(-\frac{\mathbf{s}_A^2 + z_A^2}{2\sigma_A^2}\right). \tag{7.11}$$

iii) Use the result of point **ii)** to show that for the gaussian baryon distributions the nucleus-nucleus thickness function is a Gaussian characterized by the width

$$\sigma_{AB} = \sqrt{\sigma_A^2 + \sigma_B^2}. \tag{7.12}$$

Exercise 7.11. *Binomial distribution.*
For the binomial distribution of the form

$$P(n) = \binom{N}{n} p^n (1-p)^{N-n} \tag{7.13}$$

calculate the average value $\bar{n}$ and the variance $\overline{n^2} - \bar{n}^2$. Use the obtained results to get Eqs. (3.28) and (3.30).

Exercise 7.12. *Gauge symmetry.*
Show that Lagrangian of QCD is invariant under local SU(3) transformations.

Exercise 7.13. *Bag equation of state.*
i) For the simple thermodynamic model discussed in Sec. 5.3 derive the analytic formula for the critical temperature at $\mu = 0$,

$$T_{\rm c}^4 = \frac{90B}{\pi^2 \left(16 + 21N_f/2 - g_\pi\right)}. \tag{7.14}$$

Here $g_\pi = 3$ is the degeneracy of the pionic states. We note that the expression in the denominator of Eq. (7.14) is the difference between the number of integral degrees of freedom in the plasma phase and the pion gas. **ii)** Calculate the latent heat of this transition and show that it is equal to $4B$,

$$\varepsilon_{\rm qgp}(T_{\rm c}) - \varepsilon_\pi(T_{\rm c}) = 4B. \tag{7.15}$$

Exercise 7.14. *Breit-Wigner formula.*
Fix the normalization of the right-hand-side of Eq. (6.5) with the requirement that

$$\int_{-\infty}^{+\infty} \frac{dE}{2\pi} |\psi(E)|^2 = 1. \tag{7.16}$$

What is the corresponding value of $|\psi(0)|^2$ that yields

$$\int_0^\infty dt |\psi(t)|^2 = 1. \tag{7.17}$$

PART II
RELATIVISTIC KINETIC THEORY

Chapter 8

Definitions

The second Part of this book is a short introduction to the relativistic kinetic theory. Our presentation remains mostly at the classical level — the particles are treated as well isolated excitations which are on the mass shell. However, we show how to include the quantum effects connected with **i)** the statistical properties of particles and **ii)** the internal degrees of freedom such as spin or color.

The general approach introduced in this Part will be used later to analyze specific physical situations. For example, in Chap. 11 we shall use the framework of the relativistic kinetic theory to describe the quark-gluon plasma production at the initial stage of the nuclear collisions, and in Chap. 28 we shall discuss the production rate of dileptons from a hot hadronic medium.

The importance of the kinetic theory relies on its ability to describe non-equilibrium dynamics of many body systems. Many models of nuclear reactions in different energy regimes are based on the relativistic kinetic theory. Moreover, the relativistic kinetic theory constitutes the foundation for other less general approaches like, e.g., relativistic hydrodynamics.

In this Chapter we give definitions of the basic quantities used in the relativistic kinetic theory. In particular, we define the particle current, the energy-momentum tensor, the entropy current, and different forms of the equilibrium distribution functions. We show in more detail how the quantum statistical properties of particles are reflected in the definition of their entropy current.

8.1 Distribution function

The basic quantity used in the kinetic theory is the *one-particle distribution function* $f(x,p) = f(t,\mathbf{x},\mathbf{p})$. It gives the number of particles ΔN in the phase-space volume element $\Delta^3 x \Delta^3 p$ placed at the phase-space point $(\mathbf{x},\mathbf{p})$ and the time t,

$$\Delta N = f(x,p)\, \Delta^3 x \Delta^3 p. \tag{8.1}$$

The main task of the kinetic theory is to formulate the time evolution equation for $f(x,p)$. In the non-relativistic case $f(x,p)$ satisfies the famous Boltzmann equation derived in 1872 (for a modern presentation see [1]).

The one-particle distribution function may be regarded as a classical limit (Planck's constant $h \longrightarrow 0,\ h = 2\pi\hbar$) of the *Wigner function*

$$W_{\alpha\beta}(x,p) = \frac{1}{(2\pi\hbar)^4} \int d^4y \, \exp\left(\frac{i\, p \cdot y}{\hbar}\right) \left\langle \psi^*_\beta \left(x + \frac{y}{2}\right) \psi_\alpha \left(x - \frac{y}{2}\right)\right\rangle, \tag{8.2}$$

namely

$$f(x,p) = \lim_{\hbar \to 0} \sum_\alpha W_{\alpha\alpha}(x,p). \tag{8.3}$$

In Eq. (8.2) $\psi_\alpha(x)$ is the wave function of a particle with the internal degrees of freedom specified by the index α, and the angular brackets denote averaging over various ψ^*_β and ψ_α states, i.e., they denote a transition from the complete quantum-mechanical description based on the wave function to the incomplete (statistical) quantum-mechanical description based on the density matrix

$$\left\langle \psi^*_\beta \left(x + \frac{y}{2}\right) \psi_\alpha \left(x - \frac{y}{2}\right)\right\rangle = \rho_{\alpha\beta}\left(x + \frac{y}{2}, x - \frac{y}{2}\right). \tag{8.4}$$

It is important to realize that *four* parameters p appearing in the definition of the Wigner function are independent. For example, the four-dimensional integration over p gives

$$\sum_\alpha \int d^4p \; W_{\alpha\alpha}(x,p) = \sum_\alpha \left\langle |\psi_\alpha(x)|^2 \right\rangle = \sum_\alpha \rho_{\alpha\alpha}(x,x), \tag{8.5}$$

which is the probability to find the particle at a spacetime point x. In the classical limit, however, the only non-zero contributions to $W_{\alpha\beta}(x,p)$ come from the region where $p^0 = E_p = \sqrt{m^2 + \mathbf{p}^2}$. Thus, the classical distribution function depends on the three-momentum $\mathbf{p}$ only, as was displayed explicitly above.

8.2 Particle current

The knowledge of $f(x,p)$ allows us to calculate several important macroscopic quantities. In particular, the density of particles and their three-current are given by the expressions

$$\begin{aligned} n^0(x) &= \int d^3p f(x,p), \\ \mathbf{n}(x) &= \int d^3p \, \frac{\mathbf{p}}{p^0} \, f(x,p). \end{aligned} \tag{8.6}$$

We note that the integration measure $d^3p/p^0 = d^3p/E$ is Lorentz invariant [1]. This property is most easily seen if we perform a Lorentz boost transformation with the velocity v along the x-axis. In this case the following transformation rules apply

$$\begin{aligned} E' &= \frac{E - p_x v}{\sqrt{1-v^2}}, \quad & p'_x &= \frac{p_x - E v}{\sqrt{1-v^2}}, \\ p'_y &= p_y, & p'_z &= p_z, \end{aligned} \tag{8.7}$$

[1] Speaking more precisely, it is invariant under proper, ortochronous Lorentz transformations that include spacetime translations, rotations, and Lorentz boosts.

and the new momentum integration measure is

$$d^3p' = dp'_x dp_y dp_z = \left(\frac{1}{\sqrt{1-v^2}} - \frac{p_x v}{E\sqrt{1-v^2}} \right) dp_x dp_y dp_z = \frac{E'}{E}\, d^3p. \tag{8.8}$$

Since the direction of the Lorentz boost transformation may be always chosen as the direction of the x-axis we conclude that

$$\frac{d^3p}{E} = \frac{d^3p'}{E'}. \tag{8.9}$$

Using Eq. (8.9) and the Lorentz transformation properties of the volume d^3x, we may check that the phase-space volume element $d^3p\, d^3x$ is Lorentz invariant. Therefore, we conclude from Eq. (8.1) that the phase-space distribution function $f(x,p)$ is a Lorentz scalar, see Ex. 10.1. These properties allow us to write the density of particles and the three-current in the covariant form as the four-vector (the particle four-current)

$$N^\mu(x) = \left(n^0(x), \mathbf{n}(x)\right) = \int \frac{d^3p}{p^0}\, p^\mu f(x,p). \tag{8.10}$$

Conservation of the particle number is given by the equation

$$\partial_\mu N^\mu(x) = 0. \tag{8.11}$$

We close this Section with a remark concerning our terminology. We use the name *particle current* understanding that it is the flux of particles crossing the surface of the unit area that has been placed perpendicularly to the flux. Hence, strictly speaking, N^μ is the *density of the particle current.* If particles carry conserved charges such as the baryon number b or the electric charge e, the four-currents $B^\mu = bN^\mu$ and $J^\mu = eN^\mu$ are the densities of baryon and electric current, respectively. Similar meaning has the entropy current introduced below in Sec. 8.4.

8.3 Energy-momentum tensor

The second moment of $f(x,p)$ determines the *energy-momentum tensor*

$$T^{\mu\nu}(x) = \int \frac{d^3p}{p^0}\, p^\mu p^\nu\, f(x,p). \tag{8.12}$$

The components of $T^{\mu\nu}$ have the following physical interpretation: T^{00} is the energy density, T^{0i} is the energy flow, T^{i0} is the momentum density, and T^{ij} is the pressure tensor.

The energy and momentum conservation laws have the form

$$\partial_\mu T^{\mu\nu}(x) = 0. \tag{8.13}$$

Expression (8.12) includes only the rest mass and the kinetic energy of particles. It does not include the potential energy. If the physical system consists of both particles and fields, the conservation laws are

$$\partial_\mu T^{\mu\nu}(x) + \partial_\mu T^{\mu\nu}_{\rm field}(x) = 0, \tag{8.14}$$

where $T^{\mu\nu}_{\rm field}$ is the energy-momentum tensor of the corresponding field.

8.4 Entropy current

One can also use $f(x,p)$ to calculate the entropy current of particles

$$S^\mu(x) = -\frac{1}{h^3}\int \frac{d^3p}{p^0}\, p^\mu \Phi\left[f(x,p)\right], \tag{8.15}$$

where

$$\Phi[f(x,p)] = h^3 f(x,p) \ln h^3 f(x,p) - \frac{1}{\epsilon}\left[1+\epsilon\, h^3 f(x,p)\right] \ln\left[1+\epsilon\, h^3 f(x,p)\right]. \tag{8.16}$$

Here $\epsilon = +1$ for bosons and $\epsilon = -1$ for fermions. The classical Boltzmann definition corresponds to the limit $\epsilon \to 0$

$$\Phi_{\rm cl}[f(x,p)] = h^3 f(x,p)\left[\ln h^3 f(x,p) - 1\right]. \tag{8.17}$$

The second law of thermodynamics in the relativistic notation has the form

$$\partial_\mu\, S^\mu(x) \geq 0. \tag{8.18}$$

We have placed explicitly Planck's constant in the definition of entropy (8.15) in order to have the correct normalization of the entropy density in thermal equilibrium. In the natural system of units used in the previous Chapters $h = 2\pi$ or $\hbar = 1$. To simplify the expression for the entropy, in the kinetic theory one uses often the system of units where $h = 1$. Alternatively, one may introduce the distribution functions $\tilde{f}$ which correspond to our original definition multiplied by h^3, namely

$$\tilde{f}(x,p) = h^3 f(x,p). \tag{8.19}$$

8.4.1 *Fermions*

What is the origin of the complicated form of Eq. (8.16)? As we shall see below, this form is related directly to the statistical properties of the particles. Following Landau and Lifshitz [2] we assume that the available quantum states of the particles are divided into the groups labeled by the index i ($i = 1, 2, ...$). Each group contains G_i states and a certain number N_i of the particles that are distributed among these states. The statistical weight of the whole system, $\Delta\Gamma$, is the product of the statistical weights calculated for each group,

$$\Delta\Gamma = \prod_i \Delta\Gamma_i, \tag{8.20}$$

and the entropy is given by the famous formula [2]

$$S = \ln \Delta\Gamma. \tag{8.21}$$

[2] Equation (8.21), originally written in the form displaying explicitly the Boltzmann constant, $S = k \ln W$, was first given by Planck in Ref. [3]. Nota bene, the constant k was also introduced first by Planck and many scientists at the beginning of the twentieth century (among them Lorentz as lately as in 1911) called k the Planck constant [4].

The number of the available states in the ith group is given by the volume of the phase space divided by the Planck constant,

$$G_i = \frac{\Delta^3 x_{(i)}\, \Delta^3 p_{(i)}}{h^3}, \tag{8.22}$$

and the ratio N_i/G_i is related to the phase space distribution function

$$\frac{N_i}{G_i} = \frac{\Delta N}{\Delta^3 x_{(i)}\, \Delta^3 p_{(i)}}\, h^3 = f_i h^3. \tag{8.23}$$

In the case of fermions, the statistical weight of the group is equal to the total number of ways of distributing N_i identical particles in G_i boxes with not more than one particle in each box ($N_i \leq G_i$), hence it is given by the combination

$$\Delta\Gamma_i = \frac{G_i!}{N_i!(G_i - N_i)!}. \tag{8.24}$$

With the help of the entropy definition (8.21) and the Stirling formula,

$$n! \approx (n/e)^n, \tag{8.25}$$

we find

$$S = \sum_i \left[G_i \ln G_i - N_i \ln N_i - (G_i - N_i)\ln(G_i - N_i)\right]. \tag{8.26}$$

In the next step, using Eq. (8.23) we find

$$\begin{aligned} S &= -\sum_i G_i \left[h^3 f_i \ln h^3 f_i + (1 - h^3 f_i)\ln(1 - h^3 f_i)\right] \\ &= -\int \frac{d^3p\, d^3x}{h^3} \left[h^3 f(x,p) \ln h^3 f(x,p) + \left(1 - h^3 f(x,p)\right) \ln\left(1 - h^3 f(x,p)\right)\right] \end{aligned} \tag{8.27}$$

or

$$\frac{dS}{d^3x} = -\frac{1}{h^3}\int \frac{d^3p}{p^0}\, p^0 \left[h^3 f(x,p) \ln h^3 f(x,p) + \left(1 - h^3 f(x,p)\right) \ln\left(1 - h^3 f(x,p)\right)\right]. \tag{8.28}$$

A covariant form of Eq. (8.28) is given by Eq. (8.15) with $\epsilon = -1$.

8.4.2 *Bosons*

In the case of bosons, the statistical weight of the group is equal to the total number of ways of distributing N_i identical particles in G_i boxes with multiple occupations of the same box possible,

$$\Delta\Gamma_i = \frac{(G_i + N_i - 1)!}{(G_i - 1)!N_i!}. \tag{8.29}$$

The entropy definition (8.21) gives

$$S = \sum_i \left[(G_i + N_i) \ln(G_i + N_i) - N_i \ln N_i - G_i \ln G_i \right]. \tag{8.30}$$

With the help of Eq. (8.23) we obtain

$$\begin{aligned} S &= \sum_i G_i \left[(1 + h^3 f_i) \ln(1 + h^3 f_i) - h^3 f_i \ln h^3 f_i \right] \\ &= \int \frac{d^3p\, d^3x}{h^3} \left[(1 + h^3 f(x,p)) \ln(1 + h^3 f(x,p)) - h^3 f(x,p) \ln h^3 f(x,p) \right] \end{aligned} \tag{8.31}$$

or

$$\frac{dS}{d^3x} = -\frac{1}{h^3} \int \frac{d^3p}{p^0}\, p^0 \left[h^3 f(x,p) \ln h^3 f(x,p) - \left(1 + h^3 f(x,p)\right) \ln\left(1 + h^3 f(x,p)\right) \right]. \tag{8.32}$$

A covariant form of Eq. (8.32) is given by Eq. (8.15) with $\epsilon = +1$.

8.5 Internal degrees of freedom

For particles with extra internal degrees of freedom such as spin, isospin or color, the notion of the distribution function $f(x,p)$ may be generalized to the concept of the phase-space dependent matrix $\hat{f}(x,p)$, namely

$$f(x,p) \longrightarrow \hat{f}(x,p) = \begin{bmatrix} f_{11}(x,p) & f_{12}(x,p) & \dots \\ f_{21}(x,p) & f_{22}(x,p) & \dots \\ \dots & \dots & \dots \end{bmatrix}. \tag{8.33}$$

The coefficients of $\hat{f}(x,p)$ correspond to the classical limit of the Wigner function (8.2). The diagonal elements $f_{\alpha\alpha}(x,p)$ define the phase-space densities of particles with internal quantum number specified by the label α.

For example, for fermions with spin $\frac{1}{2}$ we may introduce the 2×2 matrix

$$\hat{f}(x,p) = \begin{bmatrix} f_{\uparrow\uparrow}(x,p) & f_{\uparrow\downarrow}(x,p) \\ f_{\downarrow\uparrow}(x,p) & f_{\downarrow\downarrow}(x,p) \end{bmatrix}, \tag{8.34}$$

where $f_{\uparrow\uparrow}(x,p)$ gives the number of particles with spin "up" (oriented along an arbitrary quantization axis, let say, $\hat{z}$), and $f_{\downarrow\downarrow}(x,p)$ gives then the number of particles with spin "down". Similarly, we may treat the color degrees of freedom of quarks. In this case we use the 3×3 matrix

$$\hat{f}(x,p) = \begin{bmatrix} f_{RR}(x,p) & f_{RG}(x,p) & f_{RB}(x,p) \\ f_{GR}(x,p) & f_{GG}(x,p) & f_{GB}(x,p) \\ f_{BR}(x,p) & f_{BG}(x,p) & f_{BB}(x,p) \end{bmatrix}, \tag{8.35}$$

where $f_{RR}(x,p)$, $f_{GG}(x,p)$ and $f_{BB}(x,p)$ are the phase-space densities of red, green and blue quarks, respectively.

In all discussed cases the number of particles with *arbitrary* internal quantum numbers is given by the trace of $\hat{f}(x,p)$. This suggests an immediate generalization of our definitions (8.10), (8.12) and (8.15), namely

$$N^\mu(x) = \int \frac{d^3p}{p^0} p^\mu \ \mathrm{Tr}\left[\hat{f}(x,p)\right], \tag{8.36}$$

$$T^{\mu\nu}(x) = \int \frac{d^3p}{p^0} p^\mu p^\nu \ \mathrm{Tr}\left[\hat{f}(x,p)\right], \tag{8.37}$$

$$S^\mu(x) = -\frac{1}{h^3}\int \frac{d^3p}{p^0} p^\mu \ \mathrm{Tr}\left[\Phi\left[\hat{f}(x,p)\right]\right]. \tag{8.38}$$

In the definition of the entropy flow (8.38), the function $\Phi\left[\hat{f}\right]$ is defined by the series expansion of the function (8.16). In this way Eqs. (8.10), (8.12) and (8.15) are generalized to the sums over all independent internal degrees of freedom, as it should be for any extensive thermodynamic quantity. To see this point we may diagonalize the distribution function

$$\hat{f}(x,p) \longrightarrow \hat{f}_D(x,p) = U\hat{f}(x,p)U^{-1} = \begin{bmatrix} f_{D,1}(x,p) & 0 & \ldots \\ 0 & f_{D,2}(x,p) & \ldots \\ \ldots & \ldots & \ldots \end{bmatrix}, \tag{8.39}$$

and write

$$N^\mu(x) = \sum_{\alpha=1}^{g}\int \frac{d^3p}{p^0} p^\mu \ f_{D,\alpha}(x,p), \tag{8.40}$$

$$T^{\mu\nu}(x) = \sum_{\alpha=1}^{g}\int \frac{d^3p}{p^0} p^\mu p^\nu \ f_{D,\alpha}(x,p), \tag{8.41}$$

$$S^\mu(x) = -\frac{1}{h^3}\sum_{\alpha=1}^{g}\int \frac{d^3p}{p^0} p^\mu \ \Phi\left[f_{D,\alpha}(x,p)\right]. \tag{8.42}$$

The diagonalization (8.39) is always possible since $\hat{f}(x,p)$ is a hermitian matrix (see our definition of the Wigner function (8.2)). Moreover, the unitary transformation (8.39) does not change the physical quantities defined by Eqs. (8.36), (8.37) and (8.38).

8.6 Fluid velocity

For systems of particles carrying a conserved quantum number (for example, electric charge or baryon number) it is straightforward to define the fluid velocity through the equation

$$u^\mu(x) = \frac{N^\mu(x)}{\sqrt{N^\mu(x) N_\mu(x)}}. \tag{8.43}$$

This definition (called later *Eckart's definition* [5]) is, however, not useful for systems dominated by radiation or mesonic degrees of freedom (with zero electric charge or zero baryon number, respectively). In this case one can use *Landau's definition* of the fluid velocity [6]. It defines $u^\mu(x)$ by the equation

$$T^{\mu\nu}(x) u_\nu(x) = \lambda(x)\, u^\mu(x). \tag{8.44}$$

The normalization condition of the four-velocity, $u^\mu u_\mu = 1$, gives $\lambda = u_\alpha u_\beta T^{\alpha\beta}$. The local rest frame (LRF) is defined by the condition $u^\mu = (1,0,0,0)$. In the Landau approach, the momentum density vanishes in such a reference frame, i.e., $T^{oi}_{\rm LRF} = T^{io}_{\rm LRF} = 0$.

8.7 Relativistic equilibrium distributions

Let us now give an example of the distribution function which is the relativistic analog of the Boltzmann distribution function,

$$\begin{aligned} f(p) &= \frac{1}{h^3} \exp\left(\frac{\mu - p^\alpha u_\alpha}{T}\right) \\ &= \frac{1}{h^3} \exp\left(\frac{\mu - u^0\sqrt{m^2 + \mathbf{p}^2} + \mathbf{u}\cdot\mathbf{p}}{T}\right). \end{aligned} \tag{8.45}$$

The distribution (8.45) was introduced by *Jüttner* in 1911 [7]. It reduces to the Boltzmann equilibrium distribution in the rest frame of the fluid, where $u^\mu = (1,0,0,0)$. We note that u^μ as well as the temperature T and the chemical potential μ are independent of x. Otherwise, various transport processes would be present in the system (heat flow, diffusion) and Eq. (8.45) could not describe the true equilibrium [3]. The relativistic form of the equilibrium distribution functions valid for systems of *bosons* and *fermions* is

$$f(p) = \frac{1}{h^3}\left[\exp\left(\frac{p^\alpha u_\alpha - \mu}{T}\right) - \epsilon\right]^{-1}, \tag{8.46}$$

where $\epsilon = +1$ for bosons, and $\epsilon = -1$ for fermions. Equation (8.46) was given by Jüttner in 1928 [8].

[3] Strictly speaking, certain combinations of the gradients of μ, $T$and u_α are allowed to be different from zero even in global equilibrium. We shall discuss this point in more detail later.

Equations (8.45) and (8.46) imply the following form of the particle flow, energy-momentum tensor, and entropy flow in equilibrium

$$N^\mu = nu^\mu, \tag{8.47}$$

$$T^{\mu\nu} = (\varepsilon + P)\ u^\mu u^\nu - P g^{\mu\nu}, \tag{8.48}$$

$$S^\mu = \sigma u^\mu. \tag{8.49}$$

Here n, ε, P, and σ are the particle density, energy density, pressure, and entropy density defined in the local rest frame of the fluid element. In this special frame we have

$$N^0_{\rm LRF} = n, \quad N^i_{\rm LRF} = 0, \tag{8.50}$$

$$T^{00}_{\rm LRF} = \varepsilon, \quad T^{11}_{\rm LRF} = T^{22}_{\rm LRF} = T^{33}_{\rm LRF} = P, \tag{8.51}$$

$$S^0_{\rm LRF} = \sigma, \quad S^i_{\rm LRF} = 0. \tag{8.52}$$

Other (non-diagonal) components of the energy-momentum tensor vanish in the local rest frame, thus we obtain the following structure

$$T^{\mu\nu}_{\rm LRF} = \begin{pmatrix} \varepsilon & 0 & 0 & 0 \\ 0 & P & 0 & 0 \\ 0 & 0 & P & 0 \\ 0 & 0 & 0 & P \end{pmatrix}. \tag{8.53}$$

We note that the thermodynamic quantitites defined in the rest frame of the fluid element are Lorentz scalars. The density of particles in an arbitrary frame is related to the density defined in the rest frame by the formula

$$n^0 = nu^0 = n\gamma, \tag{8.54}$$

where $\gamma = \left(1 - v^2\right)^{-1/2}$ is the Lorentz factor. In the rest frame $\gamma = 1$ and we recover the relation $n_0 = n$. Similar equations may be written for the energy density and pressure.

The Jüttner distribution functions (8.45) and (8.46) may be also substituted into the definition of the entropy flow (8.15). In this case we find

$$S^\mu u_\mu \equiv \sigma = \frac{\varepsilon + P - \mu n}{T}. \tag{8.55}$$

Thus, if the system is in equilibrium our definition of the entropy density σ is consistent with the thermodynamic relation (5.19). The formula (8.15) is, however, also applicable for non-equilibrium situations.

8.7.1 *Classical massive particles*

Substitution of (8.45) into expression (8.10) gives the density of classical relativistic particles

$$n_{\rm cl} = \frac{4\pi}{h^3} e^{\mu/T} m^2 T K_2\left(\frac{m}{T}\right), \tag{8.56}$$

where $K_2(x)$ is the modified Bessel function of the second type. Similarly, from the definition of the energy-momentum tensor (8.12) we obtain

$$\begin{aligned} \varepsilon_{\rm cl} + P_{\rm cl} &= \frac{4\pi}{h^3} e^{\mu/T} m^3 T K_3\left(\frac{m}{T}\right), \\ P_{\rm cl} &= \frac{4\pi}{h^3} e^{\mu/T} m^2 T^2 K_2\left(\frac{m}{T}\right). \end{aligned} \tag{8.57}$$

Here $K_3(x)$ is another modified Bessel function. Definitions of the functions $K_n(x)$ as well as their asymptotic expansions and recurrence relations are given in Chap. 31. Expressions (8.56) and (8.57) yield the equation of state of the relativistic ideal gas of classical particles ($k_B = 1$)

$$P_{\rm cl} = n_{\rm cl} T. \tag{8.58}$$

8.7.2 *Relativistic massless bosons*

Thermodynamic properties of the relativistic gas of massless bosons follow from Eq. (8.46) with $\epsilon = 1$. In the special case where $m = 0$ and $\mu = 0$ we find

$$n_{\rm b} = \int \frac{d^3p}{h^3} \frac{1}{e^{p/T} - 1} = \frac{8\pi\zeta(3)}{h^3} T^3 = \frac{\zeta(3)}{\pi^2 \hbar^3} T^3. \tag{8.59}$$

The function ζ appearing in Eq. (8.59) is the Riemann zeta function defined by the series

$$\zeta(x) = \sum_{n=1}^{\infty} \frac{1}{n^x}. \tag{8.60}$$

The frequently used values are:

$$\begin{aligned} \zeta(2) &= \frac{\pi^2}{6} \approx 1.645, \\ \zeta(3) &\approx 1.202, \\ \zeta(4) &= \frac{\pi^4}{90} \approx 1.082. \end{aligned} \tag{8.61}$$

The analogs of Eq. (8.57) read

$$\varepsilon_{\rm b} = \int \frac{d^3p}{h^3} \frac{p}{e^{p/T} - 1} = \frac{4\pi^5}{15h^3} T^4 = \frac{\pi^2}{30\hbar^3} T^4, \tag{8.62}$$

$$P_{\rm b} = \int \frac{d^3p}{h^3} \frac{p}{3(e^{p/T} - 1)} = \frac{4\pi^5}{45h^3} T^4 = \frac{\pi^2}{90\hbar^3} T^4. \tag{8.63}$$

Thus, the form of the equation of state is

$$P_{\rm b} = \frac{1}{3} \varepsilon_{\rm b}. \tag{8.64}$$

Moreover, our definition of the entropy flow gives in this case

$$S^\mu_{\rm b} u_\mu \equiv \sigma_{\rm b} = \frac{\varepsilon_{\rm b} + P_{\rm b}}{T} = \frac{16\pi^5}{45h^3} T^3 = \frac{2\pi^2}{45\hbar^3} T^3, \tag{8.65}$$

which implies the constant entropy per particle,

$$\frac{\sigma_{\rm b}}{n_{\rm b}} = \frac{2\pi^4}{45\zeta(3)} \approx 3.6\,. \tag{8.66}$$

We note that Eq. (8.62) was used by us in Sec. 5.3, where we discussed the equation of state of the weakly-interacting quark-gluon plasma.

Equations (8.59), (8.62), and (8.63) follow from the generic formula

$$\int_0^\infty \frac{z^{x-1}\,dz}{e^z - 1} = \zeta(x)\Gamma(x), \tag{8.67}$$

where $\zeta(x)$ is the Riemann zeta function defined above by Eq. (8.60) and $\Gamma(x)$ is the Euler gamma function

$$\Gamma(x) = \int_0^\infty e^{-t} t^{x-1} dt. \tag{8.68}$$

To derive Eq. (8.67) we perform the following manipulations

$$\int_0^\infty \frac{z^{x-1}\,dz}{e^z - 1} = \int_0^\infty e^{-z} z^{x-1} \frac{dz}{1 - e^{-z}} = \sum_{n=0}^\infty \int_0^\infty e^{-(n+1)z} z^{x-1}\,dz$$
$$= \sum_{n=0}^\infty \int_0^\infty e^{-t} t^{x-1} (n+1)^{-x}\,dt = \sum_{n=1}^\infty \frac{1}{n^x} \int_0^\infty e^{-t} t^{x-1}\,dt = \zeta(x)\Gamma(x). \tag{8.69}$$

The special case where x is an even integer ($x = 2n$) may be expressed in the form

$$\int_0^\infty \frac{z^{2n-1}\,dz}{e^z - 1} = \frac{(2\pi)^{2n} B_n^*}{4n}, \tag{8.70}$$

where B_n^* are the Bernoulli numbers [2, 9]

$$B_n^* = \frac{2(2n)!}{(2\pi)^{2n}}\,\zeta(2n), \tag{8.71}$$

$$B_1^* = \frac{1}{6}, \quad B_2^* = \frac{1}{30}, \quad B_3^* = \frac{1}{42}, \quad B_4^* = \frac{1}{30}, \quad \ldots\;. \tag{8.72}$$

8.7.3 *Relativistic massless fermions*

8.7.3.1 *Zero chemical potential*

In this case one may use the formula which is the analog of Eq. (8.67) used for bosons,

$$\int_0^\infty \frac{z^{x-1}\,dz}{e^z + 1} = (1 - 2^{1-x})\zeta(x)\Gamma(x). \tag{8.73}$$

In order to derive Eq. (8.73) we notice that repeating the same steps as those in Eq. (8.69) we arrive at the formula

$$\int_0^\infty \frac{z^{x-1}\,dz}{e^z+1} = \Gamma(x)\sum_{n=0}^{\infty}(-1)^n\frac{1}{(n+1)^x}. \tag{8.74}$$

Collecting separately the positive and negative terms in (8.74) we find

$$\begin{aligned}\sum_{n=0}^{\infty}(-1)^n\frac{1}{(n+1)^x} &= \sum_{m=1,3,5,\ldots}^{\infty}\frac{1}{m^x} - \sum_{m=2,4,6,\ldots}^{\infty}\frac{1}{m^x}\\ &= \sum_{m=1,2,3,\ldots}^{\infty}\frac{1}{m^x} - 2\sum_{m=2,4,6,\ldots}^{\infty}\frac{1}{m^x}\\ &= \sum_{m=1,2,3,\ldots}^{\infty}\frac{1}{m^x} - 2\sum_{m=1,2,3,\ldots}^{\infty}\frac{1}{(2m)^x}\\ &= (1-2^{1-x})\zeta(x).\end{aligned} \tag{8.75}$$

This identity together with Eq. (8.74) lead directly to Eq. (8.73). The undetermined result for $x = 1$ is obtained as the integral

$$\int_0^\infty \frac{dz}{e^z+1} = \ln 2. \tag{8.76}$$

We may now conclude that the energy density and pressure of fermions at zero value of the chemical potential is related to the appropriate boson densities by a simple factor of 7/8 ($x = 4$ in Eq. (8.73)),

$$\begin{aligned}\varepsilon_{\rm f} &= \frac{7}{8}\varepsilon_{\rm b} = \frac{7\pi^2}{240\hbar^3}T^4,\\ P_{\rm f} &= \frac{7}{8}P_{\rm b} = \frac{7\pi^2}{720\hbar^3}T^4.\end{aligned} \tag{8.77}$$

On the other hand, the particle densities are related by a factor of 3/4 ($x = 3$ in Eq. (8.73)),

$$n_{\rm f} = \frac{3}{4}n_{\rm b}. \tag{8.78}$$

The last two equations imply, that the ratio of the entropy density to the particle density for massless fermions is given by the formula

$$\frac{\sigma_{\rm f}}{n_{\rm f}} = \frac{7}{6}\frac{\sigma_{\rm b}}{n_{\rm b}} \approx 4.2. \tag{8.79}$$

8.7.3.2 *Finite chemical potential*

At finite values of the chemical potential it is possible to calculate the sum of the energies of fermions and antifermions

$$
\begin{aligned}
\varepsilon_{\mathrm{f}} + \varepsilon_{\bar{\mathrm{f}}} &= \int \frac{d^3p}{h^3} \frac{p}{e^{(p-\mu)/T}+1} + \int \frac{d^3p}{h^3} \frac{p}{e^{(p+\mu)/T}+1} \\
&= \frac{4\pi T^4}{h^3} \left(\int_0^\infty dx \, \frac{x^3}{e^{x-y}+1} + \int_0^\infty dx \, \frac{x^3}{e^{x+y}+1} \right) \\
&= \frac{4\pi T^4}{h^3} \left(\int_0^\infty dx \, \frac{(x+y)^3}{e^x+1} + \int_0^\infty dx \, \frac{(x-y)^3}{e^x+1} \right. \\
&\quad \left. + \int_0^y dx (y-x)^3 \left[\frac{1}{e^x+1} + \frac{1}{e^{-x}+1} \right] \right) \qquad (8.80)
\end{aligned}
$$

The first two integrals in the last line of Eq. (8.80) may be obtained with the help of the expressions worked out in the previous Sections,

$$
\int_0^\infty \frac{dx\,(x \pm y)^3}{e^x+1} = \frac{7\pi^4}{120} + \frac{\pi^2 y^2}{4} \pm y^3 \ln 2 \pm \frac{9}{2}\, y\, \zeta(3). \qquad (8.81)
$$

The last integral in Eq. (8.80) is trivial since

$$
\left[\frac{1}{e^x+1} + \frac{1}{e^{-x}+1} \right] = 1. \qquad (8.82)
$$

In this way, we finally get

$$
\begin{aligned}
\varepsilon_{\mathrm{f}} + \varepsilon_{\bar{\mathrm{f}}} &= \frac{1}{h^3} \left(\frac{14\pi^5 T^4}{30} + 2\pi^3 T^2 \mu^2 + \pi \mu^4 \right) \\
&= \frac{1}{\hbar^3} \left(\frac{7\pi^2 T^4}{120} + \frac{T^2\mu^2}{4} + \frac{\mu^4}{8\pi^2} \right). \qquad (8.83)
\end{aligned}
$$

8.7.4 *Ideal relativistic gases — summary*

The expressions derived in the previous Section may be used to obtain thermodynamic quantities characterizing the weakly-interacting quark-gluon plasma. For example, Eq. (8.83) was used by us in Sec. 5.3, where it was multiplied by the factor $6N_f$. The factor 6 reflects the spin and color degeneracy of quarks, while N_f is the number of quark flavors. With the inclusion of gluons, the number of the internal degrees of freedom in the weakly interacting quark-gluon plasma is given by the expression

$$
\begin{aligned}
g_{\mathrm{qgp}} &= 8 \times 2 \ (\text{gluon color} \times \text{spin}) \\
&\quad + \frac{7}{8}(2 \times 3 \times 2 \times N_f) \ (\text{quarks and antiquarks} \times \text{color} \times \text{spin} \times \text{flavor}) \\
&= 16 + \frac{21}{2} N_f. \qquad (8.84)
\end{aligned}
$$

Table 8.1 Summary of the expressions defining the thermodynamic variables (mean number of particles, entropy, pressure, and energy) in the case of *quantum statistics:* $\epsilon = +1$ for bosons and $\epsilon = -1$ for fermions. The quantity g is the degeneracy factor connected with internal quantum numbers.

	$m > 0$	$m = 0$
N	$\frac{g}{2\pi^2} T V m^2 \epsilon \sum_{\kappa=1}^{\infty} \frac{\epsilon^\kappa}{\kappa} e^{\frac{\mu}{T}\kappa} K_2\left(\frac{m}{T}\kappa\right)$	$\frac{g}{\pi^2} T^3 V \epsilon \,\mathrm{Li}_3\left(\epsilon e^{\frac{\mu}{T}}\right)$
S	$\frac{g}{2\pi^2} V m^2 \epsilon \sum_{\kappa=1}^{\infty} \frac{\epsilon^\kappa}{\kappa^2} e^{\frac{\mu}{T}\kappa}$ $\times \left[(4T - \mu\kappa) K_2\left(\frac{m}{T}\kappa\right) + m\kappa K_1\left(\frac{m}{T}\kappa\right)\right]$	$\frac{g}{\pi^2} T^2 V \epsilon$ $\times \left[4T\,\mathrm{Li}_4\left(\epsilon e^{\frac{\mu}{T}}\right) - \mu\,\mathrm{Li}_3\left(\epsilon e^{\frac{\mu}{T}}\right)\right]$
P	$\frac{g}{2\pi^2} T^2 m^2 \epsilon \sum_{\kappa=1}^{\infty} \frac{\epsilon^\kappa}{\kappa^2} e^{\frac{\mu}{T}\kappa} K_2\left(\frac{m}{T}\kappa\right)$	$\frac{g}{\pi^2} T^4 \epsilon \,\mathrm{Li}_4\left(\epsilon e^{\frac{\mu}{T}}\right)$
E	$\frac{g}{2\pi^2} T V m^2 \epsilon \sum_{\kappa=1}^{\infty} \frac{\epsilon^\kappa}{\kappa^2} e^{\frac{\mu}{T}\kappa}$ $\times \left[3T K_2\left(\frac{m}{T}\kappa\right) + m\kappa K_1\left(\frac{m}{T}\kappa\right)\right]$	$\frac{3g}{\pi^2} T^4 V \epsilon \,\mathrm{Li}_4\left(\epsilon e^{\frac{\mu}{T}}\right)$

Table 8.2 The same as Table 8.1 but for *classical statistics.*

	$m > 0$	$m = 0$
N	$\frac{g}{2\pi^2} T V m^2 e^{\frac{\mu}{T}} K_2\left(\frac{m}{T}\right)$	$\frac{g}{\pi^2} T^3 V e^{\frac{\mu}{T}}$
S	$\frac{g}{2\pi^2} V m^2 e^{\frac{\mu}{T}}$ $\times \left[(4T - \mu) K_2\left(\frac{m}{T}\right) + m K_1\left(\frac{m}{T}\right)\right]$	$\frac{g}{\pi^2} T^2 V e^{\frac{\mu}{T}} (4T - \mu)$
P	$\frac{g}{2\pi^2} T^2 m^2 e^{\frac{\mu}{T}} K_2\left(\frac{m}{T}\right)$	$\frac{g}{\pi^2} T^4 e^{\frac{\mu}{T}}$
E	$\frac{g}{2\pi^2} T V m^2 e^{\frac{\mu}{T}}$ $\times \left[3T K_2\left(\frac{m}{T}\right) + m K_1\left(\frac{m}{T}\right)\right]$	$\frac{3g}{\pi^2} T^4 V e^{\frac{\mu}{T}}$

Here the factor 7/8 accounts for the difference between Bose-Einstein and Fermi-Dirac statistics. With N_f in the range 2 – 3, we find

$$37 \leq g_{\mathrm{qgp}} \leq 47.5, \tag{8.85}$$

hence, in the limit of vanishing baryon chemical potential, the plasma pressure (with the neglected bag constant contribution) is given by the formula

$$P_{\mathrm{qgp}} = g_{\mathrm{qgp}} \frac{\pi^2}{90} T^4. \tag{8.86}$$

The summary of the expressions defining thermodynamic variables of ideal relativistic gases in different limits is given in Tables 8.1 and 8.2, where we have introduced the polylogarithm function defined by the series

$$\mathrm{Li}_n(z) = \sum_{k=1}^{\infty} \frac{z^k}{k^n}. \tag{8.87}$$

The important values include:

$$\mathrm{Li}_3(1) = \zeta(3), \qquad \mathrm{Li}_3(-1) = -\frac{3}{4}\zeta(3),$$
$$\mathrm{Li}_4(1) = \frac{\pi^4}{90}, \qquad \mathrm{Li}_4(-1) = -\frac{7\pi^4}{720}. \tag{8.88}$$

Bibliography to Chapter 8

[1] K. Huang, "Statistical Mechanics," (John Wiley, New York, 1963).
[2] L. D. Landau and E. M. Lifshitz, "Statistical Physics, Part 1," (Butterworth-Heinemann, Oxford, 2001).
[3] M. Planck, *Annalen der Physik* **4** (1910) 553.
[4] A. Pais, "Subtle is the Lord: The Science and the Life of Albert Einstein," (Oxford University Press, Oxford, 1982).
[5] C. Eckart, "The Thermodynamics of irreversible processes. 3. Relativistic theory of the simple fluid," *Phys. Rev.* **58** (1940) 919–924.
[6] L. D. Landau and E. M. Lifshitz, "Fluid Mechanics," (Butterworth-Heinemann, Oxford, 2000).
[7] F. Juttner, *Ann. Phys. und Chemie* **34** (1911) 856.
[8] F. Juttner, *Zeitschr. Phys.* **47** (1928) 542.
[9] `http://mathworld.wolfram.com/BernoulliNumber.html`, eq. 40.

Chapter 9

Relativistic Kinetic Equations

In the second Chapter of Part II we discuss different forms of the kinetic equations that are used in many branches of the relativistic heavy-ion physics. In Sec. 9.1 we introduce the *Boltzmann-Vlasov equations* describing *collisionless systems*. In this case the interactions are mediated by the self-consistent mean fields. In Sec. 9.2 we analyze in greater detail the effects of collisions. The form of the *collision terms* describing binary collisions is derived. The quantum statistical properties of particles are included in the collision terms through the *Uehling-Uhlenbeck* phase-space corrections. The results obtained for simple systems (one type of particles, elastic collisions only) are generalized to more complicated systems (different types of particles, elastic and inelastic collisions) in Sec. 9.3. In Sec. 9.4 the conservation laws are analyzed. The proof of the *Boltzmann H-theorem* is given in Sec. 9.5. The role of the conservation laws in the construction of the local and global *equilibrium distribution functions* is elucidated in Sec. 9.6. In the end of the Chapter, in Sec. 9.7, we discuss a popular approximate treatment of the collision terms, which is called the *relaxation-time approximation.*

9.1 Systems without collisions

At the beginning of our discussion about the relativistic kinetic equations we consider the simplest case, i.e., the system where the effects of collisions are negligible. In this case the *relativistic Boltzmann* equation is reduced to the continuity equation expressing the conservation of the number of particles

$$p^\mu \partial_\mu \, f(x,p) = 0. \tag{9.1}$$

In the three-vector notation Eq. (9.1) has the structure

$$\left(\frac{\partial}{\partial t} + \mathbf{v}_p \cdot \nabla \right) f(t, \mathbf{x}, \mathbf{p}) = 0. \tag{9.2}$$

Integration of Eq. (9.1) over momentum leads to the conservation law (8.11). We note that the form of Eq. (9.2) is valid also for non-relativistic systems. The only difference is that the relativistic relation between the particle's velocity and momentum

$$\mathbf{v}_p = \frac{\mathbf{p}}{E_p} = \frac{\mathbf{p}}{\sqrt{\mathbf{p}^2 + m^2}} \tag{9.3}$$

should be replaced by the non-relativistic limit

$$\mathbf{v}_p = \frac{\mathbf{p}}{m}. \tag{9.4}$$

.

Equation (9.1) does not take into account the external forces. Their effect on the motion of particles may be included by adding an extra term

$$p^\mu \partial_\mu \, f(x,p) + mF^\mu \partial^p_\mu \, f(x,p) = 0. \tag{9.5}$$

Since $f(x,p)$ does not depend on p^0, Eq. (9.5) may be rewritten in the form

$$\left(\frac{\partial}{\partial t} + \mathbf{v}_p \cdot \nabla + \frac{m}{E_p} \mathbf{F} \cdot \nabla_p \right) f(t,\mathbf{x},\mathbf{p}) = 0, \tag{9.6}$$

where

$$\mathbf{F} = (F^1, F^2, F^2) = (F_x, F_y, F_z) \tag{9.7}$$

and ∇_p denotes the gradient in the momentum space

$$\nabla_p = \left(\frac{\partial}{\partial p_x}, \frac{\partial}{\partial p_y}, \frac{\partial}{\partial p_z} \right). \tag{9.8}$$

The requirement of the particle conservation in the presence of the external force $\mathbf{F}$ imposes the condition, see Ex. 10.7,

$$E_p \nabla_p \cdot \mathbf{F} = \mathbf{v}_p \cdot \mathbf{F}. \tag{9.9}$$

One may check that this condition is satisfied for the electromagnetic force, where $\mathbf{F}$ is given by Eq. (9.11). Moreover, in the presence of the external force the divergence of the energy-momentum tensor does not vanish in the general case. If Eq. (9.9) is satisfied, it is given by the expression

$$\partial_\mu T^{\mu\nu} = m \int \frac{d^3p}{E_p} \, f \left(\mathbf{F} \cdot \nabla_p \right) p^\nu. \tag{9.10}$$

In the case of the electromagnetic force, the right-hand-side of Eq. (9.10) is reduced to the product of the field tensor and the electric four-current.

9.1.1 *Boltzmann-Vlasov equations for QED plasma*

The explicit form of the relativistic force F^μ depends on the type of interaction. In the case of electrically charged particles

$$mF^\mu = qF^{\mu\nu}p_\nu, \tag{9.11}$$

where q is the electric charge of the particle and $F^{\mu\nu}$ is the electromagnetic tensor containing the electric and magnetic fields,

$$F^{\mu\nu} = -F^{\nu\mu} = \begin{pmatrix} 0 & -E_x & -E_y & -E_z \\ E_x & 0 & -B_z & B_y \\ E_y & B_z & 0 & -B_x \\ E_z & -B_y & B_x & 0 \end{pmatrix}. \tag{9.12}$$

Self-consistent equations constructed from Eq. (9.5) and the additional formula which determines $F^{\mu\nu}$ in terms of the distribution function $f(x,p)$ are called the *Boltzmann-Vlasov* equations.

Originally, Vlasov introduced such equations in 1937 to describe a collisionless electron-ion plasma. In this case we have to consider two distribution functions — one for electrons and the second one for ions. The coupled Vlasov equations have the following form

$$p^\mu \partial_\mu \, f_e(x,p) - eF^{\mu\nu}(x)p_\nu \partial^p_\mu \, f_e(x,p) = 0, \tag{9.13}$$

$$p^\mu \partial_\mu \, f_i(x,p) + ZeF^{\mu\nu}(x)p_\nu \partial^p_\mu \, f_i(x,p) = 0, \tag{9.14}$$

$$\partial_\mu F^{\mu\nu}(x) = j^\nu(x). \tag{9.15}$$

Here $f_e(x,p)$ and $f_i(x,p)$ are the distribution functions of electrons and ions, respectively. The quantity Z is the atom number of an ion, e denotes the elementary charge, and

$$j^\nu = (\rho, \mathbf{j}) \tag{9.16}$$

is the electromagnetic four-current, where

$$\rho\,(x) = e \int [Zf_i(x,p) - f_e(x,p)] \; d^3p, \tag{9.17}$$

$$\mathbf{j}\,(x) = e \int [Zf_i(x,p) - f_e(x,p)] \; \mathbf{v}_p d^3p. \tag{9.18}$$

Of course, Eq. (9.15) is nothing other but the inhomogeneous Maxwell equations [1]. The system of equations (9.13)–(9.18) has been successfully used in studies of different phenomena in ordinary plasma physics [1].

[1]The homogeneous Maxwell equations $\partial_\mu F^{*\,\mu\nu} = 0$, where $F^{*\,\mu\nu} = \varepsilon^{\mu\nu\alpha\beta}F_{\alpha\beta}$ is the dual field tensor, are automatically fulfilled if one expresses the electric and magnetic fields in terms of the scalar and vector potentials.

9.2 Systems with collisions

In the previous Section we considered the case where the interactions between particles are mediated by the self-consistent mean field. However, in many realistic cases, the correct description of the evolution of a physical system requires that the effects of collisions are taken into account explicitly. In this Section we shall present the standard treatment of such effects, which is based on the following three assumptions: **i)** only elastic binary collisions are important, **ii)** the number of the elastic binary collisions between the particles with momenta p and p_1 at the space-time point x is proportional to the product of the distribution functions $f(x,p)f(x,p_1)$, and **iii)** at distances compared to the mean free path of particles the phase-space distribution $f(x,p)$ is a smoothly varying function of x and p.

The point **ii)** is definitely the most important assumption. It is known as the *assumption of molecular chaos*. In the writings of Boltzmann it was called the *Stosszahlansatz*. It means that the colliding particles are always regarded as independent and their velocities are uncorrelated. Such a treatment of the collision process replaces the purely deterministic dynamics by the probabilistic description. In this way, an element of time asymmetry is introduced into the kinetic theory, which explains why it may be so successful in dealing with irreversible phenomena. We illustrate this aspect in Sec. 9.5 where we show how the kinetic-theory framework is used to explain the entropy growth.

9.2.1 *Loss and gain terms*

Let us consider the particles with momenta p placed at the space-time position x. By definition, their number in the phase-space element $\Delta^3 x \Delta^3 p$ is $\Delta N = f(x,p) V \Delta^3 p$ ($d^3x = V$). Since these particles scatter on other particles, their momentum changes and the number ΔN decreases. For example, a particle with momentum p interacts with a particle having momentum p_1, which as an effect gives two particles with momenta p' and p'_1. On the other hand, there exists an inverse process. Collisions of particles with momenta p' and p'_1, may result in the production of particles with momenta p and p_1. In this way the number of particles with momentum p increases. The two competing processes are depicted in Fig. 9.1. The first one should appear in the kinetic equation for the function $f(x,p)$ as the loss (sink) term, whereas the second one plays a role of the gain (source) term. The number of the lost particles may be expressed with the help of the differential transition rate $\Delta\Gamma(p,p_1|p',p'_1)$ or the differential cross section $\Delta\sigma(p,p_1|p',p'_1)$. Assuming that with each collision one particle leaves the considered phase-space element one finds

$$\begin{aligned}\Delta N_{\rm loss} &= V\Delta^3 p f(x,p) V \Delta^3 p_1 f(x,p_1) \Delta t \Delta\Gamma(p,p_1|p',p'_1) \\ &= V^2 \Delta^3 p \Delta^3 p_1 \, f(x,p) f(x,p_1) \, \Delta t \Phi_i \Delta\sigma(p,p_1|p',p'_1). \end{aligned} \tag{9.19}$$

Here the information about transition rates and cross sections collected in Chap. 30 may be helpful, in particular Eqs. (30.29) and (30.36). The quantity Φ_i in Eq. (9.19)

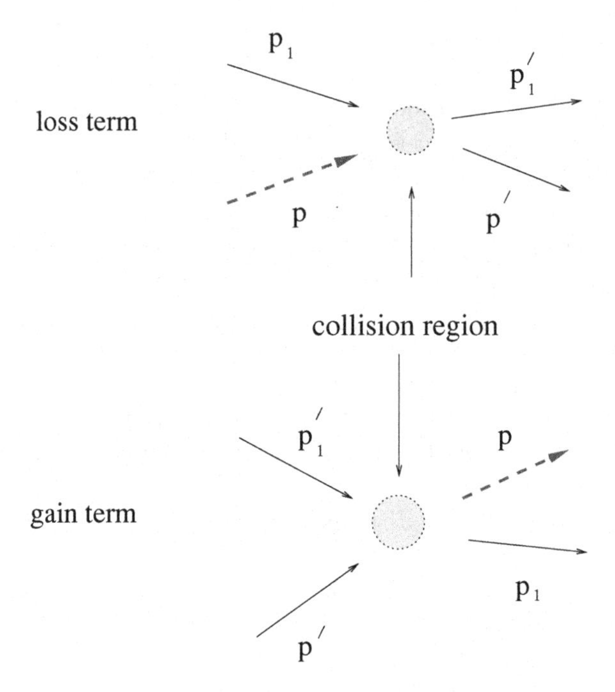

Fig. 9.1 Binary collisions or particles with momenta: p, p', p_1 and p'_1.

is the flux of the incoming particles,

$$\Phi_i = \frac{v_{\text{M}}}{V} = \frac{F_i}{Vp^0p_1^0}, \tag{9.20}$$

where F_i in the invariant flux

$$F_i = \sqrt{(p \cdot p_1)^2 - m_1^2 m_2^2} \tag{9.21}$$

and v_{M} is the Møller velocity. As Ex. 10.8 we leave the problem to show that

$$v_{\text{M}} = \frac{F_i}{p^0 p_1^0} = \sqrt{(\mathbf{v} - \mathbf{v}_1)^2 - (\mathbf{v} \times \mathbf{v}_1)^2}. \tag{9.22}$$

For the head on collisions (parallel velocities) the quantity (9.20) is reduced to the simple definition of the particle flux,

$$\Phi_i = \frac{|\mathbf{v} - \mathbf{v}_1|}{V}. \tag{9.23}$$

In the next step we rewrite Eq. (9.19) in the form

$$\begin{aligned} \Delta N_{\text{loss}} &= V^2 \Delta t \Delta^3 p \Delta^3 p_1 \, f(x,p) f(x,p_1) \, \frac{F_i}{Vp^0 p_1^0} \frac{\Delta\sigma(p, p_1 | p', p'_1)}{\Delta^3 p' \Delta^3 p'_1} \Delta^3 p' \Delta^3 p'_1 \\ &= \frac{V \Delta t \Delta^3 p}{p^0} \frac{\Delta^3 p_1}{p_1^0} \frac{\Delta^3 p'}{p'^0} \frac{\Delta^3 p'_1}{p_1'^0} \, f(x,p) f(x,p_1) \; F_i \, p'^0 p_1'^0 \frac{\Delta\sigma(p, p_1 | p', p'_1)}{\Delta^3 p' \Delta^3 p'_1} \end{aligned} \tag{9.24}$$

To simplify our notation, following Groot, Leeuwen, and Groot, see Ref. [2], we introduce the transition rate W defined by the formula

$$W(p, p_1|p', p_1') \equiv F_i\, p'^0 p_1'^0 \frac{\Delta\sigma(p, p_1|p', p_1')}{\Delta^3 p' \Delta^3 p_1'} \tag{9.25}$$

and the Lorentz invariant element of the phase space

$$d\Gamma_{\rm inv} \equiv d^3x\, dt\, \frac{d^3p}{p^0}. \tag{9.26}$$

In this way we obtain the compact expression

$$\Delta N_{\rm loss} = \Delta\Gamma_{\rm inv} \frac{\Delta^3 p_1}{p_1^0} \frac{\Delta^3 p'}{p'^0} \frac{\Delta^3 p_1'}{p_1'^0}\, f(x,p) f(x,p_1)\, W(p, p_1|p', p_1'). \tag{9.27}$$

The total number of particles *scattered from* the phase-space element $\Delta^4 x\, \Delta^3 p$ is obtained by integration over all possible momenta p_1, p' and p_1', so we redefine the quantity $\Delta N_{\rm loss}$ as the integral

$$\Delta N_{\rm loss} = \frac{1}{2}\, \Delta\Gamma_{\rm inv} \int \frac{d^3p_1}{p_1^0} \frac{d^3p'}{p'^0} \frac{d^3p_1'}{p_1'^0} f(x,p)\; f(x,p_1) W(p, p_1|p', p_1'). \tag{9.28}$$

In the exactly analogous way we derive the formula for the number of the particles *scattered into* the phase-space element $\Delta^4 x\, \Delta^3 p$,

$$\Delta N_{\rm gain} = \frac{1}{2}\, \Delta\Gamma_{\rm inv} \int \frac{d^3p_1}{p_1^0} \frac{d^3p'}{p'^0} \frac{d^3p_1'}{p_1'^0} f(x,p')\; f(x,p_1') W(p', p_1'|p, p_1). \tag{9.29}$$

In Eqs. (9.28) and (9.29) we have introduced an additional factor of 1/2 to account for the effects connected with the scattering of identical particles. The integration over d^3p' and d^3p_1' in (9.28) and (9.29) is not restricted, hence the final/initial states of identical particles are double counted if this factor is absent.

9.2.2 *Boltzmann equation*

Since the kinetic equation describes the change of the number of the particles in the Lorentz-invariant element of the phase-space, $p^\mu \partial_\mu f = p^0 dN \,/\, dt d^3x d^3p$, the kinetic equation describing the evolution of the distribution function due to the collisions has the form

$$p^\mu \partial_\mu f(x,p) = \left(\frac{\Delta N_{\rm gain}}{\Delta\Gamma_{\rm inv}} - \frac{\Delta N_{\rm loss}}{\Delta\Gamma_{\rm inv}} \right) \equiv C(x,p), \tag{9.30}$$

where the collision term (integral) $C(x,p)$ on the right-hand-side of Eq. (9.30) is

$$C(x,p) = \frac{1}{2}\int \frac{d^3p_1}{p_1^0}\frac{d^3p'}{p'^{\,0}}\frac{d^3p_1'}{p_1'^{\,0}}\left[f'f_1'W(p',p_1'|p,p_1) - f\,f_1W(p,p_1|p',p_1')\right]. \tag{9.31}$$

In Eq. (9.31) we have introduced the convenient notation

$$f' = f(x,p'),\ f_1' = f(x,p_1'),\ f = f(x,p),\ f_1 = f(x,p_1). \tag{9.32}$$

This type of notation is commonly used in the textbooks and articles discussing the kinetic equations. The relativistic Boltzmann equation (9.30) is the non-linear integro-differential equation for the one-particle distribution function $f(x,p)$. It is a straightforward generalization of the kinetic equation derived by Boltzmann in 1872.

9.2.3 *Boltzmann-Uehling-Uhlenbeck equation*

The first kinetic equation which included quantum effects was proposed by *Nordheim* in 1928 [3] and later by *Uehling* and *Uhlenbeck* in 1933 [4]. Their equation differs from the classical Boltzmann equation by the form of the collision term. The new form of the collision term for fermions takes into account the effect that the final states in the scattering processes can be occupied and, consequently, blocked by the Pauli exclusion principle. Therefore, the *collision term for fermions* may be written in the form

$$\begin{aligned}C_{\text{fermions}}(x,p) = \frac{1}{2}\int \frac{d^3p_1}{p_1^0}\frac{d^3p'}{p'^{\,0}}\frac{d^3p_1'}{p_1'^{\,0}}&\left[f'f_1'(1-h^3f\,)(1-h^3f_1)W(p',p_1'|p,p_1)\right.\\ &\left.-ff_1(1-h^3f')(1-h^3f_1')W(p,p_1|p',p_1')\right].\end{aligned} \tag{9.33}$$

In the analogous way we can treat the *scattering of bosons*. In this case we have

$$\begin{aligned}C_{\text{bosons}}(x,p) = \frac{1}{2}\int \frac{d^3p_1}{p_1^0}\frac{d^3p'}{p'^{\,0}}\frac{d^3p_1'}{p_1'^{\,0}}&\left[f'f_1'(1+h^3f\,)(1+h^3f_1)W(p',p_1'|p,p_1)\right.\\ &\left.-ff_1(1+h^3f')(1+h^3f_1')W(p,p_1|p',p_1')\right],\end{aligned} \tag{9.34}$$

where the $1+h^3f$ terms enhance the probability of scattering into already occupied states.

9.2.4 *Vlasov-Uehling-Uhlenbeck equation*

The phase-space corrections in the collision term are especially important in description of nuclear reactions at intermediate and low energies. In this case many scattering states are occupied due to the Pauli blocking effect. At such energies one should also include the mean fields which keep nucleons together. Inclusion of both effects leads to the Vlasov-Uehling-Uhlenbeck equation. Its non-relativistic form is

$$\left(\frac{\partial}{\partial t} + \mathbf{v}_p \cdot \nabla + \nabla U(\mathbf{x}, t) \cdot \nabla_p\right) f(t, \mathbf{x}, \mathbf{p}) = C(t, \mathbf{x}, \mathbf{p}), \tag{9.35}$$

where the mean field $U(\mathbf{x}, t)$ is obtained from the nucleon-nucleon potential $V(|\mathbf{x} - \mathbf{x}^{'}|)$, namely

$$U(\mathbf{x}, t) = \int d^3p' d^3x' V(|\mathbf{x} - \mathbf{x}'|)\, f(t, \mathbf{x}', \mathbf{p}'). \tag{9.36}$$

The collision term $C(t, \mathbf{x}, \mathbf{p})$ on the right-hand side of Eq. (9.35) is a non-relativistic form of the collision term (9.33), see Ex. 10.10.

9.2.5 *Transition rate W*

Expressing the cross section by the invariant matrix $\mathcal{M}_{fi} = \mathcal{M}(p, p_1|p', p_1')$, see our discussion preceding Eq. (30.36) in Chap. 30, we find (for $\kappa = 4$)

$$W(p, p_1|p', p_1') = \frac{1}{64\pi^2} \left|\mathcal{M}(p, p_1|p', p_1')\right|^2 \delta^{(4)}\left(p + p_1 - p' - p_1'\right). \tag{9.37}$$

Since $W(p, p_1|p', p_1')$ is a Lorentz scalar, it may depend on 10 invariants that may be constructed out of the four-vectors p, p_1, p' and p_1'. Four invariants are simply the squares of those four-vectors. They introduce the mass of the colliding particles. The next four invariants may be eliminated with the help of the energy-momentum conservation laws,

$$p^\mu + p_1^\mu = p'^\mu + p_1'^\mu. \tag{9.38}$$

In this way one is left with the two independent invariants, which are usually taken as the Mandelstam variables s and t,

$$\begin{aligned} s &= (p + p_1)^2 = (p' + p_1')^2, \\ t &= (p - p')^2 = (p_1 - p_1')^2, \end{aligned} \tag{9.39}$$

The dependence of the transition rate $W(p, p_1|p', p_1')$ on s and t implies that it has the symmetry connected with the simultaneous exchange of p with p_1 and p' with $p_1'^\mu$, namely

$$W(p, p_1|p', p_1') = W(p_1, p|p_1', p'). \tag{9.40}$$

This property will be crucial in our analysis of the conservation laws. We emphasize that it follows only from the Lorentz invariance and holds for any two-body collisions (also inelastic).

In the special case of the elastic collisions of identical particles, the transition rate $W(p, p_1|p', p'_1)$ may be written in the form

$$W(p, p_1|p', p'_1\,) = s\, \frac{d\sigma(s,t)}{d\Omega}\bigg|_{\mathrm{cms}}\, \delta^{(4)}(p + p_1 - p' - p'_1), \tag{9.41}$$

where $d\sigma(s,t)/d\Omega|_{\mathrm{cms}}$ is the *center-of-mass differential cross section*, s and t are Mandelstam variables defined above, and where the Dirac delta function expresses conservation of the energy and momentum, see Ex. 10.10. We note that the parameter t is related to the scattering angle θ in the center-of-mass frame, see Eq. (3.3). We also note that our approach uses the convention of Ref. [2], i.e., the quantity $d\sigma(s,t)/d\Omega|_{\mathrm{cms}}$ corresponds to the *measured* differential cross section. In the case when identical particles are scattered the *total cross section* is obtained by integrating $d\sigma(s,t)/d\Omega|_{\mathrm{cms}}$ over all directions and by dividing the result of the integration by two.

9.3 Mixtures

In the previous Section we considered a simple physical system where all particles were of the same type and they interacted elastically. In this Section we generalize our results to the situation where different types of the particles are present in the system and they may interact inelastically. Our presentation follows the treatment of Ref. [2], however, it is generalized to the case of quantum statistics.

Suppose that we have N different types of particles described by the distributions $f_k(x, p_k)$, where $k = 1, ..., N$. A straightforward generalization of Eqs. (9.30) and (9.31) is (in the case $F^\mu = 0$)

$$p_k^\mu \partial_\mu\, f_k(x, p_k) = \sum_{l=1}^{N} C_{kl}(x, p_k), \tag{9.42}$$

where the classical collision term equals [2]

$$C_{kl}(x, p_k) = \left(1 - \frac{1}{2}\delta_{kl}\right) \int \frac{d^3p_l}{p_l^0} \frac{d^3p'_k}{{p'_k}^0} \frac{d^3p'_l}{{p'_l}^0} \times \left[f'_k f'_l W_{kl}(p'_k, p'_l|p_k, p_l) - f_k\, f_l W_{kl}(p_k, p_l|p'_k, p'_l\,)\right]. \tag{9.43}$$

Equation (9.43) is still valid for the elastic collisions only. For inelastic collisions the final states may belong to different components. In this case we may write

$$C_{kl}(x, p_k) = \frac{1}{2} \sum_{i,j=1}^{N} \int \frac{d^3p_l}{p_l^0} \frac{d^3p_i}{p_i^0} \frac{d^3p_j}{p_j^0} \times \left[f_i f_j W_{ij|kl}(p_i, p_j|p_k, p_l) - f_k\, f_l W_{kl|ij}(p_k, p_l|p_i, p_j)\right], \tag{9.44}$$

[2]An extra distinction should be made between the momenta p'_l and p'_k in the case $l = k$. For simplicity of notation we skip additional indices here.

where $W_{ij|kl}$ is the transition rate for the process $k + l \rightarrow i + j$. If only elastic collisions take place in the system, the reaction rate $W_{ij|kl}$ is reduced to the form

$$W_{kl|ij} = \left(1 - \frac{1}{2}\delta_{kl}\right)\left(\delta_{ki}\delta_{lj} + \delta_{kj}\delta_{li}\right) W_{kl}. \tag{9.45}$$

Equation (9.44) can be generalized to account for quantum statistics of particles. According to the comments of the last Section we may write (for systems consisting of *either* bosons *or* fermions)

$$C_{kl}(x, p_k) = \frac{1}{2}\sum_{i,j=1}^{N} \int \frac{d^3p_l}{p_l^0}\frac{d^3p_i}{p_i^0}\frac{d^3p_j}{p_j^0}\left[f_i f_j \bar{f}_k\ \bar{f}_l W_{ij|kl} - f_k\ f_l \bar{f}_i \bar{f}_j W_{kl|ij}\right], \tag{9.46}$$

where

$$\bar{f} \equiv 1 + \epsilon\, h^3\, f\,. \tag{9.47}$$

As before, we have $\epsilon = -1$ for fermions (the phase-space available for scattered particles is reduced) and $\epsilon = 1$ for bosons (scattering into already occupied states is enhanced). We shall see that the phase-space corrections determine the forms of the equilibrium distributions for fermions and bosons.

9.4 Conservation laws

9.4.1 *Quantities conserved in elementary collisions*

Suppose that $\Psi_k(x)$ is a microscopic property of k-th particle placed at the space-time point x and having the four-momentum p_k^μ. Let us assume that $\Psi_k(x)$ is conserved in an elementary collision $k + l \rightarrow i + j$, namely

$$\Psi_k + \Psi_l = \Psi_i + \Psi_j. \tag{9.48}$$

In this case we find

$$I = \sum_{k,l=1}^{N} \int dP_k \Psi_k C_{kl}(x, p_k) = 0, \tag{9.49}$$

where

$$dP_k = \frac{d^3p_k}{p_k^0}, \tag{9.50}$$

and the collision term $C_{kl}(x, p_k)$ is given by Eq. (9.46). An example of Ψ_k can be a number of particles ($\Psi_k = 1$), any conserved charge ($\Psi_k = q_k$), or the energy and momentum ($\Psi_k = p_k^\mu$).

In order to prove relation (9.49) we write explicitly

$$I = \frac{1}{2}\sum_{i,j,k,l=1}^{N} \int dP_k dP_l dP_i dP_j \Psi_k \left[f_i f_j \bar{f}_k\ \bar{f}_l W_{ij|kl} - f_k\ f_l \bar{f}_i \bar{f}_j W_{kl|ij}\right]. \tag{9.51}$$

One can now change the arbitrary summation indices in the loss term ($k \leftrightarrow i, l \leftrightarrow j$), and then make another change ($k \leftrightarrow l, i \leftrightarrow j$) followed by the use of Eq. (9.40). These two manipulations yield

$$I = \frac{1}{4} \sum_{i,j,k,l=1}^{N} \int dP_k dP_l dP_i dP_j \left[\Psi_k + \Psi_l - \Psi_i - \Psi_j\right] f_i f_j \bar{f}_k \bar{f}_l W_{ij|kl}, \quad (9.52)$$

which is zero according to the assumption (9.48).

9.4.2 *Basic conservation laws*

The first type of the conservation law one may think of is the conservation of the particle number. Of course, in relativistic physics the number of particles is not conserved in the general case. However, if we restrict ourselves to the elastic collisions only, this conservation law should follow from the kinetic equation. In this case we take $\Psi_k = 1$. Multiplication of the Boltzmann equation (9.42) by Ψ_k, integration over p_k, and summation over k gives

$$\sum_{k=1}^{N} \int dP_k p_k^\mu \partial_\mu f_k(x, p_k) = \sum_{k=1}^{N} \partial^\mu N_\mu^k = \partial^\mu N_\mu = 0. \quad (9.53)$$

Thus the conservation law (8.11) is indeed recovered in the case of elastic collisions.

In the analogous way, we may take into consideration any conserved quantum number q_k, e.g., electric charge, baryon number or strangeness. In this case $\Psi_k = q_k$ and the appropriate four-current is

$$Q_k^\mu = \int dP_k q_k p_k^\mu f_k(x, p_k). \quad (9.54)$$

Similarly as in the previous case we find

$$\sum_{k=1}^{N} \partial^\mu Q_\mu^k = \partial^\mu Q_\mu = 0. \quad (9.55)$$

The third category is the energy-momentum conservation law. In this case we take $\Psi_k = p_k^\nu$ and find

$$\sum_{k=1}^{N} \int dP_k p_k^\nu p_k^\mu \partial_\mu f_k(x, p_k) = \sum_{k=1}^{N} \partial_\mu T_k^{\mu\nu} = \partial_\mu T^{\mu\nu} = 0. \quad (9.56)$$

Of course, in this way we reproduce the differential form of the energy-momentum conservation given by Eq. (8.13).

9.5 Boltzmann H-theorem

In this Section we give a relativistic generalization of the Boltzmann H-theorem [3], which represents the second law of thermodynamics. The divergence of the entropy flow is

$$\partial_\mu S^\mu(x) = -\sum_k \int dP_k \, p_k^\mu \partial_\mu \Phi\left[f_k\right] = -\sum_k \int dP_k \ln\left(\frac{f_k}{\bar{f}_k}\right) p_k^\mu \partial_\mu f_k, \quad (9.57)$$

where $\Phi(f_k)$ and $\bar{f}_k$ are defined in (8.16) and (9.47), respectively. We stress that in order to simplify our notation, in this and the next Sections we use the system of units where $h = 1$.

We use now the kinetic equation (9.42) with the collision term (9.46) accounting for quantum statistics of particles

$$\partial_\mu S^\mu(x) = -\frac{1}{2} \sum_{i,j,k,l=1}^{N} \int dP_k dP_l dP_i dP_j \ln\left(\frac{f_k}{\bar{f}_k}\right) \left[f_i f_j \bar{f}_k \bar{f}_l W_{ij|kl} - f_k \, f_l \bar{f}_i \bar{f}_j W_{kl|ij}\right]. \quad (9.58)$$

Change of the indices $i, j \leftrightarrow k, l$ in the loss term gives

$$\partial_\mu S^\mu(x) = -\frac{1}{2} \sum_{i,j,k,l=1}^{N} \int dP_k dP_l dP_i dP_j \left[\ln\left(\frac{f_k}{\bar{f}_k}\right) - \ln\left(\frac{f_i}{\bar{f}_i}\right)\right] f_i f_j \bar{f}_k \bar{f}_l W_{ij|kl}. \quad (9.59)$$

Subsequently, we employ the property (9.40), which leads us to the expression

$$\partial_\mu S^\mu(x) = -\frac{1}{4} \sum_{i,j,k,l=1}^{N} \int dP_k dP_l dP_i dP_j \left[\ln\left(\frac{f_k f_l}{\bar{f}_k \bar{f}_l}\right) - \ln\left(\frac{f_i f_j}{\bar{f}_i \bar{f}_j}\right)\right] f_i f_j \bar{f}_k \bar{f}_l W_{ij|kl}. \quad (9.60)$$

To proceed further we assume the *detailed balance* property

$$W_{kl|ij} = W_{ij|kl}. \quad (9.61)$$

We note that more detailed proofs of the H-theorem use the so called bilateral normalization condition, that follows directly from the unitarity of the scattering matrix. In our simplified approach, that does not include the spin description, we

[3] In the original Boltzmann notation H is the capital greek letter eta!

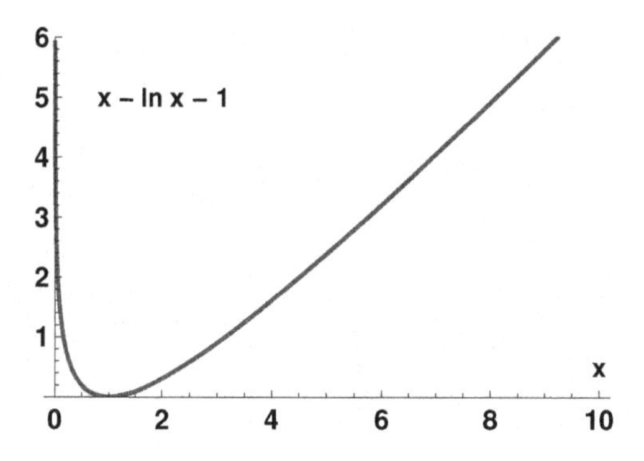

Fig. 9.2 Plot of the function $x - \ln x - 1$.

use Eq. (9.61) as the main assumption. Multiplying Eq. (9.61) by $f_k\ f_l \bar{f}_i \bar{f}_j$ and integrating over momenta one gets

$$\sum_{i,j,k,l=1}^{N} \int dP_k dP_l dP_i dP_j \left[f_k\ f_l \bar{f}_i \bar{f}_j W_{kl|ij} - f_k\ f_l \bar{f}_i \bar{f}_j W_{ij|kl} \right] = 0. \tag{9.62}$$

In the second term in Eq. (9.62) we make the change of indices $i, j \leftrightarrow k, l$

$$\sum_{i,j,k,l=1}^{N} \int dP_k dP_l dP_i dP_j \left[f_k\ f_l \bar{f}_i \bar{f}_j - f_i\ f_j \bar{f}_k \bar{f}_l \right] W_{ij|kl} = 0. \tag{9.63}$$

Putting together Eqs. (9.60) and (9.63) we obtain

$$\partial_\mu S^\mu(x) = \frac{1}{4} \sum_{i,j,k,l=1}^{N} \int dP_k dP_l dP_i dP_j \times \left[\frac{f_k f_l \bar{f}_i \bar{f}_j}{\bar{f}_k \bar{f}_l f_i f_j} - \ln \left(\frac{f_k f_l \bar{f}_i \bar{f}_j}{\bar{f}_k \bar{f}_l f_i f_j} \right) - 1 \right] f_i f_j \bar{f}_k \bar{f}_l W_{ij|kl}. \tag{9.64}$$

The expression in the square bracket has the structure $x - \ln x - 1$ with $x = f_k f_l \bar{f}_i \bar{f}_j / \bar{f}_k \bar{f}_l f_i f_j$. For positive arguments the function $x - \ln x - 1$ is positive and vanishes only at $x = 1$, see Fig. 9.2. This means that we have shown

$$\partial_\mu S^\mu \geq 0. \tag{9.65}$$

Equation (9.65) is nothing other than the statement that entropy does not decrease.

9.6 Local and global equilibrium distributions

The special case

$$\partial_\mu S^\mu = 0 \tag{9.66}$$

holds if the distribution functions satisfy the condition

$$\frac{f_k f_l}{\bar{f}_k \bar{f}_l} = \frac{f_i f_j}{\bar{f}_i \bar{f}_j} \tag{9.67}$$

or

$$\ln \frac{f_k}{\bar{f}_k} + \ln \frac{f_l}{\bar{f}_l} = \ln \frac{f_i}{\bar{f}_i} + \ln \frac{f_j}{\bar{f}_j}. \tag{9.68}$$

The last equation implies that $\ln\left(f_k/\bar{f}_k\right)$ should be expressed in terms of the collision invariants Ψ_k. In the systems with one type of particles, where the particle number, energy and momentum are conserved, the general form of Ψ_k is

$$\Psi_k = a(x) + b_\mu(x) p_k^\mu. \tag{9.69}$$

In this case we obtain

$$f_{\rm eq}(x,p) = \frac{1}{h^3}\left[\exp\left(-a(x) - b_\alpha(x) p^\alpha\right) - \epsilon\right]^{-1}. \tag{9.70}$$

Through the identifications $a(x) = \mu(x)/T(x)$ and $b_\alpha(x) = -u_\alpha(x)/T(x)$ we recover the form of the equilibrium distribution function (8.46)

$$f_{\rm eq}(x,p) = \frac{1}{h^3}\left[\exp\left(\frac{p^\alpha u_\alpha(x) - \mu(x)}{T(x)}\right) - \epsilon\right]^{-1}. \tag{9.71}$$

In formula (9.71) the thermodynamic quantities as well as the fluid velocity depend on the space-time position. This is in contrast to expression (8.46) where they were assumed to be independent of x. We say that expression (9.71) defines *local thermal equilibrium*. Substitution of the condition (9.67) into formula (9.46) gives

$$C_{kl}(x, p_k) = 0. \tag{9.72}$$

Hence, the collision term vanishes in the local equilibrium and the right-hand-side of the kinetic equation (9.42) is zero in this case. The left-hand-side of Eq. (9.42) is, however, not identically zero. It vanishes only if the following equation is satisfied

$$p^\mu \partial_\mu f_{\rm eq}(x,p) = 0, \tag{9.73}$$

which leads to the two conditions

$$\partial_\alpha\left(\frac{\mu(x)}{T(x)}\right) = 0, \qquad \partial_\alpha\left[\frac{u_\beta(x)}{T(x)}\right] + \partial_\beta\left[\frac{u_\alpha(x)}{T(x)}\right] = 0. \tag{9.74}$$

Expressions (9.71) and (9.74) define together the *global thermal equilibrium*. We note that conditions (9.74) relax our initial requirement that μ, T and u_α should be all constant in global equilibrium.

One way of dealing with Eq. (9.73) is to assume that it is approximately valid in the sense that its first and second moment vanish [5], namely

$$\partial_\mu \int \frac{d^3p}{E_p} p^\mu \, f_{\rm eq}(x,p) = 0,$$
$$\partial_\mu \int \frac{d^3p}{E_p} p^\mu p^\nu \, f_{\rm eq}(x,p) = 0. \tag{9.75}$$

In this way we obtain the equations of the form $\partial_\mu N^\mu_{\rm eq} = 0$ and $\partial_\mu T^{\mu\nu}_{\rm eq} = 0$, which are nothing else than the equations of the perfect-fluid hydrodynamics.

9.7 Relaxation-time approximation

In many practical applications one can use the following approximate form of the kinetic equation [5]

$$\hat{L} f(x,p) = -u^\alpha(x) p_\alpha \frac{f(x,p) - f_{\rm eq}(x,p)}{\tau_R}. \tag{9.76}$$

Here $\hat{L}$ is a linear differential operator which typically appears on the left-hand-side of the kinetic equation. In the simplest case of non-interacting particles one has $\hat{L} = p^\mu \partial_\mu$, see Eq. (9.1). For charged particles moving in the electromagnetic field $\hat{L} = p^\mu \partial_\mu + qF^{\mu\nu} p_\nu \partial^p_\mu$. The right-hand-side of Eq. (9.76) is an approximation for the collision term. Here $f_{\rm eq}$ is the local equilibrium distribution function, u^α is the fluid four velocity, and τ_R is a parameter describing how fast the actual distribution function $f(x,p)$ approaches $f_{\rm eq}(x,p)$. For homogeneous systems (no spatial variation of f), in the local rest frame of the fluid we have

$$p^0 \frac{\partial f}{\partial t} = -p^0 \frac{f - f_{\rm eq}}{\tau_R}, \tag{9.77}$$

so p^0 is canceled and we get the solution

$$f(t) = f_{\rm eq} + \left[f\,(t=0) - f_{\rm eq} \right] e^{-t/\tau_R}. \tag{9.78}$$

Thus τ_R is indeed seen to play the role of the *relaxation time*.

To fix the local rest frame of the fluid and to define the local temperature and density which characterize the local equilibrium distribution function f_{eq} we use the conservation laws. The conservation of the particle number (9.53) gives

$$u_\alpha(x) \int dP \; p^\alpha \left[f(x,p) - f_{\rm eq}(x,p) \right] = u_\alpha(x) \left[N^\alpha(x) - N^\alpha_{\rm eq}(x) \right] = 0, \tag{9.79}$$

and the energy-momentum conservation law (9.56) gives

$$u_\alpha(x) \int dP \; p^\alpha p^\beta \left[f(x,p) - f_{\rm eq}(x,p) \right] = u_\alpha(x) \left[T^{\alpha\beta}(x) - T^{\alpha\beta}_{\rm eq}(x) \right] = 0. \tag{9.80}$$

Equations (9.79) and (9.80) are five equations which are sufficient to determine all the needed thermodynamic parameters: the temperature $T(x)$, the chemical potential $\mu(x)$, and three independent components of the fluid four-velocity $u^\alpha(x)$. We note that these conditions are the same as those used by Landau and Lifschitz in the phenomenological treatment of transport processes in a relativistic gas.

For small relaxation time τ_R we may seek a solution of Eq. (9.76) in a form of the expansion

$$f = f_{\rm eq} + \tau_R\, f_{(1)} + \tau_R^2\, f_{(2)} + \quad \cdots \quad . \tag{9.81}$$

Substitution of the series (9.81) into Eq. (9.76) and comparison of the coefficients at various powers of τ_R allows us to calculate $f_{(i)}$ in terms of $f_{(i-1)}$ (we identify $f_{(0)} = f_{\rm eq}$). In particular we have

$$f_{(1)}(x,p) = -\frac{1}{u^\alpha(x) p_\alpha} \hat{L}\, f_{\rm eq}(x,p). \tag{9.82}$$

The function $f_{(1)}(x,p)$ can be used to calculate the corrections to the flows and the energy-momentum tensor which go beyond the perfect-fluid hydrodynamics.

Bibliography to Chapter 9

[1] N. A. Krall and A. W. Trivelpiece, "Principles of Plasma Physics," (McGraw-Hill, New York, 1973).
[2] S. R. de Groot, W. A. van Leeuwen, and C. G. van Weert, "Relativistic Kinetic Theory," (North-Holland, Amsterdam, 1980).
[3] L. Nordheim *Proc. Roy. Soc.* **A119** (1928) 689.
[4] E. A. Uehling and G. E. Uhlenbeck *Phys. Rev.* **43** (1933) 552.
[5] K. Huang, "Statistical Mechanics," (John Wiley, New York, 1963).

Chapter 10

Exercises to PART II

Exercise 10.1. *Transformation properties of the phase-space distribution function.*
Show that the integration measure $d^3p d^3x$ is invariant under Lorentz transformations and conclude that the phase-space distribution function $f(x,p)$ is a scalar.

Exercise 10.2. *Particle four-current.*
Show that the particle current (8.10) may be represented by the integral

$$N^\mu(x) = 2\int d^4p\,\theta(p^0)\,\delta(p^2-m^2)p^\mu f(x,p), \tag{10.1}$$

where $\theta(x)$ is the step function and $\delta(x)$ is the Dirac delta function. Similarly, the energy-momentum tensor is

$$T^{\mu\nu}(x) = 2\int d^4p\,\theta(p^0)\,\delta(p^2-m^2)p^\mu p^\nu f(x,p), \tag{10.2}$$

Exercise 10.3. *Thermodynamic extensive and intensive variables.*
Using the fact that the energy is an extensive function of the parameters S, V, and N derive the formula

$$E + PV = TS + \mu N. \tag{10.3}$$

Hint: Use the relation $E(\lambda S, \lambda V, \lambda N) = \lambda E(S,V,N)$, where λ is an arbitrary parameter, and the thermodynamic definitions of T, P, and μ.

Exercise 10.4. *Partition function.*
Calculate the partition function Z of the ideal relativistic Bose-Einstein/Fermi-Dirac gas. Subsequently, use the relation

$$\Omega = -T\ln Z \tag{10.4}$$

to derive the formula

$$\Omega = \epsilon V T\int \frac{d^3p}{(2\pi\hbar)^3}\ln\left[1-\epsilon\,\exp\left(\frac{\mu-\sqrt{p^2+m^2}}{T}\right)\right]. \tag{10.5}$$

Here $\epsilon = +1$ for bosons and $\epsilon = -1$ for fermions.

Exercise 10.5. *Classical relativistic distribution function.*
i) With the help of the mathematical identity

$$\Gamma\left(n+\frac{1}{2}\right) = \frac{(2n)!\sqrt{\pi}}{n!2^{2n}} \tag{10.6}$$

and the definitions given in Sec. 31.2 show that the modified Bessel functions of the second kind may be defined by the expression

$$K_n(x) = \frac{2^n n!}{(2n)!} x^{-n} \int_x^\infty d\tau \, \left(\tau^2 - x^2\right)^{n-1/2} e^{-\tau}. \tag{10.7}$$

By the integration by parts show that Eq. (10.7) is equivalent to the formula

$$K_n(x) = \frac{2^{n-1}(n-1)!}{(2n-2)!} x^{-n} \int_x^\infty d\tau \, \left(\tau^2 - x^2\right)^{n-3/2} \tau \, e^{-\tau}. \tag{10.8}$$

Check that one may also use the equation

$$K_{n+1}(x) = \frac{x^n}{(2n-1)!!} \int\limits_0^\infty e^{-x \cosh y} \sinh^{2n} y \cosh y \, dy, \tag{10.9}$$

which in the special cases gives

$$K_1(x) = \frac{1}{2} \int\limits_{-\infty}^{\infty} e^{-x \cosh y} \cosh y \, dy \tag{10.10}$$

and

$$K_0(x) = \frac{1}{2} \int\limits_{-\infty}^{\infty} e^{-x \cosh y} \, dy. \tag{10.11}$$

Note that Eqs. (10.10) and (10.11) immediately yield

$$-\frac{\partial}{\partial x} K_0(x) = K_1(x). \tag{10.12}$$

ii) Calculate the thermodynamic variables for a system described by the classical Jüttner distribution function. *Hint:* Reduce the appropriate phase-space integrals to the expressions defining the modified Bessel functions. Use the recurrence relations between the Bessel functions $K_n(x)$ with different indices n, see Sec. 31.2.

Exercise 10.6. *EOS of the weakly-interacting quark-gluon plasma.*
Calculate the energy density $\varepsilon = \varepsilon(T, \mu)$ of the weakly-interacting quark-gluon plasma consisting of massless partons. *Hint:* Treat gluons and quarks as an ideal Bose-Einstein and Fermi-Dirac gas, respectively. In case of problems see Secs. 8.7.2 and 8.7.3.

Exercise 10.7. *Collisionless systems.*
By integration of the kinetic equation (9.5) over momentum show that the number of particles is conserved only if the condition (9.9) is fulfilled. In the case of the electromagnetic interactions show that

$$\partial_\mu T^{\mu\nu} = F^{\nu\mu} j_\mu, \tag{10.13}$$

where $F^{\nu\mu}$ is the electromagnetic field tensor and j^μ is the electric current. *Hint:* Use the integration measure d^3p/E_p and assume that f vanishes at sufficiently large values of momenta.

Exercise 10.8. *Invariant flux.*
Consider a collision of two particles with masses m_1 and m_2. Show that the invariant flux is given by the formula

$$F_\mathrm{i} = \sqrt{(p_1 \cdot p_2)^2 - m_1^2 m_2^2} = p_\mathrm{i}\sqrt{s}, \tag{10.14}$$

where p_i is the initial momentum of the colliding particles in the center-of-mass frame

$$p_\mathrm{i} = \frac{\sqrt{[s-(m_1+m_2)^2][s-(m_1-m_2)^2]}}{2\sqrt{s}}. \tag{10.15}$$

In the special case $m_1 = m_2 = m$ show that

$$F_\mathrm{i} = \frac{1}{2}\sqrt{s(s-4m^2)}. \tag{10.16}$$

Prove also that the Møller velocity is given by the formula

$$v_\mathrm{M} = \frac{F_\mathrm{i}}{p^0 p_1^0} = \sqrt{(\mathbf{v}-\mathbf{v}_1)^2 - (\mathbf{v}\times\mathbf{v}_1)^2}. \tag{10.17}$$

Exercise 10.9. *Center-of-mass differential cross section.*
Consider again a collision of two particles with masses m_1 and m_2. By integration over the energy and momentum of one of the outgoing particles in the center-of-mass frame show that

$$d\sigma = \frac{1}{64\pi^2 s}\frac{q_\mathrm{f}}{p_\mathrm{i}}|\mathcal{M}_{fi}|^2 d\Omega, \tag{10.18}$$

where p_i and q_f are the momenta of the incoming and outgoing particles in the center-of-mass frame, respectively, and $d\Omega$ is the element of the solid angle in the same frame. Thus, in the special case of the elastic collision of the identical particles we may write

$$\left.\frac{d\sigma}{d\Omega}\right|_\mathrm{cms} = \frac{1}{64\pi^2 s}|\mathcal{M}_{fi}|^2. \tag{10.19}$$

Exercise 10.10. *Non-relativistic limit of the Boltzmann equation.*
Find the non-relativistic limit of the relativistic Boltzmann equation with the collision term

$$C(x,p) = \frac{1}{2}\int \frac{d^3p_1}{p_1^0}\frac{d^3p'}{p'^{\,0}}\frac{d^3p_1'}{p_1'^{\,0}}\left[f' f_1' W(p',p_1'|p,p_1) - f\, f_1 W(p,p_1|p',p_1')\right]. \tag{10.20}$$

Hints: **i)** use the formula for the transition rate

$$W(p,p_1|p',p_1') = s\left.\frac{d\sigma(s,t)}{d\Omega}\right|_{\rm cms}\delta^{(4)}(p+p_1-p'-p_1'),$$

ii) calculate the integral

$$I = \int \frac{d^3p'}{p'^{\,0}}\frac{d^3p_1'}{p_1'^{\,0}}\left[f' f_1' W(p',p_1'|p,p_1) - f\, f_1 W(p,p_1|p',p_1')\right]$$

in the center-of-mass frame where $\mathbf{p}+\mathbf{p}_1 = 0$, the result is

$$I = \frac{1}{2}\int d\Omega\,(f' f_1' - f\, f_1)\sqrt{s\,(s-4m^2)}\left.\frac{d\sigma(s,t)}{d\Omega}\right|_{\rm cms}$$

with $d\Omega$ being the element of the solid angle in the center-of-mass frame. **iii)** write the relativistic Boltzmann equation in the form

$$p^\mu\partial_\mu f = \frac{1}{2}\int\frac{d^3p_1}{p_1^0}\int d\Omega\,(f' f_1' - f f_1)\,F_{\rm i}(s)\left.\frac{d\sigma(s,t)}{d\Omega}\right|_{\rm cms},$$

where $F_{\rm i}$ is the invariant flux discussed above. **iv)** Using the non-relativistic notation show that we can write

$$(\partial_t + \mathbf{v}_p\cdot\nabla)\, f = \frac{1}{2}\int d^3p_1\int d\Omega\,(f' f_1' - f f_1)\, v_{\rm M}\left.\frac{d\sigma(s,t)}{d\Omega}\right|_{\rm cms},$$

where $v_{\rm M}$ is the Møller velocity (compare Eq. (9.22)). **v)** Analyze the non-relativistic limit of the last equation.

PART III

KINETIC DESCRIPTION OF PLASMA PRODUCTION

Chapter 11

Color-Flux-Tube Model

In this Chapter we introduce a model of the early stages of ultra-relativistic heavy-ion collisions, that is suitable for description of transparent collisions. We assume that the two receding nuclei get color-charged by random color exchange. Separation of the color charge leads to the formation of a very strong *chromoelectric field* between the nuclei. The field lines form *color-flux tubes* with uniform energy density. Under the influence of the chromoelectric field, the $q\bar{q}$ and gluon pairs are created through the tunneling process. The initial energy of the field is converted into the energy of a rapidly expanding quark-gluon plasma whose subsequent behavior is described by the kinetic Boltzmann-Vlasov equations.

The main reason for introducing the color-flux-tube model in this book is the simplicity of this approach, combined with the use of clear physical concepts related to confinement. Another reason is that the color-flux-tube model shares common features with the theory of glasma — a stage following the color glass condensate, see our discussion in Sec. 4.1.3. This similarity concerns the presence of the longitudinal color fields in the two frameworks. The difference is that the glasma is built of both chromoelectric and chromomagnetic fields, while the color-flux-tube model includes the chromoelectric field only. Another difference is that the transverse range of the color fields in glasma is determined by the saturation scale. In the color-flux-tube model the transverse radius of the tubes is a free parameter that is usually set equal to 1 fm.

In the first Section of this Chapter we describe in more detail the process of pair creation in an external constant field. Our discussion is motivated by the phenomena expected in QED. In the next Section we introduce the abelian dominance approximation of QCD and analyze the creation of $q\bar{q}$ and gluon pairs in the color-flux tubes. Subsequently, we introduce the boost-invariant Boltzmann-Vlasov equations for the quark-gluon plasma.

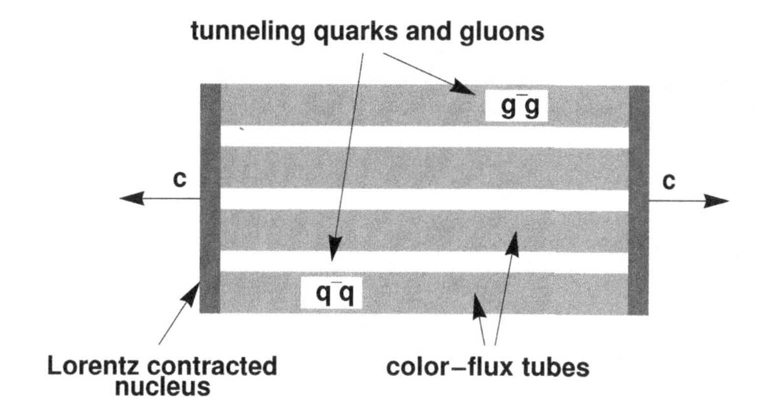

Fig. 11.1 Quark-antiquark and gluon-antigluon tunneling in strong chromoelectric fields possibly created at the initial stage of an ultra-relativistic heavy-ion collision. The blank area surrounding the tunneling pairs illustrates the effect of screening of the original field. The nuclei are Lorentz contracted and recede from each other with the velocity of light c.

11.1 Schwinger mechanism

The problem of a quantum Dirac field interacting with an external uniform classical electric field was solved exactly by Schwinger more than 50 years ago [1]. Among other results Schwinger gave the formula for the *vacuum persistence probability.* This result can be used to read off the probability of the e^+e^- pair creation (per unit time and unit volume) in an external constant electric field $\mathcal{E} > 0$ [1,2],

$$\frac{dN}{dt\,dV} = \frac{e^2\mathcal{E}^2}{4\pi^3}\sum_{n=1}^{\infty}\frac{1}{n^2}\exp\left(-\frac{\pi\, m^2 n}{e\mathcal{E}}\right). \tag{11.1}$$

Here m is the electron mass and e is the elementary charge. Schwinger's result was derived later by Casher, Neuberger, and Nussinov [3,4], who treated the decay process as tunneling and applied the WKB method to calculate the decay probability.

Casher et al. investigated further Eq. (11.1) in the context of strong interactions, interpreting $\mathcal{E}$ as the chromoelectric field, m as the quark mass, and e as the strong coupling constant. They argued that confinement may be implemented by the generation of chromoelectric color flux tubes with uniform energy density, see Fig. 11.1. Tunneling of quark pairs in such tubes represents the mechanism responsible for *multiple hadron production.*

Following the approach of Refs. [3–5] we derive Schwinger's pair-production formula (11.1) in a semiclassical tunneling calculation. The longitudinal momentum of each member of a pair, at the point $z = 0$, where the virtual particles emerge from the vacuum, must satisfy the condition

$$p_{\|}^2 + p_{\perp}^2 + m^2 = 0, \quad p_{\|} = i m_{\perp}. \tag{11.2}$$

We note that tunneling can be treated as a process which takes place in imaginary

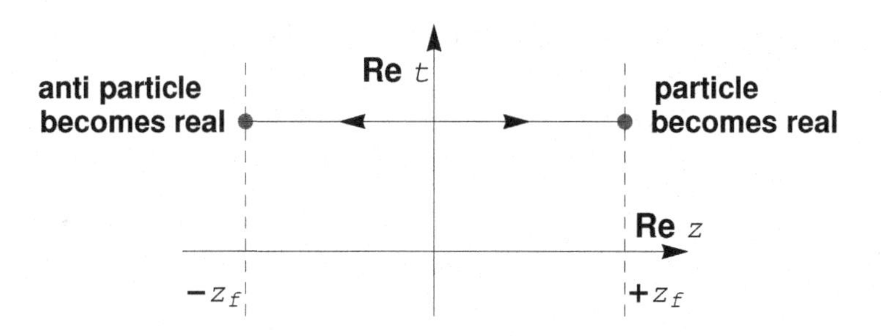

Fig. 11.2 Spacetime trajectories of the tunneling particles. The particles materialize at the points $\pm z_f$, where $z_f = m_\perp/F$.

time, see Fig. 11.2. This is the reason for imaginary momenta of the tunneling particles. The energy conservation gives

$$\sqrt{p_\parallel^2(z) + p_\perp^2 + m^2} = Fz, \tag{11.3}$$

where $F > 0$ is the constant force acting on the particle and z is the distance from the point of its first appearance. The longitudinal momentum is therefore

$$p_\parallel(z) = i\sqrt{m_\perp^2 - (Fz)^2}. \tag{11.4}$$

The action of one particle, integrated from the initial point to the point where it materializes, is [1]

$$S = i\int_0^{m_\perp/F} |p_\parallel(z)|\, dz = \frac{i\pi m_\perp^2}{4F}. \tag{11.5}$$

The probability that a virtual pair can tunnel to a real state, with each component of a pair having transverse momentum $p_\perp$, is

$$P(p_\perp) = |\, e^{2iS}|^2 = \exp\left(-\frac{\pi(m^2 + p_\perp^2)}{F}\right). \tag{11.6}$$

[1] The Lagrangian of a relativistic particle moving in an external constant field is [6]

$$L = -m\sqrt{1 - v^2} + F\, z$$

and the Euler-Lagrange equations

$$\frac{d}{dt}\frac{\partial L}{\partial v} = \frac{\partial L}{\partial z}$$

give

$$\frac{d}{dt}\left(\frac{mv}{\sqrt{1 - v^2}}\right) = F.$$

The calculation of the action yields

$$S = \int L\, dt = \int L\frac{dz}{v} = \int \left(-m\sqrt{1 - v^2} + \frac{m}{\sqrt{1 - v^2}}\right)\frac{dz}{v} = \int p_\parallel dz = i\int |p_\parallel|\, dz.$$

Knowing the tunneling probability we can calculate the probability that a pair is actually created. Following Refs. [3–5], we compute the vacuum persistence probability, which is the probability that *no* such tunneling process takes place

$$
\begin{aligned}
|\langle 0_+|0_-\rangle|^2 &= \prod_{\text{flavor}} \prod_{\text{spin}} \prod_{\mathbf{p}_\perp} \prod_{z} \prod_{t} [1 - P(p_\perp)] \\
&= \exp\left\{ \sum_{\text{flavor}} \sum_{\text{spin}} \sum_{\mathbf{p}_\perp} \sum_{z} \sum_{t} \ln [1 - P(p_\perp)] \right\}.
\end{aligned} \tag{11.7}
$$

Suppose that the spacetime volume available for tunneling is $L_x L_y L_z T$. The length required for materialization of a pair is $\Delta z = 2m_\perp/F$, and the time interval needed for tunneling is determined from the uncertainty principle $\Delta t = \pi/m_\perp$. Thus, we may write

$$
\begin{aligned}
|\langle 0_+|0_-\rangle|^2 &= \exp\left\{ 2\frac{T}{\Delta t}\frac{L_z}{\Delta z} \sum_{\text{flavor}} L_x L_y \int \frac{d^2 p_\perp}{(2\pi)^2} \ln [1 - P(p_\perp)] \right\} \\
&= \exp\left(-L_x L_y L_z T \, p\right),
\end{aligned} \tag{11.8}
$$

where 2 is the spin degeneracy factor and p is the rate at which pairs are created per unit volume

$$
p = -\frac{F}{4\pi^2} \sum_{\text{flavor}} \int_{m_f^2}^{\infty} dm_\perp^2 \, \ln [1 - P(p_\perp)] = \frac{F^2}{4\pi^3} \sum_{\text{flavor}} \sum_{n=1}^{\infty} \frac{1}{n^2} \exp\left(-\frac{\pi m_f^2 \, n}{F} \right). \tag{11.9}
$$

We note that for the electrons tunneling in a constant electric field the sum over flavors is reduced to the single term with $m_f = m$ and $F = e\mathcal{E}$. In this case Eq. (11.9) coincides with Schwinger's formula (11.1).

A nice feature of the semiclassical approach is that it gives the $p_\perp$ distribution of the tunneling pairs. Indeed, Eq. (11.9) can be written in the form

$$
\frac{dN_f}{d^4x \, d^2 p_\perp} = \frac{F}{4\pi^3} \left| \ln \left(1 - \exp\left(-\frac{\pi m_{f\perp}^2}{F} \right) \right) \right|. \tag{11.10}
$$

In the reference frame where the longitudinal momentum of the particles is zero (at the moment when they become real) we have

$$
\frac{dN_f}{d\Gamma_{\text{inv}}} = \frac{F}{4\pi^3} \left| \ln \left(1 - \exp\left(-\frac{\pi m_{f\perp}^2}{F} \right) \right) \right| \delta\left(p_{\|}\right) p^0, \tag{11.11}
$$

where $d\Gamma_{\text{inv}}$ is the Lorentz invariant element of the phase space defined above in Eq. (9.26).

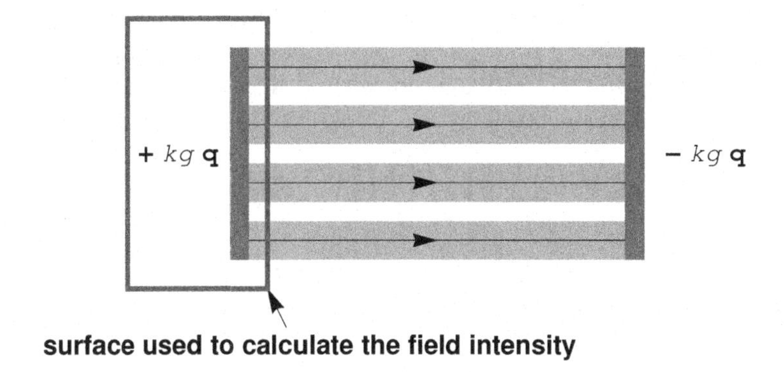

Fig. 11.3 A color-flux tube spanned by k color charges $g\mathbf{q}$. Simple geometry of the system allows us to use the Gauss law.

11.2 Abelian dominance approximation

The complete treatment of tunneling in the framework of QCD is very complicated, so we do several approximations. In this way our problem can be solved with the method outlined above. We assume that the gluon field can be separated into two parts. The first one represents the *coherent* part of the field which can be also interpreted as the *classical mean field*. The second one describes the *incoherent* part

$$F_{\mu\nu} = \langle F_{\mu\nu} \rangle + \delta F_{\mu\nu}.$$

We assume that the mean-field part can be diagonalized in color space, i.e., there is a gauge transformation U which "rotates" $\langle F_{\mu\nu} \rangle$ into the abelian subgroup of the SU(3) gauge group

$$F_{\mu\nu} \to U F_{\mu\nu} U^{-1} = F'_{\mu\nu} = \left\langle F'^{3}_{\mu\nu} \right\rangle \frac{\lambda_3}{2} + \left\langle F'^{\,8}_{\mu\nu} \right\rangle \frac{\lambda_8}{2} + \delta F'_{\mu\nu}.$$

One can see that $\left\langle F'_{\mu\nu} \right\rangle$ has two independent components, which can be represented by a two-component vector

$$\mathbf{F}_{\mu\nu} = \left(\left\langle F'^{\,3}_{\mu\nu} \right\rangle, \left\langle F'^{\,8}_{\mu\nu} \right\rangle \right). \tag{11.12}$$

The fluctuating part $\delta F'_{\mu\nu}$contains only non-diagonal terms (its diagonal components are smaller than $\mathbf{F}_{\mu\nu}$ and neglected). According to the classification introduced in Sec. 5.5, the mean-field part and the fluctuating part describe neutral and charged gluons, respectively. Quarks couple to the field $\mathbf{F}_{\mu\nu}$ through the charges $\boldsymbol{\epsilon}_i$, and the charged gluons couple to $\mathbf{F}_{\mu\nu}$ through the charges $\boldsymbol{\eta}_{ij}$, see Eqs. (5.25) and (5.33). It is further assumed that the color-field configuration formed at the initial stage of a collision corresponds to the chromoelectric field $\boldsymbol{\mathcal{E}} = \mathbf{F}^{30}$.

The Gauss law, see Fig. 11.3, applied to a color flux tube gives

$$\boldsymbol{\mathcal{E}} \mathcal{A} = kg\mathbf{q}, \tag{11.13}$$

where $\mathcal{A} = \pi r^2$ denotes the area of the transverse cross section of the tube, k is the number of color charges at the end of the tube and $g\mathbf{q} = g(q^3, q^8)$ is the color charge of a quark or a gluon. From the definition of the string tension σ (*the energy of an elementary tube per unit length*) and Eq. (11.13) we get

$$\sigma = \frac{1}{2}\mathcal{A}\boldsymbol{\mathcal{E}}\cdot\boldsymbol{\mathcal{E}} = \frac{g^2}{2\mathcal{A}}\mathbf{q}\cdot\mathbf{q}. \tag{11.14}$$

Substitution of the quark charges (5.25) into Eq. (11.14) yields (independently of the index i)

$$\sigma_q = \frac{g^2}{6\mathcal{A}}. \tag{11.15}$$

Similarly, the gluon charges (5.33) give (independently of the indices i, j)

$$\sigma_g = \frac{g^2}{2\mathcal{A}}. \tag{11.16}$$

We conclude that the string tension of a tube spanned by gluons is three times larger than that of a quark tube. The Gauss law (11.13) can be rewritten in the following form

$$\boldsymbol{\mathcal{E}} = \sqrt{\frac{2\sigma_g}{\pi r^2}}k\mathbf{q} = \sqrt{\frac{6\sigma_q}{\pi r^2}}k\mathbf{q}. \tag{11.17}$$

This equation determines the value of the initial chromoelectric field spanned by the two receding nuclei.

We have three parameters which characterize an elementary tube: the string tension σ, the strong coupling constant g, and the tube radius r. For the standard value $\sigma_q = 1$ GeV/fm ($\sigma_g = 3$ GeV/fm) we find the following relation between g and r,

$$g^2 = 6\,\mathcal{A}\,\mathrm{GeV/fm} \approx 30\,\pi\frac{r^2}{\mathrm{fm}^2}. \tag{11.18}$$

We expect the tube radius to be of the order of 1 fm. We thus see that the coupling constant is quite large, which excludes any perturbative treatment of the tunneling process. The number of color charges k may be obtained from the hypothesis of random walk in color space [7]

$$k = \sqrt{\text{number of collisions}} = \sqrt{\frac{d\nu}{d^2s}\pi r^2},$$

where $d\nu/d^2s$ is the number of collisions per unit transverse area. One expects $k \approx 3$ for central lead on lead collisions [7].

In the reference frame where the particles emerge from the vacuum with the vanishing longitudinal momentum, the production rate of quarks in the chromoelectric field (11.17) is

$$\frac{dN_{if}}{d\Gamma_{\mathrm{inv}}} = \frac{\Lambda_i}{4\pi^3}\left|\ln\left(1 - \exp\left(-\frac{\pi m_{f\,\perp}^2}{\Lambda_i}\right)\right)\right| \delta\left(p_{\parallel}\right)p^0 \equiv \mathcal{R}_{if}\;\delta\left(p_{\parallel}\right)p^0, \tag{11.19}$$

where

$$\Lambda_i = \left(g\,|\boldsymbol{\epsilon}_i \cdot \boldsymbol{\mathcal{E}}| - \sigma_q\right)\theta\left(g\,|\boldsymbol{\epsilon}_i \cdot \boldsymbol{\mathcal{E}}| - \sigma_q\right) \tag{11.20}$$

and θ is the step function, see Eq. (31.1). The quantity Λ_i describes the effective force acting on the tunneling quarks. The effect of the screening of the initial field by the tunneling particles is taken into account by the subtraction of the "elementary force" characterized by the string tension [5,8,9]. Similarly, for gluons one can find

$$\frac{d\widetilde{N}_{ij}}{d\Gamma_{\rm inv}} = \frac{\Lambda_{ij}}{4\pi^3}\left|\ln\left(1+\exp\left(-\frac{\pi p_\perp^2}{\Lambda_{ij}}\right)\right)\right|\delta\left(p_\parallel\right)p^0 \equiv \widetilde{\mathcal{R}}_{ij}\,\delta\left(p_\parallel\right)p^0, \tag{11.21}$$

where the effective force is

$$\Lambda_{ij} = \left(g\left|\boldsymbol{\eta}_{ij}\cdot\boldsymbol{\mathcal{E}}\right| - \sigma_g\right)\theta\left(g\left|\boldsymbol{\eta}_{ij}\cdot\boldsymbol{\mathcal{E}}\right| - \sigma_g\right). \tag{11.22}$$

We note that the rates (11.19) and (11.21) differ by the sign appearing before the exponent function. This difference has the origin in different statistics obeyed by quarks and gluons. A rigorous derivation of the production rates, that takes into account the statistical properties of particles, requires the use of the methods of quantum field theory. The effects of the pair production of the gauge bosons by the background fields in the case of non-abelian gauge field theories were studied, for example, in Refs. [10,11]. We also note that the rates (11.19) and (11.21) include the spin degeneracy factor, that is equal to 2 for both quarks and massless gluons.

11.3 Kinetics of the plasma formation

In the abelian dominance approximation, the kinetic equations for quarks, antiquarks, and gluons can be written as follows [12]

$$\left(p^\mu\partial_\mu + g\boldsymbol{\epsilon}_i\cdot\mathbf{F}^{\mu\nu}p_\nu\partial^p_\mu\right)G_{if}(x,p) = \frac{dN_{if}}{d\Gamma_{\rm inv}}, \tag{11.23}$$

$$\left(p^\mu\partial_\mu - g\boldsymbol{\epsilon}_i\cdot\mathbf{F}^{\mu\nu}p_\nu\partial^p_\mu\right)\overline{G}_{if}(x,p) = \frac{dN_{if}}{d\Gamma_{\rm inv}}, \tag{11.24}$$

$$\left(p^\mu\partial_\mu + g\boldsymbol{\eta}_{ij}\cdot\mathbf{F}^{\mu\nu}p_\nu\partial^p_\mu\right)\widetilde{G}_{ij}(x,p) = \frac{d\widetilde{N}_{ij}}{d\Gamma_{\rm inv}}, \tag{11.25}$$

where $G_{if}(x,p)$, $\overline{G}_{if}(x,p)$ and $G_{ij}(x,p)$ are the phase-space densities of quarks, antiquarks and gluons, respectively. Here $i,j=(1,2,3)$ are color indices, whereas $f=(u,d,s,c)$ is the flavor index. The terms on the left-hand-side describe the free motion of particles and the interaction of particles with the mean field $\mathbf{F}_{\mu\nu}$. The terms on the right-hand-side describe production of quarks and gluons due to the decay of the field.

Equations (11.23)–(11.25) determine how the particles behave under the influence of the field. In order to obtain a self-consistent set of equations we should have also the dynamic equation for the field. It can be written in the following Maxwell form

$$\partial_\mu \mathbf{F}^{\mu\nu}(x) = \mathbf{j}^\nu(x) + \mathbf{j}^\nu_D(x), \tag{11.26}$$

where

$$\mathbf{j}^\nu(x) = g \int dP\, p^\nu \left[\sum_{i=1}^{3} \boldsymbol{\epsilon}_i \sum_f \left(G_{if}(x,p) - \overline{G}_{if}(x,p)\right) + \sum_{i,j=1}^{3} \boldsymbol{\eta}_{ij} \widetilde{G}_{ij}(x,p) \right] \tag{11.27}$$

and

$$\mathbf{j}^\nu_D(x) = \int dP \left[\sum_{i=1}^{3} \sum_f \frac{dN_{if}}{d\Gamma_{\text{inv}}} \mathbf{d}^\nu_{if} + \sum_{i>j}^{3} \frac{d\widetilde{N}_{ij}}{d\Gamma_{\text{inv}}} \mathbf{d}^\nu_{ij} \right]. \tag{11.28}$$

Here $dP = d^3p/E_p$ is the momentum integration measure, see Eq. (9.50) [2]. The current in Eq. (11.26) has two components. The first one, Eq. (11.27), is related to the simple fact that particles carry color charges $\boldsymbol{\epsilon}_i$ and $\boldsymbol{\eta}_{ij}$ (*conductive current*). The second one, Eq. (11.28), has the origin in the tunneling of quarks and gluons, hence, in the creation of color charges from the vacuum (*displacement current*). The quantities $\mathbf{d}^\nu$ are the dipole moments of the produced pairs.

Let us now analyze the tunneling of the $q\overline{q}$ and gluon pairs from the plane $z = 0$. The energy balance (11.3) gives [3]

$$\sqrt{p_\parallel^2 + p_\perp^2 + m_f^2} = \Delta z_{if}\, g \left|\boldsymbol{\epsilon}_i \cdot \boldsymbol{\mathcal{E}}\right| \tag{11.29}$$

$$\sqrt{p_\parallel^2 + p_\perp^2} = \Delta z_{ij}\, g \left|\boldsymbol{\eta}_{ij} \cdot \boldsymbol{\mathcal{E}}\right|. \tag{11.30}$$

Here Δz denotes the distance from the plane $z = 0$ to the place where the particles are created, and $\boldsymbol{\mathcal{E}} = \mathbf{F}^{30}$. Thus, the dipole moments of the emerging pairs are

$$\mathbf{d}^3_{if} = g\boldsymbol{\epsilon}_i \left(2\Delta z_{if}\right) \operatorname{sign}\left(\boldsymbol{\epsilon}_i \cdot \boldsymbol{\mathcal{E}}\right) = 2\boldsymbol{\epsilon}_i \frac{\sqrt{p_\perp^2 + m_f^2}}{\boldsymbol{\epsilon}_i \cdot \boldsymbol{\mathcal{E}}}, \tag{11.31}$$

$$\mathbf{d}^3_{ij} = g\boldsymbol{\eta}_{ij} \left(2\Delta z_{ij}\right) \operatorname{sign}\left(\boldsymbol{\eta}_{ij} \cdot \boldsymbol{\mathcal{E}}\right) = 2\boldsymbol{\eta}_{ij} \frac{p_\perp}{\boldsymbol{\eta}_{ij} \cdot \boldsymbol{\mathcal{E}}}. \tag{11.32}$$

Other spacetime components of the dipole moments vanish at $z = 0$. The functions $\operatorname{sign}(\boldsymbol{\epsilon}_i \cdot \boldsymbol{\mathcal{E}})$ and $\operatorname{sign}(\boldsymbol{\eta}_{ij} \cdot \boldsymbol{\mathcal{E}})$ indicate the direction of the induced moments. The form of $\mathbf{d}^\nu_{if}$ and $\mathbf{d}^\nu_{ij}$ at $z \neq 0$ can be found by making the Lorentz transformation.

[2] The spin degeneracy factor is included in the production rates and definitions of the quark and gluon distribution functions.

[3] Here we neglect the effect of screening of the initial field by the tunneling particles. In this way, our definition of the displacement current (11.28) with the dipole moments (11.29) and (11.30) is consistent with the energy-momentum conservation laws obtained from the kinetic equations (11.23)–(11.25). A naive treatment of the screening effect, such as in Eqs. (11.20) and (11.22), fails here because the tunneling is a non-local phenomenon [13, 14].

11.4 Energy-momentum conservation laws

The energy-momentum conservation law for the system of quarks, gluons and the chromoelectric field is (8.14)

$$\partial_\mu T^{\mu\nu}(x) + \partial_\mu T^{\mu\nu}_{\rm field}(x) = 0. \tag{11.33}$$

The $\nu = 0$ component of this equation includes the terms

$$\partial_\mu T^{\mu 0}_{\rm field} = \frac{\partial}{\partial t}\left(\frac{1}{2}\mathcal{E}^2\right) = \mathcal{E}\cdot\frac{\partial \mathcal{E}}{\partial t} = -\mathbf{F}^{30}\cdot\frac{\partial \mathbf{F}^{03}}{\partial t} \tag{11.34}$$

and

$$\begin{aligned}
\partial_\mu T^{\mu 0} &= \int dP p^0 p^\mu \partial_\mu \left[\sum_{i=1}^{3}\sum_f (G_{if} + \overline{G}_{if}) + \sum_{i,j=1}^{3} \widetilde{G}_{ij}\right] \\
&= -g\mathbf{F}^{\mu\nu}\cdot \int dP\, p^0\, p_\nu\, \partial^p_\mu \left[\sum_{i=1}^{3}\epsilon_i \sum_f (G_{if} - \overline{G}_{if}) + \sum_{i,j=1}^{3} \boldsymbol{\eta}_{ij}\widetilde{G}_{ij}\right] \\
&\quad +2\int dP\, p^0 \left[\sum_{i=1}^{3}\sum_f \frac{dN_{if}}{d\Gamma_{\rm inv}} + \sum_{i>j}^{3}\frac{d\widetilde{N}_{ij}}{d\Gamma_{\rm inv}}\right],
\end{aligned} \tag{11.35}$$

where we used Eqs. (11.23)–(11.25). Using the last results we may write

$$\begin{aligned}
\mathbf{F}^{30}\cdot\frac{\partial \mathbf{F}^{03}}{\partial t} &= \mathbf{F}^{30}\cdot g\int dP p^3 \left[\sum_{i=1}^{3}\epsilon_i \sum_f (G_{if} - \overline{G}_{if}) + \sum_{i,j=1}^{3} \boldsymbol{\eta}_{ij}\widetilde{G}_{ij}\right] \\
&\quad +\mathbf{F}^{30}\cdot \int dP \left[\sum_{i=1}^{3}\frac{2p^0\epsilon_i}{\epsilon_i\cdot\mathbf{F}^{30}}\sum_f \frac{dN_{if}}{d\Gamma_{\rm inv}} + \sum_{i>j}^{3}\frac{2p^0\boldsymbol{\eta}_{ij}}{\boldsymbol{\eta}_{ij}\cdot\mathbf{F}^{30}}\frac{d\widetilde{N}_{ij}}{d\Gamma_{\rm inv}}\right].
\end{aligned} \tag{11.36}$$

We thus see, that the field equations (11.26)–(11.28) with the definitions of the dipole moments (11.31) and (11.32) represent a sufficient condition to have the total energy of the system conserved (note that the delta functions present in the production rates generate zero longitudinal momenta at $z = 0$). Similarly, we may analyze the $\nu = 3$ component of Eq. (11.33) and obtain the same conclusion.

11.5 Implementation of boost-invariance

Following the approach worked out by Białas and Czyż in Refs. [8, 12, 15–17] we impose on our Lorentz-covariant equations an additional symmetry: *longitudinal boost invariance.* This invariance means that the description of a system must have the same form in all frames boosted in the longitudinal direction [4].

[4] We recall that the general aspects related to boost-invariance were introduced and discussed in more detail in Sec. 2.7.

11.5.1 *Boost-invariant variables*

We use the following boost-invariant variables [15]

$$u = t^2 - z^2, \quad w = tp_{\parallel} - zE, \quad \mathbf{p}_{\perp}, \tag{11.37}$$

and also

$$\tau = \sqrt{u}, \quad v = Et - p_{\parallel}\, z = \sqrt{w^2 + \left(m_f^2 + \mathbf{p}_{\perp}^2\right) u}. \tag{11.38}$$

From these two equations one can easily find the energy and the longitudinal momentum of a particle

$$E = p^0 = \frac{vt + wz}{u}, \quad p_{\parallel} = \frac{wt + vz}{u}. \tag{11.39}$$

The phase-space distribution functions behave like scalars under Lorentz transformations. The requirement of boost invariance implies that they may depend only on the variables τ, w and $\mathbf{p}_{\perp}$, namely

$$f(x,p) \to f(\tau, w, \mathbf{p}_{\perp}). \tag{11.40}$$

The integration measure in the momentum sector of the phase-space is

$$dP = 2\, d^4p\, \delta\left(p^2 - m_f^2\right) \theta(p^0) = \frac{dp_{\parallel}}{p^0} d^2p_{\perp} = \frac{dw}{v} d^2p_{\perp}. \tag{11.41}$$

A chromoelectric field $\boldsymbol{\mathcal{E}} = \mathbf{F}^{30}$ does not change under Lorentz transformations along the z-axis, thus it may be written as

$$\boldsymbol{\mathcal{E}}(\tau) = -2\,\frac{d\mathbf{h}(\tau)}{du} = -\frac{1}{\tau}\frac{d\mathbf{h}(\tau)}{d\tau}, \tag{11.42}$$

where $\mathbf{h}$ is a function of the variable τ only. The production rates appearing on the right-hand-side of the kinetic equations (11.23)–(11.25) have the form

$$\frac{dN_{if}}{d\Gamma_{\rm inv}} = v\delta\left(w\right)\, \mathcal{R}_{if}, \quad \frac{d\widetilde{N}_{ij}}{d\Gamma_{\rm inv}} = v\delta\left(w\right) \widetilde{\mathcal{R}}_{ij}. \tag{11.43}$$

These two expressions are explicitly boost-invariant and reduce themselves to Eqs. (11.19) and (11.21) in the case $z = 0$.

11.5.2 *Boost-invariant transport equations*

The use of symmetries (11.37)–(11.43) in Eqs. (11.23)–(11.25) gives us the following boost-invariant transport equations [12]

$$\frac{\partial G_{if}}{\partial \tau} - g\epsilon_i \cdot \frac{d\mathbf{h}}{d\tau}\frac{\partial G_{if}}{\partial w} = \tau \mathcal{R}_{if}(\tau, p_\perp)\,\delta(w), \tag{11.44}$$

$$\frac{\partial \overline{G}_{if}}{\partial \tau} + g\epsilon_i \cdot \frac{d\mathbf{h}}{d\tau}\frac{\partial \overline{G}_{if}}{\partial w} = \tau \mathcal{R}_{if}(\tau, p_\perp)\,\delta(w), \tag{11.45}$$

$$\frac{\partial \widetilde{G}_{ij}}{\partial \tau} - g\eta_{ij} \cdot \frac{d\mathbf{h}}{d\tau}\frac{\partial \widetilde{G}_{ij}}{\partial w} = \tau \widetilde{\mathcal{R}}_{ij}(\tau, p_\perp)\,\delta(w). \tag{11.46}$$

Their formal solution is

$$G_{if}(\tau, w, p_\perp) = \int_0^\tau d\tau'\, \tau'\, \mathcal{R}_{if}(\tau', p_\perp)\,\delta(\Delta h_i(\tau, \tau') + w), \tag{11.47}$$

$$\overline{G}_{if}(\tau, w, p_\perp) = \int_0^\tau d\tau'\, \tau'\, \mathcal{R}_{if}(\tau', p_\perp)\,\delta(\Delta h_i(\tau, \tau') - w), \tag{11.48}$$

$$\widetilde{G}_{ij}(\tau, w, p_\perp) = \int_0^\tau d\tau'\, \tau'\, \widetilde{\mathcal{R}}_{ij}(\tau', p_\perp)\,\delta(\Delta h_{ij}(\tau, \tau') + w), \tag{11.49}$$

where

$$\begin{aligned} \Delta h_i(\tau, \tau') &= g\epsilon_i \cdot [\mathbf{h}(\tau) - \mathbf{h}(\tau')], \\ \Delta h_{ij}(\tau, \tau') &= g\eta_{ij} \cdot [\mathbf{h}(\tau) - \mathbf{h}(\tau')]. \end{aligned} \tag{11.50}$$

One may notice that the distribution functions (11.47)–(11.49) satisfy the following symmetry relations

$$\begin{aligned} \overline{G}_{if}(\tau, w, p_\perp) &= G_{if}(\tau, -w, p_\perp), \\ \widetilde{G}_{ij}(\tau, w, p_\perp) &= \widetilde{G}_{ji}(\tau, -w, p_\perp). \end{aligned} \tag{11.51}$$

We note also that the time integrals in Eqs. (11.47)–(11.49) reveal the non-Markovian character of the particle production mechanism: the behavior of the system at a time τ is determined by the whole evolution of the system in the time interval $0 \le \tau' \le \tau$ (for a more general discussion of the issue of memory effects in heavy-ion collisions see [18, 19]).

11.5.3 *Conductive current*

To proceed further with the calculations we substitute the distribution functions (11.47)–(11.49) into the definition of the conductive current (11.27). Equations

(11.39) and (11.51) imply that $\mathbf{j}^\nu(x)$ has the following spacetime structure

$$\mathbf{j}^\nu(x) = \left[\mathbf{j}^0(x), 0, 0, \mathbf{j}^3(x)\right] = [z, 0, 0, t]\mathcal{J}(\tau), \tag{11.52}$$

where

$$\mathcal{J}(\tau) = \frac{2g}{u}\int \frac{w\,dw}{v} d^2p_\perp \left[\sum_{i=1}^{3} \epsilon_i \sum_f G_{if}(\tau, w, p_\perp) + \sum_{i>j}^{3} \eta_{ij} \widetilde{G}_{ij}(\tau, w, p_\perp)\right]. \tag{11.53}$$

Equation (11.52) implies that the conductive current is conserved

$$\partial_\nu \mathbf{j}^\nu(x) = 0. \tag{11.54}$$

Substitution of the explicit form of the distribution functions G_{if} and $\widetilde{G}_{ij}$ in Eq. (11.53) gives

$$\begin{aligned}
\mathcal{J}(\tau) = & \\
& -\frac{g}{2\pi^3 u}\sum_{i=1}^{3}\epsilon_i \sum_f \int_0^\tau d\tau'\,\tau' \int d^2p_\perp \frac{\Delta h_i \Lambda_i(\tau')}{\sqrt{\Delta h_i^2 + m_{f\perp}^2 u}} \left|\ln\left(1 - \exp\left(-\frac{\pi m_{f\perp}^2}{\Lambda_i(\tau')}\right)\right)\right| \\
& -\frac{g}{2\pi^3 u}\sum_{i>j}^{3}\eta_{ij} \int_0^\tau d\tau'\,\tau' \int d^2p_\perp \frac{\Delta h_{ij} \Lambda_{ij}(\tau')}{\sqrt{\Delta h_{ij}^2 + p_\perp^2 u}} \left|\ln\left(1 + \exp\left(-\frac{\pi p_\perp^2}{\Lambda_{ij}(\tau')}\right)\right)\right|,
\end{aligned} \tag{11.55}$$

where we used the definitions of the production rates $\mathcal{R}_{if}$ and $\widetilde{\mathcal{R}}_{ij}$ given by Eqs. (11.19) and (11.21).

11.5.4 *Displacement current*

The boost-invariant displacement current (11.28) should have the same spacetime structure as the conductive current (11.52), so we may write

$$\mathbf{j}_D^\nu(x) = \left[\mathbf{j}_D^0(x), 0, 0, \mathbf{j}_D^3(x)\right] = [z, 0, 0, t]\mathcal{J}_D(\tau). \tag{11.56}$$

This form automatically yields the conservation law

$$\partial_\nu \mathbf{j}_D^\nu(x) = 0. \tag{11.57}$$

We know the expressions for the dipole moments induced in the plane $z = 0$, given by Eqs. (11.31) and (11.32), hence we can compute the third component of the displacement current $\mathbf{j}_D^3(t, z = 0)$ and find that

$$\mathcal{J}_D(\tau) = \frac{1}{\tau}\,\mathbf{j}_D^3(\tau, z = 0). \tag{11.58}$$

Equations (11.28), (11.29) and (11.30) give

$$\mathcal{J}_D(\tau) = \frac{2}{\tau}\sum_{i=1}^{3}\frac{\boldsymbol{\epsilon}_i}{\boldsymbol{\epsilon}_i\cdot\boldsymbol{\mathcal{E}}}\sum_f\int\frac{dw}{v}d^2p_\perp\frac{dN_{if}}{d\Gamma_{\text{inv}}}m_{f\perp} + \frac{2}{\tau}\sum_{i>j}^{3}\frac{\boldsymbol{\eta}_{ij}}{\boldsymbol{\eta}_{ij}\cdot\boldsymbol{\mathcal{E}}}\int\frac{dw}{v}d^2p_\perp\frac{d\widetilde{N}_{ij}}{d\Gamma_{\text{inv}}}p_\perp. \tag{11.59}$$

Integration over w in Eq. (11.59) is trivial and yields

$$\mathcal{J}_D(\tau) = \frac{\Lambda_i(\tau)}{2\pi^3\tau}\sum_{i=1}^{3}\frac{\boldsymbol{\epsilon}_i}{\boldsymbol{\epsilon}_i\cdot\boldsymbol{\mathcal{E}}}\sum_f\int d^2p_\perp\left|\ln\left(1-\exp\left(-\frac{\pi m_{f\perp}^2}{\Lambda_i(\tau)}\right)\right)\right|m_{f\perp} + \frac{\Lambda_{ij}(\tau)}{2\pi^3\tau}\sum_{i>j}^{3}\frac{\boldsymbol{\eta}_{ij}}{\boldsymbol{\eta}_{ij}\cdot\boldsymbol{\mathcal{E}}}\int d^2p_\perp\left|\ln\left(1+\exp\left(-\frac{\pi p_\perp^2}{\Lambda_{ij}(\tau)}\right)\right)\right|p_\perp. \tag{11.60}$$

11.5.5 *Oscillations of the quark-gluon plasma*

With the all substitutions required by the boost invariance, the field equation (11.26) may be written as

$$\frac{d^2\mathbf{h}(\tau)}{d\tau^2} = \frac{1}{\tau}\frac{d\mathbf{h}(\tau)}{d\tau} + \tau^2\left[\mathcal{J}(\tau) + \mathcal{J}_D(\tau)\right]. \tag{11.61}$$

This is an integro-differential equation for the function $\mathbf{h}(\tau)$ because the conductive current $\mathcal{J}(\tau)$ depends not only on $\mathbf{h}(\tau)$ but also on the values of $\mathbf{h}(\tau')$ for $0 \le \tau' \le \tau$. Equation (11.61) has to be solved numerically step by step for given initial values. These are taken in the form discussed earlier

$$\mathbf{h}(0) = 0, \quad \frac{1}{\tau}\frac{d\mathbf{h}}{d\tau}(0) = -\boldsymbol{\mathcal{E}}_0 = -\sqrt{\frac{2\sigma_g}{\pi r^2}}k\mathbf{q}. \tag{11.62}$$

The solution of Eq. (11.61) is independent of the initial condition for $\mathbf{h}(\tau)$ because of the cancellations connected with the gauge transformation which leaves $\boldsymbol{\mathcal{E}}$ unchanged. Since the exchange of color charges at the initial stage of a heavy-ion collision leads to the color fields spanned by gluons, we assume that $\mathbf{q}$ is one of the gluon color charges $\boldsymbol{\eta}_{ij}$. In reality, after a collision the color distribution of nuclear discs may be strongly fluctuating in the transverse direction. In this approach, however, such fluctuations are smoothed out.

The field equation (11.61) can be written explicitly in a compact form in the

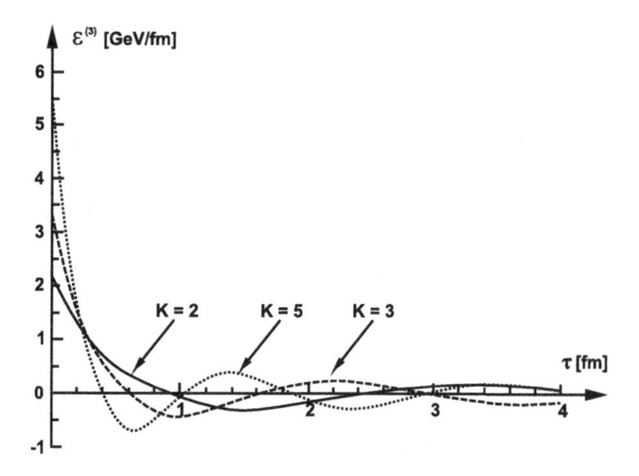

Fig. 11.4 Time dependence of the chromoelectric field obtained in the boost-invariant color-flux-tube model discussed in Sec. 11.5 (formulated originally in [12]). The results are shown for different initial conditions characterized by the parameter k ($N_f = 2$).

case where quarks can be treated as massless particles

$$
\begin{aligned}
\frac{d^2\mathbf{h}(\tau)}{d\tau^2} = & -\frac{1}{\tau}\frac{d\mathbf{h}(\tau)}{d\tau} - \frac{gN_f}{2\pi^3}\int_0^\tau d\tau'\ \tau' \sum_{i=1}^{3} \epsilon_i \Lambda_i^2(\tau')\,\beta_i C^-(\beta_i) \\
& -\frac{g}{2\pi^3}\int_0^\tau d\tau'\ \tau' \sum_{i>j}^{3} \boldsymbol{\eta}_{ij}\Lambda_{ij}^2(\tau')\,\beta_{ij} C^+(\beta_{ij}) \\
& +\frac{gN_f\,\tau}{2\pi^3}\sum_{i=1}^{3}\epsilon_i \frac{\Lambda_i^2(\tau)}{\Lambda_i(\tau)+\sigma_q}\sqrt{\frac{\Lambda_i(\tau)}{\pi}}\,\mathrm{sign}\left(\epsilon_i\cdot\boldsymbol{\mathcal{E}}\right)D^-(0) \\
& +\frac{g\,\tau}{2\pi^3}\sum_{i>j}^{3}\boldsymbol{\eta}_{ij}\frac{\Lambda_{ij}^2(\tau)}{\Lambda_{ij}(\tau)+\sigma_g}\sqrt{\frac{\Lambda_j(\tau)}{\pi}}\,\mathrm{sign}\left(\boldsymbol{\eta}_{ij}\cdot\boldsymbol{\mathcal{E}}\right)D^+(0)\,. \quad (11.63)
\end{aligned}
$$

Here N_f is the number of flavors and

$$
\beta = \frac{\Delta h(\tau,\tau')}{\sqrt{\Delta h^2(\tau,\tau') + \Lambda(\tau')\,\tau^2/\pi}}. \quad (11.64)
$$

We use here also the two functions defined by the formulas

$$
\begin{aligned}
C^\pm(y) &= \int_0^\infty d\xi \frac{\left|\ln\left(1\pm e^{-\xi}\right)\right|}{\sqrt{y^2+(1-y^2)\,\xi}}, \\
D^\pm(y) &= \int_0^\infty d\xi \sqrt{y^2+(1-y^2)\,\xi}\,\left|\ln\left(1\pm e^{-\xi}\right)\right|. \quad (11.65)
\end{aligned}
$$

The time dependence of the chromoelectric field, obtained from Eq. (11.63), is shown in Fig. 11.4. The results are presented for three different values of the initial chromoelectric field characterized by the parameter k (with $N_f = 2$). We can see that the solutions have generally an oscillatory character with a period that is shorter for larger initial fields (larger values of k). Such oscillatory behavior reminds us phenomena known from the physics of the "ordinary" QED plasma. In the considered model the amplitude of the oscillations decreases with time. Early, this effect is caused by the tunneling processes that transform the initial classical field configuration into a system of interacting quarks and gluons. Later, the decrease of the amplitude is caused by the longitudinal expansion of the system (included implicitly by the boost-invariance).

We observe that the tunneling process leads to the very fast decay of the initial field. A characteristic time of this process is a fraction of fermi. In view of the RHIC phenomenology suggesting short timescales of the production processes, this is an attractive feature of the color-flux-tube model.

Knowing $\mathbf{h}(\tau)$ as a function of the proper time τ, we may calculate the time dependence of all other interesting quantities characterizing the plasma, such as the energy density and average momenta. In Fig. 11.5 we show the time dependence of the energy density, dE/dV. One can notice that the energy transferred from the

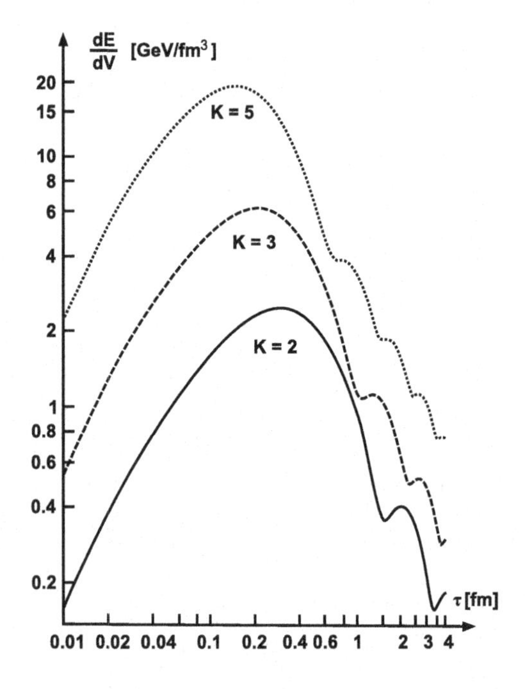

Fig. 11.5 Time dependence of the energy density of quarks and charged gluons obtained in the boost-invariant color-flux-tube model discussed in Sec. 11.5 (formulated originally in [12]). The larger is the initial value of the chromoelectric field, the larger is the maximum energy density of the produced plasma. For large values of time, the energy density drops down due to the strong longitudinal expansion.

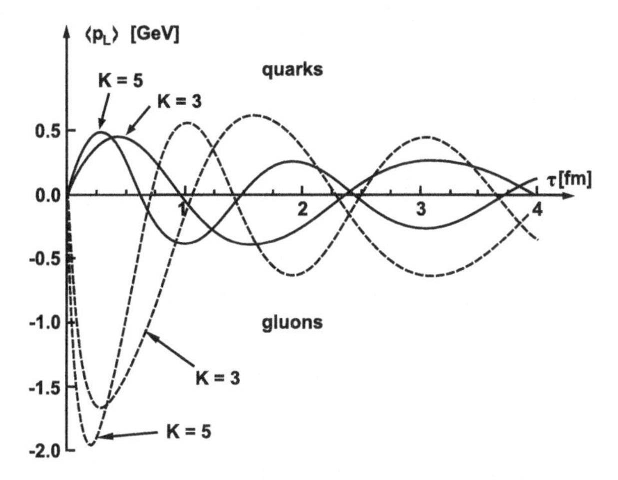

Fig. 11.6 Time dependence of the average longitudinal momentum obtained in the boost-invariant color-flux-tube model discussed in Sec. 11.5 (formulated originally in [12]). The changing signs of the average longitudinal momentum reflect the oscillatory character of the plasma motion.

field to the plasma increases strongly with k. The maximum values of the energy density reached during the evolution are: 2.5 GeV/fm^3 for $k = 2$, 6.3 GeV/fm^3 for $k = 3$, and 19.5 GeV/fm^3 for $k = 5$. All these values exceed the critical energy density for the deconfinement phase transition. This suggests that the considered model is consistent with the idea of the plasma formation.

It is interesting to discuss the origin of the bumps in the energy-density plots presented in Fig. 11.5. Their appearance is connected with the threshold effects in the particle production, see Eqs. (11.20) and (11.22). Since the energy is transferred between the field and particles back and forth, it happens that the field at later times becomes sufficiently strong and the tunneling processes reoccur. This kind of processes lead to extra particle production at later times and manifest themselves as the local maxima in Fig. 11.5.

Figure 11.6 shows the average longitudinal momenta of quarks and gluons, plotted as functions of the proper time τ for different values of the parameter k. We find that they oscillate out of phase and that gluons reach larger longitudinal momenta than quarks (this behavior is explained by the fact that gluons have larger color charges). The frequency of the oscillations increases with k.

The average transverse momenta are shown in Fig. 11.7. They also increase with k. At the very beginning of the process the tunneling of quarks and gluons with large transverse momentum is possible. When the field decreases, the smaller values of $p_\perp$ are preferred. Finally, when the field is below the threshold value incorporated in Eqs. (11.20) and (11.22), no tunneling occurs. This sequence explains the time dependence of $\langle p_\perp \rangle$ plotted in Fig. 11.7.

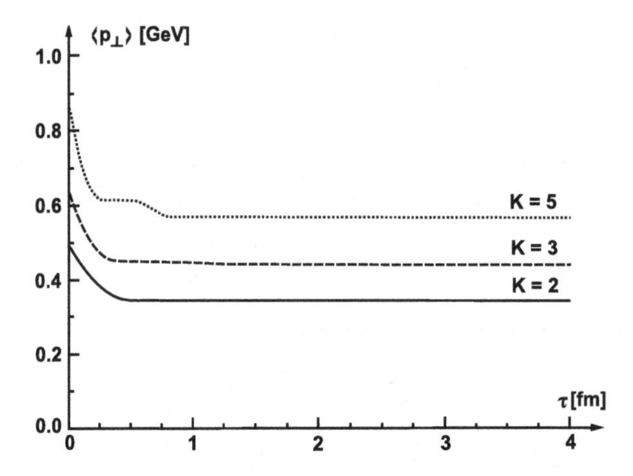

Fig. 11.7 Time dependence of the average transverse momentum of quarks and gluons obtained in the boost-invariant color-flux-tube model discussed in Sec. 11.5 (formulated originally in [12]).

11.6 Apparent thermalization

The production of particles in the fluctuating color fields may be responsible for the thermal character of the transverse-momentum spectra. To see this point, following Bialas [24], we consider the Schwinger formula for an elementary flux tube. Using Eq. (11.10), and keeping only the first term in the expansion of the logarithm, we find

$$\frac{dN_\sigma}{d^4x\, d^2p_\perp} = \frac{\sigma^2}{4\pi^3} \exp\left(-\frac{\pi m_\perp^2}{\sigma^2}\right). \tag{11.66}$$

Here we identified the force F acting on the tunneling particles with the string tension σ (see Ex. 12.3, note also that Eqs. (11.20) and (11.22) yield $\Lambda_i = \sigma_q$ and $\Lambda_{ij} = \sigma_g$ for $k = 1$ in Eq. (11.13), below we neglect the difference between σ_q and σ_g).

The main point of Ref. [24] is that the string tension may fluctuate with the probability distribution given by the Gaussian

$$P(\sigma) = \sqrt{\frac{2}{\pi\langle\sigma^2\rangle}} \exp\left(-\frac{\sigma^2}{2\langle\sigma^2\rangle}\right) \tag{11.67}$$

where

$$\langle\sigma^2\rangle = \int_0^\infty d\sigma\, \sigma^2\, P(\sigma). \tag{11.68}$$

In this case, the transverse-momentum spectrum (11.66) should be convoluted with the distribution (11.67),

$$\begin{aligned}\frac{dN}{d^4x\, d^2p_\perp} &= \int\limits_0^\infty d\sigma \, \frac{dN_\sigma}{d^4x\, d^2p_\perp} \, P(\sigma) \\ &= \frac{1}{4\pi^3}\sqrt{\frac{2}{\pi\langle\sigma^2\rangle}} \int\limits_0^\infty d\sigma \, \sigma^2 \, \exp\left(-\frac{\pi m_\perp^2}{\sigma^2} - \frac{\sigma^2}{2\langle\sigma^2\rangle}\right). \end{aligned} \tag{11.69}$$

With the help of the identity

$$\int\limits_0^\infty dx\, x^2 \, \exp(-ax^2 - bx^{-2}) = \frac{\sqrt{\pi}(1+2\sqrt{ab})}{4a^{3/2}} \exp(-2\sqrt{ab}), \tag{11.70}$$

where $a > 0$ and $b > 0$, we find

$$\frac{dN}{d^4x\, d^2p_\perp} = \frac{\sqrt{\langle\sigma^2\rangle}\left(\sqrt{\langle\sigma^2\rangle} + m_\perp\sqrt{2\pi}\right)}{4\pi^3} \exp\left(-\frac{m_\perp\sqrt{2\pi}}{\sqrt{\langle\sigma^2\rangle}}\right). \tag{11.71}$$

Neglecting the slowly varying prefactor we obtain the "thermal" formula

$$\frac{dN}{d^4x\, d^2p_\perp} \sim \exp\left(-\frac{m_\perp}{\lambda}\right) \tag{11.72}$$

with

$$\lambda = \sqrt{\frac{\langle\sigma^2\rangle}{2\pi}}. \tag{11.73}$$

Using the standard value of the string tension $\langle\sigma^2\rangle = 1$ GeV/fm we obtain $\lambda = 178$ MeV, a value that is very close to the typical chemical freeze-out temperature $T_{\rm chem} \sim 165$ MeV.

In Ref. [24] the fluctuations of the string tension were connected to the stochastic picture of the QCD vacuum [25–27]. In the previous sections we have shown that the color fields in the plasma oscillate. It turns out, that this type of behavior also leads to thermal shapes of the transverse-momentum spectra [28]. Hence, the Schwinger tunneling mechanism in strong varying fields offers an appealing explanation of a very fast formation of the transversally thermalized system in heavy-ion collisions.

Let us close this Chapter with a few general remarks. The color-flux-tube model has several attractive features. In particular, it explains large energy densities attainable in relativistic heavy-ion collisions and short formation times. It may explain also the thermal character of the transverse-momentum spectra. We must remember, however, that it is a simple model based on strong assumptions, for example: the non-linear features of QCD are eliminated, the interaction of quarks and gluons is mediated by the mean fields only, and no transverse dynamics is included.

One of the problems of the formalism described above is that it uses the Schwinger formula that was derived for uniform and static system. In heavy-ion collisions the particle production is concentrated in finite regions of space, hence, certain modifications of the Schwinger formula due to the finite transverse size of the color flux tubes are expected. The modified expressions which take into account such effects were elaborated in Refs. [20, 21].

Being a model, the color-flux-tube approach allows for different practical realizations. Of course, our presentation followed one special streamline of the arguments but other approaches are also possible, e.g., see Refs. [22, 23].

Bibliography to Chapter 11

[1] J. S. Schwinger, "On gauge invariance and vacuum polarization," *Phys. Rev.* **82** (1951) 664–679.
[2] E. Brezin and C. Itzykson, "Pair production in vacuum by an alternating field," *Phys. Rev.* **D2** (1970) 1191–1199.
[3] A. Casher, H. Neuberger, and S. Nussinov, "Chromoelectric flux tube model of particle production," *Phys. Rev.* **D20** (1979) 179–188.
[4] A. Casher, H. Neuberger, and S. Nussinov, "Multiparticle production by bubbling flux tubes," *Phys. Rev.* **D21** (1980) 1966.
[5] N. K. Glendenning and T. Matsui, "Creation of $\bar{q}q$ pair in a chromoelectric flux tube," *Phys. Rev.* **D28** (1983) 2890–2891.
[6] L. D. Landau and E. M. Lifshitz, "The Classical Theory of Fields," (Butterworth-Heinemann, Oxford, 1980).
[7] T. S. Biro, H. B. Nielsen, and J. Knoll, "Color rope model for extreme relativistic heavy ion collisions," *Nucl. Phys.* **B245** (1984) 449–468.
[8] A. Bialas and W. Czyz, "Chromoelectric flux tubes and the transverse momentum distribution in high-energy nucleus-nucleus collisions," *Phys. Rev.* **D31** (1985) 198.
[9] M. Gyulassy and A. Iwazaki, "Quark and gluon pair production in SU(n) covariant constant fields," *Phys. Lett.* **165B** (1985) 157–161.
[10] I. A. Batalin, S. G. Matinyan, and G. K. Savvidy, "Vacuum polarization by a source-free gauge field," *Sov. J. Nucl. Phys.* **26** (1977) 214.
[11] J. Ambjorn and R. J. Hughes, "Canonical quantization in non-abelian background fields. 1," *Ann. Phys.* **145** (1983) 340.
[12] A. Bialas, W. Czyz, A. Dyrek, and W. Florkowski, "Oscillations of quark-gluon plasma generated in strong color fields," *Nucl. Phys.* **B296** (1988) 611.
[13] A. Bialas, W. Czyz, A. Dyrek, and W. Florkowski, "A semiclassical boost-invariant description of pair production in chromoelectric field," *Z. Phys.* **C46** (1990) 439–444.
[14] K. Bajan and W. Florkowski, "Boost-invariant particle production in transport equations," *Acta Phys. Polon.* **B32** (2001) 3035–3052.
[15] A. Bialas and W. Czyz, "Boost-invariant Boltzmann-Vlasov equation for relativistic quark-antiquark plasma," *Phys. Rev.* **D30** (1984) 2371.
[16] A. Bialas and W. Czyz, "Oscillations of relativistic, boost-invariant quark-antiquark

plasma," *Z. Phys.* **C28** (1985) 255.

[17] A. Bialas and W. Czyz, "Production and collective motion of $q\bar{q}$ plasma in heavy-ion collisions," *Acta Phys. Polon.* **B17** (1986) 635.

[18] C. Greiner, K. Wagner, and P. G. Reinhard, "Memory effects in relativistic heavy ion collisions," *Phys. Rev.* **C49** (1994) 1693–1701.

[19] Y. Kluger, E. Mottola, and J. M. Eisenberg, "The quantum Vlasov equation and its Markov limit," *Phys. Rev.* **D58** (1998) 125015.

[20] H. P. Pavel and D. M. Brink, "$q\bar{q}$ pair creation in a flux tube with confinement," *Z. Phys.* **C51** (1991) 119–125.

[21] T. Schenfeld *et al.*, "Pair creation in a flux tube," *Phys. Lett.* **B247** (1990) 5–12.

[22] K. Sailer, T. Schoenfeld, Z. Schram, A. Schaefer, and W. Greiner, "Strings and ropes in heavy ion collisions: Towards a semiclassical unified string flux tube model," *J. Phys.* **G17** (1991) 1005–1057.

[23] K. Sailer, T. Schoenfeld, Z. Schram, A. Schaefer, and W. Greiner, "Dynamical string model on the basis of a semiclassical unified string flux tube model," *Int. J. Mod. Phys.* **A6** (1991) 4395–4435.

[24] A. Bialas, "Fluctuations of string tension and transverse mass distribution," *Phys. Lett.* **B466** (1999) 301–304.

[25] H. G. Dosch, "Gluon condensate and effective linear potential," *Phys. Lett.* **B190** (1987) 177.

[26] H. G. Dosch and Y. A. Simonov, "The area law of the Wilson loop and vacuum field correlators," *Phys. Lett.* **B205** (1988) 339.

[27] Y. A. Simonov, "Vacuum background fields in QCD as a source of confinement," *Nucl. Phys.* **B307** (1988) 512.

[28] W. Florkowski, "Schwinger tunneling and thermal character of hadron spectra," *Acta Phys. Polon.* **B35** (2004) 799–808, `nucl-th/0309049`.

Chapter 12

Exercises to PART III

Exercise 12.1. *Tunneling along the hyperbola of constant proper time.*
The high-energy processes are governed by the proper time $\tau = \sqrt{t^2 - z^2}$, rather than by the ordinary laboratory time t. Suppose that the tunneling process takes place at a fixed value of τ, not at a fixed t. Using the semiclassical method, calculate the tunneling probability in this case. *Hints:* The particles start tunneling at $z = 0$ and become real at $z = \pm z_{\mathrm{f}}$. The energy and momentum conservation laws for a particle tunneling along the z-axis give

$$\begin{aligned} m_\perp^2 + p_\parallel^2 &= \sigma^2 z_{\mathrm{f}}^2, \\ p_\parallel &= \sigma\left(t_{\mathrm{f}} - t\right), \end{aligned} \tag{12.1}$$

where σ is the string tension. Changing to rapidity and spacetime rapidity one gets

$$\begin{aligned} \tanh \mathrm{y}_{\mathrm{f}} &= \frac{m_\perp}{2\sigma\tau\sqrt{1 + m_\perp^2/(2\sigma^2\tau^2)}}, \\ \eta_{\parallel\,\mathrm{f}} &= 2\mathrm{y}_{\mathrm{f}}. \end{aligned} \tag{12.2}$$

The tunneling probability is obtained from the formula $P = \exp(-2\mathrm{Im}S)$, where S is the sum of the classical actions for the particle and antiparticle, which should be calculated as integrals over complex and not just purely imaginary time. The final result for P is the same as in the case discussed in Sec. 11.1.

Exercise 12.2. *Lorentz-transformation of the electric field.*
The electric field $\mathbf{E}$ has only one non-vanishing component, $\mathbf{E} = (0, 0, \mathcal{E})$, and the magnetic field $\mathbf{B}$ is zero. Find the electric field in the new reference system which moves along the z-axis with a constant velocity v. What are the components of the magnetic field in the new reference frame?

Exercise 12.3. *Color-flux tube as a chromoelectric capacitor.*
A tube of the chromoelectric field (or a string) can be treated as a chromoelectric capacitor. For given values of the charge at the end of the tube and the radius of the tube, calculate the force F acting on the charge. What is the value of the string tension σ (the latter is defined as the energy per unit length)? What is the relation between F and σ? *Hint:* Use the Gauss law and neglect boundary effects.

Exercise 12.4. *Boost invariance of the central region.*
Show that a scalar boost-invariant function $f(t, x, y, z)$ may depend only on $t^2 - x^2 - y^2 - z^2$. *Hint:* The scalar function is defined by condition $f'(x') = f(x)$, and the boost-invariance requires $f'(x') = f(x')$. Prime denotes the quantities in the new reference frame, $x' = Lx$, where L is the Lorentz transformation. Consider now the infinitesimal Lorentz transformation $L = 1 + \omega$ and show that f is an arbitrary function of the invariant $t^2 - x^2 - y^2 - z^2$.

Exercise 12.5. *Boost invariant phase-space distribution.*
Suppose that the distribution $f(t, \mathbf{x}, \mathbf{p})$ is a function of the variables $u = t^2 - z^2$, $w = p_z t - z\sqrt{p_z^2 + m_\perp^2}$ and $p_\perp$, $f(t, \mathbf{x}, \mathbf{p}) = G(u, w, p_\perp)$. Derive the kinetic equation satisfied by the function $G(u, w, p_\perp)$ if $f(t, \mathbf{x}, \mathbf{p})$ satisfies the Boltzmann-Vlasov equation

$$\left(p^\mu \partial_\mu + g F_{\mu\nu} p^\nu \partial_p^\mu\right) f = 0.$$

Assume that all components of $F_{\mu\nu}$ are zero except for $F_{03} = -F_{30}$. Find solutions of this equation for the case F^{30} = constant.

PART IV

RELATIVISTIC HYDRODYNAMICS

Chapter 13

Perfect Fluid

The perfect fluid is an ideal object which behaves in the very similar way as the realistic fluids in the cases where we may neglect the phenomena such as shear stresses, viscosity, or heat conduction. In the local rest frame, the perfect fluid may be completely characterized by its rest frame energy density ε and the *isotropic* pressure P.

In this Chapter we introduce the basic concepts concerning the relativistic hydrodynamics of the perfect fluid [1–6]. Having in mind very successful applications of relativistic hydrodynamics to describe relativistic heavy-ion collisions we discuss explicitly different forms of the hydrodynamic equations. In particular, we analyze in detail the structure of the equations with cylindrical symmetry (the form appropriate for the description of central collisions) and with boost-invariance (the form appropriate for the description of the central region in the most energetic heavy-ion collisions). The formalism introduced in this Chapter will be used below in Chaps. 20–22, where different hydrodynamic models of the collisions are discussed.

From the mathematical point of view, the hydrodynamic equations are linear partial differential equations. The term *linear* refers to the first order of the derivatives which appear in the equations. The coefficients standing at the derivatives are usually very complicated expressions containing the functions we want to determine, hence the hydrodynamic equations are non-linear in the sense that a linear combination of solutions does not represent the new solution. This feature makes the hydrodynamic equations quite complicated to deal with and, in the majority of interesting cases, one has to solve them numerically. Only few cases can be treated analytically — these are the cases where due to the symmetry reasons one can reduce significantly the number of independent dimensions and variables.

13.1 Thermodynamics of fluid element

Let us first introduce the concept of a fluid element [2]. This name refers to a small part of the fluid, whose volume V is much larger than the microscopic size of the particles but, at the same time, much smaller than the volume of the whole system.

The four-velocity of the fluid element is $u^\mu = \gamma(1, \mathbf{v})$, where $\mathbf{v}$ is the three-velocity and $\gamma = (1 - v^2)^{-\frac{1}{2}}$ is the Lorentz factor. We specify the change of V by expression

$$\frac{dV}{d\tau} \equiv u^\mu \, \partial_\mu V = V \partial_\mu \, u^\mu. \tag{13.1}$$

In the local rest frame of the fluid element, defined by the condition $\mathbf{v} = 0$, Eq. (13.1) is reduced to the time derivative dV/dt. If there is a quantum number $N = nV$ conserved during the time evolution of the system, then the conservation law has the standard form

$$\partial_\mu (n u^\mu) = 0, \tag{13.2}$$

which together with (13.1) yields

$$\frac{dN}{d\tau} = 0. \tag{13.3}$$

Let us now consider the case where N corresponds to the baryon number. Since the baryon number in the fluid element is conserved, the first law of thermodynamics reads

$$dE = TdS - PdV. \tag{13.4}$$

Here E, T, S and P are the energy, temperature, entropy and pressure of the fluid element, respectively. The change from extensive to intensive variables,

$$\varepsilon = \frac{E}{V}, \quad s = \frac{S}{N}, \tag{13.5}$$

gives [2]

$$d\varepsilon = \frac{\varepsilon + P}{n} dn + nT ds. \tag{13.6}$$

This equation implies that the energy density ε can be treated as a function of the baryon density n and the entropy per baryon, s. The functional dependence $\varepsilon = \varepsilon(n, s)$ defines the equation of state of a considered physical system. If the equation of state is known, the pressure and temperature may be calculated from the formulas

$$P(n, s) = n \left(\frac{\partial \varepsilon}{\partial n} \right)_s - \varepsilon \tag{13.7}$$

and

$$T(n, s) = \frac{1}{n} \left(\frac{\partial \varepsilon}{\partial s} \right)_n . \tag{13.8}$$

Introducing enthalpy per baryon,

$$w = \frac{W}{N} = \frac{\varepsilon + P}{n}, \tag{13.9}$$

we can also write

$$d\varepsilon = w \, dn + nT \, ds, \tag{13.10}$$

$$dP = n \, dw - nT \, ds. \tag{13.11}$$

In many physical situations the baryon density in a system is negligible. This happens, for example, in the central region of the heavy-ion collisions at the RHIC top energies, where initially the gluonic and later the mesonic degrees of freedom play the dominant role. The thermodynamic relations for systems with zero baryon density are

$$\varepsilon + P = T\,\sigma, \tag{13.12}$$

$$dP = \sigma\, dT, \qquad d\varepsilon = T\, d\sigma. \tag{13.13}$$

Here σ is the entropy volume density,

$$\sigma = \frac{S}{V}. \tag{13.14}$$

13.2 Euler equation and conservation laws

13.2.1 *Systems with non-zero baryon density*

The *perfect fluid* is defined formally by the form of its energy-momentum tensor, namely

$$T^{\mu\nu} = (\varepsilon + P)u^\mu u^\nu - P g^{\mu\nu}, \tag{13.15}$$

where $g^{\mu\nu}$ is the metric tensor with $g^{00} = 1$. From our discussion of the kinetic theory we know that such a form of the energy-momentum tensor follows from the assumption of local thermal equilibrium, see Eq. (8.48). Equations of motion of the perfect fluid are obtained from the conservation laws

$$\partial_\mu T^{\mu\nu} = 0. \tag{13.16}$$

Substituting (13.15) in (13.16) and using (13.2) gives

$$\frac{d}{d\tau}(wu^\nu) \equiv u^\mu \partial_\mu (wu^\nu) = \frac{1}{n}\partial^\nu P. \tag{13.17}$$

The projection of (13.17) on the fluid four-velocity u_ν and the use of the thermodynamic identity (13.11) yields

$$\frac{ds}{d\tau} \equiv u^\mu \partial_\mu s = 0. \tag{13.18}$$

Equation (13.17) is the relativistic analog of the Euler equation used in the classical hydrodynamics, while Eq. (13.18) states that the entropy per baryon is conserved. In other words the flow is *adiabatic* (there is no heat exchange between different fluid elements). It is interesting that this condition follows directly from the assumed form of the energy-momentum tensor (13.15). In the classical hydrodynamics, the adiabaticity of the flow is one of the independent assumptions which characterize the perfect fluid [1].

With the help of Eqs. (13.11) and (13.18) one may rewrite Eq. (13.17) in the form

$$(u^\mu u^\nu - g^{\mu\nu})\,(\partial_\mu w - T\partial_\mu s) = -w\,u^\mu \partial_\mu u^\nu, \tag{13.19}$$

which is equivalent to

$$\frac{du^\nu}{d\tau} = -\frac{1}{wn}\Delta^{\mu\nu}\partial_\mu P. \tag{13.20}$$

Here we introduced the projection tensor

$$\Delta^{\mu\nu} = u^\mu u^\nu - g^{\mu\nu}. \tag{13.21}$$

The four-vectors appearing on both sides of Eq. (13.20) are explicitly perpendicular to u^μ, hence only three equations in (13.20) are independent. Together with Eqs. (13.2) and (13.18) they form a system of five equations for five unknown variables: n, s and three components of the velocity $\mathbf{v}$. We may summarize these equations in the form

$$n\frac{du^\nu}{d\tau} = -\Delta^{\mu\nu}\left[c_\mathrm{s}^2(n,s)\partial_\mu n + d_\mathrm{s}^2(n,s)\partial_\mu s\right], \tag{13.22}$$

$$\partial_\mu(nu^\mu) = 0, \tag{13.23}$$

$$u^\mu\partial_\mu s = 0, \tag{13.24}$$

where

$$c_\mathrm{s}^2 = \frac{n}{\epsilon + P}\left(\frac{\partial P}{\partial n}\right)_s = \frac{n}{w}\left(\frac{\partial w}{\partial n}\right)_s \tag{13.25}$$

and

$$d_\mathrm{s}^2 = \frac{n}{\epsilon + P}\left(\frac{\partial P}{\partial s}\right)_n. \tag{13.26}$$

We note that $c_\mathrm{s}(n,s)$ is the velocity of sound in the medium characterized by the equation of state $\varepsilon = \varepsilon(n,s)$. Clearly, the explicit form of the equation of state is required in order to close the system of the hydrodynamic equations. We also note that Eqs. (13.23)–(13.24) imply the conservation law $\partial_\mu(\sigma u^\mu) = 0$.

If the flow satisfies the condition s = const., Eqs. (13.22)–(13.24) are reduced to a simpler form, namely

$$n\frac{du^\nu}{d\tau} = -c_\mathrm{s}^2(n)\,\Delta^{\mu\nu}\partial_\mu n, \quad \partial_\mu(nu^\mu) = 0. \tag{13.27}$$

13.2.2 *Systems with vanishing baryon density*

The relativistic hydrodynamic equations may be also derived for the fluid with vanishing chemical potential. In this case, in analogy to (13.17) and (13.18) we obtain

$$\frac{d}{d\tau}(Tu^\nu) = u^\mu \partial_\mu (Tu^\nu) = \partial^\nu T, \tag{13.28}$$

$$\partial_\mu \left(\sigma u^\mu\right) = 0. \tag{13.29}$$

Further manipulations allow us to write Eq. (13.28) as

$$\frac{du^\nu}{d\tau} = -\frac{1}{T}\Delta^{\mu\nu}\partial_\mu T, \tag{13.30}$$

which is an analog of (13.20). Switching from T to σ one finds

$$\sigma\frac{du^\nu}{d\tau} = -c_{\rm s}^2(\sigma)\,\Delta^{\mu\nu}\partial_\mu\sigma, \quad \partial_\mu(\sigma u^\mu) = 0, \tag{13.31}$$

where the sound velocity of baryon free matter is defined by the formula

$$c_{\rm s}^2 = \frac{\partial P}{\partial \epsilon} = \frac{\sigma}{T}\frac{\partial T}{\partial \sigma}. \tag{13.32}$$

13.3 Euler equation in Cartesian coordinates

In this Section, using the Cartesian coordinates, we give the explicit form of the relativistic Euler equations, (13.17) and (13.28), in the non-relativistic three-vector notation. This form is convenient for the numerical analysis of the hydrodynamic equations in general situations which do not exhibit any simplifying spacetime symmetries. We shall also treat the Cartesian form as the starting point for the analysis of the cases with the cylindrical symmetry, Sec. 13.4, or with boost invariance, Sec. 13.5.

13.3.1 *Non-zero baryon density*

In Sec. 13.2.1 we showed that for systems with non-zero baryon density the relativistic Euler equation is given by the formula (13.17),

$$u^\mu \partial_\mu (wu^\nu) = \frac{1}{n}\partial^\nu P = \partial^\nu w - T\partial^\nu s. \tag{13.33}$$

One may check, see Exercise 1 to Part IV, that this equation may be rewritten in the equivalent three-vector form as

$$\frac{\partial}{\partial t}\left(w\gamma\mathbf{v}\right) + \nabla\left(w\gamma\right) = \mathbf{v}\times\left(\nabla\times w\gamma\mathbf{v}\right) + \frac{T}{\gamma}\nabla s. \tag{13.34}$$

The three-vector form is the good starting point to analyze each component of the Euler equation separately. First, we calculate the differentials of the Lorentz gamma factor

$$d\gamma = d\left(1-v^2\right)^{-1/2} = \gamma^3 v\, dv, \qquad d\left(\gamma v\right) = \gamma^3 dv. \tag{13.35}$$

With the help of these two equations one finds the acceleration equation for v_x

$$\frac{\left(1-v_y^2-v_z^2\right)}{(1-v^2)}\frac{dv_x}{dt}+\frac{v_x v_y}{(1-v^2)}\frac{dv_y}{dt}+\frac{v_x v_z}{(1-v^2)}\frac{dv_z}{dt}-\frac{T}{w}(1-v^2)\frac{\partial s}{\partial x}$$
$$+v_x\frac{\partial \ln w}{\partial t}+\left(1-v_y^2-v_z^2\right)\frac{\partial \ln w}{\partial x}+v_x\left(v_y\frac{\partial \ln w}{\partial y}+v_z\frac{\partial \ln w}{\partial z}\right)=0. \tag{13.36}$$

To simplify the notation, we have introduced here the total time derivative

$$\frac{d}{dt}=\frac{\partial}{\partial t}+v_x\frac{\partial}{\partial x}+v_y\frac{\partial}{\partial y}+v_z\frac{\partial}{\partial z}. \tag{13.37}$$

The final form of the y- and z-component of the Euler equation is obtained by the consecutive changes of the indices $(x\to y, y\to z, z\to x)$ in Eq. (13.36). In this way we obtain

$$\frac{\left(1-v_z^2-v_x^2\right)}{(1-v^2)}\frac{dv_y}{dt}+\frac{v_y v_z}{(1-v^2)}\frac{dv_z}{dt}+\frac{v_y v_x}{(1-v^2)}\frac{dv_x}{dt}-\frac{T}{w}(1-v^2)\frac{\partial s}{\partial y}$$
$$+v_y\frac{\partial \ln w}{\partial t}+\left(1-v_z^2-v_x^2\right)\frac{\partial \ln w}{\partial y}+v_y\left(v_z\frac{\partial \ln w}{\partial z}+v_x\frac{\partial \ln w}{\partial x}\right)=0 \tag{13.38}$$

and

$$\frac{\left(1-v_x^2-v_y^2\right)}{(1-v^2)}\frac{dv_z}{dt}+\frac{v_z v_x}{(1-v^2)}\frac{dv_x}{dt}+\frac{v_z v_y}{(1-v^2)}\frac{dv_y}{dt}-\frac{T}{w}(1-v^2)\frac{\partial s}{\partial z}$$
$$+v_z\frac{\partial \ln w}{\partial t}+\left(1-v_x^2-v_y^2\right)\frac{\partial \ln w}{\partial z}+v_z\left(v_x\frac{\partial \ln w}{\partial x}+v_y\frac{\partial \ln w}{\partial y}\right)=0. \tag{13.39}$$

13.3.2 *Vanishing baryon density*

For systems with the vanishing baryon density the relativistic Euler equation has a simpler form, namely

$$u^\mu \partial_\mu (T u^\nu) = \partial^\nu T, \tag{13.40}$$

and in the equivalent three-vector notation may be rewritten as [7]

$$\frac{\partial}{\partial t}(T\gamma\mathbf{v})+\nabla(T\gamma)=\mathbf{v}\times(\nabla\times T\gamma\mathbf{v}). \tag{13.41}$$

One can see that these equations may be formally obtained from the equations derived for the systems with non-zero baryon density if we replace enthalpy by the temperature and assume that the entropy per baryon is constant, $s=$ const. Thus, the Cartesian form of the Euler equation for baryon-free matter follows directly from Eqs. (13.36), (13.38) and (13.39)

$$\frac{\left(1-v_y^2-v_z^2\right)}{(1-v^2)}\frac{dv_x}{dt}+\frac{v_x v_y}{(1-v^2)}\frac{dv_y}{dt}+\frac{v_x v_z}{(1-v^2)}\frac{dv_z}{dt}$$
$$+v_x\frac{\partial \ln T}{\partial t}+\left(1-v_y^2-v_z^2\right)\frac{\partial \ln T}{\partial x}+v_x\left(v_y\frac{\partial \ln T}{\partial y}+v_z\frac{\partial \ln T}{\partial z}\right)=0, \tag{13.42}$$

$$\frac{\left(1-v_z^2-v_x^2\right)}{(1-v^2)}\frac{dv_y}{dt}+\frac{v_y v_z}{(1-v^2)}\frac{dv_z}{dt}+\frac{v_y v_x}{(1-v^2)}\frac{dv_x}{dt}$$
$$+v_y\frac{\partial \ln T}{\partial t}+\left(1-v_z^2-v_x^2\right)\frac{\partial \ln T}{\partial y}+v_y\left(v_z\frac{\partial \ln T}{\partial z}+v_x\frac{\partial \ln T}{\partial x}\right)=0, \tag{13.43}$$

$$\frac{\left(1-v_x^2-v_y^2\right)}{(1-v^2)}\frac{dv_z}{dt}+\frac{v_z v_x}{(1-v^2)}\frac{dv_x}{dt}+\frac{v_z v_y}{(1-v^2)}\frac{dv_y}{dt}$$
$$+v_z\frac{\partial \ln T}{\partial t}+\left(1-v_x^2-v_y^2\right)\frac{\partial \ln T}{\partial z}+v_z\left(v_x\frac{\partial \ln T}{\partial x}+v_y\frac{\partial \ln T}{\partial y}\right)=0. \tag{13.44}$$

13.4 Euler equation with cylindrical symmetry

In the cases of central or semi-central collisions and also when one analyzes the minimum-bias data the expansion of matter may be approximately described as

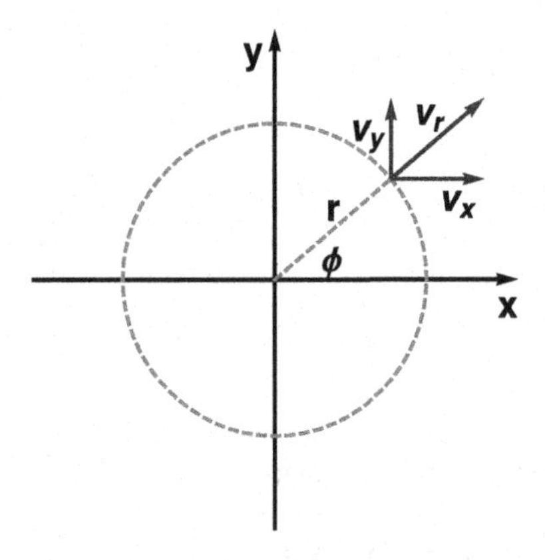

Fig. 13.1 In the case of cylindrical symmetry we use the standard parameterization, see Eq. (13.45). Only the radial component of the transverse flow is present. The azimuthal component is zero since we exclude the possibility that the whole system rotates around the z axis.

cylindrically symmetric. In such situations the cylindrical symmetry may be imposed on the equations of relativistic hydrodynamics and their analysis becomes simpler. In the cylindrically symmetric case the scalar thermodynamic quantities such as temperature or entropy density depend only on the variables $t, r = \sqrt{x^2 + y^2}$ and z. On the other hand, the velocity field has the following radial structure, see Fig. 13.1,

$$v_x = v_r\,(t,r)\cos\phi, \qquad v_y = v_r\,(t,r)\sin\phi, \qquad v_z = v_z(t,r). \tag{13.45}$$

The change of the variables from x, y to r, ϕ, see Eq. (31.17), implies that the following relations should be used:

$$\begin{aligned}
\frac{\partial v_x}{\partial x} &= \frac{\partial v_r}{\partial r}\cos^2\phi + v_r\frac{\sin^2\phi}{r}, \\
\frac{\partial v_y}{\partial x} &= \frac{\partial v_r}{\partial r}\sin\phi\cos\phi - v_r\cos\phi\frac{\sin\phi}{r}, \\
\frac{\partial v_x}{\partial y} &= \frac{\partial v_r}{\partial r}\cos\phi\sin\phi - v_r\sin\phi\frac{\cos\phi}{r}, \\
\frac{\partial v_y}{\partial y} &= \frac{\partial v_r}{\partial r}\sin^2\phi + v_r\frac{\cos^2\phi}{r}.
\end{aligned} \tag{13.46}$$

Implementing Eqs. (13.45) and (13.46) in Eqs. (13.36)–(13.39) one gets the *radial Euler equation* for systems with non-zero baryon density

$$\begin{aligned}
&\frac{(1-v_z^2)}{(1-v^2)}\left(\frac{\partial v_r}{\partial t} + v_r\frac{\partial v_r}{\partial r} + v_z\frac{\partial v_r}{\partial z}\right) + \frac{v_r\,v_z}{(1-v^2)}\left(\frac{\partial v_z}{\partial t} + v_r\frac{\partial v_z}{\partial r} + v_z\frac{\partial v_z}{\partial z}\right) \\
&+v_r\frac{\partial \ln w}{\partial t} + (1-v_z^2)\frac{\partial \ln w}{\partial r} + v_r\,v_z\frac{\partial \ln w}{\partial z} - \frac{T}{w}(1-v^2)\frac{\partial s}{\partial r} = 0
\end{aligned} \tag{13.47}$$

and the *longitudinal Euler* equation

$$\begin{aligned}
&\frac{(1-v_r^2)}{(1-v^2)}\left(\frac{\partial v_z}{\partial t} + v_r\frac{\partial v_z}{\partial r} + v_z\frac{\partial v_z}{\partial z}\right) + \frac{v_z\,v_r}{(1-v^2)}\left(\frac{\partial v_r}{\partial t} + v_r\frac{\partial v_r}{\partial r} + v_z\frac{\partial v_r}{\partial z}\right) \\
&+v_z\frac{\partial \ln w}{\partial t} + v_z\,v_r\frac{\partial \ln w}{\partial r} + (1-v_r^2)\frac{\partial \ln w}{\partial z} - \frac{T}{w}(1-v^2)\frac{\partial s}{\partial z} = 0.
\end{aligned} \tag{13.48}$$

Here $v = \sqrt{v_r^2 + v_z^2}$ is the magnitude of the velocity.

The analogs of Eqs. (13.47) and (13.48) for the systems with vanishing baryon density are obtained with the help of the formal substitutions $s =$ const. and

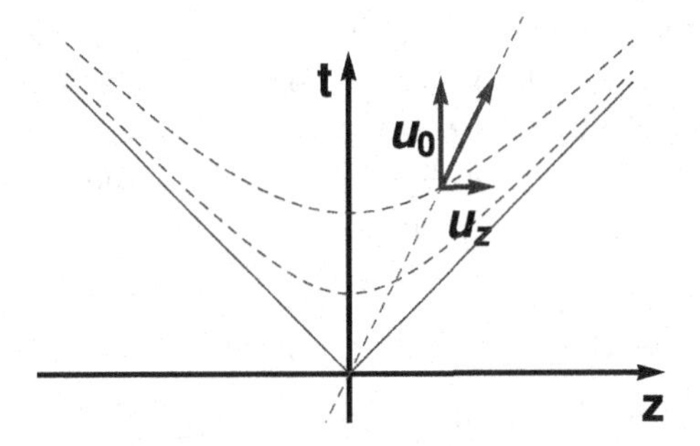

Fig. 13.2 For the boost invariant systems one has $v_z = u_z/u_0 = z/t$, see Eq. (13.51). The hyperbolas show the lines of the constant longitudinal proper time, $\tau = \sqrt{t^2 - z^2} = \text{const}$. The evolution of matter is confined to the interior of the light cone where $|z| < t$.

$w \to T$. In this way we find

$$\frac{(1-v_z^2)}{(1-v^2)}\left(\frac{\partial v_r}{\partial t} + v_r \frac{\partial v_r}{\partial r} + v_z \frac{\partial v_r}{\partial z}\right) + \frac{v_r\, v_z}{(1-v^2)}\left(\frac{\partial v_z}{\partial t} + v_r \frac{\partial v_z}{\partial r} + v_z \frac{\partial v_z}{\partial z}\right)$$
$$+ v_r \frac{\partial \ln T}{\partial t} + \left(1 - v_z^2\right)\frac{\partial \ln T}{\partial r} + v_r\, v_z \frac{\partial \ln T}{\partial z} = 0 \tag{13.49}$$

and

$$\frac{(1-v_r^2)}{(1-v^2)}\left(\frac{\partial v_z}{\partial t} + v_r \frac{\partial v_z}{\partial r} + v_z \frac{\partial v_z}{\partial z}\right) + \frac{v_z\, v_r}{(1-v^2)}\left(\frac{\partial v_r}{\partial t} + v_r \frac{\partial v_r}{\partial r} + v_z \frac{\partial v_r}{\partial z}\right)$$
$$+ v_z \frac{\partial \ln T}{\partial t} + v_z\, v_r \frac{\partial \ln T}{\partial r} + \left(1 - v_r^2\right)\frac{\partial \ln T}{\partial z} = 0. \tag{13.50}$$

13.5 Euler equation with boost-invariance

The influential papers by Feynmann and Bjorken suggested that the particle production at very high energies has the boost-invariant character, see our discussion in Secs. 2.6 and 2.7. The RHIC data did not confirm this picture, however, if one restricts the analysis to the midrapidity region, $-1 \leq \mathrm{y} \leq 1$, the assumption of boost invariance represents a quite reasonable approximation.

Similarly to the case of cylindrical symmetry, also the requirement of boost-invariance may be implemented in the relativistic hydrodynamic equations. In this case the thermodynamic quantities depend only on the variables $\tau = \sqrt{t^2 - z^2}$, x

and y, whereas the velocity field has the structure (see our discussion in Sec. 2.7 and Fig. 13.2)

$$u^\mu = \gamma(1, v_x, v_y, v_z) = \bar{\gamma}(\tau, x, y)\, \frac{t}{\tau} \left(1, \frac{\tau}{t}\bar{v}_x(\tau, x, y), \frac{\tau}{t}\bar{v}_y(\tau, x, y), \frac{z}{t}\right). \qquad (13.51)$$

Here the functions $\bar{v}_x, \bar{v}_y$ depend only on $\tau = \sqrt{t^2 - z^2}$, x and y, and coincide with v_x, v_y for $z = 0$. Similarly, at $z = 0$ the quantity $\bar{\gamma}$ coincides with the Lorentz gamma factor γ. By direct differentiation one finds

$$\frac{\partial v_x}{\partial t} + v_z \frac{\partial v_x}{\partial z} = \frac{\tau^2}{t^2}\frac{\partial \bar{v}_x}{\partial \tau}, \qquad v_x \frac{\partial v_x}{\partial x} = \frac{\tau^2}{t^2}\bar{v}_x \frac{\partial \bar{v}_x}{\partial x}, \qquad v_y \frac{\partial v_x}{\partial y} = \frac{\tau^2}{t^2}\bar{v}_y \frac{\partial \bar{v}_x}{\partial y}. \quad (13.52)$$

We also find

$$\frac{1 - v_y^2 - v_z^2}{1 - v^2} = \frac{1 - \bar{v}_y^2}{1 - \bar{v}^2}, \qquad \frac{v_x v_y}{1 - v^2} = \frac{\bar{v}_x \bar{v}_y}{1 - \bar{v}^2}, \qquad (13.53)$$

where $\bar{v} = \sqrt{\bar{v}_x^2 + \bar{v}_y^2}$. Similar formulas may be found for other expressions involving v_x, v_y and v_z. They allow us to rewrite the two transverse components of the Euler equation, Eqs. (13.36) and (13.38), in the form

$$\begin{aligned} &\frac{(1 - \bar{v}_y^2)}{(1 - \bar{v}^2)} \left(\frac{\partial \bar{v}_x}{\partial \tau} + \bar{v}_x \frac{\partial \bar{v}_x}{\partial x} + \bar{v}_y \frac{\partial \bar{v}_x}{\partial y}\right) + \frac{\bar{v}_x\, \bar{v}_y}{(1 - \bar{v}^2)} \left(\frac{\partial \bar{v}_y}{\partial \tau} + \bar{v}_x \frac{\partial \bar{v}_y}{\partial x} + \bar{v}_y \frac{\partial \bar{v}_y}{\partial y}\right) \\ &+ \bar{v}_x \frac{\partial \ln w}{\partial \tau} + \left(1 - \bar{v}_y^2\right) \frac{\partial \ln w}{\partial x} + \bar{v}_x\, \bar{v}_y \frac{\partial \ln w}{\partial y} - \frac{T}{w}(1 - \bar{v}^2)\frac{\partial s}{\partial x} = 0, \qquad (13.54) \end{aligned}$$

$$\begin{aligned} &\frac{(1 - \bar{v}_x^2)}{(1 - \bar{v}^2)} \left(\frac{\partial \bar{v}_y}{\partial \tau} + \bar{v}_x \frac{\partial \bar{v}_y}{\partial x} + \bar{v}_y \frac{\partial \bar{v}_y}{\partial y}\right) + \frac{\bar{v}_y\, \bar{v}_x}{(1 - \bar{v}^2)} \left(\frac{\partial \bar{v}_x}{\partial \tau} + \bar{v}_x \frac{\partial \bar{v}_x}{\partial x} + \bar{v}_y \frac{\partial \bar{v}_x}{\partial y}\right) \\ &+ \bar{v}_y \frac{\partial \ln w}{\partial \tau} + \left(1 - \bar{v}_x^2\right) \frac{\partial \ln w}{\partial y} + \bar{v}_x\, \bar{v}_y \frac{\partial \ln w}{\partial x} - \frac{T}{w}(1 - \bar{v}^2)\frac{\partial s}{\partial y} = 0. \qquad (13.55) \end{aligned}$$

We note that for boost-invariant systems it is sufficient to consider the space-time evolution of the system in the plane $z = 0$, where $\tau = t$ and $\bar{v}_{x,y} = v_{x,y}$. The values of the thermodynamic potentials and the flow in the regions away from this plane may be obtained by the appropriate Lorentz boosts. With these remarks in mind, it is obvious that the boost-invariant equations may be obtained directly from Eqs. (13.36) and (13.38) if one makes first the substitution $v_z = z/t$ and subsequently sets $z = 0$. Proceeding in this way we find

$$\frac{(1-v_y^2)}{(1-v^2)}\left(\frac{\partial v_x}{\partial t}+v_x\frac{\partial v_x}{\partial x}+v_y\frac{\partial v_x}{\partial y}\right)+\frac{v_x\,v_y}{(1-v^2)}\left(\frac{\partial v_y}{\partial t}+v_x\frac{\partial v_y}{\partial x}+v_y\frac{\partial v_y}{\partial y}\right)$$
$$+v_x\frac{\partial \ln w}{\partial t}+\left(1-v_y^2\right)\frac{\partial \ln w}{\partial x}+v_x\,v_y\frac{\partial \ln w}{\partial y}-\frac{T}{w}(1-v^2)\frac{\partial s}{\partial x}=0, \tag{13.56}$$

$$\frac{(1-v_x^2)}{(1-v^2)}\left(\frac{\partial v_y}{\partial t}+v_x\frac{\partial v_y}{\partial x}+v_y\frac{\partial v_y}{\partial y}\right)+\frac{v_y\,v_x}{(1-v^2)}\left(\frac{\partial v_x}{\partial t}+v_x\frac{\partial v_x}{\partial x}+v_y\frac{\partial v_x}{\partial y}\right)$$
$$+v_y\frac{\partial \ln w}{\partial t}+\left(1-v_x^2\right)\frac{\partial \ln w}{\partial y}+v_x\,v_y\frac{\partial \ln w}{\partial x}-\frac{T}{w}(1-v^2)\frac{\partial s}{\partial y}=0. \tag{13.57}$$

We observe that the structure of Eqs. (13.54)–(13.55) and Eqs. (13.56)–(13.57) is indeed equivalent. For the boost-invariant systems we find that Eq. (13.39) is automatically fulfilled. Hence, as expected, the requirement of boost invariance reduces the number of the original equations to two.

It is also interesting to note that the form of the boost-invariant equations, (13.54) and (13.55), and the structure of the cylindrically symmetric equations, (13.47) and (13.48), is identical.

13.5.1 *Combining boost-invariance with cylindrical symmetry*

For description of the midrapidity region in central collisions we may use the hydrodynamic equations which include both cylindrical symmetry and boost-invariance. This may be achieved by either implementing boost-invariance in Eqs. (13.47) and (13.48) or by imposing cylindrical symmetry in Eqs. (13.56) and (13.57). Of course, in both cases, independently of the way we proceed, we find the same single equation which plays a role of the radial/transverse Euler equation in this case

$$v_r\frac{\partial \ln w}{\partial t}+\frac{\partial \ln w}{\partial r}+\frac{1}{1-v_r^2}\left(\frac{\partial v_r}{\partial t}+v_r\frac{\partial v_r}{\partial r}\right)-\frac{T}{w}(1-v_r^2)\frac{\partial s}{\partial r}=0. \tag{13.58}$$

Neglecting the baryon density we may rewrite Eq. (13.58) as follows

$$v_r\frac{\partial \ln T}{\partial t}+\frac{\partial \ln T}{\partial r}+\frac{1}{1-v_r^2}\left(\frac{\partial v_r}{\partial t}+v_r\frac{\partial v_r}{\partial r}\right)=0. \tag{13.59}$$

We note that this form agrees with Eq. (2.22) of Ref. [7]. We also note that Eqs. (13.58) and (13.59) are identical in form with the equations describing a longitudinal one-dimensional expansion, see Eqs. (13.70) and (13.71) discussed below.

13.6 Entropy conservation

In the previous Sections we discussed different forms of the relativistic Euler equation. We know, however, that the closed system of the hydrodynamic equations consist of the Euler equation, the conservation laws, and the equation of state. Thus, in this Section we turn to the discussion of the entropy conservation law (other conservation laws have the same structure). Similarly to the discussion of the Euler equation we analyze separately the cases with non-zero and zero baryon density.

13.6.1 *Non-zero baryon density*

For the systems with non-zero baryon density the entropy conservation law is included in the simple form by the continuity equation (13.18). For cylindrically symmetric systems we have

$$u^\mu \partial_\mu s = \gamma \left(\frac{\partial s}{\partial t} + v_r \frac{\partial s}{\partial r} + v_z \frac{\partial s}{\partial z} \right) = 0. \tag{13.60}$$

On the other hand, for the boost-invariant systems the entropy depends on τ, x, and y, where $\tau = \sqrt{t^2 - z^2}$. In this case we find

$$u^\mu \partial_\mu s = \gamma \left(\frac{\partial s}{\partial \tau} \frac{t}{\tau} + v_x \frac{\partial s}{\partial x} + v_y \frac{\partial s}{\partial y} - v_z \frac{\partial s}{\partial \tau} \frac{z}{\tau} \right) = 0. \tag{13.61}$$

For the boost-invariant systems we also have $v_z = z/t$, hence we may write

$$\frac{\partial s}{\partial \tau} \frac{\tau}{t} + v_x \frac{\partial s}{\partial x} + v_y \frac{\partial s}{\partial y} = 0 \tag{13.62}$$

or

$$\frac{\partial s}{\partial \tau} + \bar{v}_x \frac{\partial s}{\partial x} + \bar{v}_y \frac{\partial s}{\partial y} = 0, \tag{13.63}$$

where the fluid transverse velocities $\bar{v}_x, \bar{v}_y$ were introduced in Sec. 13.5. Again, we observe the formal equivalence of the equations valid for the cylindrically symmetric and boost-invariant systems.

It is important to note that for the systems with non-zero baryon density we also have to include the baryon number conservation, see Eq. (13.23). Since the form of this equation is identical to the entropy conservation law used for the systems with zero baryon number we directly switch to the discussion of the equation $\partial_\mu(\sigma u^\mu) = 0$. We shall see that this equation has different form for cylindrically symmetric and boost-invariant systems.

13.6.2 *Vanishing baryon density*

In the case of vanishing baryon density the entropy conservation is given by Eq. (13.29). In the Cartesian coordinates it has the form

$$\frac{\partial \ln \sigma}{\partial t} + \mathbf{v} \cdot \nabla \ln \sigma + \frac{v}{1 - v^2} \left(\frac{\partial v}{\partial t} + \mathbf{v} \cdot \nabla v \right) + \nabla \cdot \mathbf{v} = 0. \tag{13.64}$$

In the case of the cylindrical symmetry, Eq. (13.64) may be transformed to the expression

$$\frac{\partial \ln \sigma}{\partial t} + v_r \frac{\partial \ln \sigma}{\partial r} + v_z \frac{\partial \ln \sigma}{\partial z} + \frac{v_r}{1-v^2}\left(\frac{\partial v_r}{\partial t} + v_r \frac{\partial v_r}{\partial r} + v_z \frac{\partial v_r}{\partial z}\right)$$
$$+\frac{v_z}{1-v^2}\left(\frac{\partial v_z}{\partial t} + v_r \frac{\partial v_z}{\partial r} + v_z \frac{\partial v_z}{\partial z}\right) + \frac{\partial v_r}{\partial r} + \frac{v_r}{r} + \frac{\partial v_z}{\partial z} = 0. \tag{13.65}$$

Equation (13.65) together with Eqs. (13.49) and (13.50) form three equations for four unknown variables: v_r, v_z, T and σ. To close this system we have to provide the equation of state specifying σ in terms of T. Then, a closed system of equations is formed which may be used to study the hydrodynamic behavior of cylindrically symmetric baryon-free matter (e.g., the matter formed in the central collisions in the midrapidity range).

Similarly, for the boost-invariant systems one gets

$$\frac{\partial \ln \sigma}{\partial \tau} + \bar{v}_x \frac{\partial \ln \sigma}{\partial x} + \bar{v}_y \frac{\partial \ln \sigma}{\partial y} + \frac{\bar{v}_x}{1-\bar{v}^2}\left(\frac{\partial \bar{v}_x}{\partial \tau} + \bar{v}_x \frac{\partial \bar{v}_x}{\partial x} + \bar{v}_y \frac{\partial \bar{v}_x}{\partial y}\right)$$
$$+\frac{\bar{v}_y}{1-\bar{v}^2}\left(\frac{\partial \bar{v}_y}{\partial \tau} + \bar{v}_x \frac{\partial \bar{v}_y}{\partial x} + \bar{v}_y \frac{\partial \bar{v}_y}{\partial y}\right) + \frac{\partial \bar{v}_x}{\partial x} + \frac{\partial \bar{v}_y}{\partial y} + \frac{1}{\tau} = 0, \tag{13.66}$$

where $\bar{v} = \sqrt{\bar{v}_x^2 + \bar{v}_y^2}$. Equation (13.66) together with Eqs. (13.54) and (13.55), where the formal substitutions $s =$ const. and $w \to T$ are done to switch to the baryon free case, form the system of three equations for four unknown variables: $\bar{v}_x, \bar{v}_y, T$ and σ. Similarly to the cylindrical case, in order to close the system we have to provide the equation of state. Finally, one can notice that the equations (13.65) and (13.66) have different structure. *Thus, we conclude that the basic difference between the form of cylindrically symmetric and boost-invariant system resides in the form of the conservation law $\partial_\mu(\sigma u^\mu) = 0$ or $\partial_\mu(n u^\mu) = 0$.*

Of the special interest is the boost-invariant cylindrically symmetric expansion. In this case one obtains the following equation (at $z = 0$)

$$\frac{\partial \ln \sigma}{\partial t} + v_r \frac{\partial \ln \sigma}{\partial r} + \frac{1}{1-v^2}\left(v_r \frac{\partial v_r}{\partial t} + \frac{\partial v_r}{\partial r}\right) + \frac{v_r}{r} + \frac{1}{t} = 0. \tag{13.67}$$

The last two terms in Eq. (13.67) follow from the divergence of the three-velocity of the fluid rewritten in the cylindrical coordinates. One may check that Eqs. (13.59) and (13.67) reproduce the system of equations studied by Baym et al. in Ref. [7].

Neglecting the transverse expansion in (13.67) one obtains the simple equation

$$\frac{\partial \ln \sigma}{\partial t} + \frac{1}{t} = 0, \tag{13.68}$$

which leads to the famous Bjorken scaling solution [8]

$$\sigma(t) = \sigma_{\rm i}\frac{t_{\rm i}}{t}. \tag{13.69}$$

We note that in this case Eq. (13.59) is trivially fulfilled since $v_r = 0$ and T is independent of r. The important consequences of Eq. (13.69) will be studied in Chap. 21.

13.7 One-dimensional expansion

Very often in the model studies one neglects the transverse expansion of matter and concentrates on the longitudinal expansion of the system (we must stress here, however, that the realistic description of the evolution of matter requires the incorporation of the transverse expansion which is quite substantial at RHIC). In such a simplified case Eqs. (13.36) and (13.38) are automatically fulfilled, while Eq. (13.39) is reduced to the form

$$v_z \frac{\partial \ln w}{\partial t} + \frac{\partial \ln w}{\partial z} + \frac{1}{1-{v_z}^2}\left(\frac{\partial v_z}{\partial t} + v_z \frac{\partial v_z}{\partial z}\right) - \frac{T}{w}(1-{v_z}^2)\frac{\partial s}{\partial z} = 0. \tag{13.70}$$

The analog of Eq. (13.70) for baryon-free matter is the following

$$v_z \frac{\partial \ln T}{\partial t} + \frac{\partial \ln T}{\partial z} + \frac{1}{1-{v_z}^2}\left(\frac{\partial v_z}{\partial t} + v_z \frac{\partial v_z}{\partial z}\right) = 0. \tag{13.71}$$

To form a complete set of hydrodynamic equations in the case $\mu_B = 0$, Eq. (13.71) should be considered together with the entropy conservation, see Eq. (13.65), which in the case of purely longitudinal expansion has the form

$$\frac{\partial \ln \sigma}{\partial t} + v_z \frac{\partial \ln \sigma}{\partial z} + \frac{v_z}{1-v_z^2}\left(\frac{\partial v_z}{\partial t} + v_z \frac{\partial v_z}{\partial z}\right) + \frac{\partial v_z}{\partial z} = 0 \tag{13.72}$$

For boost-invariant systems $v_z = z/t$ and the temperature T depends on the combination $t^2 - z^2$. Hence, in this case Eq. (13.71) is automatically fulfilled. This property is used in the Bjorken model [8], where the only non-trival equation is related to the entropy conservation, see Eq. (13.69). The Bjorken model is discussed by us in greater detail below in Chap. 21.

13.7.1 *Simple Riemann solution*

If the equation of state is of the form $P = c_s^2\, \varepsilon$, where the sound velocity c_s is a constant, the system of equations (13.71) and (13.72) admits a simple solution where

$$v(t,z) = \frac{c_s\, t + z}{t + c_s\, z} \tag{13.73}$$

and

$$T(t,z) = T_i \left(\frac{1 - c_s\, t - z}{1 + c_s\, t + z}\right)^{c_s/2}. \tag{13.74}$$

The validity of this solution may be checked by the direct substitution of (13.73) and (13.74) in Eqs. (13.71) and (13.72). For any given time t the range of z is

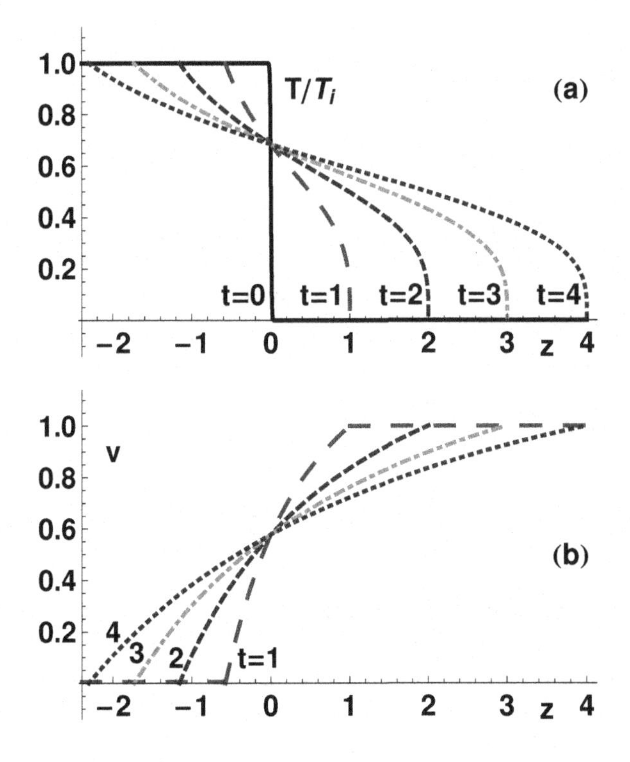

Fig. 13.3 The temperature (**a**) and velocity (**b**) profiles describing the rarefaction wave entering the region of hot matter. Initially, the matter is uniformly distributed in the region $z \leq 0$ and its temperature is $T_{\rm i}$. The rarefaction wave enters the system from right to left at the speed of sound. The value $c_{\rm s}^2 = 1/3$ has been used in the plots. The long-dashed, dashed, dotted-dashed and dotted lines describe the profiles for $t = 1$, 2, 3, and 4 fm, respectively. The solid line in the left panel describes the initial step-like profile of the temperature. On the right-hand-side, the edge of matter expands in vacuum at the speed of light. For each value of time, the Riemann solution is restricted to the interval $-c_{\rm s}\, t < z < c\, t$.

restricted to the interval $-c_{\rm s} t < z < t$. Equations (13.73) and (13.74) describe the rarefaction wave entering the region of uniformly distributed hot matter initially placed in the region $z \leq 0$ and characterized by the constant initial temperature $T_{\rm i}$. The front of the rarefaction wave travels to the left at the speed of sound $c_{\rm s}$. At the other end, the matters expands in vacuum at the speed of light, see Fig. 13.3.

Characteristic features of the simple Riemann solution appear in more sophisticated models of heavy-ion collisions. For instance, in the Landau model the Riemann waves describe the expansion of the outer/fragmentation regions. The non-trivial Khalatnikov solutions describing the behavior of the central part of matter are matched to such simple waves, see Chap. 20. Similarly, the initial transverse expansion of matter may be also described approximately by the simple waves if the initial distribution of matter is uniform (step-like) in the transverse plane. For further use, in particular for the analysis of the Landau model, it is convenient to introduce the relation between the velocity and the temperature in the simple waves. Simple manipulations with formulas (13.73) and (13.74) yield

$$c_s \vartheta = -\ln \frac{T}{T_i}, \quad v = \tanh \vartheta. \tag{13.75}$$

We note that Eq. (13.73) may be also rewritten in the form

$$z = t \frac{v - c_s}{1 + v\, c_s}. \tag{13.76}$$

13.8 Boost-invariant cylindrically asymmetric case

13.8.1 *Introduction of cylindrical coordinates*

Equation (13.66) should be analyzed together with Eqs. (13.56) and (13.57), and also with the equation of state connecting T and σ. Such a closed system of equations may be used to study cylindrically asymmetric expansion of matter formed in the midrapidity region during the most energetic heavy-ion collisions (e.g., top RHIC energies where the effects of the baryon number on the evolution of matter may be neglected). In particular, this set of equation may be used to address the problem of the formation of the elliptic flow.

Due to the physical importance of this case, we discuss it in more detail now. At first we show that the hydrodynamic equations may be cast in a quite elegant form [9] if one uses the cylindrical coordinates and uses the following parameterization of the transverse velocities, see Fig. 13.4,

$$v_x = v \cos(\phi + \alpha), \quad v_y = v \sin(\phi + \alpha). \tag{13.77}$$

Here $v = \sqrt{v_x^2 + v_y^2}$ and α is the dynamic angle describing deviations of the transverse flow from the radial direction.

Adding together Eq. (13.56) multiplied by v_x and Eq. (13.57) multiplied by v_y we obtain (if necessary, to describe the central baryon-free region we make the formal subsitutions: $s =$ const., $w \to T$, $z = 0$)

$$\frac{\partial}{\partial t}(rT\gamma v) + r\cos\alpha \frac{\partial}{\partial r}(T\gamma) + \sin\alpha \frac{\partial}{\partial \phi}(T\gamma) = 0. \tag{13.78}$$

Similarly, subtracting Eq. (13.56) multiplied by v_y from Eq. (13.57) multiplied by v_x we find

$$T\gamma^2 v \left(\frac{d\alpha}{dt} + \frac{v \sin\alpha}{r} \right) - \sin\alpha \frac{\partial T}{\partial r} + \frac{\cos\alpha}{r} \frac{\partial T}{\partial \phi} = 0, \tag{13.79}$$

where

$$\frac{d}{dt} = \frac{\partial}{\partial t} + v_x \frac{\partial}{\partial x} + v_y \frac{\partial}{\partial y} = \frac{\partial}{\partial t} + v\cos\alpha \frac{\partial}{\partial r} + \frac{v \sin\alpha}{r} \frac{\partial}{\partial \phi}. \tag{13.80}$$

Finally, the entropy conservation gives

$$\frac{\partial}{\partial t}(rt\sigma\gamma) + \frac{\partial}{\partial r}(rt\sigma\gamma v \cos\alpha) + \frac{\partial}{\partial \phi}(t\sigma\gamma v \sin\alpha) = 0. \tag{13.81}$$

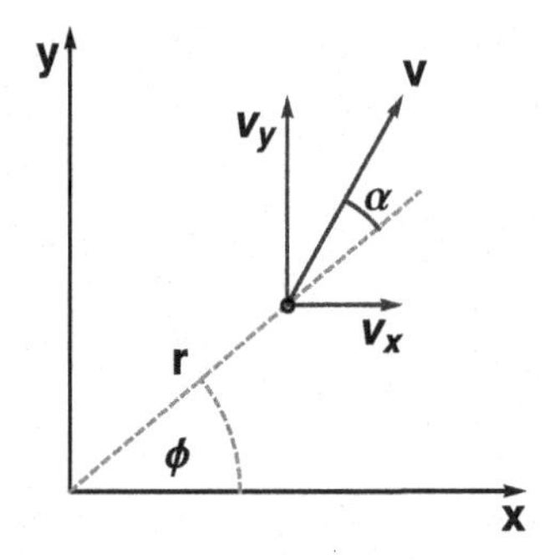

Fig. 13.4 The notation used to define the components of the fluid velocity in the transverse plane, $z = 0$. The magnitude of the flow v and the angle α may be treated as two dynamical variables equivalent to v_x and v_y.

13.8.2 *Global entropy conservation*

By integration of Eq. (13.81) over the space coordinates r and ϕ we obtain the following constraint

$$\frac{\partial}{\partial t}\left[\int_0^\infty drr \int_0^{2\pi} d\phi\, t\, \sigma(t,r,\phi)\, \gamma(t,r,\phi)\right] = 0, \tag{13.82}$$

which yields

$$\int_0^\infty drr \int_0^{2\pi} d\phi\, \sigma(t,r,\phi)\, \gamma(t,r,\phi) = \frac{\text{const.}}{t}. \tag{13.83}$$

We note that the integration of the second term in Eq. (13.81) gives zero since $\sigma \to 0$ for $r \to \infty$. On the other hand, the integration of the third term in Eq. (13.81) vanishes due to the periodic conditions of the type $\sigma(t,r,0) = \sigma(t,r,2\pi)$.

Equation (13.83) represents the generalization of the famous Bjorken scaling law for the entropy density, see Eq. (13.69). If no transverse flow is present, the Lorentz gamma factor is equal to unity and the entropy density is a function of the time only. In this case Eq. (13.83) is reduced to Eq. (13.69). In the realistic situations including the presence of transverse flow, Eq. (13.83) suggests that the decrease of the entropy is faster than in the Bjorken model — it is triggered by both the longitudinal and transverse expansion of the system. Another nice feature of Eq. (13.83) is that it may serve as a convenient check of the numerical method used to solve the hydrodynamic equations.

13.8.3 *Characteristic form*

Equations (13.78), (13.79) and (13.81) are three equations for four unknown functions: T, σ, v, and α. To close this system of equations we should add the equation of state, i.e., the functional relation connecting T and σ. In this Section, following the method introduced in Ref. [7], we show that the information about the equation of state may be conveniently encoded in the hydrodynamic equations through the use of the temperature dependent sound velocity. In the general form, this property appeared already in Sec. 13.2.2, see Eq. (13.31). In our specific situation (the boost-invariant and cylindrically asymmetric expansion) the method of Ref. [7] has further advantages: the number of the independent equations is reduced by one (at the expense of the formal extension of the radial coordinate r to the negative values) and the initial conditions at the origin of the system are automatically fulfilled.

We refer to the formulation proposed in [7] as to the characteristic form of the hydrodynamic equations. Besides the sound velocity $c_s(T)$ this form makes use of the potential Φ defined by the differential equation

$$d\Phi = \frac{d\ln T}{c_s} = c_s d\ln\sigma. \tag{13.84}$$

The integration of Eq. (13.84) allows us to express Φ in terms of the temperature or to express temperature in terms of Φ. In this way we obtain the two functions: $\Phi_T(T)$ and $T_\Phi(\Phi)$. The functions of this type turn out to be very useful in the further analysis of the hydrodynamic equations [1].

In addition to the potential Φ we introduce also the quantities ϑ and $a_\pm$ defined by the relations

$$v = \tanh\vartheta, \quad a_\pm = \exp(\Phi \pm \vartheta). \tag{13.85}$$

Multiplication of Eq. (13.81) by the expression $c_s/(rt\sigma\cosh\vartheta)$ and the use of Eq. (13.84) leads us to the formula

$$\begin{aligned} 0 = {} & \frac{\partial\Phi}{\partial t} + v\cos\alpha\frac{\partial\Phi}{\partial r} + v\frac{\sin\alpha}{r}\frac{\partial\Phi}{\partial\phi} + c_s v\frac{\partial\vartheta}{\partial t} + c_s\cos\alpha\frac{\partial\vartheta}{\partial r} + c_s\frac{\sin\alpha}{r}\frac{\partial\vartheta}{\partial\phi} \\ & + c_s v\frac{\cos\alpha}{r}\frac{\partial\alpha}{\partial\phi} - c_s v\sin\alpha\frac{\partial\alpha}{\partial r} + c_s\left[\frac{1}{t} + \frac{v\cos\alpha}{r}\right]. \end{aligned} \tag{13.86}$$

Similarly, Eq. (13.78) may be rewritten in the form

$$0 = c_s v\frac{\partial\Phi}{\partial t} + c_s\cos\alpha\frac{\partial\Phi}{\partial r} + c_s\frac{\sin\alpha}{r}\frac{\partial\Phi}{\partial\phi} + \frac{\partial\vartheta}{\partial t} + v\cos\alpha\frac{\partial\vartheta}{\partial r} + v\frac{\sin\alpha}{r}\frac{\partial\vartheta}{\partial\phi}. \tag{13.87}$$

The sum and the difference of Eqs. (13.86) and (13.87) gives two equations for the functions $a_\pm$

$$\begin{aligned} & \frac{\partial a_\pm}{\partial t} + \frac{(v\pm c_s)}{(1\pm c_s v)}\cos\alpha\frac{\partial a_\pm}{\partial r} + \frac{(v\pm c_s)}{(1\pm c_s v)}\frac{\sin\alpha}{r}\frac{\partial a_\pm}{\partial\phi} \\ & - \frac{c_s v}{(1\pm c_s v)}\left(\sin\alpha\frac{\partial\alpha}{\partial r} - \frac{\cos\alpha}{r}\frac{\partial\alpha}{\partial\phi}\right)a_\pm \\ & + \frac{c_s}{(1\pm c_s v)}\left[\frac{1}{t} + \frac{v\cos\alpha}{r}\right]a_\pm = 0, \end{aligned} \tag{13.88}$$

[1] We introduce the subscripts to make clear what kind of the argument is expected for a given function. For example, the temperature may be considered as a function of entropy or Φ. In those two cases one should use the functions T_σ or T_Φ, respectively.

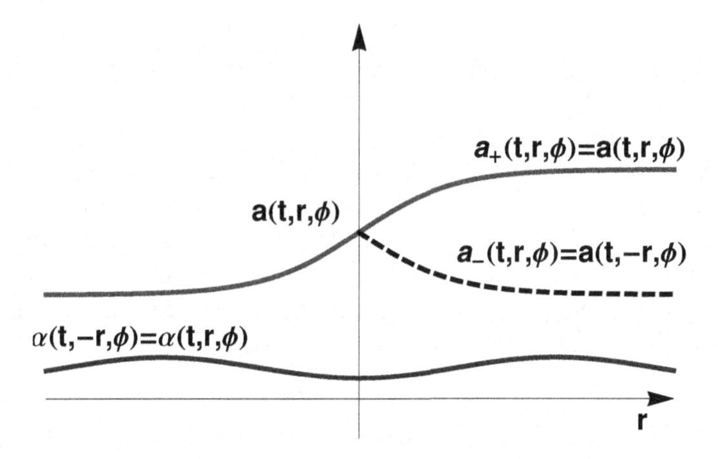

Fig. 13.5 The example illustrating the definition of the functions $a_+(t, r, \phi)$ and $a_-(t, r, \phi)$ in terms of the single function $a(t, r, \phi)$ defined in the extended region of r. The function $\alpha(t, r, \phi)$ is symmetrically extended to the negative values of r.

while Eq. (13.79) takes the form

$$\frac{\partial \alpha}{\partial t} = \frac{\left(1 - v^2\right) c_s}{v} \left(\sin\alpha \frac{\partial \Phi}{\partial r} - \frac{\cos\alpha}{r} \frac{\partial \Phi}{\partial \phi} \right) - v \left(\cos\alpha \frac{\partial \alpha}{\partial r} + \frac{\sin\alpha}{r} \frac{\partial \alpha}{\partial \phi} + \frac{\sin\alpha}{r} \right). \tag{13.89}$$

Equations (13.88) and (13.89) are three equations for three unknown functions: $a_+(t, r, \phi)$, $a_-(t, r, \phi)$, and $\alpha(t, r, \phi)$. We note, that the velocity v and the potential Φ are functions of a_+ and a_-,

$$v = \frac{a_+ - a_-}{a_+ + a_-}, \quad \Phi = \frac{1}{2} \ln(a_+ a_-). \tag{13.90}$$

Similarly, the sound velocity appearing in Eqs. (13.88) and (13.89) may be also represented as a function of a_+ and a_-,

$$c_s(T) = c_s \left\{ T_\Phi \left[\frac{1}{2} \ln(a_+ a_-) \right] \right\}. \tag{13.91}$$

Clearly, this system of equations is closed if the function T_Φ is known. It may be obtained directly from the equation of state.

13.8.4 *Boundary conditions*

For the moment we shall restrict our considerations to the cylindrically symmetric case which was studied originally in [7]. For such a symmetric situation all terms in Eq. (13.89) vanish ($\partial\Phi/\partial\phi = 0$ and $\alpha = 0$) whereas Eqs. (13.88) are reduced to Eqs. (2.24) of Ref. [7]

$$\frac{\partial a_\pm}{\partial t} + \frac{(v \pm c_s)}{(1 \pm c_s v)} \frac{\partial a_\pm}{\partial r} + \frac{c_s}{(1 \pm c_s v)} \left[\frac{1}{t} + \frac{v}{r} \right] a_\pm = 0. \tag{13.92}$$

Due to the symmetry reasons, the velocity v should vanish at $r = 0$,

$$v(t, r = 0) = 0. \tag{13.93}$$

This condition is achieved by demanding that the two functions $a_+(t,r)$ and $a_-(t,r)$ may be expressed in terms of a single function $a(t,r)$ with the help of the prescription, see Fig. 13.5,

$$\begin{aligned} a_+(t,r) &= a(t,r), \quad r > 0, \\ a_-(t,r) &= a(t,-r). \end{aligned} \tag{13.94}$$

The ansatz (13.94) and the structure of Eqs. (13.92) indicate that the equation for $a_-(t,r)$ may be obtained from the equation for $a_+(t,r)$ by the substitution: $r \to -r$. This observation further means that two equations (13.92) may be reduced to a single equation for the function $a(t,r)$ with the range of the variable r extended to negative values,

$$0 = \frac{\partial a}{\partial t} + \frac{(v + c_s)}{(1 + c_s v)} \frac{\partial a}{\partial r} + \frac{c_s}{(1 + c_s v)} \left[\frac{1}{t} + \frac{v}{r} \right] a. \tag{13.95}$$

Furthermore, the ansatz (13.94) and Eq. (13.90) automatically yield the desired boundary condition for the temperature

$$\left. \frac{\partial T(t,r)}{\partial r} \right|_{r=0} = 0. \tag{13.96}$$

The presented method of dealing with cylindrically symmetric hydrodynamic equations was introduced in Ref. [7]. The analogous method for dealing with cylindrically non-symmetric systems was given in Ref. [10]. Similarly to the cylindrically symmetric case we combine the two functions a_+ and a_- into one function a defined in the extended r-space

$$\begin{aligned} a_+(t,r,\phi) &= a(t,r,\phi), \quad r > 0, \\ a_-(t,r,\phi) &= a(t,-r,\phi). \end{aligned} \tag{13.97}$$

The inclusion of negative values of r implies that we also have to define the function $\alpha(t,r,\phi)$ in the range $r < 0$. This is done by the assumption

$$\alpha(t,-r,\phi) = \alpha(t,r,\phi), \quad r > 0. \tag{13.98}$$

Using definitions (13.97) and (13.98), Eqs. (13.88) are reduced to a single equation for the function $a(t,r,\phi)$,

$$\begin{aligned} &\frac{\partial a}{\partial t} + \frac{(v + c_s)}{(1 + c_s v)} \cos\alpha \frac{\partial a}{\partial r} + \frac{(v + c_s)}{(1 + c_s v)} \frac{\sin\alpha}{r} \frac{\partial a}{\partial \phi} \\ &- \frac{c_s v}{(1 + c_s v)} \left(\sin\alpha \frac{\partial \alpha}{\partial r} - \frac{\cos\alpha}{r} \frac{\partial \alpha}{\partial \phi} \right) a \\ &+ \frac{c_s}{(1 + c_s v)} \left[\frac{1}{t} + \frac{v \cos\alpha}{r} \right] a = 0, \end{aligned} \tag{13.99}$$

Equation (13.99) should be solved together with Eq. (13.89), where the range of r may be extended naturally to negative values. The use of the cylindrical coordinates leads to the periodic boundary conditions

$$\begin{aligned} a(t,r,0) &= a(t,r,2\pi), \\ \alpha(t,r,0) &= \alpha(t,r,2\pi). \end{aligned} \tag{13.100}$$

We also note that Eqs. (13.97) and (13.98) yield the following boundary conditions for the temperature and the function $\alpha(t,r,\phi)$,

$$\left.\frac{\partial T(t,r,\phi)}{\partial r}\right|_{r=0} = 0, \quad \left.\frac{\partial \alpha(t,r,\phi)}{\partial r}\right|_{r=0} = 0. \tag{13.101}$$

In addition, multiplying the equation for a_+ in (13.88) by a_- and subtracting the equation for a_- multiplied by a_+ one finds

$$\left.\frac{1}{r}\frac{\partial T(t,r,\phi)}{\partial \phi}\right|_{r=0} = 0. \tag{13.102}$$

Bibliography to Chapter 13

[1] L. D. Landau and E. M. Lifshitz, "Fluid Mechanics," (Butterworth-Heinemann, Oxford, 2000).

[2] C. W. Misner, K. S. Thorne, and J. A. Wheeler, "Gravitation," (W.H. Freeman, San Francisco, 1973) 1279 p.

[3] R. B. Clare and D. Strottman, "Relativistic hydrodynamics and heavy-ion reactions," *Phys. Rept.* **141** (1986) 177–280.

[4] J.-P. Blaizot and J.-Y. Ollitrault, "Hydrodynamics of quark-gluon plasmas," *Adv. Ser. Direct. High Energy Phys.* **6** (1990) 393–470.

[5] D. H. Rischke, S. Bernard, and J. A. Maruhn, "Relativistic hydrodynamics for heavy ion collisions. 1. General aspects and expansion into vacuum," *Nucl. Phys.* **A595** (1995) 346–382.

[6] D. H. Rischke, Y. Pursun, and J. A. Maruhn, "Relativistic hydrodynamics for heavy-ion collisions. 2. Compression of nuclear matter and the phase transition to the quark-gluon plasma," *Nucl. Phys.* **A595** (1995) 383–408.

[7] G. Baym, B. L. Friman, J. P. Blaizot, M. Soyeur, and W. Czyz, "Hydrodynamics of ultrarelativistic heavy-ion collisions," *Nucl. Phys.* **A407** (1983) 541–570.

[8] J. D. Bjorken, "Highly relativistic nucleus-nucleus collisions: The central rapidity region," *Phys. Rev.* **D27** (1983) 140–151.

[9] A. Dyrek and W. Florkowski, "Boost-invariant motion of relativistic perfect fluid," *Acta Phys. Polon.* **B15** (1984) 653–666.

[10] M. Chojnacki and W. Florkowski, "Characteristic form of boost-invariant and cylindrically non-symmetric hydrodynamic equations," *Phys. Rev.* **C74** (2006) 034905.

Chapter 14

General Aspects of Perfect-Fluid Dynamics

In this Chapter we discuss general properties of the relativistic flow of the perfect fluid. Our starting point are the hydrodynamic equations introduced in the previous Chapter. In particular, we discuss here the non-relativistic limit, the Bernoulli equation, and the conservation of circulation.

14.1 Non-relativistic limit

In the classical physics the number of particles in the fluid element is conserved. Hence, discussing the non-relativistic limit, we may simply identify the conserved quantity N introduced in Sec. 13.1 with the particle number. Proceeding in the same way as in Sec. 13.1 we find the following non-relativistic thermodynamic relation

$$\frac{1}{\rho}dP + T\, ds_{\rm nr} = dw_{\rm nr}, \tag{14.1}$$

where ρ is the mass density (M/V) and $s_{\rm nr}$ is the entropy per unit mass (S/M). The quantity $w_{\rm nr}$ is the non-relativistic enthalpy per unit mass $(W_{\rm nr}/M)$. The latter is connected with its relativistic counterpart by the formula

$$w = \frac{W}{N} = \frac{M + W_{\rm nr}}{N} = m + \frac{M}{N}\frac{W_{\rm nr}}{M} = m + m w_{\rm nr}, \tag{14.2}$$

where m denotes the particle mass. The mass gives contribution to w since the relativistic enthalpy, as well as the relativistic energy, include the rest energy of particles. In the transition from the relativistic to the non-relativistic hydrodynamics we extract this part. In fact we observe that m is the dominant contribution to w if the non-relativistic limit is taken. Having also in mind that the Lorentz factor γ approaches unity in this limit, the spatial components of the relativistic Euler equation (13.17)

$$\frac{d}{d\tau}(w\gamma v^i) = -\frac{1}{n}\nabla^i P, \tag{14.3}$$

may be approximated by the expression

$$\frac{dv^i}{dt} = -\frac{1}{mn}\nabla^i P = -\frac{1}{\rho}\nabla^i P, \tag{14.4}$$

which is nothing other but the Euler equation known from the classical hydrodynamics.

Similarly, the non-relativistic limit may be obtained from Eqs. (13.36)–(13.39). For example, in the non-relativistic limit Eq. (13.36) does not include the time derivatives of v_y and v_z (in the notation containing explicitly the speed of light the products of the velocities are supressed by the factor c^2). One may check that this equation is then reduced to the x-component of the non-relativistic Euler equation $dv_x/dt = -(1/\rho)(\partial P/\partial x)$.

14.2 Bernoulli equation

Let us consider the stationary flow of the relativistic perfect fluid described by Eq. (13.34). The stationarity of the flow means that the term with the time derivative vanishes. In addition, we assume that the flow satisfies the condition $s =$ const. In this case Eq. (13.34) takes the form

$$\nabla (w\gamma) = \mathbf{v} \times (\nabla \times w\gamma\mathbf{v}) \,. \tag{14.5}$$

By myltiplying Eq. (14.5) by $\mathbf{v}$ we obtain the condition

$$\mathbf{v} \cdot \nabla (w\gamma) = 0, \tag{14.6}$$

which states that the quantity $w\gamma$ is constant along the stream lines. Considering the non-relativistic limit we find

$$w\gamma \approx (m + m w_{\rm nr}) \left(1 + \frac{v^2}{2}\right) \approx m \left(1 + \frac{v^2}{2} + w_{\rm nr}\right) = \text{const.}\,, \tag{14.7}$$

which is equivalent to the well-known non-relativistic form of the Bernoulli equation [1]

$$\frac{v^2}{2} + w_{\rm nr} = \text{const.} \tag{14.8}$$

14.3 Relativistic circulation

The example of the Bernoulli equation shows that it is useful to define other physical quantities which are relativistic generalizations of the well established non-relativistic concepts. In particular, the relativistic *circulation* is defined by the expression [2]

$$\mathcal{C} = \frac{1}{m} \oint_C w \; u^\mu \; \delta x_\mu, \tag{14.9}$$

where w is the enthalpy per baryon with mass m (per particle in the non-relativistic case), and C is an arbitrary, closed integration contour in the Minkowski space. For relativistic fluids with zero baryon number the appropriate definition is

$$\mathcal{C} = \oint_C T \; u^\mu \; \delta x_\mu. \tag{14.10}$$

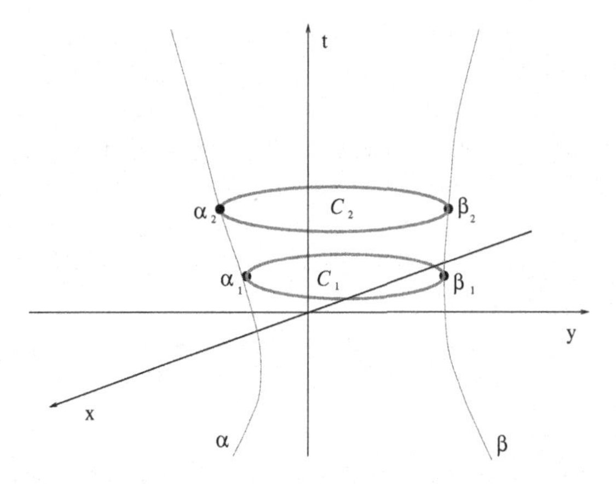

Fig. 14.1 The world lines of the fluid elements α and β. The space-time interval between the points α_1 and α_2 is the same as the interval between the points $\beta_1$and β_2, namely $(\alpha_2 - \alpha_1)^2 = (\beta_2 - \beta_1)^2 = (\delta\tau)^2$. The conservation of circulation discussed in Sec. 14.3 implies that the calculation of circulation for the curve $\mathcal{C}_1$ gives the same result as the calculation for the curve $\mathcal{C}_2$.

We can specify now the conditions under which the circulation is not changed during the hydrodynamic evolution of the fluid. The total proper-time derivative of (14.9) can be written as a sum of two terms

$$\frac{d\mathcal{C}}{d\tau} = \frac{1}{m} \oint_C \frac{d}{d\tau} (w \; u^\nu) \; \delta x_\nu + \frac{1}{m} \oint_C w \; u^\nu \; \frac{d\delta x_\nu}{d\tau}. \tag{14.11}$$

The second term in (14.11) is connected with the change of the integration contour, as determined by the flow, see Fig. 14.1. The Euler equation (13.17) gives

$$\frac{d\mathcal{C}}{d\tau} = \frac{1}{m} \oint_C \frac{\partial^\nu P}{n} \delta x_\nu + \frac{1}{m} \oint_C w \; u^\nu \; \delta u_\nu. \tag{14.12}$$

The second term on the right-hand-side of Eq. (14.12) vanishes, since it contains the term $u^\nu \delta u_\nu = 0$. The first term on the right-hand-side of Eq. (14.12) vanishes for each closed contour C if and only if

$$dP = n \, d\Phi_{\rm rel}, \tag{14.13}$$

where $\Phi_{\rm rel}$ is an arbitrary, sufficiently smooth function of the space-time coordinates. The equality of the mixed derivatives, $\partial^\alpha \partial^\beta \Phi_{\rm rel} = \partial^\beta \partial^\alpha \Phi_{\rm rel}$, leads us to the condition

$$\frac{d\mathcal{C}}{d\tau} = 0 \Leftrightarrow \partial^\alpha \left(\frac{1}{n} \partial^\beta P \right) - \partial^\beta \left(\frac{1}{n} \partial^\alpha P \right) = 0. \tag{14.14}$$

The approach presented above shows strong similarities to the case of the classical, non-relativistic hydrodynamics. In the latter case, circulation is defined by the expression

$$\mathcal{C} = \oint_C \mathbf{v} \cdot \delta \, \mathbf{l}, \tag{14.15}$$

where $\mathbf{v}$ is the three-velocity field and C is a closed integration contour in space. The non-relativistic circulation does not change in time if

$$dP = \rho d\Phi_{\mathrm{cl}}, \tag{14.16}$$

where ρ is the mass density. We note that the relativistic definition (14.9) is reduced to the non-relativistic expression (14.15) in the limit $w \to m$ and $\gamma \to 1$. This is the reason why we introduced the factor $1/m$ in Eq. (14.9). On the other hand, the relativistic definition (14.10), valid for baryon-free matter, has no classical limit.

The conditions (14.13) and (14.16) are fulfilled if the flow is characterized by a single thermodynamic parameter. In general, P, T, ε are functions of two parameters, see Eqs. (13.7) and (13.8). The constraints of the type $T(n,s) = \text{const.}$ or $n = \text{const.}$, which may be suitable in different physical situations, reduce the number of independent thermodynamic parameters to one. In such cases the circulation is conserved.

14.3.1 *Vortices in relativistic fluids*

In Sec. 13.3.1 we showed that the relativistic Euler equation for the fluid with non-zero baryon density may be written with the help of the three-vector notation as

$$\frac{\partial}{\partial t}\left(w\gamma\mathbf{v}\right) + \nabla\left(w\gamma\right) = \mathbf{v} \times \left(\nabla \times w\gamma\mathbf{v}\right) + \frac{T}{\gamma}\nabla s. \tag{14.17}$$

For the flow satisfying the condition $s = \text{const.}$ it is convenient to introduce the quantity

$$\mathbf{\Omega} = \nabla \times w\gamma\mathbf{v}. \tag{14.18}$$

Calculating of the rotation of Eq. (14.17) gives

$$\frac{\partial \mathbf{\Omega}}{\partial t} = \nabla \times \left(\mathbf{v} \times \mathbf{\Omega}\right). \tag{14.19}$$

The equation of the same form appears in the classical hydrodynamics. It describes the change in time of the rotation of the three-velocity

$$\omega = \nabla \times \mathbf{v} \tag{14.20}$$

and has the form

$$\frac{\partial \omega}{\partial t} = \nabla \times \left(\mathbf{v} \times \omega\right). \tag{14.21}$$

The form of Eqs. (14.19) and (14.21) implies the following property: if $\mathbf{\Omega}$ or ω is zero in the whole space at a given time t_0, then it remains zero for any later

Table 14.1 Comparison between classical and relativistic hydrodynamics of the perfect fluid.

classical hydrodynamics	relativistic hydrodynamics
mass (baryon-number) conservation	
$\frac{\partial \rho}{\partial t} + \nabla \cdot (\rho \mathbf{v}) = 0, \quad \rho = \frac{M}{V}$	$\frac{\partial (n\gamma)}{\partial t} + \nabla \cdot (n\gamma \mathbf{v}) = 0, \quad n = \frac{N}{V}$
adiabaticity	
$\frac{ds_{\rm nr}}{dt} = 0, \quad s_{\rm nr} = \frac{S}{M}$	$\frac{ds}{d\tau} = u^\mu \partial_\mu s = 0, \quad s = \frac{S}{N}$
Euler equation	
$\frac{d}{dt}\mathbf{v} = -\frac{1}{\rho}\nabla P$	$\frac{d}{d\tau}(w\gamma \mathbf{v}) = -\frac{1}{n}\nabla P$
circulation	
$\mathcal{C} = \oint \mathbf{v} \cdot \delta \mathbf{l}$	$\mathcal{C} = \oint w u^\mu \delta x_\mu$
vorticity	
$\omega = \nabla \times \mathbf{v}$	$\mathbf{\Omega} = \nabla \times w\gamma \mathbf{v}$

time $t > t_0$. If $\mathbf{\Omega} = 0$ in a certain region of *space*, then there are no vortices in this region and the circulation vanishes, namely

$$\mathcal{C} = \oint_{\partial A} w\gamma \mathbf{v} \cdot \delta\, \mathbf{l} = \int_A (\nabla \times w\gamma \mathbf{v})\, d\mathbf{a} = \int_A \mathbf{\Omega} \cdot d\mathbf{a} = 0. \tag{14.22}$$

Here ∂A is the boundary of an arbitrary surface A inside the discussed region. Thus, the correspondence of Eqs. (14.19) and (14.21) indicates that $\mathbf{\Omega}$ can be treated as the relativistic generalization of the classical vorticity ω. The comparison between this and other quantities used in the classical and relativistic hydrodynamics of the perfect fluid is presented in Table 14.1.

14.4 Relativistic shock waves

During the hydrodynamic evolution, discontinuities of various physical quantities (velocity, pressure, etc.) may occur. Surfaces where these quantities change suddenly are called *surfaces of discontinuity.* The conservation laws applied in the neighborhood of these surfaces lead to the special matching conditions for the tensors describing the fluid behavior. Usually such matching conditions are defined in the rest frame of the discontinuity (we discuss here only space-like discontinuities) and one chooses a Lorentz frame where the x-axis is along the normal to the surface of discontinuity. With such a choice of the reference frame, the energy-momentum conservation laws require that the energy and momentum fluxes should be continuous, hence we obtain

$$\begin{aligned}
\tilde{w}_l\, u_{0,l}\, u_{x,l} &= \tilde{w}_r\, u_{0,r}\, u_{x,r}, \\
\tilde{w}_l\, u_{x,l}^2 + P_l &= \tilde{w}_r\, u_{x,r}^2 + P_r, \\
\tilde{w}_l\, u_{y,l}\, u_{x,l} &= \tilde{w}_r\, u_{y,r}\, u_{x,r}, \\
\tilde{w}_l\, u_{z,l}\, u_{x,l} &= \tilde{w}_r\, u_{z,r}\, u_{x,r}.
\end{aligned} \tag{14.23}$$

Here the indices l and r refer to the left and right sides of the discontinuity surface, respectively, and $\tilde{w} = \varepsilon + P$ is the enthalpy volume density. Similarly, the baryon number conservation law yields

$$n_l u_{x,l} = n_r u_{x,r}. \tag{14.24}$$

Generally, one distinguishes between two types of discontinuities. In the first case $u_{x,r} = 0$ and consequently $u_{x,l} = u_{x,r} = 0$ and $P_l = P_r$. In this type of discontinuities there is no energy flux across the surface. However, the tangential components of the velocity and the thermodynamic variables other than pressure may be discontinuous. Because of this behavior we call this type of the discontinuity a *tangential discontinuity* or a *contact discontinuity.* In the second case $u_{x,r} \neq 0$. This type of the discontinuity is called a *shock wave.* The relativistic theory of shock waves was initialized by Taub in 1948 [3] and then was developed by many authors, for example, see Refs. [4–8].

Assuming $u_{x,r} \neq 0$ and dividing the last two equations in Eq. (14.23) by the first equation we find that the tangential components of the velocity are continuous. Therefore, we may restrict our considerations to the case $u_y = u_z = 0$. The second equation in (14.23) and Eq. (14.24) lead to the formula

$$j \equiv n_l u_{x,l} = n_r u_{x,r} = \pm \left(\frac{P_r - P_l}{\tilde{V}_l - \tilde{V}_r} \right)^{1/2} \tag{14.25}$$

where we have defined the specific volume

$$\tilde{V} = \frac{w}{n} = \frac{\tilde{w}}{n^2}. \tag{14.26}$$

In the next step we calculate the square of the first equation in (14.23) and eliminate the velocities using Eq. (14.25). In this way we obtain the formula

$$w_l^2 - w_r^2 = (P_l - P_r) \left(\tilde{V}_l - \tilde{V}_r \right). \tag{14.27}$$

This relation is called the *shock adiabat.* It was first derived by Taub [3] and represents a relativistic generalization of the Rankine-Hugoniot relation [1].

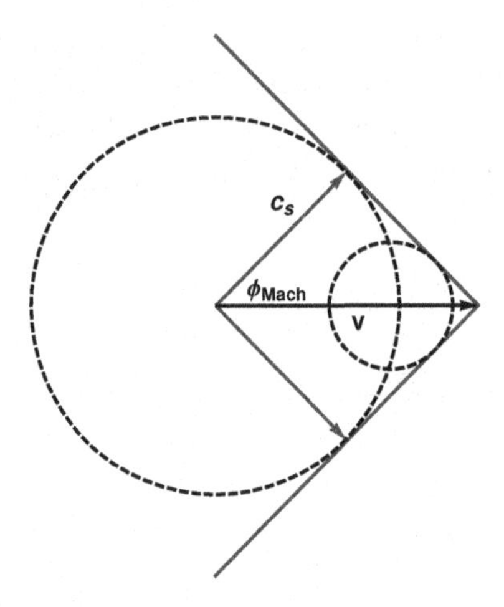

Fig. 14.2 Creation of the Mach cone by a source moving with the velocity v.

The shock adiabat defines a curve representing thermodynamic states $(\tilde{V}_r, P_r)$ in the (V, P) space which are related to the state $(\tilde{V}_l, P_l)$. For a given value of $u_{x,l}$, the solutions of Eqs. (14.23) and (14.24) are given as the intersections of the shock adiabat with the straight line defined by Eq. (14.25).

14.5 Mach cones

As we mentioned in Sec. 1.4, one of the most striking RHIC results is the observation of the jet quenching, i.e., the suppression of highly energetic particles. This phenomenon suggests that the medium created at RHIC has very high density. On the other hand, the large values of the elliptic flow suggest that the soft degrees of freedom are thermalized very fast and the created system behaves like a perfect fluid, see Sec. 2.5.2. These two observations taken together indicate that the energy deposited by the jet into the system is also thermalized, absorbed by the fluid, and its further behavior may be described as the propagation of the sound waves.

From the elementary physics we know that when a probe moves through a medium with the velocity larger than the speed of sound, the energy deposited by the probe forms a forward moving conical shock wave. The simple reason for this behavior is the interference of the spherical sound waves created at the subsequent points, where the energy is deposited to the medium by the moving particle. Simple geometric arguments, see Fig. 14.2, give the formula connecting the cone angle, $\phi_{\rm Mach}$, the sound velocity, $c_{\rm s}$, and the velocity of the jet, $v \approx c$,

$$\cos \phi_{\rm Mach} = \frac{c_{\rm s}}{v}. \tag{14.28}$$

The experimental observation of the Mach cone would be an important confirmation that the system produced at RHIC behaves like a fluid. Moreover, the measurement of the cone angle gives us the information about the sound velocity and, indirectly, about the equation of state of the produced matter.

The appearance of the Mach cones in low energy nuclear collisions was suggested already in 1970s [9–11], however, they have not been observed. In agreement with the expectations listed at the beginning of this Section, the RHIC physics brought the evidence that the Mach cones may have been indeed formed [12]. The evidence comes from the observation of a double peaked structure in two-particle correlations on the away-side from the trigerred high-energy particle [13, 14]. Nevertheless, the more recent studies generated some skepticism, attributing the double peaked structure to the method of subtraction of the background two-particle correlation (ZYAM method). At the moment, the problem of creation of Mach cones in heavy-ion collisions attracts much attention, for example, see Refs. [15–17].

14.6 Convective stability

During a hydrodynamic expansion, the particles in an accelerated fluid element experience an inertial force in the direction opposite to the acceleration. In this case, a convective instability, analogous to that in an atmosphere, is possible [18,19]. To derive the condition for convective stability we may first consider the case of the static fluid placed in a uniform gravitational field. Suppose that the fluid element is displaced adiabatically upwards (from height z to height $z+\Delta z$, with fixed entropy per baryon, s = const., see Fig. 14.3). The convective stability requires that the fluid element is heavier than the fluid it replaces, hence

$$g\,\tilde{w}(P',s) > g\,\tilde{w}(P',s'). \tag{14.29}$$

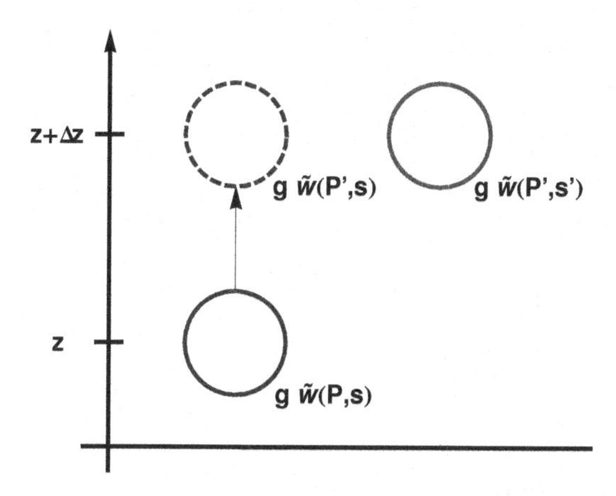

Fig. 14.3 Virtual displacement of the fluid element from height z to height $z+\Delta z$. If the condition (14.29) is satisfied, the fluid is stable against convection.

Here we use the fact that in a relativistic fluid the gravity g couples to the enthalpy density $\tilde{w} = \varepsilon + P$. We also use the notation $s' = s(z + \Delta z)$ and $P' = P(z + \Delta z)$. For small Δz the stability conditions is

$$g\left(\frac{\partial \tilde{w}}{\partial s}\right)_P \frac{ds}{dz} < 0. \tag{14.30}$$

Using the thermodynamic relation

$$\left(\frac{\partial \tilde{w}}{\partial s}\right)_P = \frac{T}{c_P}\left(\frac{\partial \varepsilon}{\partial T}\right)_P \tag{14.31}$$

where c_P is the specific heat per net baryon at constant pressure, and the fact that the specific heat is positive we obtain

$$g\left(\frac{\partial \varepsilon}{\partial T}\right)_P \frac{ds}{dz} < 0. \tag{14.32}$$

For non-relativistic matter the energy density is replaced by the mass density ρ and $(\partial\rho/\partial T)_P < 0$. Thus, one obtains the standard condition for an atmosphere to be stable, $ds/dz > 0$ (the entropy per particle increases with increasing altitude). To generalize Eq. (14.32) to the case of an expanding fluid without gravity we replace, according to the equivalence principle, the gravitational acceleration, $-g\hat{z}$, by minus the acceleration of the fluid element in its local rest frame (denoted below by primes)

$$\left(\frac{\partial \varepsilon}{\partial T}\right)_P \frac{ds}{dz'}\frac{dv'}{dt'} < 0. \tag{14.33}$$

This formula may be rewritten in the explicitly Lorentz covariant way as

$$\left(\frac{\partial \varepsilon}{\partial T}\right)_P a^\mu \partial_\mu s < 0, \tag{14.34}$$

where a^μ is the four-acceleration, $a^\mu = u^\nu \partial_\nu u^\mu$.

We note that the derivative $(\partial\varepsilon/\partial T)_P$ vanishes in the case where the fluid has zero net baryon number (or more generally does not have a composition degree of freedom). It also vanishes in the case where the energy density is proportional to the pressure. A gas of non-interacting massless particles is in neutral equilibrium with respect to convention.

Bibliography to Chapter 14

[1] L. D. Landau and E. M. Lifshitz, "Fluid Mechanics," (Butterworth-Heinemann, Oxford, 2000).

[2] A. H. Taub, "On circulation in relativistic hydrodynamics," *Arch. Rational Mech. Anal.* **3** (1959) 312–324.
[3] A. H. Taub, "Relativistic Rankine-Hugoniot equations," *Phys. Rev.* **74** (1948) 328–334.
[4] W. Scheid, H. Muller, and W. Greiner, "Nuclear shock waves in heavy-ion collisions," *Phys. Rev. Lett.* **32** (1974) 741–745.
[5] G. F. Bertsch, "Comment on nuclear shock waves in heavy-ion collisions," *Phys. Rev. Lett.* **34** (1975) 697–698.
[6] P. J. Steinhardt, "Relativistic detonation waves and bubble growth in false vacuum decay," *Phys. Rev.* **D25** (1982) 2074.
[7] P. Danielewicz and P. V. Ruuskanen, "Shock phenomena in baryonless strongly interacting matter," *Phys. Rev.* **D35** (1987) 344.
[8] J. P. Blaizot and J.-Y. Ollitrault, "The structure of hydrodynamic flows in expanding quark-gluon plasmas," *Phys. Rev.* **D36** (1987) 916.
[9] H. G. Baumgardt *et al.*, "Shock waves and Mach cones in fast nucleus-nucleus collisions," *Z. Phys.* **A273** (1975) 359–371.
[10] J. Hofmann, H. Stoecker, U. W. Heinz, W. Scheid, and W. Greiner, "Possibility of detecting density isomers in high density nuclear Mach shock waves," *Phys. Rev. Lett.* **36** (1976) 88–91.
[11] H. Stoecker, J. F. Hofmann, W. Scheid, and W. Greiner, "High density nuclear Mach shock waves in central high- energy heavy-ion collisions," *Fizika* **9** (1977) 671–706.
[12] J. Casalderrey-Solana, E. V. Shuryak, and D. Teaney, "Conical flow induced by quenched QCD jets," *J. Phys. Conf. Ser.* **27** (2005) 22–31.
[13] **STAR** Collaboration, J. Adams *et al.*, "Distributions of charged hadrons associated with high transverse momentum particles in p p and Au + Au collisions at $\sqrt{s_{\rm NN}} = 200$ GeV," *Phys. Rev. Lett.* **95** (2005) 152301.
[14] **PHENIX** Collaboration, S. S. Adler *et al.*, "Modifications to di-jet hadron pair correlations in Au + Au collisions at $\sqrt{s_{\rm NN}} = 200$ GeV," *Phys. Rev. Lett.* **97** (2006) 052301.
[15] T. Renk and J. Ruppert, "Mach cones in an evolving medium," *Phys. Rev.* **C73** (2006) 011901.
[16] L. M. Satarov, H. Stoecker, and I. N. Mishustin, "Mach shocks induced by partonic jets in expanding quark- gluon plasma," *Phys. Lett.* **B627** (2005) 64–70.
[17] G. Torrieri, B. Betz, J. Noronha, and M. Gyulassy, "Mach cones in heavy ion collisions," *Acta Phys. Polon.* **B39** (2008) 3281–3308.
[18] B. L. Friman, G. Baym, and P. V. Ruuskanen, "Stability of hydrodynamic flow in ultra-relativistic nucleus nucleus collisions," Proceedings of the Int. Workshop on Gross Properties of Nuclei and Nuclear Excitations, Hirschegg, 1987, 53-60.
[19] W. Florkowski, B. L. Friman, G. Baym, and P. V. Ruuskanen, "Convective stability of hot matter in ultrarelativistic heavy ion collisions," *Nucl. Phys.* **A540** (1992) 659–674.

Chapter 15

Initial Conditions and Freeze-Out

Similarly to other partial differential equations describing physical systems, the hydrodynamic equations should be supplied with the appropriate initial and boundary conditions. The boundary conditions require that the energy density and other thermodynamic variables drop to zero at large distances from the collision center (from the collision axis, if the boost-invariant systems are considered). On the other hand, there is no obvious form of the initial conditions. The latter should be obtained from the kinetic calculations, such as the parton cascades, but the microscopic processes leading to local equilibrium are still poorly understood. In this situation one uses simple parameterizations of the initial entropy or energy density profiles to start the hydrodynamic evolution. In Sec. 15.1 we briefly review popular choices applied in the ultra-relativistic heavy-ion physics.

The uncertainty of the initial conditions has its counterpart in the uncertainty of the final conditions for the hydrodynamic evolution. By the final conditions we understand the instructions when and where we should stop the hydrodynamic expansion. This problem is closely related to the issue of modeling the freeze-out stage. The most common prescription for switching from the hydrodynamic regime to the hadron free-streaming stage is presented in Secs. 15.2 and 15.3. The final Section of this Chapter introduces a very useful concept of the *emission function*. It is used frequently in the studies of the correlation functions.

15.1 Initial conditions

15.1.1 *Boost-invariant systems*

For the boost-invariant systems with vanishing baryon chemical potential one usually assumes that either the initial entropy density, $\sigma_{\rm i}(\mathbf{x}_\perp) = \sigma(\tau_{\rm i}, \mathbf{x}_\perp)$, or the initial energy density, $\varepsilon_{\rm i}(\mathbf{x}_\perp) = \varepsilon(\tau_{\rm i}, \mathbf{x}_\perp)$, are directly related to the density of *sources of particle production*, $\rho_{\rm sr}(\mathbf{x}_\perp)$. The sources considered in this context are *wounded nucleons* or *binary collisions*. The symmetry with respect to the Lorentz boosts along the collision axis means that it is sufficient to consider all these quantities in the plane $z = 0$. In general, a mixed model is used, see Eq. (3.59), with a lin-

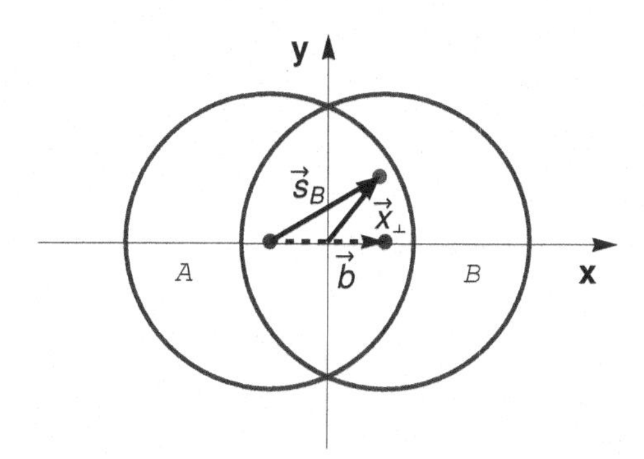

Fig. 15.1 The typical arrangement of the coordinate system in the transverse plane. The impact vector, denoted by the dashed arrow, lies in the reaction plane along the x-axis, $\mathbf{b} = (b, 0)$. The center of the nucleus B has the coordinates $(b/2, 0)$, while the center of the nucleus A is located at $(-b/2, 0)$. The position of the wounded nucleon is given by the two-dimensional vector $\mathbf{x}_\perp = (x, y)$.

ear combination of the wounded-nucleon density $\overline{w}(\mathbf{x}_\perp)$ and the density of binary collisions $\overline{n}(\mathbf{x}_\perp)$. This leads to the two popular choices:

$$\sigma_{\rm i}(\mathbf{x}_\perp) \propto \rho_{\rm sr}(\mathbf{x}_\perp) = \frac{1-\kappa}{2}\,\overline{w}(\mathbf{x}_\perp) + \kappa\,\overline{n}(\mathbf{x}_\perp) \tag{15.1}$$

or

$$\varepsilon_{\rm i}(\mathbf{x}_\perp) \propto \rho_{\rm sr}(\mathbf{x}_\perp) = \frac{1-\kappa}{2}\,\overline{w}(\mathbf{x}_\perp) + \kappa\,\overline{n}(\mathbf{x}_\perp)\,. \tag{15.2}$$

The wounded-nucleon and binary-collision densities in Eqs. (15.1) and (15.2) are obtained from the Glauber model discussed thoroughly in Chap. 3. In order to obtain the average density of the wounded nucleons at the transverse position $\mathbf{x}_\perp$ we use Eq. (3.52). One has to be careful, however, to correctly interpret the geometric meaning of the vector $\mathbf{s}_B$ appearing in Eq. (3.52). Initially, in the calculations presented in Sec. 3.5 the vectors $\mathbf{s}_B$ denoted the positions of the nucleons in the nucleus B with reference to the center of that nucleus, see Fig. 3.2. However, in the second line of Eq. (3.49) a change of the integration variables is made which effectively means that in the next equations $\mathbf{s}_B$ denotes the positions of the nucleons in the nucleus B with reference to the center of the nucleus A, see Fig. 15.1. Thus, we may write

$$\mathbf{s}_B = \frac{\mathbf{b}}{2} + \mathbf{x}_\perp \tag{15.3}$$

and find that the average density of the wounded nucleons in the nucleus B at the transverse position $\mathbf{x}_\perp$ is

$$\overline{w}_B(\mathbf{x}_\perp) = B\,T_B\left(-\frac{\mathbf{b}}{2} + \mathbf{x}_\perp\right)\left\{1 - \left[1 - \sigma_{\rm in}\,T_A\left(\frac{\mathbf{b}}{2} + \mathbf{x}_\perp\right)\right]^A\right\}. \tag{15.4}$$

The calculation of the density of the wounded nucleons in the nucleus A is analogous and yields

$$\overline{w}_A(\mathbf{x}_\perp) = A\,T_A\left(\frac{\mathbf{b}}{2}+\mathbf{x}_\perp\right)\left\{1-\left[1-\sigma_{\rm in}\,T_B\left(-\frac{\mathbf{b}}{2}+\mathbf{x}_\perp\right)\right]^B\right\}. \tag{15.5}$$

Therefore, for the collision of two nuclei, $A+B$, one may use the final expression in the form

$$\overline{w}(\mathbf{x}_\perp) = \overline{w}_A(\mathbf{x}_\perp) + \overline{w}_B(\mathbf{x}_\perp). \tag{15.6}$$

In the case of the binary collisions, similar geometrical considerations lead to the formula

$$\overline{n}(\mathbf{x}_\perp) = \sigma_{\rm in}\,A\,B\,T_A\left(\frac{\mathbf{b}}{2}+\mathbf{x}_\perp\right)T_B\left(-\frac{\mathbf{b}}{2}+\mathbf{x}_\perp\right). \tag{15.7}$$

We recall that $\sigma_{\rm in}$ in Eqs. (15.4), (15.5), and (15.7) is the nucleon-nucleon inelastic cross section.

In the hydrodynamic codes with zero baryon chemical potential, the initial conditions may be specified for an arbitrary thermodynamic variable, not necessarily for the entropy or energy density. The knowledge of the equation of state allows for easy changes of those variables. For example, one may use the temperature as the independent variable and assume the initial conditions in the form

$$T(\tau_{\rm i},\mathbf{x}_\perp) = T_\sigma\left[\sigma_{\rm i}\frac{\rho_{\rm sr}(\mathbf{x}_\perp)}{\rho_{\rm sr}(0)}\right]. \tag{15.8}$$

Here $T_\sigma(\sigma)$ is the inverse function to the function $\sigma(T)$, and $\sigma_{\rm i}$ is the initial entropy at the center of the system. The initial central temperature $T_{\rm i}$ equals $T_\sigma(\sigma_{\rm i})$. Besides the initial profile of one thermodynamic variable, one has to specify also the initial transverse flow profile. In most cases, however, it is set equal to zero.

15.1.2 *Non-boost-invariant systems*

If the physical system is not boost-invariant, one prefers to use the coordinates τ and $\eta_\parallel$ to parameterize its spacetime evolution rather than to stay with the Cartesian coordinates t and z. There are two main reasons for this choice: Firstly, if one uses the Cartesian coordinates and the initial conditions are set at the given laboratory time $t=t_{\rm i}$, the thermalization process seems to be independent of rapidity. This is in contrast with the natural expectations that the thermalization of fast particles should be delayed by the time dilation effects. Secondly, the use of the coordinates τ and $\eta_\parallel$ is convenient from the numerical point of view [1].

Thus, the initial conditions for non-boost-invariant systems should specify the entropy/energy density and the longitudinal fluid rapidity as functions of x, y and $\eta_\parallel$ for a fixed proper time $\tau=\tau_{\rm i}$. The natural generalization of Eqs. (15.1) and (15.2) are the formulas

$$\sigma_{\rm i}(\mathbf{x}_\perp,\eta_\parallel) \propto H\left(\eta_\parallel-\eta_\parallel^{\rm sh}(x,y)\right)\left[\frac{1-\kappa}{2}\,\overline{w}(\mathbf{x}_\perp)+\kappa\,\overline{n}(\mathbf{x}_\perp)\right] \tag{15.9}$$

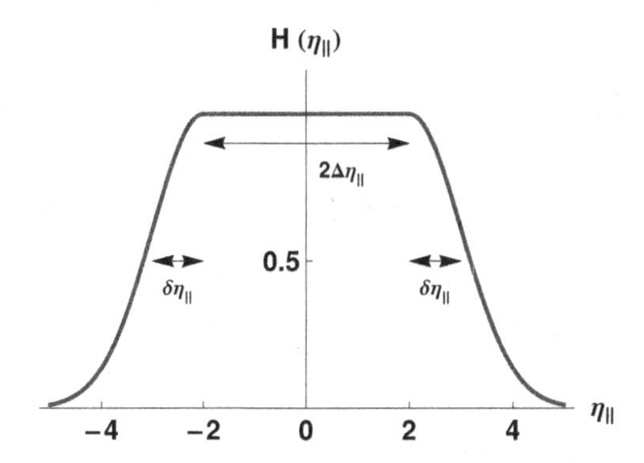

Fig. 15.2 The function $H(\eta_\parallel)$ defined by Eq. (15.11) with $\Delta\eta_\parallel = 2$ and $\delta\eta_\parallel = 1$.

and

$$\varepsilon_{\rm i}(\mathbf{x}_\perp, \eta_\parallel) \propto H\left(\eta_\parallel - \eta_\parallel^{\rm sh}(x,y)\right)\left[\frac{1-\kappa}{2}\,\overline{w}\,(\mathbf{x}_\perp) + \kappa\,\overline{n}\,(\mathbf{x}_\perp)\right]. \tag{15.10}$$

The function H defines the longitudinal profile of the entropy/energy density. It may be taken in the form [2]

$$H(\eta_\parallel) = \exp\left[-\frac{(|\eta_\parallel| - \Delta\eta_\parallel)^2}{2(\delta\eta_\parallel)^2}\,\theta(|\eta_\parallel| - \Delta\eta_\parallel)\right], \tag{15.11}$$

where θ is the step function, the parameter $\Delta\eta_\parallel$ defines the half width of the initial plateau in spacetime rapidity, and $\delta\eta_\parallel$ defines the half width of the Gaussian tails on both sides of the plateau, see Fig. 15.2. By the appropriate changes of the parameters $\Delta\eta_\parallel$ and $\delta\eta_\parallel$ we may vary between the boost-invariant-like and Gaussian-like initial conditions.

The most popular choice for the initial fluid rapidity is that it is equal to spacetime rapidity, namely

$$\vartheta_\parallel(\tau_{\rm i}, x, y, \eta_\parallel) = \eta_\parallel. \tag{15.12}$$

Correspondingly, the transverse fluid rapidity is set equal to zero,

$$\vartheta_\perp(\tau_{\rm i}, x, y, \eta_\parallel) = 0. \tag{15.13}$$

We note that during the hydrodynamic evolution, due to the action of the longitudinal and transverse pressures, the flow rapidities increase. Thus we have $\vartheta_\parallel > \eta_\parallel$ and $\vartheta_\perp > 0$ for $\tau > \tau_{\rm i}$.

The quantity $\eta_\parallel^{\rm sh}$ in Eq. (15.9) describes a shift in the spacetime rapidity of matter due to the difference of the momenta of the nucleons in the projectile and target. It is given by the expression [2,3]

$$\eta_\parallel^{\rm sh} = \frac{1}{2}\ln\left(\frac{\overline{w}_A + \overline{w}_B + v_{\rm proj}(\overline{w}_A - \overline{w}_B)}{\overline{w}_A + \overline{w}_B - v_{\rm proj}(\overline{w}_A - \overline{w}_B)}\right), \tag{15.14}$$

where $v_{\rm proj}$ is the velocity of the projectile in the center-of-mass frame.

15.2 Freeze-out

As the hadronic fireball formed in ultra-relativistic heavy-ion collisions expands, its density decreases and the mean free path of particles increases. Eventually, this process leads to the decoupling of hadrons which become non-interacting objects traveling freely to detectors. Since after this stage the momenta of particles do not change anymore [1], we refer to the decoupling mechanism as to the thermal freeze-out process (the momenta of particles become frozen). Besides the decrease of density also the growing rate of the collective expansion favors the process of decoupling. If the expansion rate is much larger than the scattering rate, the freeze-out may happen even at relatively large densities. Generally speaking, the process of decoupling is a complicated non-equilibrium process which should be studied with the help of the kinetic equations. In particular, different processes and/or different types of particles may decouple at different times, thus one introduces very often the hierarchy of different freeze-outs — see our discussion in Secs. 4.4 and 4.5 where a distinction between the chemical and thermal freeze-out was made.

In many practical applications simple criteria for decoupling/freeze-out are applied. The most popular criterion for the *thermal freeze-out* uses the concept of the fixed final temperature $T_{\rm f}$. When the local temperature drops to $T_{\rm f}$ all kinds of the processes in the fluid element stop. This condition may be very easily adopted in the hydrodynamic description of the evolution of matter formed in heavy-ion collisions. In fact, this is the reason why we discuss the freeze-out in the Chapter devoted to the relativistic hydrodynamics. The condition $T = T_{\rm f}$ defines a three-dimensional hypersurface in the Minkowski space. From the hydrodynamic calculation we know the values of the other thermodynamic parameters and flow on this hypersurface. This information allows us to calculate many interesting observables and compare them with the existing data. The appropriate formalism is based on the Cooper-Frye formula discussed below in Sec. 15.3.

The more realistic treatment of the freeze-out may be achieved if the hydrodynamic description is combined with the transport model. In this case the hydrodynamical fields are mapped to hadron distributions and typically the UrQMD simulations are performed to describe freeze-out in the dynamic way [4–8]. Interestingly, this mapping is also done with the help of the Cooper-Frye formula, which in this case defines a "switching hypersurface" between the hydrodynamic regime and the hadron kinetic regime.

If the freeze-out process is very fast, the emitted particles carry information about their earlier thermal distributions. Thus the observation of hadrons may bring us information about the thermodynamic conditions of matter in the late stages just before the thermal freeze-out. This kind of phenomenon is similar to the decoupling of the microwave radiation in the Early Universe.

---

[1] In more detailed analysis one should include the weak decays of certain particles.

15.3 Cooper-Frye formula

The number of particles that decouple in a small volume dV and have an equilibrium distribution $f_{\rm eq}(E_p)$ at freeze-out is given by the standard expression

$$dN = dV \int d^3p f_{\rm eq}(E_p). \qquad (15.15)$$

The problem exists, however, that this equation is valid in the rest frame of the fluid element. We have to generalize it in such a way that we may add distributions from different fluid elements that move independently. The first step in this direction is to write the previous formula in the covariant way [9]

$$dN = dV_\mu \int p^\mu \frac{d^3p}{E_p} f_{\rm eq}(p \cdot u). \qquad (15.16)$$

Here u^μ is the four-velocity of the fluid element and $dV^\mu = dV u^\mu$. Clearly, in the rest frame of the fluid element $u^\mu = (1,0,0,0)$ and Eq. (15.16) is reduced to Eq. (15.15). The total number of particles, emitted from all fluid elements, is obtained by integration over the volume elements dV^μ. Since dV^μ and u^μ depend on the spacetime position x, we obtain

$$N = \int dV_\mu(x) \int p^\mu \frac{d^3p}{E_p} f_{\rm eq}\left(p \cdot u(x)\right). \qquad (15.17)$$

In order to find the final answer to our problem of determining the distribution of particles emitted at the thermal freeze-out one important modification should be still done. It turns out that the freeze-out volume elements cannot be generally written in the form $dV^\mu = dV u^\mu$. For example, at the early stages of the collisions many particles are emitted from the edge of the system. This emission has a surface character and the appropriate "fluid element" is obtained by the multiplication of the area of the emission region by the time this emission takes. In this case dV^μ is space-like. Thus, the final formula that we seek has the structure of Eq. (15.17) but the fluid element $dV^\mu(x)$ should be taken as independent of $u^\mu(x)$. The form of $dV^\mu(x)$ should follow from the model/theory used to describe the spacetime evolution of matter.

With these remarks in mind we write the expression for the number of particles N which decouple on the freeze-out hypersurface Σ in the form

$$N = \int \frac{d^3p}{E_p} \int d\Sigma_\mu(x) p^\mu f(x,p). \qquad (15.18)$$

This is the famous Cooper-Frye formula [10] used in the hydrodynamic codes. The three-dimensional element of the freeze-out hypersurface, $d\Sigma^\mu$, replaced here the fluid element dV moving with the four-velocity u^μ. The quantity $d\Sigma^\mu$ may be calculated with the help of the formula known from the differential geometry [11]

$$d\Sigma_\mu = \varepsilon_{\mu\alpha\beta\gamma} \frac{dx^\alpha}{d\alpha} \frac{dx^\beta}{d\beta} \frac{dx^\gamma}{d\gamma} d\alpha d\beta d\gamma, \qquad (15.19)$$

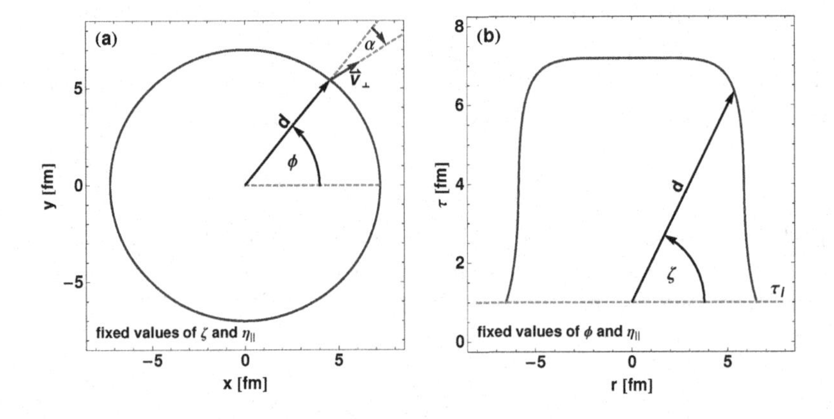

Fig. 15.3 Parameterization of the freeze-out hypersurface. The part **(a)** shows the view in the $x-y$ plane with the fixed values of ζ and $\eta_\parallel$. The part **(b)** represents the view in the $\tau - r$ plane with the fixed values of ϕ and $\eta_\parallel$.

where $\varepsilon_{\mu\alpha\beta\gamma}$ is the Levi-Civita tensor and the parameters α, β, and γ are used to parameterize the three-dimensional freeze-out manifold in the four-dimensional Minkowski space. Similarly to the three-dimensional case, $d\Sigma^\mu$ may be interpreted as the four-vector that is perpendicular to the hypersurface and its norm is equal to the "volume element" of the hypersurface.

Equation (15.19) defines the components of the four-vector $d\Sigma^\mu$ in terms of three-dimensional determinants (we adopt the convention $\varepsilon_{0123} = +1$):

$$\begin{aligned}
d\Sigma^0 &= d\Sigma_0 = \varepsilon_{0ijk}\frac{dx^i}{d\alpha}\frac{dx^j}{d\beta}\frac{dx^k}{d\gamma}d\alpha d\beta d\gamma = \frac{\partial(x^1,x^2,x^3)}{\partial(\alpha,\beta,\gamma)}d\alpha d\beta d\gamma,\\
d\Sigma^1 &= -d\Sigma_1 = -\varepsilon_{1\alpha\beta\gamma}\frac{dx^\alpha}{d\alpha}\frac{dx^\beta}{d\beta}\frac{dx^\gamma}{d\gamma}d\alpha d\beta d\gamma = \frac{\partial(x^2,x^3,x^0)}{\partial(\alpha,\beta,\gamma)}d\alpha d\beta d\gamma,\\
d\Sigma^2 &= -d\Sigma_2 = -\varepsilon_{2\alpha\beta\gamma}\frac{dx^\alpha}{d\alpha}\frac{dx^\beta}{d\beta}\frac{dx^\gamma}{d\gamma}d\alpha d\beta d\gamma = -\frac{\partial(x^3,x^0,x^1)}{\partial(\alpha,\beta,\gamma)}d\alpha d\beta d\gamma,\\
d\Sigma^3 &= -d\Sigma_3 = -\varepsilon_{3\alpha\beta\gamma}\frac{dx^\alpha}{d\alpha}\frac{dx^\beta}{d\beta}\frac{dx^\gamma}{d\gamma}d\alpha d\beta d\gamma = \frac{\partial(x^0,x^1,x^2)}{\partial(\alpha,\beta,\gamma)}d\alpha d\beta d\gamma. \quad (15.20)
\end{aligned}$$

Here we introduced the short-hand notation for the determinants,

$$\frac{\partial(x,y,z)}{\partial(\alpha,\beta,\gamma)} = \begin{vmatrix} \frac{\partial x}{\partial\alpha} & \frac{\partial x}{\partial\beta} & \frac{\partial x}{\partial\gamma} \\ \frac{\partial y}{\partial\alpha} & \frac{\partial y}{\partial\beta} & \frac{\partial y}{\partial\gamma} \\ \frac{\partial z}{\partial\alpha} & \frac{\partial z}{\partial\beta} & \frac{\partial z}{\partial\gamma} \end{vmatrix}. \quad (15.21)$$

Most of the three-dimensional freeze-out hypersurfaces that come out from the hydrodynamic calculations may be parameterized in the following way

$$\begin{aligned}
t &= \left[\tau_{\rm i} + d\left(\phi,\zeta,\eta_\parallel\right)\sin\zeta\right]\,\cosh\eta_\parallel, \\
z &= \left[\tau_{\rm i} + d\left(\phi,\zeta,\eta_\parallel\right)\sin\zeta\right]\,\sinh\eta_\parallel, \\
x &= d\left(\phi,\zeta,\eta_\parallel\right)\cos\zeta\,\cos\phi, \\
y &= d\left(\phi,\zeta,\eta_\parallel\right)\cos\zeta\,\sin\phi.
\end{aligned} \tag{15.22}$$

Here ϕ is the azimuthal angle in the $y-x$ plane, whereas $\eta_\parallel$ is the spacetime rapidity

$$\eta_\parallel = \frac{1}{2}\ln\frac{t+z}{t-z} = \tanh^{-1}\left(\frac{z}{t}\right). \tag{15.23}$$

Equations (15.22) lead to the compact formulas for the longitudinal proper time τ and the transverse distance r,

$$\begin{aligned}
\tau - \tau_{\rm i} &= \sqrt{t^2 - z^2} - \tau_{\rm i} = d\left(\phi,\zeta,\eta_\parallel\right)\sin\zeta, \\
r &= \sqrt{x^2+y^2} = d\left(\phi,\zeta,\eta_\parallel\right)\cos\zeta.
\end{aligned} \tag{15.24}$$

The parameter $\tau_{\rm i}$ is the initial proper time — the value of τ when the initial conditions for the hydrodynamic evolution are specified.

For fixed ϕ and $\eta_\parallel$ the quantity $d\left(\phi,\zeta,\eta_\parallel\right)$ is the distance between the hypersurface point with coordinates $(\phi,\zeta,\eta_\parallel)$ and the spacetime point $(\tau=\tau_{\rm i}, x=0, y=0)$, see Fig. 15.3. The variable ζ, restricted to the range $0 \le \zeta \le \pi/2$, is an angle in the $\tau - r$ space. We have introduced the angle ζ because in most cases the freeze-out curves in the $\tau - r$ plane may be treated as functions of this parameter. The use of the transverse distance r is inconvenient, since very often two freeze-out points correspond to one value of r.

Clearly, the three variables ϕ, $\eta_\parallel$, and ζ may be regarded as the special choice of the parameters α, β, and γ used in Eqs. (15.19)–(15.21). If the function d is independent of ϕ, the freeze-out hypersurface is cylindrically symmetric. Similarly, if d is independent of $\eta_\parallel$, the freeze-out hypersurface is boost-invariant. The cylindrically symmetric and boost-invariant freeze-out hypersurfaces are defined by the function d depending solely on the variable ζ. The substitution of the parameterization (15.22) in Eq. (15.20) gives

$$\begin{aligned}
d\Sigma^0 &= d\,\cos\zeta\left[(\tau_{\rm i} + d\sin\zeta)\left(d\,\sin\zeta - \tfrac{\partial d}{\partial\zeta}\,\cos\zeta\right)\cosh\eta_\parallel + \tfrac{\partial d}{\partial_{\eta_\parallel}}\,\sinh\eta_\parallel\right], \\
d\Sigma^1 &= -d\,(\tau_{\rm i} + d\,\sin\zeta)\left[\left(d\,\cos\zeta + \tfrac{\partial d}{\partial\zeta}\,\sin\zeta\right)\cos\zeta\cos\phi + \tfrac{\partial d}{\partial\phi}\,\sin\phi\right], \\
d\Sigma^2 &= -d\,(\tau_{\rm i} + d\,\sin\zeta)\left[\left(d\,\cos\zeta + \tfrac{\partial d}{\partial\zeta}\,\sin\zeta\right)\cos\zeta\sin\phi + \tfrac{\partial d}{\partial\phi}\,\cos\phi\right], \\
d\Sigma^3 &= d\,\cos\zeta\left[(\tau_{\rm i} + d\,\sin\zeta)\left(d\,\sin\zeta - \tfrac{\partial d}{\partial\zeta}\,\cos\zeta\right)\sinh\eta_\parallel + \tfrac{\partial d}{\partial_{\eta_\parallel}}\,\cosh\eta_\parallel\right].
\end{aligned} \tag{15.25}$$

With Eqs. (15.25) and the standard definition of the four-momentum expressed in terms of the rapidity and transverse momentum,

$$p^\mu = (m_\perp \cosh \mathrm{y}, p_\perp \cos \phi_p, p_\perp \sin \phi_p, m_\perp \sinh \mathrm{y}) \,, \tag{15.26}$$

where $m_\perp = \sqrt{m^2 + p_\perp^2}$ is the transverse mass, we find the explicit form of the Cooper-Frye integration measure [10]

$$\begin{aligned} d\Sigma_\mu \, p^\mu = {} & d\,(\tau_{\rm i} + d\,\sin\zeta)\,\Big[d\cos\zeta\,\big(m_\perp \sin\zeta \cosh\left(\eta_\parallel - \mathrm{y}\right) + p_\perp \cos\zeta \cos\left(\phi - \phi_p\right)\big) \\ & + \frac{\partial d}{\partial \zeta}\cos\zeta\,\big(-m_\perp \cos\zeta \cosh\left(\eta_\parallel - \mathrm{y}\right) + p_\perp \sin\zeta \cos\left(\phi - \phi_p\right)\big) \\ & + \frac{\partial d}{\partial \phi} p_\perp \sin\left(\phi - \phi_p\right) + \frac{\partial d}{\partial \eta_\parallel} m_\perp \cot\zeta \sinh\left(\eta_\parallel - \mathrm{y}\right)\Big]\, d\eta_\parallel d\phi d\zeta. \end{aligned} \tag{15.27}$$

For boost-invariant systems the function d depends only on ϕ and ζ. The term $\partial d/\partial \eta_\parallel$ may be omitted in this case and the Cooper-Frye formula leads to the six-dimensional particle distribution at the thermal freeze-out

$$\begin{aligned} & \frac{dN}{d\mathrm{y} d\phi_p p_\perp dp_\perp d\eta_\parallel d\phi d\zeta} \\ & = \frac{g}{(2\pi)^3}\, d\,(\tau_{\rm i} + d\,\sin\zeta)\,\Big[d\cos\zeta\,\big(m_\perp \sin\zeta \cosh\left(\eta_\parallel - \mathrm{y}\right) + p_\perp \cos\zeta \cos\left(\phi - \phi_p\right)\big) \\ & + \frac{\partial d}{\partial \zeta}\cos\zeta\,\big(-m_\perp \cos\zeta \cosh\left(\eta_\parallel - \mathrm{y}\right) + p_\perp \sin\zeta \cos\left(\phi - \phi_p\right)\big) \\ & + \frac{\partial d}{\partial \phi} p_\perp \sin\left(\phi - \phi_p\right)\Big] \\ & \times \left\{ \exp\left[\frac{\beta}{\sqrt{1 - v_\perp^2}} \left(m_\perp \cosh(\mathrm{y} - \eta_\parallel) - p_\perp v_\perp \cos(\phi + \alpha - \phi_p)\right) - \beta\mu \right] \pm 1 \right\}^{-1}. \end{aligned} \tag{15.28}$$

The formulas derived in this Section will be used frequently in the next Chapters, in particular in Chap. 24, where different freeze-out geometries are analyzed.

The momentum distribution following from Eq. (15.18) has the form

$$E_p \frac{dN}{d^3p} = \frac{dN}{d\mathrm{y} d^2 p_\perp} = \int d\Sigma_\mu(x) p^\mu f(x,p) \tag{15.29}$$

For the systems of particles which are in local thermodynamic equilibrium we have

$$E_p \, \frac{dN}{d^3p} = \int d\Sigma_\mu(x)\, p^\mu f_{\rm eq}\left(p^\alpha\, u_\alpha(x)\right), \tag{15.30}$$

where the function $f_{\rm eq}$ is the equilibrium distribution function

$$f_{\rm eq}\left(p^\alpha\, u_\alpha\right) = \frac{1}{(2\pi)^3}\left[\exp\left(\frac{p^\alpha\, u_\alpha - \mu}{T}\right) - \epsilon\right]^{-1}. \qquad (15.31)$$

Here, as always, the case $\epsilon = -1\,(+1)$ corresponds to the Fermi-Dirac (Bose-Einstein) statistics, and the limit $\epsilon \to 0$ yields the classical (Boltzmann) statistics, see Eq. (8.46) in Sec. 8.7.

15.3.1 *Space-like emission*

In the transition from Eq. (15.17) to (15.18) we argued that the freeze-out hypersurface contains the space-like parts, see Fig. 15.4 — we use the convention that in the space-like (time-like) region the vector normal to the freeze-out hypersurface is space-like (time-like). Now, we want to emphasize that the treatment of the space-like parts might lead to conceptual problems since particles emitted from such regions of the hypersurface enter again the system and the hydrodynamic description of such regions (combined with the use of the Cooper-Frye formula) becomes inadequate. Recently much work has been done to develop a consistent description of the freeze-out process from the space-like parts [12–21]. At the same time, quantitative arguments have been also presented [6] that the "wrong" contributions from the space-like parts are very small and may be neglected in the realistic hydrodynamic calculations.

To see the last point we have to consider the covariant form of the distribution function. For the particles going outwards we have a factor

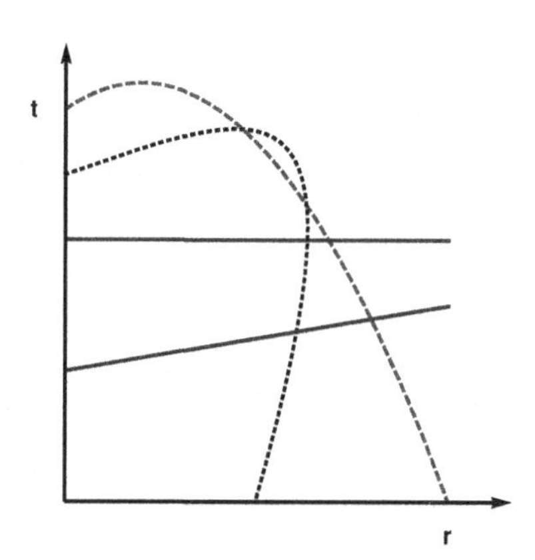

Fig. 15.4 A priori possible, different freeze-out curves in Minkowski $t - r$ space (for some fixed azimuthal angle ϕ). The dotted and dashed lines describe the cases where both the space-like and time-like parts are present. The solid lines describe the cases where only the time-like part is present.

$\exp(-E_p\gamma/T + p_\perp v_\perp \gamma/T)$, whereas for the particles going inwards the momentum changes sign and we have $\exp(-E_p\gamma/T - p_\perp v_\perp \gamma/T)$. Hence, the ratio of the "in" and "out" contributions behaves like $\exp(-2p_\perp v_\perp \gamma/T)$. For $p_\perp \sim 400$ MeV, $v_\perp \sim 0.5$, and $T \sim 130$ MeV one finds $\exp(-2p_\perp v_\perp \gamma/T) \sim 0.03$. Clearly, if the strong radial flow is present at freeze-out, the emitted particles pick up some part of the momentum of the fluid element and move outwards. The problems with the inward emission may appear, however, if the flow is not strong and/or the freeze-out hypersurface bends to the origin of the coordinate system, as the dotted curve in Fig. 15.4.

15.4 Emission function

Equation (15.29) suggests the introduction of the concept of the *emission function* (sometimes also called the *source function*),

$$S(x, \mathbf{p}) = \int d\Sigma_\mu(x')\, p^\mu\, \delta^{(4)}(x' - x) f(x', p), \tag{15.32}$$

The physical interpretation of the emission function is very straightforward — it gives the distribution of the spacetime positions and momenta of the emitted particles. In other words, the emission function gives the number of the particles emitted in the phase-space element $\Delta^3 x\, \Delta^3 p$ per unit time (in order to obtain the Lorentz invariant quantity this ratio is also multiplied by the energy of the emitted particles)

$$S(x, \mathbf{p}) = E_p \frac{\Delta N}{\Delta t\, \Delta^3 x\, \Delta^3 p}. \tag{15.33}$$

The integration of the emission function over time and space yields the invariant momentum distribution

$$\frac{dN}{dy d^2 p_\perp} = \int d^4x\, S(x, \mathbf{p}) = \int d\Sigma_\mu(x)\, p^\mu\, f(x, p). \tag{15.34}$$

The concept of the emission function turns out to be very useful in modeling of the physical conditions at freeze-out (in quite general situations not restricted to the hydrodynamic picture used in this Section). The emission function is used to calculate not only the particle spectra but to study other physical observables such as, for example, the HBT correlation radii. In this case one calculates the Fourier transform defined by the expression

$$S(\mathbf{k}, \mathbf{q}) = \int d^4x\, e^{iq\cdot x}\, S(x, \mathbf{k}) = \int d\Sigma_\mu(x)\, k^\mu\, e^{iq\cdot x}\, f(x, k). \tag{15.35}$$

Bibliography to Chapter 15

[1] T. Hirano, "Is early thermalization achieved only near midrapidity in Au + Au collisions at $\sqrt{s_{NN}} = 130$ GeV?," *Phys. Rev.* **C65** (2002) 011901.

[2] T. Hirano and K. Tsuda, "Collective flow and two pion correlations from a relativistic hydrodynamic model with early chemical freeze out," *Phys. Rev.* **C66** (2002) 054905.

[3] P. Bozek and I. Wyskiel, "Rapid hydrodynamic expansion in relativistic heavy-ion collisions," *Phys. Rev.* **C79** (2009) 044916.

[4] A. Dumitru, S. A. Bass, M. Bleicher, H. Stoecker, and W. Greiner, "Direct emission of multiple strange baryons in ultra-relativistic heavy-ion collisions from the phase boundary," *Phys. Lett.* **B460** (1999) 411–416.

[5] S. A. Bass *et al.*, "Hadronic freeze-out following a first order hadronization phase transition in ultrarelativistic heavy-ion collisions," *Phys. Rev.* **C60** (1999) 021902.

[6] D. Teaney, J. Lauret, and E. V. Shuryak, "A hydrodynamic description of heavy ion collisions at the sps and rhic," `nucl-th/0110037`.

[7] C. Nonaka and S. A. Bass, "Space-time evolution of bulk QCD matter," *Phys. Rev.* **C75** (2007) 014902.

[8] H. Petersen, J. Steinheimer, G. Burau, M. Bleicher, and H. Stocker, "A Fully Integrated Transport Approach to Heavy Ion Reactions with an Intermediate Hydrodynamic Stage," *Phys. Rev.* **C78** (2008) 044901.

[9] B. Touschek, "Covariant statistical mechanics," *Nuovo Cimento* **B58** (1933) 295.

[10] F. Cooper and G. Frye, "Comment on the single particle distribution in the hydrodynamic and statistical thermodynamic models of multiparticle production," *Phys. Rev.* **D10** (1974) 186.

[11] C. W. Misner, K. S. Thorne, and J. A. Wheeler, "Gravitation," (W.H. Freeman, San Francisco, 1973) 1279 p.

[12] C. Anderlik *et al.*, "Freeze out in hydrodynamical models," *Phys. Rev.* **C59** (1999) 3309–3316.

[13] V. K. Magas *et al.*, "Freeze-out in hydrodynamical models in relativistic heavy ion collisions," *Nucl. Phys.* **A661** (1999) 596–599.

[14] Y. M. Sinyukov, S. V. Akkelin, and Y. Hama, "On freeze-out problem in hydro-kinetic approach to a+a collisions," *Phys. Rev. Lett.* **89** (2002) 052301.

[15] K. A. Bugaev, "Relativistic kinetic equations for finite domains and freeze-out problem," *Phys. Rev. Lett.* **90** (2003) 252301.

[16] V. K. Magas, A. Anderlik, C. Anderlik, and L. P. Csernai, "Non-equilibrated post freeze out distributions," *Eur. Phys. J.* **C30** (2003) 255–261.

[17] K. A. Bugaev, "Boundary conditions of the hydro-cascade model and relativistic kinetic equations for finite domains," *Phys. Rev.* **C70** (2004) 034903.

[18] F. Grassi, "Particle emission in hydrodynamics: a problem needing a solution," *Braz. J. Phys.* **35** (2005) 52–69.

[19] E. Molnar *et al.*, "Covariant description of kinetic freeze out through a finite time-like layer," *J. Phys.* **G34** (2007) 1901–1916.

[20] E. Molnar, L. P. Csernai, V. K. Magas, A. Nyiri, and K. Tamosiunas, "Covariant description of kinetic freeze out through a finite space-like layer," `nucl-th/0503047`.

[21] K. Tamosiunas, L. P. Csernai, E. Molnar, A. Nyiri, and V. K. Magas, "Modelling of boltzmann transport equation for freeze-out," *J. Phys.* **G31** (2005) S1001–S1004.

Chapter 16

Exercises to PART IV

Exercise 16.1. *Sound velocity:*
i) Starting from the standard definition of the sound velocity,

$$c_s^2 = \left(\frac{\partial P}{\partial \varepsilon}\right)_s, \tag{16.1}$$

check that it is equivalent to Eq. (13.25). **ii)** Show that for an adiabatic transformation one has

$$\left(\frac{\partial P}{\partial \tilde{V}}\right)_s = -\frac{c_s^2 n^2}{1-c_s^2} \leq 0, \tag{16.2}$$

where $\tilde{V} = w/n$. **iii)** Derive the low temperature limit of the sound velocity of the massive pion gas

$$c_s \approx \sqrt{\frac{T}{m_\pi}}. \tag{16.3}$$

Exercise 16.2. *Relativistic hydrodynamic equations in the non-relativistic three-vector notation.*
Show that the relativistic Euler equation (13.33) may be written in the form given by Eq. (13.34),

$$\frac{\partial}{\partial t}(w\gamma\mathbf{v}) + \nabla(w\gamma) = \mathbf{v}\times(\nabla\times w\gamma\mathbf{v}) + \frac{T}{\gamma}\nabla s. \tag{16.4}$$

Similarly, show that Eq. (13.40) is equivalent to (13.41),

$$\frac{\partial}{\partial t}(T\gamma\mathbf{v}) + \nabla(T\gamma) = \mathbf{v}\times(\nabla\times T\gamma\mathbf{v}). \tag{16.5}$$

Exercise 16.3. *One-dimensional relativistic hydrodynamic equations and the light-cone variables.*

i) Show that the one-dimensional relativistic hydrodynamic equations (13.16) with the energy-momentum tensor (13.15) where $\varepsilon = 3P$ may be written in the form

$$\frac{\partial \varepsilon}{\partial x_+} + 2\frac{\partial \left(\varepsilon e^{-2\vartheta}\right)}{\partial x_-} = 0,$$
$$\frac{\partial \varepsilon}{\partial x_-} + 2\frac{\partial \left(\varepsilon e^{2\vartheta}\right)}{\partial x_+} = 0, \tag{16.6}$$

where $x_\pm$ are the light cones variables,

$$x_\pm = x^0 \pm x^3 = t \pm z,$$

and the parameter ϑ is defined by the relations

$$u^0 = \cosh\vartheta, u^3 = \sinh\vartheta$$

(note that $u^1 = u^2 = 0$ for the pure longitudinal expansion). **ii)** Prove that Eqs. (16.6) are equivalent to the equations

$$(\cosh 2\vartheta + 2)\frac{\partial \ln \varepsilon}{\partial t} + \sinh 2\vartheta \frac{\partial \ln \varepsilon}{\partial z} + 4\frac{\partial \vartheta}{\partial z} = 0,$$
$$\sinh 2\vartheta \frac{\partial \ln \varepsilon}{\partial t} + (\cosh 2\vartheta - 2)\frac{\partial \ln \varepsilon}{\partial z} - 4\frac{\partial \vartheta}{\partial t} = 0. \tag{16.7}$$

iii) Show also that for the one-dimensional longitudinal expansion Eqs. (13.71) and (13.72), derived by us in Chap. 13, are equivalent to Eqs. (16.6) and (16.7). *Hint:* Use the substitutions: $v_z = \tanh\vartheta$, $d\ln\varepsilon = 4d\ln T$, $d\ln\sigma = 3d\ln T$.

Exercise 16.4. *Hydrodynamic potential* Φ *(Baym's formalism).*
Starting from the definition of the temperature dependent sound velocity

$$c_s(T) = \frac{1}{\sqrt{3}}\left[1 - \frac{1}{2}\left(\frac{1}{1 + (T/T_0)^{2n}}\right)\right], \tag{16.8}$$

where T_0 and n are parameters, integrate Eq. (13.84) and show that the following relations hold:

$$\Phi_T(T) = \frac{\sqrt{3}}{2n}\ln\frac{(T/T_0)^{4n}}{1 + 2(T/T_0)^{2n}}$$

and

$$T_\Phi(\Phi) = T_0\left[e^{\frac{2n\Phi}{\sqrt{3}}} + \sqrt{e^{\frac{2n\Phi}{\sqrt{3}}} + e^{\frac{4n\Phi}{\sqrt{3}}}}\right]^{\frac{1}{2n}}.$$

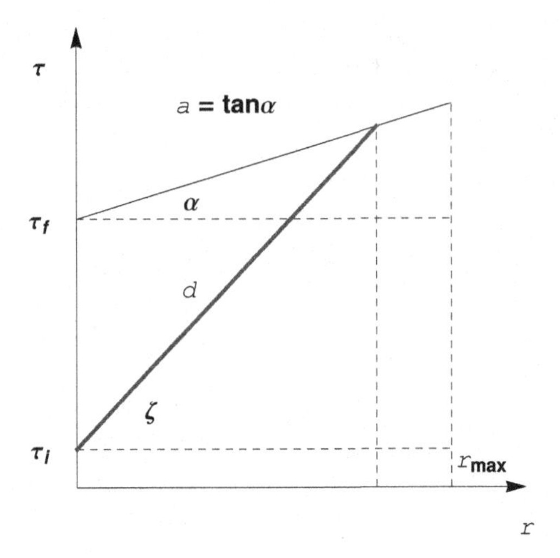

Fig. 16.1 Parameterization of the freeze-out curve (the projection of the freeze-out hypersurface on the $\tau - r$ plane) considered in Ex. 16.6.

Exercise 16.5. *Rapidity dependent initial conditions.*
Derive Eq. (15.14).

Exercise 16.6. *Freeze-out hypersurface.*
i) Consider the cylindrically symmetric and boost-invariant freeze-out hypersurface Σ_μ. In the case where the freeze-out takes place at the fixed proper time, $\tau = \tau_\mathrm{f}$, we have

$$d(\zeta) = \frac{\tau_\mathrm{f} - \tau_\mathrm{i}}{\sin\zeta}, \qquad r = d(\zeta)\, \cos\zeta.$$

Use Eq. (15.27) and replace the integration variable ζ by r . Show that the integration measure in the Cooper-Frye formula equals

$$d\Sigma_\mu p^\mu = r\tau_\mathrm{f}\, m_\perp \cosh(\mathrm{y} - \eta_\parallel)\, d\eta_\parallel\, dr\, d\phi. \tag{16.9}$$

Note that the change of the integration range from $\zeta_\mathrm{min} \leq \zeta \leq \pi/2$ to $0 \leq r \leq r_\mathrm{max}$ introduces an extra change of sign.

ii) Generalize the previous case to the freeze-out defined by the condition $\tau = \tau_\mathrm{f} + ar$, see Fig. 16.1. In this case

$$d(\zeta) = \frac{\tau_\mathrm{f} - \tau_\mathrm{i}}{\sin\zeta - a\cos\zeta}$$

and

$$d\Sigma_\mu p^\mu = r\,(\tau_\mathrm{f} + ar)\, \left[m_\perp \cosh(\mathrm{y} - \eta_\parallel) - ap_\perp \cos(\phi - \phi_p)\right]\, d\eta_\parallel\, dr\, d\phi. \tag{16.10}$$

Of course, in the simple cases considered in this Exercise, the parameterization with the help of the variable r is much more convenient than the parameterization using ζ. This is not the case for more complicated freeze-out hypersurfaces obtained from the hydrodynamical calculations, where different freeze-out points may correspond to the same value of r (with fixed ϕ and $\eta_\parallel$).

PART V

PARTICLE INTERFEROMETRY

Chapter 17

Pion Correlation Functions

The correlations between identical pions have different physical reasons: **i)** conservation laws, **ii)** specific features of the production mechanism and final state interactions, and **iii)** Bose-Einstein statistics [1].

The correlations due to the conservation laws are of the two types — the energy-momentum conservation leads to the kinematical constraints while the conservation of quantum numbers such as isospin leads to the correlation between the numbers of different types of pions produced in a given reaction. The typical example of the correlation induced by the production mechanism is the momentum correlation caused by the resonance production — more than one half of pions produced in the ultra-relativistic heavy-ion collisions comes from the resonance decays. On the other hand, the typical correlation induced by the final state interactions is that caused by the long-range Coulomb interaction. The last source of the correlations listed above is the Bose-Einstein statistics. This is an effect connected with the required symmetrization of the one-particle pion wave functions.

Because the available phase space for pion production in ultra-relativistic heavy-ion collisions is very large (hundreds or even thousands of pions are produced in the typical heavy-ion collision) the correlations induced by the conservation laws are negligible. On the contrary, the correlations due to the resonance decays and Coulomb repulsion are substantial and should be always taken into account together with the statistical effects. In most cases we are interested in extracting the correlation effects caused by the Bose-Einstein statistics, hence, the experimental data are frequently corrected for the Coulomb interaction. Of course, the data cannot be corrected for all possible effects and the study of correlations involves always a detailed comparison of the specific model predictions with the data.

The study of the pion correlations, especially those due to the Bose-Einstein statistics, is also known as *pion interferometry*, *pion femtoscopy*, or *HBT analysis*. The last term emphasizes the analogy to the well known technique of second order *intensity interferometry* developed by Hanbury-Brown and Twiss (HBT) [2, 3] to measure stellar angular sizes [1]. In high-energy physics, the ideas of pion interfer-

[1]This analogy is sometimes misunderstood and the momentum correlations are confused with the spacetime HBT correlations. Although the two types of the correlations have the common

ometry were used to deduce the space-time structure of the interaction regions [7–9]. Experimentally, pion interferometry was first used by Goldhaber, Goldhaber, Lee, and Pais in Ref. [10] to determine the size of the pion production region in $p\bar{p}$ collisions. That is why the increase of the correlation function for small relative momenta is sometimes called the GGLP effect.

In this Chapter we introduce the main concepts used in the studies of identical particle correlations. In particular, we present in more detail the most popular framework used to study and interpret the pion HBT data. This framework is based on the assumption that the emission function may be well approximated by the Gaussian shape.

17.1 General definitions

17.1.1 *Connection to inclusive distributions*

Let us start with the introduction of the general concepts concerning the pion interferometry. The central object discussed in this context is the pion correlation function defined as the inclusive two-particle distribution $\mathcal{P}_2(\mathbf{p}_1, \mathbf{p}_2)$ divided by the product of the inclusive one-particle distributions $\mathcal{P}_1(\mathbf{p})$, namely

$$C(\mathbf{p}_1, \mathbf{p}_2) = \frac{\mathcal{P}_2(\mathbf{p}_1, \mathbf{p}_2)}{\mathcal{P}_1(\mathbf{p}_1)\mathcal{P}_1(\mathbf{p}_2)}. \tag{17.1}$$

Here $\mathbf{p}_1$ and $\mathbf{p}_2$ are the three-momenta of the two pions. The one- and two-particle distributions in Eq. (17.1) correspond to the single and double inclusive cross sections. For example, in the case of the positive pions we may write

$$C(\mathbf{p}_1, \mathbf{p}_2) = \frac{\sigma_{\rm tot} d\sigma_{\rm incl}(\pi^+\pi^+)/d^3p_1 d^3p_2}{(d\sigma_{\rm incl}(\pi^+)/d^3p_1)(d\sigma_{\rm incl}(\pi^+)/d^3p_2)}, \tag{17.2}$$

where $\sigma_{\rm tot}$ is the total cross section (see our discussion of the inclusive cross sections in Sec. 30.5, compare also Eq. (3.60) in Sec. 3.8). Equation (17.2) gives the straightforward prescription for the experimental determination of the $\pi^+\pi^+$ correlation function.

17.1.2 *Connection to density matrix*

The one-particle and two-particle pion distribution functions may be expressed as the expectation values of the products of the pion creation and annihilation opera-

---

quantum statistical origin, the momentum correlations of identical particles yield the spacetime picture of the source, whereas the spacetime HBT correlations (dependence of the number of coincident two-photon counts on the distance between two detectors) provide the information on the characteristic relative three-momenta of emitted photons, which gives the angular size of a star without the knowledge of its radius and lifetime [4–6].

tors in a multiparticle state denoted in the following by the brackets $\langle ... \rangle$,

$$\mathcal{P}_1(\mathbf{p}) = E_p \frac{dN}{d^3p} = E_p \langle \hat{a}^+_{\mathbf{p}} \hat{a}_{\mathbf{p}} \rangle, \tag{17.3}$$

$$\mathcal{P}_2(\mathbf{p}_1, \mathbf{p}_2) = E_{p_1} E_{p_2} \frac{dN}{d^3p_1 d^3p_2} = E_{p_1} E_{p_2} \langle \hat{a}^+_{\mathbf{p}_1} \hat{a}^+_{\mathbf{p}_2} \hat{a}_{\mathbf{p}_2} \hat{a}_{\mathbf{p}_1} \rangle. \tag{17.4}$$

If the Wick theorem can be applied, i.e., if we may make use of the relation

$$\langle \hat{a}^+_{\mathbf{p}_1} \hat{a}^+_{\mathbf{p}_2} \hat{a}_{\mathbf{p}_2} \hat{a}_{\mathbf{p}_1} \rangle = \langle \hat{a}^+_{\mathbf{p}_1} \hat{a}_{\mathbf{p}_1} \rangle \langle \hat{a}^+_{\mathbf{p}_2} \hat{a}_{\mathbf{p}_2} \rangle + \langle \hat{a}^+_{\mathbf{p}_1} \hat{a}_{\mathbf{p}_2} \rangle \langle \hat{a}^+_{\mathbf{p}_2} \hat{a}_{\mathbf{p}_1} \rangle, \tag{17.5}$$

then the correlation function may be rewritten in the form

$$C(\mathbf{p}_1, \mathbf{p}_2) = 1 + \frac{|\rho(\mathbf{p}_1, \mathbf{p}_2)|^2}{\rho(\mathbf{p}_1, \mathbf{p}_1)\rho(\mathbf{p}_2, \mathbf{p}_2)}, \tag{17.6}$$

where the quantity

$$\rho(\mathbf{p}_1, \mathbf{p}_2) = \sqrt{E_{p_1} E_{p_2}} \langle \hat{a}^+_{\mathbf{p}_1} \hat{a}_{\mathbf{p}_2} \rangle \tag{17.7}$$

is the density matrix.

17.1.3 *Very simple model*

The effect of the Bose statistics on the correlation function may be illustrated in a very simple model including the symmetrization of the pion wave function. Let us assume that pions are produced at points $\mathbf{x}_1$ and $\mathbf{x}_2$ which are randomly distributed in space. The distribution of the points is given by the density distribution $\rho(\mathbf{x})$. The two-particle distribution of identical pions with momenta $\mathbf{p}_1$ and $\mathbf{p}_2$ is given by the expression

$$\mathcal{P}_2(\mathbf{p}_1, \mathbf{p}_2) = \int d^3x_1 d^3x_2 \rho(\mathbf{x}_1)\rho(\mathbf{x}_2)|\Psi_{12}|^2, \tag{17.8}$$

where Ψ_{12} is the wave function

$$\Psi_{12} = \frac{1}{\sqrt{2}(2\pi)^3} \left(e^{i\,\mathbf{p}_1\cdot\mathbf{x}_1 + i\,\mathbf{p}_2\cdot\mathbf{x}_2} + e^{i\,\mathbf{p}_1\cdot\mathbf{x}_2 + i\,\mathbf{p}_2\cdot\mathbf{x}_1} \right). \tag{17.9}$$

The second term arises from the symmetrization required by the Bose-Einstein statistics. To proceed further we introduce the relative momentum

$$\mathbf{q} = \mathbf{p}_1 - \mathbf{p}_2, \tag{17.10}$$

and the center-of-mass momentum

$$\mathbf{k} = \frac{1}{2}(\mathbf{p}_1 + \mathbf{p}_2). \tag{17.11}$$

Using these variables we find

$$\Psi_{12} = \frac{1}{\sqrt{2}(2\pi)^3} e^{i\,\mathbf{k}\cdot(\mathbf{x}_1+\mathbf{x}_2)} \left[e^{i\,\mathbf{q}\cdot(\mathbf{x}_1-\mathbf{x}_2)/2} + e^{-i\,\mathbf{q}\cdot(\mathbf{x}_1-\mathbf{x}_2)/2} \right] \tag{17.12}$$

and the simple calculation yields

$$\mathcal{P}_2(\mathbf{p}_1, \mathbf{p}_2) = \int \frac{d^3x_1}{(2\pi)^3} \frac{d^3x_2}{(2\pi)^3} \rho(\mathbf{x}_1)\rho(\mathbf{x}_2) \left[1 + \frac{1}{2}\left(e^{i\,\mathbf{q}\cdot(\mathbf{x}_1-\mathbf{x}_2)} + e^{-i\,\mathbf{q}\cdot(\mathbf{x}_1-\mathbf{x}_2)}\right)\right]. \tag{17.13}$$

The one-particle distribution is obtained in the similar way

$$\mathcal{P}_1(\mathbf{p}) = \int d^3x \rho(\mathbf{x}) \left|\frac{e^{i\,\mathbf{p}\cdot\mathbf{x}}}{(2\pi)^{3/2}}\right|^2 = \int \frac{d^3x}{(2\pi)^3}\rho(\mathbf{x}). \tag{17.14}$$

The fact that this distribution is independent of momentum reflects here the uncertainty relation — particles are produced in well defined places $\mathbf{x}$, hence their momenta are uniformly distributed. From the definition of the correlation function we obtain

$$C(\mathbf{k}, \mathbf{q}) = 1 + \frac{\left|\int d^3x e^{-i\,\mathbf{q}\cdot\mathbf{x}}\rho(\mathbf{x})\right|^2}{\left[\int d^3x \rho(\mathbf{x})\right]^2}. \tag{17.15}$$

For very small values of $\mathbf{q}$, the correlation function increases to 2, which is the GGLP effect.

17.1.4 *Connection to emission function*

The essential problem of the model introduced in Sec. 17.1.3 is that it does not include the time dependence of the pion source. Such dependence was introduced by Kopylov and Podgoretsky [7,8] who used the Klein-Gordon equation in the presence of several chaotic source currents to argue that the correlation function measures not only the space but also the time Fourier transform of the source distribution function depending generally on the spacetime coordinates t and $\mathbf{x}$. This concept may be realized by a simple relation between the density matrix and the emission function, namely by the ansatz

$$\rho(\mathbf{p}_1, \mathbf{p}_2) = \rho\left(\mathbf{k} + \frac{1}{2}\mathbf{q}, \mathbf{k} - \frac{1}{2}\mathbf{q}\right) = \int d^4x \, \exp(iq \cdot x) S(x, \mathbf{k}). \tag{17.16}$$

The relative and average (center-of-mass) momenta $\mathbf{q}$ and $\mathbf{k}$ were defined by Eqs. (17.10) and (17.11), respectively. The generalization to the full spacetime integrals requires that we have to define also the temporal components of the four-vectors $q = (q^0, \mathbf{q})$ and $k = (k^0, \mathbf{k})$. In most of the practical applications one assumes that particles are on the mass shell, which leads in the natural way to the definitions

$$q^0 = E_{p_1} - E_{p_2} = \sqrt{m_\pi^2 + p_1^2} - \sqrt{m_\pi^2 + p_2^2} \tag{17.17}$$

and

$$k^0 = \frac{1}{2}\left(E_{p_1} + E_{p_2}\right) = \frac{1}{2}\left(\sqrt{m_\pi^2 + p_1^2} + \sqrt{m_\pi^2 + p_2^2}\right). \tag{17.18}$$

The mass-shell constraints also yield the condition

$$q \cdot k = \frac{1}{2}(p_1 - p_2) \cdot (p_1 + p_2) = \frac{1}{2}\left(p_1^2 - p_2^2\right) = \frac{1}{2}\left(m_\pi^2 - m_\pi^2\right) = 0, \tag{17.19}$$

which can be used to eliminate q^0, namely

$$q^0 = \frac{\mathbf{k} \cdot \mathbf{q}}{k^0} \equiv \boldsymbol{\beta} \cdot \mathbf{q}, \tag{17.20}$$

where $\boldsymbol{\beta}$ describes the velocity of a pair.

In Sec. 15.4 we showed that the momentum distribution of pions is simply the space-time integral of the emission function

$$E_p \frac{dN}{d^3p} = \frac{dN}{dy\, d^2p_\perp} = \rho(\mathbf{p}, \mathbf{p}) = \int d^4x\, S(x, \mathbf{p}). \tag{17.21}$$

Similarly, we obtain the compact expression for the correlation function in terms of the emission function [2]

$$C(\mathbf{k}, \mathbf{q}) = 1 + \frac{\left|\int d^4x\, e^{iq\cdot x}\, S(x, \mathbf{k})\right|^2}{\int d^4x\, S(x, \mathbf{k} + \frac{1}{2}\mathbf{q}) \int d^4y\, S(y, \mathbf{k} - \frac{1}{2}\mathbf{q})}. \tag{17.22}$$

Very often the q-dependence of the denominator on the right-hand-side of Eq. (17.22) is rather weak, so we may use the so called *smoothness* approximation

$$S\left(x, \mathbf{k} + \frac{1}{2}\mathbf{q}\right) S\left(y, \mathbf{k} - \frac{1}{2}\mathbf{q}\right) \approx S(x, \mathbf{k})\, S(y, \mathbf{k}). \tag{17.23}$$

As long as this approximation is valid, the correlation function may be reduced to the form

$$C(\mathbf{k}, \mathbf{q}) = 1 + \frac{\left|\int d^4x\, e^{iq\cdot x}\, S(x, \mathbf{k})\right|^2}{\left[\int d^4x\, S(x, \mathbf{k})\right]^2}. \tag{17.24}$$

The differences coming from the usage of Eq. (17.24) instead of Eq. (17.22) are illustrated in Ex. 18.2, where the Zajc model is discussed [11]. Equation (17.24) suggests a simple interpretation — the correlation function is the Fourier transform of the emission function. Thus, the range of the correlation function is related to the space-time extensions of the emitting source.

17.2 Gaussian parametrization

In this Section, following Wiedemann and Heinz [12], we consider a Gaussian parametrization of the emission function

$$S(x, \mathbf{k}) = \mathcal{N}(\mathbf{k}) \exp\left[-\frac{1}{2}\tilde{x}^\mu(\mathbf{k})\, B_{\mu\nu}(\mathbf{k})\, \tilde{x}^\nu(\mathbf{k})\right]. \tag{17.25}$$

[2] We use the same letter to denote the correlation function in the $(\mathbf{p}_1, \mathbf{p}_2)$ space and the correlation function in the $(\mathbf{k}, \mathbf{q})$ space. The use of the appropriate arguments follows from the mathematical context.

Here $\mathcal{N}(\mathbf{k})$ is an arbitrary normalization, and the coordinates $\widetilde{x}^{\mu}(\mathbf{k})$ are deviations from the mean values $\overline{x}^{\mu}(\mathbf{k})$,

$$\widetilde{x}^{\mu}(\mathbf{k}) = x^{\mu} - \overline{x}^{\mu}(\mathbf{k}). \tag{17.26}$$

The formula (17.25) was frequently used to construct the framework for the interpretation of the correlation data, see for example Refs. [13–17] which we follow here in our discussion.

Without the loss of generality we may assume that the coefficients $B_{\mu\nu}(\mathbf{k})$ form a symmetric matrix B, namely $B_{\mu\nu}(\mathbf{k}) = B_{\nu\mu}(\mathbf{k})$. This property follows from the fact that any antisymmetric part of $B_{\mu\nu}(\mathbf{k})$ does not contribute to the symmetric sum over μ and ν in Eq. (17.25). The symmetric matrix B may be diagonalized with the help of the transformation

$$B = R\,D\,R^{T}, \tag{17.27}$$

where R is an orthogonal matrix satisfying the condition $RR^{T} = R^{T}R = 1$ and the matrix D has only diagonal elements (they are eigenvalues of the matrix B). The orthogonality of R implies that $\det R = \pm 1$ and $\det B = \det D$. We exclude the possibility that the matrix R represents the reflection, hence we consider only the cases with $\det R = +1$.

The diagonalization of the matrix B is the technical way which allows us to calculate the correlation function corresponding to the emission function (17.25). Because we consider four-dimensional rotations rather than Lorentz transformations in the diagonalization procedure, it is useful to switch from the covariant notation to the standard matrix notation. This is done by the replacements of the type

$$\begin{aligned} &B_{\mu\nu} \to B_{mn}, \quad B^{-1}_{mn} \to (B^{-1})^{\mu\nu}, \quad g_{\mu\nu} \to g_{mn}, \\ &\widetilde{x}^{\mu} \to \widetilde{x}_{m}, \quad \widetilde{x}_{\mu} \to g_{mn}\,\widetilde{x}_{n}, \quad q_{\alpha}\,\widetilde{x}^{\alpha} \to q_{a}\,\widetilde{x}_{b}\,g_{ab}, \end{aligned} \tag{17.28}$$

where the objects with latin indices are four by four matrices or vectors. Using the substitution rules (17.28) Eq. (17.27) may be written in the form

$$B_{mn} = R_{ms}\,D_{sr}\,R^{T}_{rn}, \tag{17.29}$$

thus the argument of the exponent in Eq. (17.25) includes the term

$$\widetilde{x}_{m} R_{ms}\,D_{sr}\,R^{T}_{rn}\,\widetilde{x}_{n}. \tag{17.30}$$

The summation over the identical indices is understood here. We also note that we do not distinguish between the upper and lower latin indices. The expression (17.30) is simplified if one introduces the vectors y defined by the formula

$$y_{r} = R^{T}{}_{rn}\widetilde{x}_{n} = \widetilde{x}_{n} R_{\,nr}. \tag{17.31}$$

The inverse transformation to Eq. (17.31) is $\widetilde{x}_{s} = R_{sr}y_{r} = y_{r}R^{T}{}_{rs}$. Since the matrix D is diagonal we obtain the desired result

$$\widetilde{x}_{m} R_{ms}\,D_{sr}\,R^{T}_{rn}\,\widetilde{x}_{n} = y_{s}D_{sr}y_{r} = \sum_{r=0}^{3} D_{rr}y_{r}^{2}. \tag{17.32}$$

We are now ready to calculate the space-time integral of the emission function,

$$\int d^4x\, S(x,\mathbf{k}) = \mathcal{N}(\mathbf{k}) \int d^4x \exp\left[-\frac{1}{2}\widetilde{x}^{\,\mu}(\mathbf{k})\, B_{\mu\nu}(\mathbf{k})\, \widetilde{x}^{\,\nu}(\mathbf{k})\right]. \tag{17.33}$$

This quantity gives us directly the momentum spectrum of pions. Changing the integration variables from x to y we obtain a product of the Gaussian integrals which can be easily done

$$\begin{aligned}\int d^4x\, S(x,\mathbf{k}) &= \mathcal{N}(\mathbf{k}) \int d^4y \left|\det\left(\frac{\partial x^\mu}{\partial y^\nu}\right)\right| \exp\left[-\frac{1}{2}\sum_{r=0}^{3} D_{rr} y_r^2\right] \\ &= \mathcal{N}(\mathbf{k}) \frac{(2\pi)^2}{\sqrt{D_{00}(\mathbf{k})D_{11}(\mathbf{k})D_{22}(\mathbf{k})D_{33}(\mathbf{k})}} \\ &= \mathcal{N}(\mathbf{k}) \frac{(2\pi)^2}{\sqrt{\det B(\mathbf{k})}}.\end{aligned} \tag{17.34}$$

Here we used the property that the determinants of the matrices B and D are equal. We note that $|\det(\partial x^\mu/\partial y^\nu)|$ is the absolute value of the Jacobian determinant, which equals unity in this case since $|\det(\partial x^\mu/\partial y^\nu)| = |\det R\,|$.

The off-diagonal elements of the density matrix, necessary to determine the correlation function, may be calculated from Eq. (17.16)

$$\rho\left(\mathbf{k}+\tfrac{1}{2}\mathbf{q}, \mathbf{k}-\tfrac{1}{2}\mathbf{q}\right) = \mathcal{N}(\mathbf{k}) \int d^4x \exp\left[iq_\alpha x^\alpha - \frac{1}{2}\widetilde{x}^\mu(\mathbf{k})B_{\mu\nu}(\mathbf{k})\,\widetilde{x}^\nu(\mathbf{k})\right]. \tag{17.35}$$

Introducing the matrix notation (17.28) we rewrite the right-hand-side of the last equation in the form

$$\mathcal{N}(\mathbf{k}) \exp\left[iq_\alpha \overline{x}^\alpha(\mathbf{k})\right] \int d^4\widetilde{x} \exp\left[i\sum_{r=0}^{3} q_r \widetilde{x}_r g_{rr} - \frac{1}{2}\sum_{r=0}^{3} D_{rr} y_r^2\right]. \tag{17.36}$$

Changing the integration variable from $\widetilde{x}$ to y we rewrite this result as

$$\mathcal{N}(\mathbf{k}) \exp\left[iq_\alpha \overline{x}^\alpha(\mathbf{k})\right] \int d^4y \exp\left[\sum_{r=0}^{3}\left(-\frac{1}{2}D_{rr}y_r^2 + ic_r y_r\right)\right], \tag{17.37}$$

where $c_r = \sum_{s=0}^{3} q_s g_{ss} R_{sr}$. By completing the square

$$-\frac{1}{2}D_{rr}y_r^2 + ic_r y_r = -\frac{1}{2}D_{rr}\left(y_r - i\frac{c_r}{D_{rr}}\right)^2 - \frac{1}{2}c_r D_{rr}^{-1} c_r \tag{17.38}$$

and performing the y-integration we find

$$\rho\left(\mathbf{k}+\tfrac{1}{2}\mathbf{q}, \mathbf{k}-\tfrac{1}{2}\mathbf{q}\right) = \mathcal{N}(\mathbf{k}) \exp\left[iq_\alpha \overline{x}^\alpha(\mathbf{k})\right] \frac{(2\pi)^2 \exp\left[-\frac{1}{2}\sum\limits_{r=0}^{3} c_r D_{rr}^{-1} c_r\right]}{\sqrt{\det B(\mathbf{k})}}. \tag{17.39}$$

Using now the explicit form of the vectors c_r we obtain

$$\begin{aligned}\sum_{r=0}^{3} c_r D_{rr}^{-1} c_r &= \sum_{r,s,t=0}^{3} q_s g_{ss} R_{sr} D_{rr}^{-1} q_t g_{tt} R_{tr} \\ &= \sum_{r,s,t=0}^{3} q_s g_{ss} R_{sr} D_{rr}^{-1} R_{rt}^{T} g_{tt} q_t \\ &= \sum_{s,t=0}^{3} q_s g_{ss} B_{st}^{-1} g_{tt} q_t \\ &= q_\sigma (B^{-1})^{\sigma\tau} q_\tau . \end{aligned} \tag{17.40}$$

In the last line we came back to the covariant notation, see Eq. (17.28). Putting the last two equations together we find

$$\rho\left(\mathbf{k}+\tfrac{1}{2}\mathbf{q},\mathbf{k}-\tfrac{1}{2}\mathbf{q}\right) = \mathcal{N}(\mathbf{k}) \exp\left[i q_\alpha \overline{x}^\alpha(\mathbf{k})\right] \frac{(2\pi)^2 \exp\left[-\frac{1}{2} q_\mu (B^{-1})^{\mu\nu} q_\nu\right]}{\sqrt{\det B(\mathbf{k})}}. \tag{17.41}$$

If the smoothness approximation (17.23) is valid, which is formally expressed by the condition

$$\frac{\mathcal{N}^2(\mathbf{k})}{\mathcal{N}(\mathbf{k}+\mathbf{q}/2)\mathcal{N}(\mathbf{k}-\mathbf{q}/2)} \left[\frac{\det(B(\mathbf{k}+\mathbf{q}/2)) \det(B(\mathbf{k}-\mathbf{q}/2))}{\det(B(\mathbf{k})) \det(B(\mathbf{k}))}\right]^{1/2} \approx 1, \tag{17.42}$$

the correlation function (17.24) corresponding to our Gaussian ansatz attains the simple form

$$C\left(\mathbf{k},\mathbf{q}\right) = 1 + \exp\left[-q_\mu (B^{-1})^{\mu\nu}(\mathbf{k})\, q_\nu\right]. \tag{17.43}$$

17.2.1 *Mass-shell projection*

Equation (17.43) may be rewritten in a form which explicitly uses the mass-shell condition (17.20). To do so, we first introduce a concept of averaging with the emission function $S(x,\mathbf{k})$ [13–17]. For $f(x,\mathbf{k})$ being an arbitrary function of the space-time coordinates x and the three-momentum $\mathbf{k}$, the average value of f is defined as

$$\langle f\rangle(\mathbf{k}) = \frac{\int d^4x\, f(x,\mathbf{k})\, S(x,\mathbf{k})}{\int d^4x\, S(x,\mathbf{k})}. \tag{17.44}$$

Going through the computational steps familiar from the previous Section we find

$$\langle \tilde{x}_\mu \tilde{x}_\nu\rangle(\mathbf{k}) = \langle \tilde{x}_\nu \tilde{x}_\mu\rangle(\mathbf{k}) = (B^{-1})_{\mu\nu}(\mathbf{k}) \tag{17.45}$$

and

$$C\left(\mathbf{k},\mathbf{q}\right) = 1 + \exp\left[-q^\mu\, q^\nu \langle \tilde{x}_\mu \tilde{x}_\nu\rangle(\mathbf{k})\right]. \tag{17.46}$$

With the help of the mass-shell condition (17.20) we obtain

$$
\begin{aligned}
q^\mu q^\nu \langle \tilde{x}_\mu \tilde{x}_\nu \rangle(\mathbf{k}) &= \sum_{i,j=1}^{3} q^i q^j \left[\beta^i \beta^j \langle \tilde{t}\, \tilde{t} \rangle + \langle \tilde{x}^i \, \tilde{x}^j \rangle - 2\beta^j \langle \tilde{t}\, \tilde{x}^i \rangle \right] (\mathbf{k}) \\
&= \sum_{i,j=1}^{3} q^i q^j \left[\beta^i \beta^j \langle \tilde{t}\, \tilde{t} \rangle + \langle \tilde{x}^i \, \tilde{x}^j \rangle - \beta^j \langle \tilde{t}\, \tilde{x}^i \rangle - \beta^i \langle \tilde{t}\, \tilde{x}^j \rangle \right] (\mathbf{k}) \\
&= \sum_{i,j=1}^{3} q^i q^j \langle (\tilde{x}^i - \beta^i \tilde{t})(\tilde{x}^j - \beta^j \tilde{t}) \rangle (\mathbf{k}). \qquad (17.47)
\end{aligned}
$$

Substituting Eq. (17.47) into Eq. (17.46) yields an equivalent form of the correlation function, namely

$$
C(\mathbf{k}, \mathbf{q}) = 1 + \exp \left[- \sum_{i,j=1}^{3} q^i q^j R^2_{ij}(\mathbf{k}) \right], \qquad (17.48)
$$

where the coefficients $R^2_{ij}(\mathbf{k})$ are defined as

$$
R^2_{ij}(\mathbf{k}) = \langle (\tilde{x}^i - \beta^i \tilde{t})(\tilde{x}^j - \beta^j \tilde{t}) \rangle (\mathbf{k}) \qquad (17.49)
$$

17.2.2 *Out-side-long coordinate system*

In the practical calculations it is useful to choose the coordinate system in such a way that the vector $\mathbf{k}$ has only two non-zero components

$$
\mathbf{k} = \left(k_\perp, 0, k_\parallel \right). \qquad (17.50)
$$

The z-axis agrees here with the beam axis and determines the *long* direction. In the transverse plane, the x-axis is chosen parallel to the vector component of the momentum of a pair, which is transverse to the beam direction. In this way the *out* direction is fixed. The remaining y-axis (of the orthogonal right-handed frame) determines the *side* direction. Similarly to Eq. (17.50), we have

$$
\boldsymbol{\beta} = \frac{\mathbf{k}}{k^0} = \left(\beta_\perp, 0, \beta_\parallel \right). \qquad (17.51)
$$

The symmetric matrix $R^2_{ij}(\mathbf{k})$ has 6 independent coefficients. In the *out-side-long* coordinate system the diagonal elements are

$$
\begin{aligned}
R^2_{11}(\mathbf{k}) &= R^2_{\text{out}}(\mathbf{k}) = \langle (\tilde{x} - \beta_\perp \tilde{t})^2 \rangle (\mathbf{k}), \\
R^2_{22}(\mathbf{k}) &= R^2_{\text{side}}(\mathbf{k}) = \langle \tilde{y}^2 \rangle (\mathbf{k}), \qquad (17.52) \\
R^2_{33}(\mathbf{k}) &= R^2_{\text{long}}(\mathbf{k}) = \langle (\tilde{z} - \beta_\parallel \tilde{t})^2 \rangle (\mathbf{k}),
\end{aligned}
$$

whereas the off-diagonal elements have the form

$$
\begin{aligned}
R^2_{12}(\mathbf{k}) &= R^2_{\text{out}-\text{side}}(\mathbf{k}) = \langle (\tilde{x} - \beta_\perp \tilde{t})\, \tilde{y} \rangle(\mathbf{k}), \\
R^2_{23}(\mathbf{k}) &= R^2_{\text{side}-\text{long}}(\mathbf{k}) = \langle \tilde{y}(\tilde{z} - \beta_\parallel \tilde{t}) \rangle(\mathbf{k}), \\
R^2_{31}(\mathbf{k}) &= R^2_{\text{long}-\text{out}}(\mathbf{k}) = \langle (\tilde{z} - \beta_\parallel \tilde{t})(\tilde{x} - \beta_\perp \tilde{t}) \rangle(\mathbf{k}).
\end{aligned}
\tag{17.53}
$$

17.2.2.1 *Boost-invariant cylindrically symmetric sources*

For cylindrically symmetric emission functions $\bar{y}(\mathbf{k}) = 0$, since we have a reflection symmetry with respect to the side direction, $y \to -y$. In this case the radii $R_{\text{out}-\text{side}}$ and $R_{\text{side}-\text{long}}$ vanish, since they are linear in $\tilde{y}$. If our system is boost-invariant, similar symmetry arguments hold for the z-axis, and the terms linear in $\tilde{z}$ vanish as well. This leads to further simplifications. In the longitudinally comoving system (LCMS), where in addition $\beta_\parallel = 0$, we find

$$
\begin{aligned}
R^2_{\text{out}}(\mathbf{k}) &= \langle (\tilde{x} - \beta_\perp \tilde{t})^2 \rangle(\mathbf{k}), \\
R^2_{\text{side}}(\mathbf{k}) &= \langle \tilde{y}^2 \rangle(\mathbf{k}), \\
R^2_{\text{long}}(\mathbf{k}) &= \langle \tilde{z}^2 \rangle(\mathbf{k}),
\end{aligned}
\tag{17.54}
$$

and

$$
R^2_{\text{out}-\text{side}}(\mathbf{k}) = R^2_{\text{side}-\text{long}}(\mathbf{k}) = R^2_{\text{long}-\text{out}}(\mathbf{k}) = 0. \tag{17.55}
$$

Moreover, in this case the correlation function becomes

$$
C(\mathbf{k}, \mathbf{q}) = 1 + \exp\left[-R^2_{\text{out}}(k_\perp) q^2_{\text{out}} - R^2_{\text{side}}(k_\perp) q^2_{\text{side}} - R^2_{\text{long}}(k_\perp) q^2_{\text{long}}\right], \tag{17.56}
$$

where $q_{\text{out}} = q^1, q_{\text{side}} = q^2$, and $q_{\text{long}} = q^3$; compare Eq. (4.12).

17.2.2.2 *Determination of HBT radii in model calculations*

Suppose that our theoretical model yields a boost-invariant cylindrically symmetric emission function $S(x, \mathbf{k})$. The formalism outlined above may be directly used to evaluate the HBT radii: $R_{\text{out}}(k_\perp), R_{\text{side}}(k_\perp)$, and $R_{\text{long}}(k_\perp)$, provided the emission function $S(x, \mathbf{k})$ is indeed well reproduced by the Gaussian function. In each case we calculate the Fourier transform, compare Eq. (15.35),

$$
S(k_\perp, q_i) = \int d^4x\, e^{i\tilde{q}_i \cdot x}\, S(x; k_\perp, 0, 0), \tag{17.57}
$$

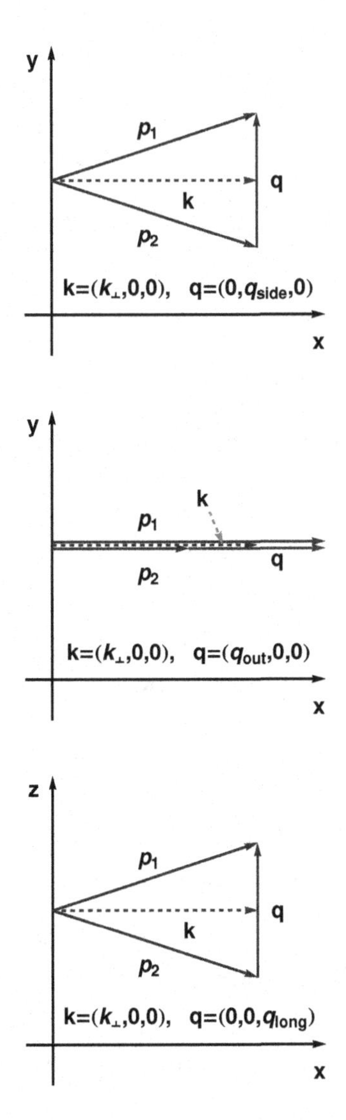

Fig. 17.1 Standard parametrizations of the vectors **k** and **q** used to determine the HBT radii $R_{\text{side}}, R_{\text{out}}$ and R_{long}.

where the three-vector **k** has been pointed along the x-axis, and the four-vector $\tilde{q}_i$ is one out of the three possible choices $i = \text{out, side, long}$,

$$
\begin{aligned}
\tilde{q}_{\text{out}} &= (\Delta E_{\text{out}}, q_{\text{out}}, 0, 0)\,, \\
\tilde{q}_{\text{side}} &= (\Delta E_{\text{side}}, 0, q_{\text{side}}, 0)\,, \\
\tilde{q}_{\text{long}} &= (\Delta E_{\text{long}}, 0, 0, q_{\text{long}})\,.
\end{aligned}
\tag{17.58}
$$

Only in the case of the out-radius, the difference between the energies of the two pions does not vanish

$$\Delta E_{\text{out}} = \sqrt{m_\pi^2 + (k_\perp + q_{\text{out}}/2)^2} - \sqrt{m_\pi^2 + (k_\perp - q_{\text{out}}/2)^2}. \tag{17.59}$$

In the two remaining cases we find, see Fig. 17.1,

$$\Delta E_{\text{side}} = \Delta E_{\text{long}} = 0. \tag{17.60}$$

Following (17.56), the HBT radii may be calculated as the derivatives

$$R_i^2 = -\left.\frac{\partial}{\partial q_i^2}\left[\frac{|S(k_\perp, q_i)|^2}{S^2(k_\perp, 0)}\right]\right|_{q_i^2=0}. \tag{17.61}$$

17.3 Monte-Carlo approach

Many theoretical models of heavy-ion collisions are coupled to the Monte-Carlo codes that may serve as event generators. With such a model at hand, the correlation function for identical pions may be obtained with the two-particle method [18, 19]. In this approach the evaluation of the correlation function is reduced to the calculation of the following expression

$$C(\mathbf{q}, \mathbf{k}) = \frac{\sum\limits_i \sum\limits_{j \neq i} \delta_\Delta(\mathbf{q} - \mathbf{p}_i + \mathbf{p}_j)\ \delta_\Delta\left(\mathbf{k} - \frac{1}{2}[\mathbf{p}_i + \mathbf{p}_j]\right)\ |\Psi(\mathbf{q}^*, \mathbf{r}^*)|^2}{\sum\limits_i \sum\limits_{j \neq i} \delta_\Delta(\mathbf{q} - \mathbf{p}_i + \mathbf{p}_j)\ \delta_\Delta\left(\mathbf{k} - \frac{1}{2}[\mathbf{p}_i + \mathbf{p}_j]\right)}, \tag{17.62}$$

where δ_Δ denotes the box function [3]

$$\delta_\Delta(\mathbf{p}) = \begin{cases} 1, & \text{if } |p_x| \le \frac{\Delta}{2}, |p_y| \le \frac{\Delta}{2}, |p_z| \le \frac{\Delta}{2} \\ 0, & \text{otherwise.} \end{cases} \tag{17.63}$$

In the numerator of Eq. (17.62) we calculate the sum of the squares of modules of the wave function determined for all pion pairs with the relative momentum $\mathbf{q}$ and the pair average momentum $\mathbf{k}$. For each pair the wave function $\Psi(\mathbf{q}^*, \mathbf{r}^*)$ is calculated in the rest frame of the pair; $\mathbf{q}^*$ and $\mathbf{r}^*$ denote the relative momentum and the relative distance in the pair rest frame, respectively. In the denominator of Eq. (17.62) we put the number of pairs with the relative momentum $\mathbf{q}$ and the average momentum $\mathbf{k}$. The correlation function (17.62) is then expressed with the help of the Bertsch-Pratt coordinates $k_T, q_{\text{out}}, q_{\text{side}}, q_{\text{long}}$ and approximated by the formula (17.56).

17.4 HBT data vs. theoretical expectations

The technique of the HBT correlations has been developed with the aim of measuring the pion emission regions. It has been soon realized, however, that the widths of

[3] In [18, 19] one uses the bin resolution $\Delta = 5$ MeV.

the correlation functions determine the sizes of the regions where the pions are correlated rather than the overall sizes of the physical systems from which the pions are emitted. For example, it is commonly observed that the correlation radii decrease with the increasing average momentum of the pair. This effect may be explained by the collective expansion of matter that collimates the emitted particles [20] — the fast pions have small relative momenta and are sensitive to Bose-Einstein statistics only if they are emitted from the neighboring spacetime points. After Makhlin and Sinyukov [21] we say that the correlation functions measure the *homogeneity lengths* of the system.

Much effort has been also put into better understanding of the repulsive Coulomb effects on the correlation function [22–24]. The corrections to the Gaussian form of the correlation function (4.12) have been worked out for this case. We refer to the modified form used in the fits as to the Bowler-Sinyukov formula. Besides the final state Coulomb interactions, also the final state strong interactions are studied [25, 26].

The potential ability of the correlation measurement to deliver information about the time extension of the system created in heavy-ion collisions was an inspiring point for many studies that connected the HBT technique with the analysis of the phase transition between the quark-gluon plasma and a hadron gas. The main argument used in those studies was that the first order phase transition should manifest itself by the long living fireball [27–30]. In the studies of the correlation functions, this behavior would be reflected in the large values of the ratio $R_{\rm out}/R_{\rm side}$.

One of the spectacular RHIC results is that the measured ratio $R_{\rm out}/R_{\rm side}$ is close to unity, see Figs. 4.7 and 4.8. This result contradicts the conjecture that the deconfinement phase transition is of the first order. There are, however, further problems with the theoretical description of the HBT radii, since most of the hydrodynamic and kinetic models are not able to reproduce the experimental HBT results [31, 32]. This problem is known as the RHIC HBT puzzle. It is worth mentioning that the failure of the kinetic and hydrodynamic models may be contrasted with the successes of simple thermal models which consistently describe one- and two-particle observations, for example, see [33].

For more information about the pion correlations we refer the reader to the review articles [5, 6, 12, 34–39].

Bibliography to Chapter 17

[1] M. Gyulassy, S. K. Kauffmann, and L. W. Wilson, "Pion interferometry of nuclear collisions. I. Theory," *Phys. Rev.* **C20** (1979) 2267–2292.

[2] R. Hanbury Brown and R. Q. Twiss, "A New type of interferometer for use in radio astronomy," *Phil. Mag.* **45** (1954) 663–682.

[3] R. Hanbury Brown and R. Q. Twiss, "A Test of a new type of stellar interferometer on Sirius," *Nature* **178** (1956) 1046–1048.

[4] G. I. Kopylov and M. I. Podgoretsky, "Interference of two-particle states in

elementary-particle physics and astronomy," *Zh. Eksp. Teor. Fiz.* **69** (1975) 414–421.

[5] G. Baym, "The physics of Hanbury Brown-Twiss intensity interferometry: From stars to nuclear collisions," *Acta Phys. Polon.* **B29** (1998) 1839–1884.

[6] R. Lednicky, "Correlation femtoscopy of multiparticle processes," *Phys. Atom. Nucl.* **67** (2004) 72–82.

[7] G. I. Kopylov and M. I. Podgoretsky, "Correlations of identical particles emitted by highly excited nuclei," *Sov. J. Nucl. Phys.* **15** (1972) 219–223.

[8] G. I. Kopylov and M. I. Podgoretsky, "Multiple production and interference of particles emitted by moving sources," *Sov. J. Nucl. Phys.* **18** (1974) 336–341.

[9] G. I. Kopylov, "Like particle correlations as a tool to study the multiple production mechanism," *Phys. Lett.* **B50** (1974) 472–474.

[10] G. Goldhaber, S. Goldhaber, W.-Y. Lee, and A. Pais, "Influence of Bose-Einstein statistics on the antiproton proton annihilation process," *Phys. Rev.* **120** (1960) 300–312.

[11] W. A. Zajc, in "Particle Production in Highly Matter", H.H. Gutbrod and J. Rafelski eds., NATO ASI Series B **303** (Plenum, New York, 1993) p. 435.

[12] U. A. Wiedemann and U. W. Heinz, "Particle interferometry for relativistic heavy-ion collisions," *Phys. Rept.* **319** (1999) 145–230.

[13] S. Chapman, P. Scotto, and U. W. Heinz, "Model independent features of the two-particle correlation function," *Heavy Ion Phys.* **1** (1995) 1–31.

[14] S. Chapman, P. Scotto, and U. W. Heinz, "A New cross term in the two particle HBT correlation function," *Phys. Rev. Lett.* **74** (1995) 4400–4403.

[15] S. Chapman, J. R. Nix, and U. W. Heinz, "Extracting source parameters from Gaussian fits to two particle correlations," *Phys. Rev.* **C52** (1995) 2694–2703.

[16] U. W. Heinz, B. Tomasik, U. A. Wiedemann, and Y. F. Wu, "Lifetimes and sizes from two-particle correlation functions," *Phys. Lett.* **B382** (1996) 181–188.

[17] U. A. Wiedemann, P. Scotto, and U. W. Heinz, "Transverse momentum dependence of Hanbury-Brown-Twiss correlation radii," *Phys. Rev.* **C53** (1996) 918–931.

[18] A. Kisiel, W. Florkowski, and W. Broniowski, "Femtoscopy in hydro-inspired models with resonances," *Phys. Rev.* **C73** (2006) 064902.

[19] A. Kisiel, "Non-identical particle femtoscopy in models with single freeze-out," *Braz. J. Phys.* **37** (2007) 917–924.

[20] S. Pratt, "Pion interferometry for exploding sources," *Phys. Rev. Lett.* **53** (1984) 1219–1221.

[21] A. N. Makhlin and Y. M. Sinyukov, "The hydrodynamics of hadron matter under a pion interferometric microscope," *Z. Phys.* **C39** (1988) 69.

[22] S. Pratt, "Coherence and Coulomb effects on pion interferometry," *Phys. Rev.* **D33** (1986) 72–79.

[23] M. G. Bowler, "Coulomb corrections to Bose-Einstein correlations have been greatly exaggerated," *Phys. Lett.* **B270** (1991) 69–74.

[24] Y. Sinyukov, R. Lednicky, S. V. Akkelin, J. Pluta, and B. Erazmus, "Coulomb corrections for interferometry analysis of expanding hadron systems," *Phys. Lett.* **B432** (1998) 248–257.

[25] S. Pratt, T. Csorgo, and J. Zimanyi, "Detailed predictions for two pion correlations in ultrarelativistic heavy ion collisions," *Phys. Rev.* **C42** (1990) 2646–2652.

[26] R. Lednicky and V. L. Lyuboshits, "Final state interaction effect on correlations in narrow particles pairs," in "Nantes 1990, Proceedings, Particle correlations and interferometry in nuclear collisions", 42-54.

[27] S. Pratt, "Pion interferometry of quark-gluon plasma," *Phys. Rev.* **D33** (1986)

1314–1327.
[28] G. F. Bertsch, "Pion interferometry as a probe of the plasma," *Nucl. Phys.* **A498** (1989) 173c–180c.
[29] D. H. Rischke and M. Gyulassy, "The time-delay signature of quark-gluon plasma formation in relativistic nuclear collisions," *Nucl. Phys.* **A608** (1996) 479–512.
[30] S. Soff, S. A. Bass, and A. Dumitru, "Pion interferometry at RHIC: Probing a thermalized quark gluon plasma?," *Phys. Rev. Lett.* **86** (2001) 3981–3984.
[31] Q. Li, M. Bleicher, and H. Stoecker, "Transport model analysis of particle correlations in relativistic heavy ion collisions at femtometer scales," *Phys. Rev.* **C73** (2006) 064908.
[32] Q. Li, M. Bleicher, and H. Stocker, "Pion freeze-out as seen through HBT correlations in heavy ion collisions from FAIR/AGS to RHIC energies," *J. Phys.* **G34** (2007) 2037–2044.
[33] F. Retiere and M. A. Lisa, "Observable implications of geometrical and dynamical aspects of freeze-out in heavy ion collisions," *Phys. Rev.* **C70** (2004) 044907.
[34] W. Bauer, C. K. Gelbke, and S. Pratt, "Hadronic interferometry in heavy ion collisions," *Ann. Rev. Nucl. Part. Sci.* **42** (1992) 77–100.
[35] U. W. Heinz and B. V. Jacak, "Two-particle correlations in relativistic heavy-ion collisions," *Ann. Rev. Nucl. Part. Sci.* **49** (1999) 529–579.
[36] R. M. Weiner, "Boson interferometry in high energy physics," *Phys. Rept.* **327** (2000) 249–346.
[37] T. Csorgo, "Particle interferometry from 40 MeV to 40 TeV," *Heavy Ion Phys.* **15** (2002) 1–80.
[38] B. Tomasik and U. A. Wiedemann, "Central and non-central HBT from AGS to RHIC," `hep-ph/0210250`.
[39] K. Zalewski, "Bose-Einstein correlations in multiple particle production from 1959 to 1989," `hep-ph/0607222`.

Chapter 18

Exercises to PART V

Exercise 18.1. *Very simple model for the correlation function.*
Show that the model introduced in Sec. 17.1.3 may be defined equivalently by the density operator of the form

$$\hat{\rho} = \int d^3x |\mathbf{x}\rangle \rho(\mathbf{x}) \langle \mathbf{x}|.$$

This form of $\hat{\rho}$ describes a single particle mixed state that is an incoherent superposition of pure states, each corresponding to a given production point. Calculate the correlation function for the two cases: **i)** $\rho(\mathbf{x}) = \rho_0 = \text{const.}$ for $|\mathbf{x}| \leq R_S$ and $\rho(\mathbf{x}) = 0$ for $|\mathbf{x}| > R_S$, **ii)** $\rho(\mathbf{x}) = \exp(-\mathbf{x}^2/\lambda)$. The parameters ρ_0, R_S and λ are real constants. Show that the models i) and ii) are equivalent for small values of the relative momentum q if $R_S = \sqrt{5\lambda}$.

Exercise 18.2. *Zajc model.*
Consider a model for the emission function proposed by Zajc,

$$S(t, \mathbf{x}, \mathbf{p}) = N_s N \exp\left[-\frac{1}{2(1-s^2)}\left(\frac{\mathbf{x}^2}{R_0^2} - 2s\frac{\mathbf{x}\cdot\mathbf{p}}{R_0 P_0} + \frac{\mathbf{p}^2}{P_0^2}\right)\right]\delta(t), \tag{18.1}$$

where N, R_0, P_0 and $0 \leq s < 1$ are parameters, and

$$N_s = \frac{E_p}{(2\pi R_s P_0)^3}, \quad R_s = R_0\sqrt{1-s^2}. \tag{18.2}$$

Calculate the correlation function using Eq. (17.22) and then Eq. (17.24). Compare the results.

Answers: If Eq. (17.22) is used one gets

$$C(\mathbf{q}, \mathbf{k}) = 1 + \exp\left(-R_{\mathrm{HBT}}^2 \mathbf{q}^2\right), \tag{18.3}$$

where

$$R_{\mathrm{HBT}}^2 = R_s^2\left(1 - \frac{1}{(2R_s P_0)^2}\right). \tag{18.4}$$

If Eq. (17.24) is used, then the form (18.3) is valid but R_{HBT} is reduced to R_s.

Exercise 18.3. *Boost-invariant emission function.*
Consider a system that is described by the boost-invariant and cylindrically symmetric emission function

$$S(x,p) = \frac{m_\perp \cosh(\eta_\parallel - \mathrm{y})}{(2\pi)^3} f(r)\, \delta(\tau - \tau_\mathrm{f}) \times \exp\left[-\beta\left(m_\perp \cosh(\eta_\parallel - \mathrm{y}) \cosh\vartheta_\perp - p_\perp \cos(\phi - \phi_p) \sinh\vartheta_\perp\right)\right], \tag{18.5}$$

where the function $f(r)$ describes the distribution of emitting sources in the transverse plane, β and τ_f are constants, and $\vartheta_\perp$ is the transverse fluid rapidity that may depend on the transverse position r.

i) Make the substitution

$$\mathbf{p} = (p_x, p_y, p_z) \rightarrow \mathbf{k} = (k_\perp, 0, 0)$$

and using Eq. (17.57) show that the Fourier transforms of the emission function (18.5) in the side, out, and long directions may be expressed by the following expressions:

$$S(k_\perp, q_\mathrm{side}) = \frac{\tau_\mathrm{f} m_\perp}{2\pi^2} \int_0^\infty r\, dr\, f(r)\, K_1(\beta m_\perp \cosh\vartheta_\perp) \times \mathcal{I}_0(\beta k_\perp \sinh\vartheta_\perp, q_\mathrm{side} r), \tag{18.6}$$

$$S(k_\perp, q_\mathrm{out}) = \frac{\tau_\mathrm{f} m_\perp}{2\pi^2} \int_0^\infty r\, dr\, f(r)\, K_1(\beta m_\perp \cosh\vartheta_\perp - i\Delta E_\mathrm{out} \tau_\mathrm{f}) \times I_0(\beta k_\perp \sinh\vartheta_\perp - i q_\mathrm{out} r), \tag{18.7}$$

$$S(k_\perp, q_\mathrm{long}) = \frac{\tau_\mathrm{f} m_\perp}{2\pi^2} \int_0^\infty r\, dr\, f(r)\, \mathcal{K}_1(\beta m_\perp \cosh\vartheta_\perp, q_\mathrm{long} \tau_\mathrm{f}) \times I_0(\beta k_\perp \sinh\vartheta_\perp). \tag{18.8}$$

The functions $\mathcal{I}_0$ and $\mathcal{K}_1$ are defined below by Eq. (25.33) and Eq. (25.42), respectively.

ii) Suppose that the function $f(r)$ is peaked at some value of r, $f(r) \sim \delta(r - R_\mathrm{geo})$. In this case, that corresponds physically to the surface emission, the integrals on the right-hand-side of Eqs. (18.6) – (18.8) may be replaced by the values of the integrands at $r = R_\mathrm{geo}$. Using Eqs. (18.6) and (17.61) show that the side radius may be calculated from the formula

$$R^2_\mathrm{side} = -R^2_\mathrm{geo} \frac{\partial}{\partial b^2} \left[\frac{\mathcal{I}_0^2(a,b)}{\mathcal{I}_0^2(a,0)}\right]\Bigg|_{b^2=0}, \tag{18.9}$$

where

$$a = \beta k_\perp \sinh\vartheta_\perp(R_\mathrm{geo}), \quad b = q_\mathrm{side} R_\mathrm{geo}. \tag{18.10}$$

In the next step, using Eq. (25.33) show that

$$R_{\rm side}^2 = R_{\rm geo}^2 \frac{I_1(a)}{a I_0(a)}. \tag{18.11}$$

Expanding this equation for small values of a one gets

$$R_{\rm side}^2 = \frac{R_{\rm geo}^2}{2}\left(1 - \frac{a^2}{8}\right). \tag{18.12}$$

Thus, if the expansion may be neglected we obtain the pure geometric result $R_{\rm side}^2 = R_{\rm geo}^2/2$. We also observe that the flow reduces the side radius.

iii) Similarly to the previous case, using Eqs. (18.8) and (17.61) show that the long radius may be calculated from the formula

$$R_{\rm long}^2 = -\tau_{\rm f}^2 \frac{\partial}{\partial b^2}\left[\frac{\mathcal{K}_1^2(a,b)}{\mathcal{K}_1^2(a,0)}\right]\Bigg|_{b^2=0}, \tag{18.13}$$

where

$$a = \beta m_\perp \cosh\vartheta_\perp(R_{\rm geo}), \quad b = q_{\rm long}\tau_{\rm f}. \tag{18.14}$$

Using Eq. (25.42) derive the Herrmann-Bertsch formula

$$R_{\rm long}^2 = \tau_{\rm f}^2 \frac{K_2(a)}{a K_1(a)}. \tag{18.15}$$

This time, expanding this equation for large values of a one gets

$$R_{\rm long}^2 = \frac{\tau_{\rm f}^2}{a} = \frac{\tau_{\rm f}^2 T}{m_\perp \cosh\vartheta_\perp(R_{\rm geo})}. \tag{18.16}$$

Even for very large transverse flow, reaching half the speed of light, we have $\cosh\vartheta_\perp \approx 1$. Then, Eq. (18.16) is reduced to the famous Makhlin–Sinyukov formula

$$R_{\rm long} = \tau_{\rm f}\sqrt{\frac{T}{m_\perp}}. \tag{18.17}$$

Equation (18.17) has been frequently used in the estimates of $R_{\rm long}$. However, we must remember that the limit $a \gg 1$ is sometimes not very well justified. In such situation Eq. (18.15) should be used.

iv) Use similar methods to obtain $R_{\rm out}$. In this case, in the limit of small transverse flow and large $\beta m_\perp$, we find the Chapman-Scotto-Heinz formula

$$R_{\rm out}^2 \approx R_{\rm side}^2 + \frac{k_\perp^2 \tau_{\rm f}^2 T^2}{2 m_\perp^4}. \tag{18.18}$$

PART VI

HYDRODYNAMIC DESCRIPTION OF NUCLEAR COLLISIONS

Chapter 19

Historical Perspective: Fermi Statistical Model

The use of relativistic hydrodynamics to describe particle production in hadronic collisions has a long history which starts with the famous work of Landau in the early 1950s [1], see also [2, 3]. Landau's considerations were preceded, however, by a few approaches that used pure statistical and thermodynamic concepts in the analysis of the hadronic collisions [4–6]. Such approaches may be regarded as pre-hydrodynamic models and we think it is important to present them before we turn to the discussion of the genuine hydrodynamic models.

Probably, the most influential early statistical model of hadron production was formulated by Fermi [6, 7]. Fermi assumed that when two relativistic nucleons collide, the energy available in their center-of-mass frame is released in a very small volume V, whose magnitude corresponds to the Lorentz contracted characteristic pion field volume V_0, i.e., $V = 2m_{\rm N}V_0/\sqrt{s}$, where $V_0 = (4/3)\,\pi R_\pi^3$ with $R_\pi = 1/m_\pi$, and $\sqrt{s}$ is the center-of-mass energy. Subsequently, such a dense system decays into one of many accessible multiparticle states. The decay probability is calculated in the framework of the standard statistical physics.

The reason for the introduction of the statistical approach was the breakdown of the perturbation theory in the attempts to describe strongly interacting systems. Clearly, the large values of the coupling constant prohibit the application of the perturbation theory. On the other hand, the large value of the coupling is responsible for the phenomenon of multiple production of particles, which is a characteristic feature of strong interactions (in QED, where the coupling is small, the production of many particle states is suppressed by the appropriate power of the fine structure constant). Generally speaking, the probability of the transition into a given state is proportional to the square of the corresponding matrix element and to the density of states. In the statistical description the matrix elements are treated as constants and the main effect comes from the phase space. Thus, the statistical approach represents a simple theoretical modeling of collisions which may be regarded as the complementary approach to the perturbation schemes which typically break down at a certain scale. The main heuristic argument for the justification of the use of the statistical approach is that the role of the phase space naturally grows with the increasing energy of the collisions.

19.1 Pion production in low-energy nucleon collisions

The first physical example considered in Ref. [6] is the pion production in low-energy nucleon-nucleon collisions. This case was chosen mainly because of its simplicity — only a few states of rather low energy are taken into account in this analysis. In fact, Fermi considers exactly two states: the state (a) where the two nucleons scatter elastically, and the state (b) in which an additional pion is formed. In the case (b) we have altogether 3 particles — two nucleons and a pion. In view of the introductory remarks presented above, the application of the statistical approach to describe low-energy collisions cannot be sufficiently well justified. The work of Fermi showed, however, that this kind of approach led to surprisingly good description of the data. This success evidently inspired Fermi and his followers to apply the statistical approach to various hadronic processes.

The general formula used in the Fermi model to find the probability for the formation of the state with n particles has the form

$$S(n) = \left[\frac{V}{(2\pi)^3}\right]^{(n-1)} \frac{dQ(W)}{dW}. \tag{19.1}$$

Here W is the total energy of the colliding system, dQ/dW is the number of states per unit energy, and V is the interaction volume. The power $n-1$ arises from the fact that in the center-of-mass frame the momenta of only $n-1$ particles are independent. We note that the probabilities $S(n)$ given by Eq. (19.1) are not normalized and the appropriate normalization should be done in the end of the calculations.

In this Section we assume that V is practically equal to V_0 (low energy) and W is the kinetic energy of the two colliding nucleons. We denote this kinetic energy as $T_{\rm kin}$. The phase space volume of the case (a) is the two-nucleon phase space corresponding to $T_{\rm kin}$. At low energy, i.e., for $T_{\rm kin}$ slightly exceeding the pion mass m_π, $Q_2(T_{\rm kin})$ may be calculated in the non-relativistic limit

$$\begin{aligned} Q_2(T_{\rm kin}) &= \int d^3q_1 d^3q_2\, \theta\left(T_{\rm kin} - \frac{q_1^2}{2m_{\rm N}} - \frac{q_2^2}{2m_{\rm N}}\right) \delta^{(3)}\left(\mathbf{q}_1 + \mathbf{q}_2\right) \\ &= \int d^3q_1 \theta\left(T_{\rm kin} - \frac{q_1^2}{m_{\rm N}}\right) = 4\pi \int_0^{\sqrt{T_{\rm kin} m_{\rm N}}} q_1^2 dq_1 = \frac{4\pi}{3}\left(T_{\rm kin} m_{\rm N}\right)^{3/2}. \end{aligned} \tag{19.2}$$

Here $\mathbf{q}_1$ and $\mathbf{q}_2$ are the final nucleon momenta and θ denotes the step function. According to Fermi's theory, see Eq. (19.1), the probability for the formation of such a state is given by the expression

$$S\,(2) = \frac{V}{(2\pi)^3} \frac{dQ_2}{dT_{\rm kin}} = \frac{V m_{\rm N}^{3/2}}{4\pi^2} T_{\rm kin}^{1/2}. \tag{19.3}$$

Similarly, the non-relativistic volume of the three-particle phase space of two nucleons and a pion is obtained from the formula

$$\begin{aligned}
Q_3(T_{\rm kin}) &= \int d^3p\, d^3q_1 d^3q_2\, \theta\left(\Delta T_{\rm kin} - \frac{p^2}{2m_\pi} - \frac{q_1^2}{2m_{\rm N}} - \frac{q_2^2}{2m_{\rm N}}\right) \delta^{(3)}\left(\mathbf{p}+\mathbf{q}_1+\mathbf{q}_2\right) \\
&= \int d^3p\, d^3P d^3q\, \theta\left(\Delta T_{\rm kin} - \frac{p^2}{2m_\pi} - \frac{\frac{1}{2}P^2 + 2\,q^2}{2m_{\rm N}}\right) \delta^{(3)}\left(\mathbf{p}+\mathbf{P}\right) \\
&= \int d^3p\, d^3q\, \theta\left(\Delta T_{\rm kin} - \frac{p^2}{a^2} - \frac{q^2}{b^2}\right) \\
&= (4\pi)^2\, a^3 b^3 \int_0^\infty x^2 dx \int_0^\infty y^2 dy\, \theta\left(\Delta T_{\rm kin} - x^2 - y^2\right). \qquad (19.4)
\end{aligned}$$

Here $\Delta T_{\rm kin} = T_{\rm kin} - m_\pi$ is the kinetic energy left over after a pion was created, $\mathbf{p}$ is the final pion momentum, $\mathbf{P} = \mathbf{q}_1 + \mathbf{q}_2$, and $\mathbf{q} = (\mathbf{q}_1 - \mathbf{q}_2)/2$. In Eq. (19.4) we introduced also the coefficients a and b defined by the equations

$$\frac{1}{a^2} = \frac{1}{2m_\pi} + \frac{1}{4m_{\rm N}}, \qquad \frac{1}{b^2} = \frac{1}{m_{\rm N}}. \qquad (19.5)$$

The last integral in Eq. (19.4) may be done in the polar coordinates where $x = r\cos\phi$ and $y = r\sin\phi$,

$$\begin{aligned}
Q_3(T_{\rm kin}) &= 16\,\pi^2 \left(\frac{4\,m_\pi\, m_{\rm N}^2}{2m_{\rm N} + m_\pi}\right)^{3/2} \int_0^{\sqrt{\Delta T_{\rm kin}}} r^5 dr \int_0^{\pi/2} \cos^2\phi\, \sin^2\phi\, d\phi \\
&= \frac{\pi^3}{6} \left(\frac{4\,m_\pi\, m_{\rm N}^2}{2m_{\rm N} + m_\pi}\right)^{3/2} (\Delta T_{\rm kin})^3. \qquad (19.6)
\end{aligned}$$

Thus, the probability of formation of the two-nucleon plus one-pion state is

$$\begin{aligned}
S(3) &= \frac{V^2}{(2\pi)^6} \frac{dQ_3}{dT_{\rm kin}} = \frac{V^2}{16\pi^3} \left(\frac{m_\pi\, m_{\rm N}^2}{2m_{\rm N} + m_\pi}\right)^{3/2} (\Delta T_{\rm kin})^2 \\
&\approx \frac{V^2 m_{\rm N}^{3/2} m_\pi^{3/2} (\Delta T_{\rm kin})^2}{32\sqrt{2}\pi^3}. \qquad (19.7)
\end{aligned}$$

We may consider the ratio $S(3)/(S(2) + S(3)) \approx S(3)/S(2)$ as the probability of the pion formation,

$$\frac{S(3)}{S(2)} = \frac{V m_\pi^{3/2}}{8\sqrt{2}\pi} \frac{(T_{\rm kin} - m_\pi)^2}{T_{\rm kin}^{1/2}} \approx \frac{V m_\pi}{8\sqrt{2}\pi} (T_{\rm kin} - m_\pi)^2. \qquad (19.8)$$

Here we assumed that $T_{\rm kin}$ is slightly larger than the threshold energy m_π. Assuming that the Lorentz contraction is negligible, $V = V_0 = (4/3)\,\pi m_\pi^{-3}$, we obtain the compact result

$$\frac{S(3)}{S(2)} = \frac{1}{6\sqrt{2}}\left(\frac{T_{\rm kin}}{m_\pi} - 1\right)^2. \tag{19.9}$$

The cross-section for pion formation is given by the product of the total cross section and the probability (19.9). Fermi compared the result of his statistical model to the data on nucleon-nucleon collisions at the beam energy of 345 MeV (the proton energy available around 1950 at Berkeley) and found surprisingly good agreement. The idea of statistical production seemed to work well (even outside the region of its potential best applicability!).

19.2 Multiple pion production

Let us generalize the considerations of the previous Section to the multiple pion production at higher energies. In order to have compact analytic expressions we neglect spin, disregard the momentum and angular momentum conservation, and treat all particles as independent. In this case the general Fermi formula reads

$$S(n) = \left[\frac{V}{(2\pi)^3}\right]^n \frac{dQ(W)}{dW}, \tag{19.10}$$

where n is the number of the produced particles. The phase space density may be easily found for the case where the final state consists of non-relativistic particles with masses m_i and for the case where the final state consists of ultra-relativistic (massless) particles. The corresponding two formulas are (see Exercises 23.1 – 23.3)

$$S(n) = \frac{V^n (m_1 ... m_n)^{3/2}}{(2\pi)^{3n/2}} \frac{(T_{\rm kin})^{3n/2-1}}{\Gamma(3n/2)} \tag{19.11}$$

and

$$S(n) = \frac{V^n}{\pi^{2n}} \frac{W^{3n-1}}{\Gamma(3n)}. \tag{19.12}$$

It is also possible to work out the formula for the case where s particles are non-relativistic, with mass $m_{\rm N}$ (nucleons), and n particles are massless (approximate treatment of pions). In this case the available kinetic energy equals $W - s\,m_{\rm N}$ hence we get

$$\begin{aligned} S(s,n) &= \frac{V^{s+n}}{(2\pi)^{3(s+n)}} \frac{dQ_{s+n}(W - s m_{\rm N})}{dW} \\ &= \frac{m_{\rm N}^{3s/2} V^{s+n}}{2^{3s/2}\pi^{2n+3s/2}} \frac{(W - s m_{\rm N})^{3n+3s/2-1}}{\Gamma(3n+3s/2)}. \end{aligned} \tag{19.13}$$

It is convenient to rewrite this equation in the equivalent form as

$$S(s,n) = \frac{m_{\rm N}^{3s/2} V^{s/2+1/3}}{2^{3s/2}\pi^{s/2+2/3}\Gamma(3n+3s/2)} \left[\frac{V^{1/3}(W - s m_{\rm N})}{\pi^{2/3}}\right]^{3n+3s/2-1}. \quad (19.14)$$

Equation (19.14) may be corrected to include the momentum conservation. Since the momenta of nucleons are much greater than the momenta of pions, the changes due to the momentum conservation may be applied to the nucleons only. Similarly to the situation considered in Sec. 19.1 such changes are included by reducing the number of outcoming particles by one. Because we concentrate only on the nucleons now, s should be changed to $s-1$ in all places except for the term $W - s m_{\rm N}$. Taking into account the events where two nucleons collide and a few new pions appear (besides the two original nucleons) we should set $s = 2$ in the corrected formula. In this way we arrive at the expression

$$S(2,n) = \frac{s_0}{\Gamma(3n+3/2)} \left[\frac{V^{1/3}(W - 2m_{\rm N})}{\pi^{2/3}}\right]^{3n+1/2}, \quad (19.15)$$

where s_0 is a constant which absorbed the s-dependent prefactor in Eq. (19.14). Following Fermi we obtain the volume V as the Lorentz contracted typical pion field volume, namely

$$V = V_0 \frac{2m_{\rm N}}{W} = \frac{4}{3}\frac{\pi}{m_\pi^3}\frac{2m_{\rm N}}{W} = \frac{8\pi}{3}\frac{m_{\rm N}}{m_\pi^3 W}. \quad (19.16)$$

Substituting Eq. (19.16) into Eq. (19.15) one obtains

$$S(2,n) = \frac{s_0}{\Gamma(3n+3/2)} \left[\left(\frac{8}{3\pi}\right)^{1/3} \frac{m_{\rm N}}{m_\pi} \frac{\bar{w}-2}{\bar{w}^{1/3}}\right]^{3n+1/2}, \quad (19.17)$$

where $\bar{w} = W/m_{\rm N}$ is the initial energy measured in the units of the nucleon mass [1]. The probabilities $\bar{p}(n,\bar{w})$ of producing n pions at the initial energy corresponding to $\bar{w}$ are obtained from the formula

$$\bar{p}(n,\bar{w}) = \frac{S(2,n,\bar{w})}{\sum_{k=0}^{\infty} S(2,k,\bar{w})}. \quad (19.18)$$

The probabilities $\bar{p}(n,\bar{w})$ obtained for different collision energies are given in Table 19.1. One observes the fast decrease of the probability for elastic nucleon-nucleon collisions. We note that the values shown here are slightly different from the original values given in Ref. [6]. The differences are caused by the slightly different input value for $m_\pi/m_{\rm N}$ and by the exact numerical calculation of the Euler gamma function defined by Eq. (8.68).

[1] In this Section we use Fermi's original notation. Note that w has no relation to wounded nucleons or enthalpy.

Table 19.1 Probabilities for production of n pions in nucleon-nucleon collsions at the energy $W = \bar{w} m_{\rm N}$.

$\bar{w}$	W [GeV]	$n = 0$	1	2	3	4	5	6	7	$\bar{n}$
2.5	2.35	48.6	47.6	3.8	0.1					0.6
3.0	2.82	8.7	56.6	30.2	4.3	0,2				1.3
3.5	3.29	1.6	30.0	46.2	18.9	3.1	0.3			1.9
4.0	3.76	0.3	12.5	39.8	33.8	11.5	1.9	0.2		2.5
4.5	4.23		4.7	26.1	38.5	22.7	6.7	1.1	0.1	3.0
5.0	4.70		1.7	14.8	33.8	31.1	14.2	3.7	0.6	3.6

19.3 From statistical to thermodynamical production

In the last chapter of his seminal paper [6] Fermi considers the collisions at extremely high energies. He argues that in this case a detailed statistical considerations may be replaced by the simple thermodynamic arguments. Assuming that the matter is thermalized one can calculate the temperature of the produced hadronic system from the thermodynamic relations valid for massless particles. Fermi took into account only the production of pions, nucleons and antinucleons. For pions we use Eq. (8.62) multiplied by a factor of 3 which accounts for three different isospin states. For nucleoons and antinucleons we use Eq. (8.83) multiplied by a factor of 4 (two different spins and isospins of nucleons). In this way we obtain the expression for the energy density, $\varepsilon_\pi + \varepsilon_{N+\bar{N}}$, which may be related to the initial beam energy W, namely

$$\left(\varepsilon_\pi + \varepsilon_{N+\bar{N}}\right) V = \frac{\pi^2 V T^4}{3} = W. \tag{19.19}$$

Using the expression for the Lorentz contracted volume we rewrite Eq. (19.19) in the following form

$$T^4 = \frac{3}{2\pi^2} \frac{W^2}{V_0 m_{\rm N}} = \frac{9}{8\pi^3} \frac{W^2 m_\pi^3}{m_{\rm N}}. \tag{19.20}$$

This equation may be used to calculate the abundances of the produced pions, nucleons and antinucleons from the thermodynamic relations giving the particle densities in terms of the temperature.

19.4 Fermi's legacy

In the last twenty years, the thermodynamic method of analyzing hadronic collisions has been developed into an important subfield of high-energy nuclear physics. At present, this kind of approach is particularly useful in the analysis of hadronic abundances which yield the information about the physical conditions at the chemical freeze-out. The main difference between the current thermodynamic models and the original Fermi model resides in the fact that nowadays one considers all known hadronic states (resonances) and various properties of particles may be treated exactly. The inclusion of all resonances leads to complications because one has to take into account the feeding from the decays of heavier resonances to the yields of the light particles that are directly detected. On the other hand, the inclusion of such decays turns out to be the most important feature responsible for the success of thermodynamic models in explaining various heavy-ion data. We shall discuss the recent thermodynamic approaches, called popularly the thermal models, in more detail in Chap. 24.

Recently, the progress has been made also in the area of the statistical models. The initial studies of the phase-space integrals changed into the advanced analysis of the microcanonical ensembles [8–11].

Bibliography to Chapter 19

[1] L. D. Landau, "On the multiparticle production in high-energy collisions," *Izv. Akad. Nauk SSSR Ser. Fiz.* **17** (1953) 51–64.

[2] I. M. Khalatnikov, "Some questions of the relativistic hydrodynamics," *Zh. Eksp. Teor. Fiz.* . **26** (1954) 529.

[3] S. Z. Belenkij and L. D. Landau, "Hydrodynamic theory of multiple production of particles," *Nuovo Cim. Suppl.* **3S10** (1956) 309.

[4] H. Koppe, "Die Mesonenausbeute beim Beschuss von leichten Kernen mit a-Teilchen," *Z. f. Naturforschung (Notizen)* **3A** (1948) 251.

[5] H. Koppe, "On the Production of Mesons," *Phys. Rev.* **76** (1949) 688–688.

[6] E. Fermi, "High-energy nuclear events," *Prog. Theor. Phys.* **5** (1950) 570–583.

[7] E. Fermi, "Angular Distribution of the Pions Produced in High Energy Nuclear Collisions," *Phys. Rev.* **81** (1951) 683–687.

[8] F. Becattini and L. Ferroni, "Statistical model and microcanonical ensemble," *J. Phys.* **G31** (2005) S1091–S1094.

[9] F. Becattini and L. Ferroni, "The microcanonical ensemble of the ideal relativistic quantum gas," *Eur. Phys. J.* **C51** (2007) 899–912.

[10] F. Becattini and L. Ferroni, "The microcanonical ensemble of the ideal relativistic quantum gas with angular momentum conservation," *Eur. Phys. J.* **C52** (2007) 597–615.

[11] F. Becattini and F. Piccinini, "The ideal relativistic spinning gas: polarization and spectra," *Annals Phys.* **323** (2008) 2452–2473.

Chapter 20

Landau Model

An important assumption of the Fermi model [1] is that the system reaches equilibrium at the stage of maximum compression and then instantaneously breaks up into multiparticle hadronic states. This assumption is relaxed in the Landau hydrodynamic model [2]: instead of assuming instantaneous break-up, Landau's approach allows for the expansion of matter before the hadron decoupling. This expansion is described by the hydrodynamics of the perfect fluid. It allows the system to cool down on the way to the freeze-out point. Lowering the freeze-out temperature causes that most of the final hadrons are light particles, mainly pions — the Fermi model would give larger probabilities of production of heavier particles.

In fact, the idea of modification of the Fermi approach was put forward by Pomeranchuk already in 1951 [3]. Pomeranchuk realized that the strong interaction between produced particles could not cease immediately, just after the particles left the considered Lorentz-contracted volume. The particles in the system should interact until the average distance between them becomes larger than the typical interaction distance, that is of the order of the inverse pion mass, m_π^{-1}.

Landau proposed his hydrodynamic approach to describe proton-proton collisions. Following Fermi, he assumed that the two colliding protons released their energy in the volume corresponding to the Lorentz-contracted size of a proton. However, he also noticed that such substantial contraction would lead to the formation of a large longitudinal pressure gradient. Under the influence of this gradient the system starts expanding. The transverse gradient is also present but initially the gradient in the longitudinal direction is much larger and the early expansion may be regarded as one-dimensional. Landau assumed that the one-dimensional expansion was governed by the laws of relativistic hydrodynamics of the perfect fluid with the equation of state of the form $\varepsilon = 3P$. At some point the transverse expansion cannot be neglected and the hydrodynamic expansion becomes three-dimensional. Since the problem of analytic treatment of the three-dimensional hydrodynamics is very complicated, Landau treated this stage in a very approximate way — he assumed that the angular deviation of the path of each fluid element reached in the one-dimensional stage would remain the same during the subsequent three-dimensional stage.

As the matter further expands, its temperature decreases. When it falls down to the sufficiently low values of the order of the pion mass, the mean free path becomes large and the fluid elements break up into individual pions, in agreement with the Pomeranchuk original idea. In our presentation of the Landau model we refer to the nucleus-nucleus collisions but, obviously, the main ingredients of the model remain the same as in the original formulation where it was applied to more elementary proton-proton collisions.

20.1 Initial one-dimensional expansion

In the case of one-dimensional expansion of matter along the collision axis z, the equations of relativistic hydrodynamics with zero baryon chemical potential, Eq. (13.28), may be written in the form

$$\begin{aligned} u^0\partial_0(Tu^0) + u^3\partial_3(Tu^0) &= \partial^0 T, \\ u^0\partial_0(Tu^3) + u^3\partial_3(Tu^3) &= \partial^3 T. \end{aligned} \tag{20.1}$$

Equations (20.1) together with the normalization condition for four-velocity lead us to the formula

$$\frac{\partial}{\partial x^0}(Tu_3) = \frac{\partial}{\partial x^3}(Tu_0), \tag{20.2}$$

which means that the quantities Tu_0 and Tu_3 may be written as the derivatives of a potential field $\Phi_{\rm L}$,

$$Tu_0 = -\partial_0\Phi_{\rm L}, \quad Tu_3 = -\partial_3\Phi_{\rm L}. \tag{20.3}$$

The total differential of the potential $\Phi_{\rm L}$ is then

$$\begin{aligned} d\Phi_{\rm L} &= \partial_0\Phi_{\rm L}\,dx^0 + \partial_3\Phi_{\rm L}\,dx^3 = -Tu^0\,dx^0 + Tu^3\,dx^3 \\ &= -Tu^0\,dt + Tu^3\,dz = -T\cosh\vartheta\,dt + T\sinh\vartheta\,dz. \end{aligned} \tag{20.4}$$

Here we have introduced our standard parameterization of the fluid velocity in terms of the fluid rapidity,

$$u^0 = \cosh\vartheta, \quad u^3 = \sinh\vartheta, \tag{20.5}$$

which automatically satisfies the normalization condition $u^\mu u_\mu = 1$. It turns out that the next convenient step is to perform the Legendre transformation and switch from the potential $\Phi_{\rm L}$ to the potential χ defined as

$$\chi = \Phi_{\rm L} + Tu^0 t - Tu^3 z. \tag{20.6}$$

The total differential of χ is

$$\begin{aligned} d\chi &= \left(u^0 t - u^3 z\right) dT + tT\,du_0 - Tz\,du^3 \\ &= (t\cosh\vartheta - z\sinh\vartheta)\,dT + (t\sinh\vartheta - z\cosh\vartheta)\,Td\vartheta, \end{aligned} \tag{20.7}$$

and the partial derivatives of χ are

$$\frac{\partial \chi}{\partial T} = t \cosh\vartheta - z \sinh\vartheta,$$
$$\frac{\partial \chi}{\partial \vartheta} = T \left(t \sinh\vartheta - z \cosh\vartheta \right). \tag{20.8}$$

Equations (20.8) may be used to determine the spacetime coordinates in terms of the temperature T and the fluid rapidity ϑ,

$$t = \frac{\partial \chi}{\partial T} \cosh\vartheta - \frac{1}{T} \frac{\partial \chi}{\partial \vartheta} \sinh\vartheta,$$
$$z = \frac{\partial \chi}{\partial T} \sinh\vartheta - \frac{1}{T} \frac{\partial \chi}{\partial \vartheta} \cosh\vartheta. \tag{20.9}$$

Of course we finally want to obtain the spacetime picture of the collision, i.e., we want to have T and ϑ expressed in terms of t and z. This aim may be achieved if the function χ is known. The equation for χ may be obtained from the two-dimensional version of the entropy conservation law, Eq. (13.29),

$$\frac{\partial}{\partial t} (\sigma \cosh\vartheta) + \frac{\partial}{\partial z} (\sigma \sinh\vartheta) = 0. \tag{20.10}$$

Changing the independent variables from (t, z) to (T, ϑ) we obtain

$$\begin{aligned} 0 &= \frac{\partial(\sigma \cosh\vartheta, z)}{\partial(t, z)} + \frac{\partial(\sigma \sinh\vartheta, t)}{\partial(z, t)} \\ &= \frac{\partial(\sigma \cosh\vartheta, z)}{\partial(t, z)} - \frac{\partial(\sigma \sinh\vartheta, t)}{\partial(t, z)} \\ &= \frac{\partial(t, z)}{\partial(T, \vartheta)} \left[\frac{\partial(\sigma \cosh\vartheta, z)}{\partial(t, z)} - \frac{\partial(\sigma \sinh\vartheta, t)}{\partial(t, z)} \right] \\ &= \frac{\partial(\sigma \cosh\vartheta, z)}{\partial(T, \vartheta)} - \frac{\partial(\sigma \sinh\vartheta, t)}{\partial(T, \vartheta)}. \end{aligned} \tag{20.11}$$

A straightforward calculation of the determinants gives

$$\begin{aligned} &\frac{1}{\sigma} \frac{d\sigma}{dT} \left[\frac{\partial}{\partial \vartheta} (z \cosh\vartheta - t \sinh\vartheta) - (z \sinh\vartheta - t \cosh\vartheta) \right] \\ &+ \frac{\partial}{\partial T} (t \cosh\vartheta - z \sinh\vartheta) = 0. \end{aligned} \tag{20.12}$$

With the help of Eq. (20.8) we rewrite Eq. (20.12) as the partial differential equation for the potential χ,

$$\frac{1}{\sigma} \frac{d\sigma}{dT} \left(\frac{\partial \chi}{\partial T} - \frac{1}{T} \frac{\partial^2 \chi}{\partial \vartheta^2} \right) + \frac{\partial^2 \chi}{\partial T^2} = 0. \tag{20.13}$$

20.1.1 *Khalatnikov equation*

Further simplifications may be achieved if we introduce the variable [1]

$$Y = \ln\left(\frac{T}{T_{\rm i}}\right). \tag{20.14}$$

In this way we obtain [4]

$$\frac{\partial^2 \chi}{\partial \vartheta^2} - c_{\rm s}^2 \frac{\partial^2 \chi}{\partial Y^2} + (c_{\rm s}^2 - 1)\frac{\partial \chi}{\partial Y} = 0. \tag{20.15}$$

In the case of the constant sound velocity Eq. (20.15) becomes a partial differential equation with constant coefficients. For example, in the case $c_{\rm s}^2 = 1/3$ we obtain

$$3\frac{\partial^2 \chi}{\partial \vartheta^2} - \frac{\partial^2 \chi}{\partial Y^2} - 2\frac{\partial \chi}{\partial Y} = 0. \tag{20.16}$$

Equation (20.15) is an important result. It shows that the one-dimensional problem of relativistic hydrodynamics with the equation of state of the form $P = c_{\rm s}^2\, \varepsilon$ (with constant sound velocity) may be reduced to the problem of solving an appropriate linear partial differential equation. In the following we shall refer to Eq. (20.15) and its special case, Eq. (20.16), as to the Khalatnikov equations [4].

Clearly, the solutions of the differential equation (20.15) are not determined solely by its form but depend also on the specific choice of the boundary conditions. In the Landau model we consider expansion of a flat disk whose initial width is

$$\Delta = 2l = 2\,R\,\frac{2m_{\rm N}}{\sqrt{s_{\rm NN}}}, \tag{20.17}$$

where R is the nucleus radius and the factor $2m_{\rm N}/\sqrt{s_{\rm NN}}$ accounts for the Lorentz contraction. We choose the coordinate system in such a way that the coordinate $z = -l$ determines the symmetry plane (i.e., for $z = -l$ the matter is always at rest, while the initial right edge of the system is placed at $z = 0$). We use the Khalatnikov equation to find the *non-trivial solutions* realized for the times $t > t_0$, where t_0 is the time needed for the rarefaction wave to reach the plane $z = -l$,

$$t_0 = \frac{l}{c_{\rm s}}. \tag{20.18}$$

For $t > t_0$, the non-trivial solution consists of a superposition of the incident rarefaction wave and the waves reflected from the wall $z = -l$, see Fig. 20.1.

Since the matter is at rest for $z = -l$, from Eq. (20.9) we obtain the first boundary condition

$$\left.\frac{\partial \chi}{\partial \vartheta}\right|_{\vartheta=0} = l\,T_{\rm i}\,e^{Y}. \tag{20.19}$$

[1] In order to keep similarity to Landau's original notation we use the letter Y to denote the logarithm of the temperature. Landau used lowercase y but this letter is in our case reserved for the spatial coordinate.

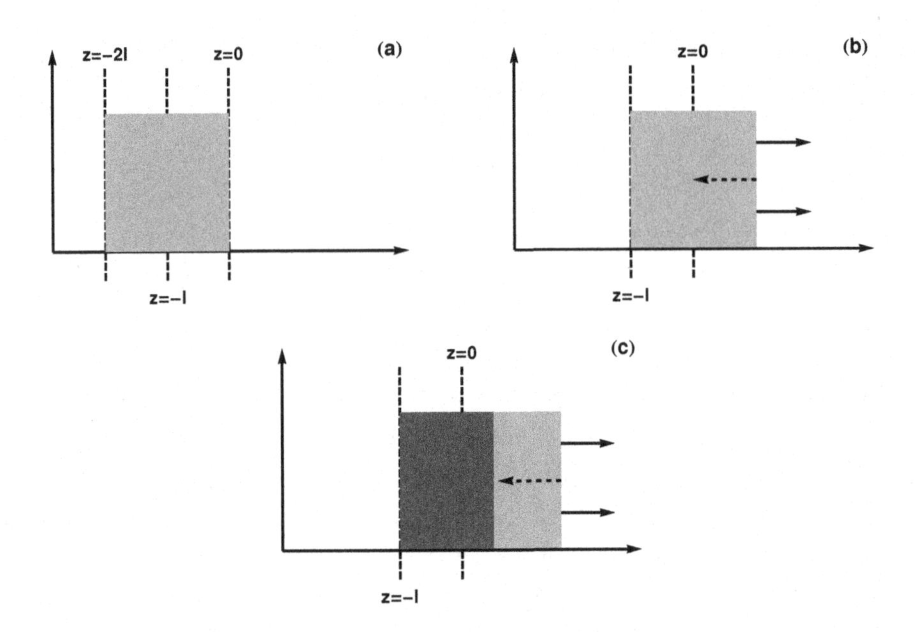

Fig. 20.1 The panel (**a**) illustrates the distribution of matter at the initial stage of the collision. The matter forms a highly compressed disk of the width $\Delta = 2l$. Because of the reflection symmetry with respect to the plane $z = -l$ we may consider only the expansion of matter for $z \geq -l$. The initial right edge of the system is placed at $z = 0$. This geometry allows us to use the results obtained for the simple Riemann solutions. The panel (**b**) illustrates the fact that initially the evolution of matter consists of the expansion of matter into vacuum (indicated by the two solid arrows) and the rarefaction wave entering the system (indicated by the dashed arrow). After the time $t_0 = l/c_s$, when the rarefaction wave hits the plane $z = -l$, the evolution of the central region becomes quite complicated. It consists of the incident rarefaction wave and the reflected waves. In the outer region the simple Riemann solution always holds, and should be matched to the non-trivial solution found by Khalatnikov. The non-trivial region is marked as the darker region in the panel (**c**).

The second boundary condition is obtained by matching the non-trivial solution with the outer simple wave solution, see our discussion in Sec. 13.7.1. In this case we have

$$z = t\,\frac{v - c_s}{1 + v\,c_s}, \quad \vartheta = -\frac{Y}{c_s}, \tag{20.20}$$

see Eqs. (13.75) and (13.76). Substituting Eq. (20.9) into Eq. (13.76) we find the relation

$$\begin{aligned}&\frac{\partial \chi}{\partial Y}\left(\sinh\vartheta - \frac{\tanh\vartheta - c_s}{1 - \tanh\vartheta c_s}\cosh\vartheta\right)\\&= \frac{\partial \chi}{\partial \vartheta}\left(\cosh\vartheta - \frac{\tanh\vartheta - c_s}{1 - \tanh\vartheta c_s}\sinh\vartheta\right),\end{aligned}$$

which after simple manipulations is reduced to

$$\frac{\partial \chi}{\partial Y} c_s - \frac{\partial \chi}{\partial \vartheta} = 0. \tag{20.21}$$

The expression on the left-hand-side of this equation represents the derivative along the line $\vartheta = -Y/c_s$, hence the potential χ should be a constant there. Since χ is anyway defined up to a constant, we choose the second boundary condition in the form

$$\chi = 0 \quad \text{for} \quad \vartheta = -\frac{Y}{c_s}. \tag{20.22}$$

Khalatnikov found that the solution of Eq. (20.15), satisfying the boundary conditions (20.19) and (20.22), may be written in the following integral form

$$\chi = -\frac{l\,T_i}{c_s}\, e^{Y} \int\limits_{c_s\vartheta}^{-Y} e^{(1+\beta_s)Y'} I_0\left(\beta_s\sqrt{Y'^2 - c_s^2\vartheta^2}\right) dY', \tag{20.23}$$

where

$$\beta_s = \frac{1-c_s^2}{2c_s^2}, \tag{20.24}$$

and I_0 is the modified Bessel function of the first kind. We note that our definition (20.14) implies that the quantity Y is negative, hence the upper limit of the integration in Eq. (20.23) is positive and $\chi \leq 0$.

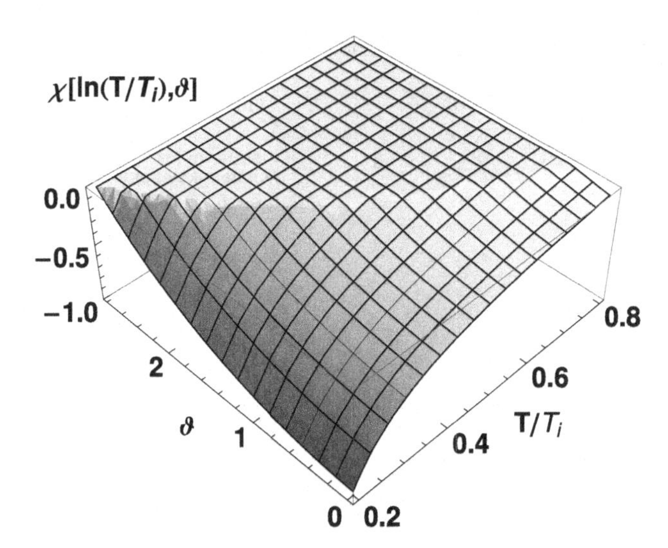

Fig. 20.2 The Khalatnikov function $\chi\,[\ln(T/T_i), \vartheta]$ plotted in the temperature range $0.2\,T_i \leq T \leq 0.8\,T_i$ and the rapidity range $0 \leq \vartheta \leq -\ln(T/T_i)/c_s$ ($c_s = 1/\sqrt{3}$, $l = 0.065$ fm, $T_i = 500$ MeV). In the region $\vartheta > -\ln(T/T_i)/c_s$, the function χ has been conventionally set equal to zero.

An example of the Khalatnikov function for $c_s = 1/\sqrt{3}\,(b_s = 1)$, $l = 0.065$ fm and $T_i = 500$ MeV is shown in Fig. 20.2. The value $l = 0.065$ fm corresponds to the radius of the gold nucleus, $R_{Au} = 6.5$ fm, divided by the Lorentz gamma factor of 100 (this is a realistic value for the top RHIC energies where $\sqrt{s_{NN}} = 200$ GeV). For $\vartheta = -\ln(T/T_i)/c_s$ the Khalatnikov function vanishes. In Fig. 20.2 we have extended the range of ϑ to the region $\vartheta > -\ln(T/T_i)/c_s$, where χ has been conventionally set equal to zero.

When dealing with the Khalatnikov solution, we should remember about the specific geometry — the longitudinal coordinate $z = -l$ is taken as the symmetry plane of the collision in the center-of-mass frame. Moreover, the solution (20.23) describes the evolution of the central part of matter for times $t > t_0 = l/c_s$. The outer regions are described always by the Riemann solutions, hence the region of the applicability of the Khalatnikov solution is always restricted to the finite interval of the longitudinal coordinate z. This feature is shown in Fig. 20.3 where we show isotherms corresponding to the Khalatnikov solution. For lower temperatures we observe larger ranges of the applicability of the Khalatnikov solution. It is also interesting to observe that the isotherms are quite well approximated by the hyperbolas of the constant proper time. However, the small differences are quite important. They are responsible for the deviations of the Landau model from the boost-invariant model of Bjorken.

The Khalatnikov potential was used in the early works on hydrodynamics [5–8]. Afterwards, other methods of dealing with the equations of relativistic hydrodynamics took over. Nevertheless, more recently the Khalatnikov method has become

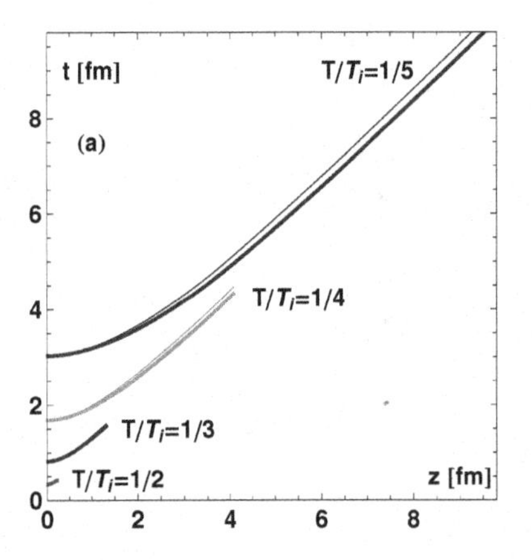

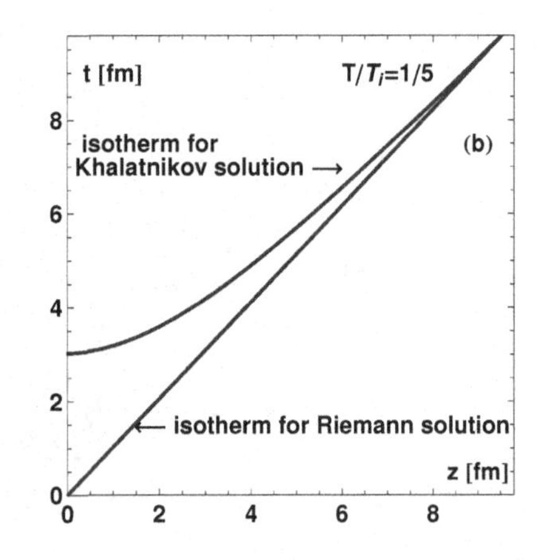

Fig. 20.3 **(a)** Isotherms in the non-trivial region described by the Khalatnikov equation (thick lines) and their hyperbolic approximations (thin lines), plotted for various values of the temperature, $T/T_i = 1/2$, $1/3$, $1/4$, and $1/5$. **(b)** The isotherm $T = T_i/5$ following from the Khalatnikov solution (upper part) combined with the isotherm corresponding to the Riemann solution (lower part). The parameters used in the plot: $c_s = 1/\sqrt{3}$, $l = 0.065$ fm, $T_i = 500$ MeV.

again the subject of interesting formal studies, see for example [9, 10].

20.2 Approximate formulas for one-dimensional expansion

In this Section, following Refs. [5, 8], we discuss the approximate forms of the Khalatnikov solution. As we shall see below, the approximate solutions lead to the Gaussian rapidity distributions. This is a quite interesting feature since the RHIC data show that the measured rapidity distributions are indeed Gaussians. Moreover the width of the experimental distributions are compatible with the Landau predictions. This type of the observations triggered new interest in the foundations and predictions of the Landau model [11]. Below we shall discuss several of those currently studied issues.

In the studies based on the Landau model one most often assumes that the final temperature is significantly lower than the initial temperature, hence, we are formally interested in the limit $|\ln(T/T_\mathrm{i})| = |Y| >> 1$. As shown in Ref. [8], in this limit the approximate formulas for the Khalatnikov function and its derivatives read (with $c_\mathrm{s}^2 = 1/3$)

$$\chi \approx -\frac{lT_\mathrm{i}}{\sqrt{6\pi}} \frac{\exp\left[-Y + \sqrt{Y^2 - c_\mathrm{s}^2\vartheta^2}\right]}{\sqrt{-Y}} \left(1 - \frac{7}{24Y} + \frac{32c_\mathrm{s}^2\vartheta^2 + 83}{384Y^2} + \cdots\right), \quad (20.25)$$

$$\frac{\partial\chi}{\partial Y} \approx \frac{2lT_\mathrm{i}}{\sqrt{6\pi}} \frac{\exp\left[-Y + \sqrt{Y^2 - c_\mathrm{s}^2\vartheta^2}\right]}{\sqrt{-Y}} \left(1 - \frac{1}{24Y} + \frac{128c_\mathrm{s}^2\vartheta^2 - 1}{384Y^2} + \cdots\right), \quad (20.26)$$

$$\frac{\partial\chi}{\partial\vartheta} \approx lT_\mathrm{i}\left[\frac{\vartheta}{3\sqrt{6\pi}} \frac{\exp\left[-Y + \sqrt{Y^2 - c_\mathrm{s}^2\vartheta^2}\right]}{(-Y)^{-3/2}} \left(1 - \frac{1}{8Y} + \cdots\right) + e^{Y + 2c_\mathrm{s}\vartheta}\right]. \quad (20.27)$$

The above equations suggest that $\partial\chi/\partial Y >> \partial\chi/\partial\vartheta$, so we may neglect the term $\partial\chi/\partial\vartheta$.

Landau in his original work makes further approximation. He argues that in the leading order of magnitude the functions χ and $\partial\chi/\partial Y$ are given by

$$\chi \approx -lT_\mathrm{i} \exp\left[-Y + \sqrt{Y^2 - c_\mathrm{s}^2\vartheta^2}\right] \quad (20.28)$$

and

$$\frac{\partial\chi}{\partial Y} \approx \Delta T_\mathrm{i} \exp\left[-Y + \sqrt{Y^2 - c_\mathrm{s}^2\vartheta^2}\right]. \quad (20.29)$$

Within such an approximation, the formulas (20.9) with $c_\mathrm{s}^2 = 1/3$ give

$$\frac{t+z}{\Delta} \approx \exp\left[\vartheta - 2Y + \sqrt{Y^2 - \vartheta^2/3}\right],$$
$$\frac{t-z}{\Delta} \approx \exp\left[-\vartheta - 2Y + \sqrt{Y^2 - \vartheta^2/3}\right]. \quad (20.30)$$

Equations (20.30) are useful since they allow us to express the fluid rapidity and temperature in terms of the spacetime coordinates t and z, namely

$$\vartheta = \frac{1}{2} \ln \frac{t+z}{t-z} \tag{20.31}$$

and

$$Y = -\frac{1}{3} \left[\ln \frac{t+z}{\Delta} + \ln \frac{t-z}{\Delta} - \sqrt{\ln \frac{t+z}{\Delta} \ln \frac{t-z}{\Delta}} \right]. \tag{20.32}$$

It is interesting to note that although the Landau model is not boost-invariant the fluid velocity following from Eq. (20.31) has a boost-invariant form. This is a consequence of the approximations made in transition from Eq. (20.23) to (20.28).

20.3 Transition to three-dimensional expansion

The one-dimensional expansion considered above may be regarded as a good approximation for the initial stage of the collision only. Clearly, after some time the motion of the fluid becomes three-dimensional and its theoretical description becomes quite complicated. In fact, only with the recent numerical codes, one is able to describe fully the three dimensional evolution of matter formed in heavy-ion collisions (this issue will be discussed below in Sec. 22.2). In the original Landau framework the three-dimensional expansion is included in a very approximate way. First, the transition line from one- to three-dimensional expansion is found. Then, it is assumed that the three-dimensional expansion looks like a break-up which does not change the rapidity distribution.

The arguments used to determine the transition line are quite heuristic. In the brief formulation of Ref. [8] they sound as follows: After time t, the distance between the relativistic fluid element and the origin of the system is about t (note that $c = 1$). At this time the longitudinal coordinate of the element is z. If $\sqrt{t^2 - z^2} > R_\mathrm{f}$, where $R_\mathrm{f} \approx 2R$ is the transverse size (diameter) of the nucleus, the transverse expansion cannot be neglected. Hence, the condition for the end of one-dimensional expansion is given by the equation

$$t^2 - z^2 = R_\mathrm{f}^2. \tag{20.33}$$

Using Eq. (20.9) we express this conditions as

$$\left(\frac{\partial \chi}{\partial Y} \right)^2 - \left(\frac{\partial \chi}{\partial \vartheta} \right)^2 = R_\mathrm{f}^2 T_\mathrm{i}^2 e^{2Y}. \tag{20.34}$$

In the approximation used by Landau this condition takes the form

$$\frac{\partial \chi}{\partial Y} = R_\mathrm{f} T_\mathrm{i} e^{Y}. \tag{20.35}$$

20.4 Rapidity distributions

20.4.1 *Original formulation*

Following the original formulation of the Landau model we calculate the rapidity distribution at the transition point from one- to three-dimensional expansion. Assuming that the three-dimensional expansion leaves the rapidity distribution unaffected, we may compare it directly to the experimental data. Of course this method, based on many crude approximations, leads to a quite rough estimate of the rapidity spectrum.

Using the results of the previous Section, we seek the solution of Eq. (20.33) in the form

$$t = R_{\rm f} \cosh \vartheta, \quad z = R_{\rm f} \sinh \vartheta. \tag{20.36}$$

Substituting Eq. (20.36) into Eq. (20.34), where only the term containing the derivative of χ with respect to Y is kept (in the Landau approximation, Eq. (20.29)), we obtain the relation connecting the fluid rapidity ϑ with the fluid temperature Y along the transition line, namely

$$L = -2Y + \sqrt{Y^2 - \vartheta^2/3}. \tag{20.37}$$

Here we have introduced the variable

$$L = \ln\left(\frac{R_{\rm f}}{\Delta}\right) = \ln\left(\frac{\sqrt{s_{\rm NN}}}{2m_{\rm N}}\right). \tag{20.38}$$

Equation (20.37) may be obtained also directly from Eq. (20.35). It has the solution for Y in the form

$$Y = -\frac{2}{3}L + \frac{1}{3}\sqrt{L^2 - \vartheta^2}. \tag{20.39}$$

Working in the Landau approximation we also find

$$\frac{t \pm z}{\Delta} = \frac{R_{\rm f}\, e^{\pm\vartheta}}{\Delta} = e^{L \pm \vartheta}, \tag{20.40}$$

and from Eq. (20.39) we obtain the rapidity profile of the temperature (along the transition line)

$$T = T_{\rm i} \exp\left[-\frac{1}{3}\left(2L - \sqrt{L^2 - \vartheta^2}\right)\right]. \tag{20.41}$$

Similarly, we find the rapidity profile of the entropy density

$$\sigma = \sigma_{\rm i} \exp\left[-2L + \sqrt{L^2 - \vartheta^2}\right]. \tag{20.42}$$

In order to calculate the rapidity distribution we first consider a thin slice of the fluid whose thickness in its rest frame is dz'. The entropy contained in this slice is $dS = \pi R^2 \sigma dz'$. Since the fluid element moves in the center-of-mass frame with the velocity $v = \tanh \vartheta$, may write

$$dS = \pi R^2 \sigma dz' = \pi R^2 \sigma (\cosh \vartheta dz - \sinh \vartheta dt). \tag{20.43}$$

The differentials dz and dt may be expressed in terms of dY and $d\vartheta$. In the considered case $dz' = R_{\rm f} d\vartheta$, hence

$$dS = \pi R^2 R_{\rm f}\, \sigma_{\rm i} \exp\left[-2L + \sqrt{L^2 - \vartheta^2}\right] d\vartheta. \tag{20.44}$$

For $\vartheta \ll L$ we find

$$\frac{dS}{d\vartheta} = \pi R^2 R_{\rm f} e^{-L} \sigma_{\rm i} \exp\left(-\frac{\vartheta^2}{2L}\right). \tag{20.45}$$

If we identify the entropy density with the particle density and the fluid rapidity ϑ with the particle rapidity y we may write

$$\frac{dN}{d{\rm y}} = \frac{N}{(2\pi L)^{1/2}} \exp\left(-\frac{{\rm y}^2}{2L}\right), \tag{20.46}$$

where N is the total multiplicity.

We note that L is a parameter fixed by the geometry of the colliding system, hence larger values of ϑ correspond to lower values of the temperature T. This means that T appearing in Eq. (20.41) cannot be identified with the constant decoupling temperature defined by the inverse pion mass. Equation (20.41) defines the temperature of the transition from one- to three-dimensional motion only and the freeze-out temperature is lower. This reasoning indicates that the fluid elements necessarily take part in the three-dimensional expansion, however this late stage cannot be precisly described in the Landau model.

It must be kept in mind that the Gaussian distribution represents very crude mathematical approximation done within the Landau model, which is in turn a crude approximation to the physical situation. On the other hand, Eq. (20.46) describes very well various rapidity distributions measured in different hadronic and heavy-ion experiments. This is the reason why the Landau model attracted so much attention in the past [12,13] and it is again discussed nowadays in the context of the RHIC data [11,14,15].

20.4.2 *Freeze-out at constant temperature*

There are several approaches based on the Landau model where the freeze-out description differs from the original formulation. In those cases one assumes that the longitudinal expansion dominates and the freeze-out happens at constant value of the temperature (in this longitudinally expanding system). Then, we may again use Eq. (20.43) expressing dz and dt in terms of dY and $d\vartheta$. To do this change in the general case we use Eq. (20.9) rewritten as

$$t = \frac{e^{-Y}}{T_{\rm i}}\left(\frac{\partial\chi}{\partial Y}\cosh\vartheta - \frac{\partial\chi}{\partial\vartheta}\sinh\vartheta\right),$$
$$z = \frac{e^{-Y}}{T_{\rm i}}\left(\frac{\partial\chi}{\partial Y}\sinh\vartheta - \frac{\partial\chi}{\partial\vartheta}\cosh\vartheta\right). \tag{20.47}$$

By direct differentiation of Eq. (20.47) one gets

$$dz' = \frac{e^{-Y}}{T_\mathrm{i}} \left[\left(\frac{\partial \chi}{\partial \vartheta} - \frac{\partial^2 \chi}{\partial Y \partial \vartheta} \right) dY + \left(\frac{\partial \chi}{\partial Y} - \frac{\partial^2 \chi}{\partial \vartheta^2} \right) d\vartheta \right]. \tag{20.48}$$

For constant value of the final temperature $Y = Y_\mathrm{f} = \ln(T_\mathrm{f}/T_\mathrm{i})$ we find

$$dz' = c_\mathrm{s}^2 \frac{e^{-Y}}{T_\mathrm{i}} \left(\frac{\partial \chi}{\partial Y} - \frac{\partial^2 \chi}{\partial Y^2} \right) d\vartheta, \tag{20.49}$$

where we used the Khalatnikov equation to make the substitution

$$\frac{\partial \chi}{\partial Y} - \frac{\partial^2 \chi}{\partial \vartheta^2} = c_\mathrm{s}^2 \left(\frac{\partial \chi}{\partial Y} - \frac{\partial^2 \chi}{\partial Y^2} \right). \tag{20.50}$$

We are ready now to write the expression for the entropy dS in a compact form as

$$dS = \pi R^2 \frac{\sigma_\mathrm{i}}{T_\mathrm{i}} c_\mathrm{s}^2 e^{2Y} \frac{\partial}{\partial Y} \left(\chi - \frac{\partial \chi}{\partial Y} \right) d\vartheta. \tag{20.51}$$

The difference $\chi - \partial\chi/\partial Y$ may be expressed by the combination of the Bessel functions only (without the integral), hence we obtain [9]

$$\frac{dS}{d\vartheta} = \pi R^2 c_\mathrm{s} \sigma_\mathrm{i}\, l\, e^{Y_\mathrm{f}} \left[I_0 \left(\sqrt{Y_\mathrm{f}^2 - c_\mathrm{s}^2 \vartheta^2} \right) - \frac{Y_\mathrm{f} I_1 \left(\sqrt{Y_\mathrm{f}^2 - c_\mathrm{s}^2 \vartheta^2} \right)}{\sqrt{Y_\mathrm{f}^2 - c_\mathrm{s}^2 \vartheta^2}} \right]. \tag{20.52}$$

With the usual assumption that the particle density is proportional to the entropy density, $dN \sim dS$, Eq. (20.52) may be used to describe the rapidity distribution of particles. Of course, one assumes here implicitly that the fluid rapidity ϑ may be identified with the particle rapidity y. As we know from our previous discussion, such identification is a quite robust approximation.

Bibliography to Chapter 20

[1] E. Fermi, "High-energy nuclear events," *Prog. Theor. Phys.* **5** (1950) 570–583.

[2] L. D. Landau, "On the multiparticle production in high-energy collisions," *Izv. Akad. Nauk SSSR Ser. Fiz.* **17** (1953) 51–64.

[3] I. Y. Pomeranchuk, "On the theory of multiple particle production in a single collision," *Dokl. Akad. Nauk Ser. Fiz.* **78** (1951) 889–891.

[4] I. M. Khalatnikov, "Some questions of the relativistic hydrodynamics," *Zh.Eksp. Teor. Fiz.* . **26** (1954) 529.

[5] S. Z. Belenkij and L. D. Landau, "Hydrodynamic theory of multiple production of particles," *Nuovo Cim. Suppl.* **3S10** (1956) 309.

[6] G. A. Milekhin, "On hydrodynamic theory of multiple production of particles," *Zh. Eksp. Teor. Fiz.* . **35** (1958) 978.

[7] G. A. Milekhin, "Hydrodynamic theory of multiple production of particles in collisions of fast nucleons on nuclei," *Zh. Eksp. Teor. Fiz.* . **35** (1958) 1185.

[8] S. Chadha, C. S. Lam, and Y. C. Leung, "On Proton Proton Collisions in the Hydrodynamic Theory," *Phys. Rev.* **D10** (1974) 2817.

[9] D. K. Srivastava, J.-e. Alam, S. Chakrabarty, B. Sinha, and S. Raha, "Hydrodynamics of ultrarelativistic heavy ion collisions: Considerations of boost noninvariance and stopping," *Ann. Phys.* **228** (1993) 104–145.

[10] G. Beuf, R. Peschanski, and E. N. Saridakis, "Entropy flow of a perfect fluid in (1+1) hydrodynamics," *Phys. Rev.* **C78** (2008) 064909.

[11] P. Steinberg, "Landau hydrodynamics and RHIC phenomena," *Acta Phys. Hung.* **A24** (2005) 51–57.

[12] P. Carruthers and M. Duong-Van, "New scaling law based on the hydrodynamical model of particle production," *Phys. Lett.* **B41** (1972) 597–601.

[13] P. Carruthers and M. Doung-van, "Rapidity and angular distributions of charged secondaries according to the hydrodynamical model of particle production," *Phys. Rev.* **D8** (1973) 859–874.

[14] C.-Y. Wong, "Landau Hydrodynamics Revisited," *Phys. Rev.* **C78** (2008) 054902.

[15] C.-Y. Wong, "Lectures on Landau Hydrodynamics," `arXiv:0809.0517`.

Chapter 21

Bjorken Model

In the Landau model the initial conditions are specified for a given laboratory time, considered in the center-of-mass frame, when the matter is highly compressed and at rest. This description seems to loose one aspect of high-energy hadronic collisions, namely the fact that fast particles are produced later and further away from the collision center than the slow ones. This feature is naturally built in the string models of multiparticle production. It is possible to account for this effect in the hydrodynamic description by imposing special initial conditions. This idea was proposed and studied by Bjorken.

The Bjorken hydrodynamic model [1] was based on the assumption that the rapidity distribution of the charged particles, $dN_{\rm ch}/dy$, is constant in the *mid-rapidity* region [1]. This fact means that the central region is invariant under Lorentz boosts along the beam axis. This in turn implies that the longitudinal flow has the form $v_z = z/t$ and all thermodynamic quantities characterizing the central region depend only on the longitudinal proper time $\tau = \sqrt{t^2 - z^2}$ and the transverse coordinates x and y, see Sec. 2.7. Ignoring the effects of transverse expansion, the dependence on the transverse coordinates vanishes and we obtain a simple one-dimensional hydrodynamic model which can be treated easily with the help of analytic methods.

The Bjorken model, as the hydrodynamic model with boost-invariance, had predecessors in Refs. [2–6]. However, the advantage of Bjorken's approach over the previous work is that Ref. [1] introduces the formula which may be used to make estimates of the initial energy density accessible in relativistic heavy-ion collisions. The calculations based on this formula indicate that high-energy nuclear collisions lead to the formation of matter whose energy density exceeds by at least one order of magnitude the energy density of normal nuclear matter. This observation suggests the formation of the quark-gluon plasma in such collisions.

[1] From our previous discussions we know that the condition $dN_{\rm ch}/dy$ = constant is realized in a relatively narrow rapidity range close to y = 0

21.1 Pure longitudinal expansion

As stated above, assuming the boost-invariance and ignoring the transverse expansion we obtain a one-dimensional hydrodynamic model where the thermodynamic variables are functions of the longitudinal proper time only,

$$\varepsilon = \varepsilon(\tau), \qquad P = P(\tau), \qquad T = T(\tau), \qquad \text{etc.} \tag{21.1}$$

The initial conditions for the hydrodynamic expansion are imposed along the hyperbola of the constant proper time

$$\sqrt{t^2 - z^2} = \tau_{\rm i}. \tag{21.2}$$

In this way one accounts for the time dilation effects characterizing the particle production. The fluid four-velocity field has the form that is the special case of Eq. (2.51),

$$u^\mu = \frac{1}{\tau}(t, 0, 0, z) = \gamma\left(1, 0, 0, \frac{z}{t}\right). \tag{21.3}$$

This form implies

$$\partial_\mu u^\mu = \frac{1}{\tau}. \tag{21.4}$$

Equation (21.3) indicates that the rapidity of the fluid is equal to the spacetime rapidity. For simplicity we identify those two quantities with the rapidity of the particles

$$\mathrm{y} = \operatorname{arctanh}\left(\mathrm{v}_\parallel\right) = \operatorname{arctanh}\left(\frac{z}{t}\right), \tag{21.5}$$

$$t = \tau \cosh \mathrm{y}, \qquad z = \tau \sinh \mathrm{y}. \tag{21.6}$$

At high energies, the baryon number density in the central region is negligible. Hence, the hydrodynamic equations (13.28) and (13.29) may be applied. The entropy conservation (13.29) gives

$$\partial_\mu \left(\sigma u^\mu\right) = \frac{d\sigma(\tau)}{d\tau} + \frac{\sigma(\tau)}{\tau} = 0. \tag{21.7}$$

The solution of this equation is

$$\sigma(\tau) = \sigma(\tau_{\rm i}) \frac{\tau_{\rm i}}{\tau}. \tag{21.8}$$

On the other hand, the thermodynamic relations (13.12) and (13.13) used in (21.7) give

$$\frac{d\varepsilon}{\varepsilon + P} = -\frac{d\tau}{\tau}. \tag{21.9}$$

This equation can be solved if the equation of state is known. For ultra-relativistic particles $P = \lambda\, \varepsilon$ (with $\lambda = c_s^2 = 1/3$) and we find

$$\varepsilon(\tau) = \varepsilon(\tau_{\rm i}) \left(\frac{\tau_{\rm i}}{\tau}\right)^{1+\lambda}. \tag{21.10}$$

21.2 Simple estimates

Equation (21.8) implies that entropy per unit rapidity is a constant of the motion. To check this property, note that in the reference frame where the fluid element is at rest we have

$$d^3x = d^2x_\perp \tau \, dy. \tag{21.11}$$

Thus the entropy contained in the interval dy around $\mathrm{y} = 0$ is

$$dS = \tau \, \sigma(\tau) \int d^2x_\perp \, dy \tag{21.12}$$

and

$$\frac{d}{d\tau}\left[\frac{dS}{dy}\right] = \int d^2x_\perp \frac{d}{d\tau}\left[\tau \, \sigma(\tau)\right] = 0. \tag{21.13}$$

This result allows us to make the simple estimate of the energy density achieved in the central region of the ultra-relativistic heavy-ion collisions. We first calculate the entropy density $\sigma(\tau_\mathrm{i})$ at the time τ_i when the hydrodynamic description starts

$$\sigma(\tau_\mathrm{i}) = \frac{1}{\tau_\mathrm{i} \mathcal{A}} \frac{dS}{dy}(\mathrm{y} = 0) \approx \frac{3.6}{\tau_\mathrm{i} \mathcal{A}} \frac{dN}{dy}(\mathrm{y} = 0). \tag{21.14}$$

Here we assumed the proportionality between the entropy density and the multiplicity of the measured pions (compare Eqs. (8.59) and (8.65), valid for relativistic massless bosons, which give $\sigma/n = 2\pi^4/(45\zeta(3)) \approx 3.6$). The transverse overlap area of the two colliding nuclei is denoted by $\mathcal{A}$. For symmetric central collisions $\mathcal{A} = \pi\left(3A/(4\pi\rho_0)\right)^{2/3}$, where A is the mass number of the nuclei and ρ_0 is the nuclear saturation density. The energy density at the time τ_i can be expressed in terms of the entropy density $\sigma(\tau_\mathrm{i})$

$$\varepsilon(\tau_\mathrm{i}) = \left[\frac{1215 \, \sigma^4(\tau_\mathrm{i})}{128 \, g \, \pi^2}\right]^{\frac{1}{3}} \approx \frac{5.4}{g^{\frac{1}{3}}}\left[\frac{1}{\tau_\mathrm{i} \mathcal{A}} \frac{dN}{dy}\right]^{\frac{4}{3}}. \tag{21.15}$$

Here we have used again the thermodynamic relations for a gas of massless bosons.

For the most energetic ($\sqrt{s_{\mathrm{NN}}} = 200$ GeV) central Au + Au collisions studied at RHIC, the multiplicity of charged particles per unit rapidity at $\mathrm{y} = 0$ is about 600. Hence, the total number of the produced particles (dominated by pions) is

$$\frac{dN}{dy} \approx \frac{3}{2} 600 \approx 900. \tag{21.16}$$

Taking $g_\pi = 3$ and the standard value $\tau_\mathrm{i} = 1$ fm, one gets

$$\varepsilon_{\mathrm{pions}}(\tau_\mathrm{i}) = 9.6 \, \frac{\mathrm{GeV}}{\mathrm{fm}^3}. \tag{21.17}$$

Such a large energy density suggests that the quark-gluon plasma might have been formed at the initial stages of the collisions. However, since the concept of the pion gas at such a high energy density breaks down, we should make another estimate, with the degeneracy factor reflecting the number of the internal degrees of freedom in the plasma phase.

Assuming that the system at $\tau = \tau_{\rm i}$ is the weakly-interacting quark-gluon plasma that may be effectively regarded as an ideal gas of massless bosons with $g_{\rm qgp}$ degrees of freedom we may again use Eq. (21.15). For $g_{\rm qgp} = 37$, see Eq. (8.85), one finds

$$\varepsilon_{\rm qgp}\left(\tau_{\rm i}\right) \approx 4.2 \, \frac{\rm GeV}{\rm fm^3}. \tag{21.18}$$

With the inclusion of strangeness, the number of the internal degrees of freedom in the plasma is larger, $g_{\rm qgp} = 47.5$, and

$$\varepsilon_{\rm qgp}\left(\tau_{\rm i}\right) \approx 3.8 \, \frac{\rm GeV}{\rm fm^3}. \tag{21.19}$$

We conclude that the two values, (21.18) and (21.19), are again consistent with the hypothesis of the plasma formation.

The estimates presented above use the connection between the particle density and entropy, which is valid, strictly speaking, only for massless bosons. To relax this assumption one may try to connect the initial energy density directly with the measured energy of the finally observed particles. Following the same steps as in the analysis of the entropy, we define the energy contained in the rapidity interval dy placed around $\rm y = 0$ by the formula

$$dE = \tau \; \varepsilon\left(\tau\right) \int d^2x_\perp \; dy. \tag{21.20}$$

With the help of Eq. (21.10) we find

$$\frac{d}{d\tau}\left[\tau^{\lambda} \frac{dE}{dy}\right] = \int d^2x_\perp \frac{d}{d\tau}\left[\tau^{\,1+\lambda} \; \varepsilon\left(\tau\right)\right] = 0. \tag{21.21}$$

Equations (21.20) and (21.21) yield

$$\varepsilon(\tau_{\rm i}) = \frac{1}{\tau_{\rm i}\,\mathcal{A}} \left(\frac{\tau}{\tau_{\rm i}}\right)^{\lambda} \frac{dE}{dy}\left(\tau, {\rm y} = 0\right). \tag{21.22}$$

If we want to use the experimental value for the energy density, the proper time τ should be identified with the freeze-out time $\tau_{\rm f}$. This, however, leads to problems, since the ratio $\tau_{\rm f}/\tau_{\rm i}$ is poorly known. The typical values are $\tau_{\rm i} \sim 1$ fm and $\tau_{\rm f} \sim 10$ fm, which gives $\left(\tau_{\rm f}/\tau_{\rm i}\right)^{\lambda} \approx 2$.

The appearance of the factor $\left(\tau_{\rm f}/\tau_{\rm i}\right)^{\lambda}$ in Eq. (21.22) reflects the fact that the energy per unit rapidity is not conserved. The longitudinal pressure in the system does not vanish (by this we mean the non-zero T^{33} component of the energy-momentum tensor) and it acts on the adjacent fluid elements. Hence, there is a "leakage" of the energy from the considered central rapidity unit. This energy is converted into the longitudinal flow. The increase of the factor $\left(\tau/\tau_{\rm i}\right)^{\lambda}$ with time compensates the decrease of the energy density $dE(\tau)/dy$ in such a way that the left-hand-side of Eq. (21.22) remains constant.

Since we expect that the ratio $(\tau_f/\tau_i)^\lambda$ is of the order of unity, one often uses Eq. (21.22) without this factor,

$$\varepsilon(\tau_i) = \frac{1}{\tau_i \mathcal{A}} \frac{dE_\perp}{dy}. \tag{21.23}$$

Here $E_\perp$ is the transverse energy of the produced particles (the sum of the transverse masses of the particles at $y = 0$). Sometimes, one also uses the equation

$$\varepsilon(\tau_i) = \frac{3}{2\tau_i \mathcal{A}} \langle m_\perp \rangle \frac{dN_{ch}}{dy}. \tag{21.24}$$

where N_{ch} is the multiplicity of charged particles. The use of Eq. (21.23) or other similar formulas by the four RHIC experiments indicates that in the central Au+Au collisions at the highest RHIC beam energy the initial energy density ε_i reaches the value of about 5 GeV/fm^3 at $\tau_i = 1$ fm [7–10]. The value of 5 GeV/fm^3 is well above the critical energy density for the deconfinement phase transition.

In the end of this Section we want to emphasize that the largest uncertainty in the estimates of the initial energy density ε_i is connected with our poor knowledge of the thermalization time τ_i.

21.3 First-order phase transition

The one-dimensional picture of the hydrodynamic expansion used in the Bjorken model allows for a simple inclusion of the first-order phase transition. Assuming that the hadronic and plasma phases are described by the bag equation of state, Eqs. (5.10)–(5.15) with $\mu = 0$, we may construct the following physical scenario: Initially, for $\tau \geq \tau_i$, the system expands and cools down in the plasma phase. The entropy decreases according to the scaling law (21.8),

$$\sigma(\tau) = \frac{\sigma_i \tau_i}{\tau}. \tag{21.25}$$

At $\tau = \tau_c$ the phase transition starts. In the time interval $\tau_c \leq \tau \leq \tau_h$ the system is a mixture of the plasma and the pion gas. The temperature of the mixed phase is fixed and equal to T_c. Its entropy is defined by the equation

$$\sigma(\tau) = \sigma_\pi(T_c)\xi(\tau) + \sigma_{qgp}(T_c)(1 - \xi(\tau)) = \frac{\sigma_i \tau_i}{\tau}, \tag{21.26}$$

where $\xi(\tau)$ defines the volume fraction of the pion phase. The last equality in Eq. (21.26) stresses the fact that the total entropy density should always obey Eq. (21.8).

Equation (21.26) with the conditions $\xi(\tau_c) = 0$ and $\xi(\tau_h) = 1$ determines the time dependence of the function $\xi(\tau)$, namely

$$\xi(\tau) = \frac{1 - \tau_c/\tau}{1 - g_\pi/g_{qgp}}, \tag{21.27}$$

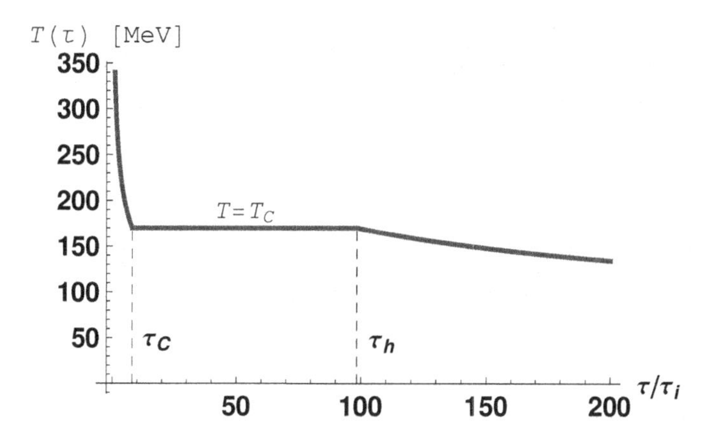

Fig. 21.1 The time dependence of the plasma temperature in the one-dimensional Bjorken scenario with the first-order phase transition; $T_c = 170$ MeV, $T_i = 2\,T_c$, $\tau_i = 1$ fm, $\tau_c = 8\,\tau_i$, and $\tau_h = 12.3\,\tau_c$.

where the parameters g_π and g_{qgp} denote the numbers of internal degrees of freedom in the pion gas and the plasma, respectively. From Eq. (21.27) we infer that the mixed plasma is ended at the time

$$\tau_h = \tau_c \frac{g_{qgp}}{g_\pi}. \tag{21.28}$$

For $N_f = 2\,(3)$ we find $g_{qgp} = 37\,(47.5)$ and $\tau_h = 12.3\,(15.8)\,\tau_c$. Further evolution of the system (in the hadronic phase, for $\tau \geq \tau_h$) is defined by the temperature changes

$$T = T_c \left(\frac{\tau_h}{\tau}\right)^{1/3}. \tag{21.29}$$

The time dependence of the temperature in the three time intervals considered above is shown in Fig. 21.1.

21.4 Bjorken vs. Landau

Very often the opinions are presented that the Landau model is applicable at lower beam energies, while the Bjorken model is suitable for the description of ultra-relativistic collisions. The real situation is more complex. In Chap. 2 we showed that the hadron rapidity distributions measured at the RHIC energies are described well by the Gaussian profiles, which suggests the usefulness of the Landau model even at the highest available energies. On the other hand, we also showed that the collisions studied at RHIC do not exhibit the baryon stopping, which suggests the validity of the Bjorken approach. Interestingly, at the highest RHIC energies we deal with a kind of mixture of Landau- and Bjorken-like behavior — the rapidity distributions are Gaussians, while the baryons in the colliding nuclei are not stopped. Clearly, the more advanced hydrodynamic codes are necessary to describe the behavior of matter produced in the ultra-relativistic collisions.

Bibliography to Chapter 21

[1] J. D. Bjorken, "Highly relativistic nucleus-nucleus collisions: The central rapidity region," *Phys. Rev.* **D27** (1983) 140–151.
[2] R. C. Hwa, "Statistical description of hadron constituents as a basis for the fluid model of high-energy collisions," *Phys. Rev.* **D10** (1974) 2260.
[3] C. B. Chiu, E. C. G. Sudarshan, and K.-H. Wang, "Hydrodynamical expansion with frame independence symmetry in high-energy multiparticle production," *Phys. Rev.* **D12** (1975) 902–908.
[4] C. B. Chiu and K.-H. Wang, "Hydrodynamical model with massless constituents," *Phys. Rev.* **D12** (1975) 272–276.
[5] M. I. Gorenstein, Y. M. Sinyukov, and V. I. Zhdanov, "On scaling solutions in the hydrodynamical theory of multiparticle production," *Phys. Lett.* **B71** (1977) 199–202.
[6] M. I. Gorenstein, Y. M. Sinyukov, and V. I. Zhdanov, "On scale invariant solutions in the hydrodynamical theory of multiparticle production," *Zh. Eksp. Teor. Fiz.* **74** (1978) 833–845.
[7] **BRAHMS** Collaboration, I. Arsene *et al.*, "Quark gluon plasma and color glass condensate at RHIC? The perspective from the BRAHMS experiment," *Nucl. Phys.* **A757** (2005) 1–27.
[8] **PHENIX** Collaboration, K. Adcox *et al.*, "Formation of dense partonic matter in relativistic nucleus nucleus collisions at RHIC: Experimental evaluation by the PHENIX collaboration," *Nucl. Phys.* **A757** (2005) 184–283.
[9] **PHOBOS** Collaboration, B. B. Back *et al.*, "The PHOBOS perspective on discoveries at RHIC," *Nucl. Phys.* **A757** (2005) 28–101.
[10] **STAR** Collaboration, J. Adams *et al.*, "Experimental and theoretical challenges in the search for the quark gluon plasma: The STAR collaboration's critical assessment of the evidence from RHIC collisions," *Nucl. Phys.* **A757** (2005) 102–183.

Chapter 22

Modeling RHIC Data

Modern hydrodynamic calculations follow general concepts introduced in the Landau and Bjorken models. However, they differ from the Landau original description in the way how they treat the initial conditions. They also go beyond the simple Bjorken approach by including transverse expansion and, eventually, by breaking the boost-invariance. Additionally, the recent hydrodynamic codes use modern equations of state inspired by the lattice simulations of QCD and advanced hadron-gas calculations.

In the Landau and Bjorken models, the thermalized matter was a gas of ultra-relativistic particles, mainly pions, satisfying the extreme relativistic equation of state $P = \varepsilon/3$. At present, more accurate equations of state for hot and dense matter are known. Expecting the phase transition to the quark-gluon plasma, one can use the plasma equation of state including the phase transition back to ordinary hadronic matter, see our discussion in Sec. 6.6. The fact that the phase transition may be easily implemented in the hydrodynamic description is one of the great advantages of using hydrodynamics as a tool to describe the evolution of matter formed in heavy-ion collisions.

Despite the differences in the form of the initial conditions or in the form of the equations of state, the hydrodynamic calculations addressing the RHIC physics are based on the following basic assumptions: **i)** a large amount of the initial kinetic energy is thermalized in a short time, $\tau_{\rm i} \leq 1$ fm, forming a locally equilibrated very dense and hot system that is called a fireball or a firecylinder, **ii)** the strongly interacting matter in the fireball is treated as a relativistic fluid, which is characterized by a few smooth functions: the fluid four-velocity field $u^\mu(x)$ and local thermodynamic variables such as temperature $T(x)$ and baryon chemical potential $\mu_B(x)$, **iii)** the initial transverse hydrodynamic flow at the thermalization point $\tau = \tau_{\rm i}$ is in most cases set equal to zero. *Nota bene*, there is a growing interest in including a non-zero initial transverse flow. In this context one argues that the processes leading to thermalization of the system are responsible also for the formation of the transverse flow which should be included as a non-trivial initial condition for the hydrodynamic equations.

In view of the simplicity of the main physical assumptions listed above, it was a striking observation that many data collected by the four RHIC experiments turned out to be in a good *quantitative* agreement with the predictions of hydrodynamic models — especially for the most central collisions at the highest RHIC beam energy of $\sqrt{s_{\rm NN}} = 200$ GeV.

The hydrodynamic codes used to describe the physical phenomena studied at RHIC may be divided into three classes: The first class consists of the boost-invariant models, whose validity is restricted to the central rapidity region [1–20]. Such models are usually called the 2+1 hydrodynamic models. The number 2 refers in this case to the two non-trivial space dimensions and the number 1 refers to one time dimension. The evolution in the third space dimension is trivial due to the assumption of boost-invariance. The second class consists of fully three-dimensional models, which are known as 3+1 codes [21–28]. Finally, the third class includes the hydrodynamic models which take into account the effects of viscosity and other transport phenomena [29–45]. The inclusion of the viscous effects represents the forefront of the present investigations in the field of ultra-relativistic heavy-ion collisions. Most of those approaches are still restricted to 2+1 versions.

Using the hydrodynamic approaches, one tries to reproduce the main soft-hadronic observables. They include: the transverse momentum spectra of different hadronic species, the rapidity distributions, the elliptic flow coefficient v_2, the directed flow coefficient v_1, and the HBT radii. The calculations of the rapidity distributions and the directed flow require the use of the 3+1 codes. The main successes of the perfect fluid hydrodynamics are connected with the very good description of the transverse-momentum spectra and the elliptic flow coefficient v_2. On the other hand, there are well known problems with reproducing the HBT radii in the hydrodynamic models. This problem is commonly known as the "RHIC HBT puzzle".

22.1 2+1 hydrodynamic models

The first 2+1 hydrodynamic models used to interpret the RHIC results (Huovinen, Kolb, Heinz, Ruuskanen, and Voloshin [7], Teaney and Shuryak [10], Kolb and Rapp [14]) assumed the bag equation of state for the plasma phase and the resonance gas model for the hadronic phase. The two phases were connected with the first-order phase transition. The latent heat ranged from 0.8 GeV/fm^3 in [10] to 1.15 GeV/fm^3 in [7, 14]. The starting time for the hydrodynamic evolution was 1 fm in [10] and 0.6 fm in [7, 14]. The models differed also in how they treated the hadronic phase. In Ref. [10] the hydrodynamic evolution was coupled to the hadronic rescattering model RQMD. In this way the hadron decoupling processes were governed by the realistic hadronic cross sections. In Ref. [14] a partial chemical equilibrium was incorporated in order to have an earlier chemical freeze-out. On the other hand, in Ref. [7] the full chemical equilibrium was assumed.

A brief review of the early hydrodynamic models was done in the PHENIX

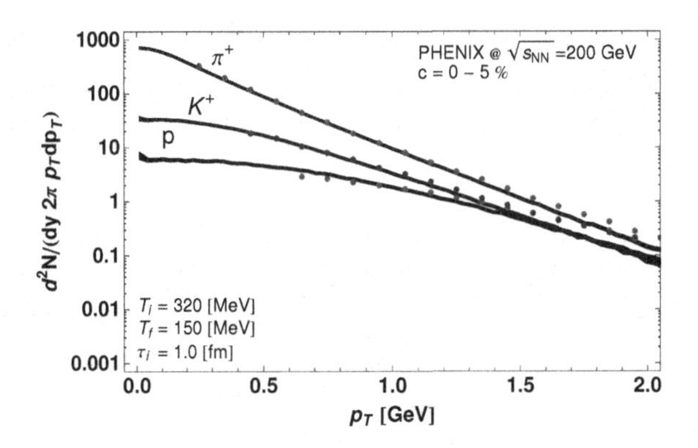

Fig. 22.1 Transverse-momentum spectra of π^+, K^+, and protons. The PHENIX experimental results [47] for Au+Au collisions at $\sqrt{s_{\rm NN}}$ = 200 GeV and the centrality class 0–5% (points) are compared to the results obtained in the framework of the Cracow 2+1 hydrodynamic model [18] discussed in Sec. 22.1.1. The model parameters are: $b = 2.26$ fm, $\tau_{\rm i} = 1.0$ fm, $T_{\rm i} = 320$ MeV, and $T_{\rm f} = 150$ MeV.

RHIC White Paper [46]: The results presented in [10] reproduce the experimental pion and proton spectra as well as the pion and proton v_2. The approach of Ref. [7] reproduces the pion spectra together with the pion and proton v_2. The results presented in [14] reproduce the pion spectra for $p_\perp \leq 1$ GeV and the proton v_2 for $p_\perp \leq 0.7$ GeV, but they do not agree with the experimental proton spectra and the pion elliptic flow. None of those models is able to reproduce the HBT radii .

In the last years several efforts have been made to improve the hydrodynamic codes. Such improvements included, for example, the construction of a more realistic equation of state, the inclusion of fluctuations of the initial conditions, and the modeling of a continuous freeze-out. Of course, the main technical improvement was a development of the 3+1 codes and inclusion of viscosity.

22.1.1 *Cracow 2+1 hydrodynamic model*

Before we start our discussion of the 3+1 codes, we shall analyze in more detail the results of the recent 2+1 hydrodynamic model which addresses both the one- and two-particle observables in Au+Au collisions at the maximal beam energy $\sqrt{s_{\rm NN}}$ = 200 GeV (Chojnacki et al. [18]). In this approach, called below the Cracow 2+1 hydrodynamic model, the initial conditions are specified according to Eq. (15.1) with $\kappa = 0.14$. The initial time for the hydrodynamic evolution is 1 fm and the baryon chemical potential in the hydrodynamic equations is zero. In the numerical calculations one uses the form of the hydrodynamic equations that was discussed in Sec. 13.8 and the equation of state introduced in Sec. 6.6. The main physical parameters are: the initial temperature at the center of the system $T_{\rm i}$ and the final (thermal freeze-out) temperature $T_{\rm f}$. No distinction between the chemical and

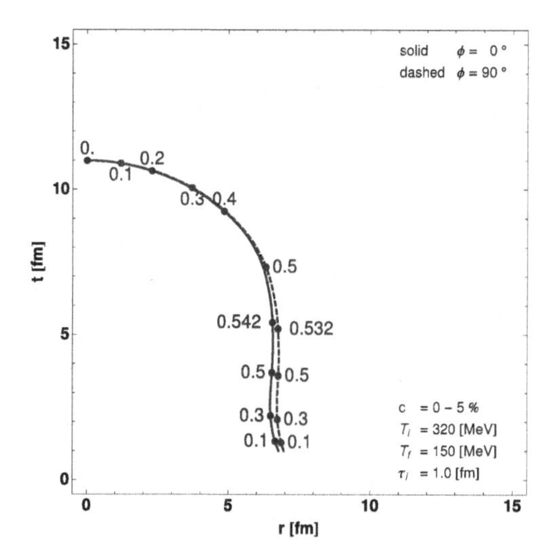

Fig. 22.2 In-plane and out-of-plane freeze-out curves, i.e., the intersections of the freeze-out hypersurface with the planes $y = 0$ ($\phi = 0^o$) and $x = 0$ ($\phi = 90^o$), obtained in the Cracow 2+1 hydrodynamic model [18]. The two curves overlap, indicating that the system at freeze-out is almost azimuthally symmetric in the transverse plane. The numbers denote the values of the transverse flow. The parameters are the same as in Fig. 22.1.

thermal freeze-out is made. The hypersurfaces extracted from the hydrodynamic model are used as the input for the thermal event generator THERMINATOR [49]. In the process of particle generation, the chemical potentials: $\mu_B = 28.5$ MeV, $\mu_S = 6.9$ MeV, and $\mu_{I_3} = -0.9$ MeV are introduced. These values are known from the thermal analysis of particle abundances [1], see Refs. [50,51] and Table 24.1.

In order to compare the model predictions with the data one has to use the appropriate value of the impact parameter in the theoretical calculations. In Ref. [18] Eq. (2.22) is used, where the value of the total inelastic cross section, $\sigma_{\rm in}^{\rm AuAu} = 6.4$ b, is determined from the Monte Carlo simulations of the Glauber model [52]. For the centrality class 0–5% this gives $b = 2.26$ fm (Eq. (2.22) with $c = 2.5\%$), while for the centrality class 20–30% one gets $b = 7.16$ fm (Eq. (2.22) with $c = 25\%$), see also Table 3.1.

22.1.1.1 *Central collisions*

In Fig. 22.1 the model results for the hadron transverse-momentum spectra in the central Au+Au collisions are presented. They are obtained with the initial starting time for hydrodynamics $\tau_{\rm i} = 1$ fm, the initial central temperature $T_{\rm i} = 320$ MeV, and the final (freeze-out) temperature $T_{\rm f} = 150$ MeV. The dots show the PHENIX data [47] for positive pions, positive kaons, and protons, whereas the solid lines show

[1] The values of the chemical potentials are much smaller than the characteristic temperature scale, hence, their effect on the hydrodynamic evolution may be neglected.

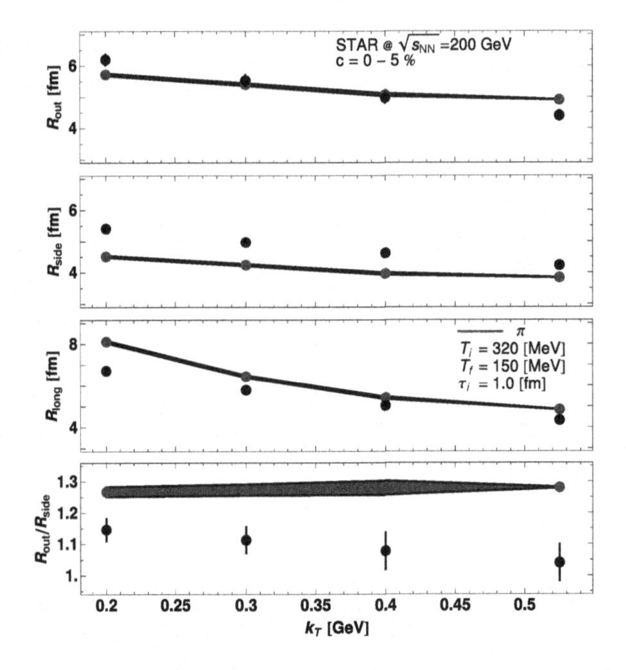

Fig. 22.3 The pionic HBT radii plotted as functions of the average transverse momentum of the pair compared to the STAR data [48] at the centrality 0–5%. The Cracow 2+1 model calculation [18] uses the two-particle method and includes the Coulomb effects. The values of the model parameters are the same as in Fig. 22.1.

the results obtained from the hydrodynamic code linked to THERMINATOR. The hydrodynamic model describes the data well up to the $p_\perp$ values of about 1.5 GeV. For larger values of $p_\perp$ the model underpredicts the data. This effect may be explained by the presence of the semi-hard processes that are not included in the hydrodynamic approach.

In Fig. 22.2 we show the freeze-out curves obtained from the hydrodynamic calculations. The in-plane and out-of-plane freeze-out curves are defined as the intersections of the freeze-out hypersurface with the planes $y = 0\,(\phi = 0^o)$ and $x = 0\,(\phi = 90^o)$. The two freeze-out curves practically overlap, showing that the expansion of the system is almost azimuthally symmetric in the transverse plane. This feature is certainly expected for the central collisions, since the impact parameter is very small.

Finally, in Fig. 22.3 the model results and the STAR data [48] for the HBT radii are presented. For the same values of the parameters, a quite reasonable agreement is found. Discrepancies at the level of 10–15% are observed in the behavior of the $R_{\rm side}$, which is too small, and $R_{\rm long}$, which is too large, most likely due to the assumption of strict boost invariance. The ratio $R_{\rm out}/R_{\rm side} \simeq 1.25$ is larger than one, which is a typical discrepancy of hydrodynamic studies. Nevertheless, when compared to other hydrodynamic calculations, the ratio $R_{\rm out}/R_{\rm side}$ obtained in [18]

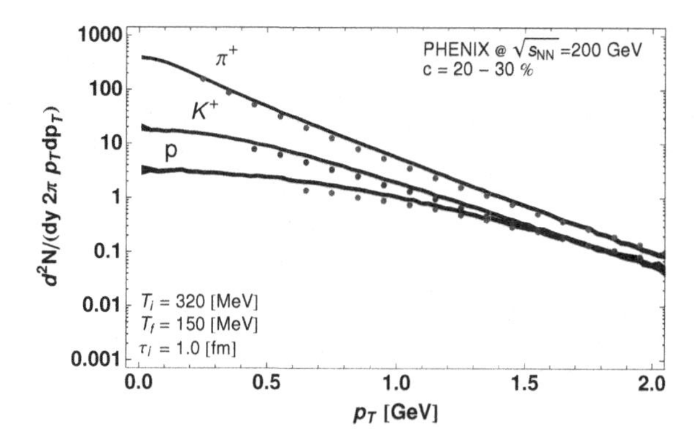

Fig. 22.4 Transverse-momentum spectra of π^+, K^+, and protons. The PHENIX experimental results [47] for Au+Au collisions at $\sqrt{s_{\rm NN}}$ = 200 GeV and the centrality class 20–30% (points) are compared to the Cracow 2+1 model calculations [18] (solid lines) with the parameters: $\tau_{\rm i}$ = 1.0 fm, $T_{\rm i}$ = 320 MeV, $T_{\rm f}$ = 150 MeV, and b = 7.16 fm.

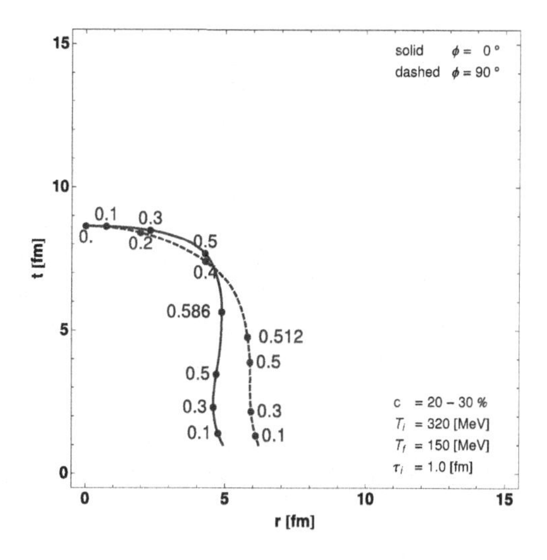

Fig. 22.5 The freeze-out curves for c = 20–30%. The solid line describes the in-plane profile, while the dashed line describes the out-of-plane profile. The values of the Cracow 2+1 model parameters [18] are the same as in Fig. 22.4.

is significantly closer to the experimental value.

22.1.1.2 *Non-central collisions*

Let us now turn to the discussion of the non-central events. We consider the centrality classes 20–30% and 20–40%. The data are compared with the model results obtained with the impact parameter b = 7.16 fm. The values of the initial central

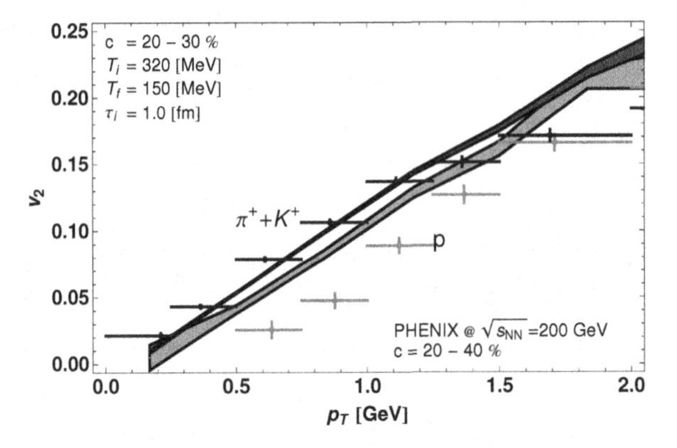

Fig. 22.6 The elliptic flow coefficient v_2. The values measured by PHENIX [53] at $\sqrt{s_{\rm NN}} = 200$ GeV and the centrality class 20–40% are indicated by the upper (pions + kaons) and lower (protons) points, with the horizontal bars indicating the $p_\perp$ bin. The corresponding Cracow 2+1 model calculations [18] are indicated by the solid lines, with the bands displaying the statistical error of the Monte-Carlo method. The parameters are the same as in Fig. 22.4.

temperature and the final temperature are the same as in the case of the central collisions, i.e., $T_{\rm i} = 320$ MeV and $T_{\rm f} = 150$ MeV.

The comparison of the experimental and model transverse-momentum spectra is presented in Fig. 22.4. Similarly to the central events, we find a reasonable agreement between the data and the model results. The two freeze-out curves obtained from the hydrodynamic calculations are shown in Fig. 22.5. We observe that the out-of-plane profile is wider than the in-plane profile. This difference indicates that the system is elongated along the y-axis at the moment of freeze-out. This feature is in qualitative agreement with the HBT measurements of the azimuthal dependence of $R_{\rm side}$.

In Fig. 22.6 the model results and the PHENIX data [53] on the elliptic flow coefficient $v_2(p_\perp)$ are compared. We observe that the v_2 of pions+kaons agrees with the data. Taking into account the uncertainly in the initial eccentricity, which is at the level of 10–20%, the obtained agreement is reasonable. On the other hand, the model prediction for proton v_2 is too large. The discrepancy is probably caused by the final-state elastic interactions, that are not included in this approach.

The HBT radii for non-central RHIC collisions are shown in Fig. 22.7. Similarly to the central collisions, we observe that $R_{\rm side}$ is slightly too small and $R_{\rm long}$ is slightly too large. Still, the ratio $R_{\rm out}/R_{\rm side}$ is very close to the data. Comparing the values of $R_{\rm side}$ with other hydrodynamic calculations one finds that the presented values are larger. This effect is caused by the halo of decaying resonances which increases the system size by about 1 fm, see Ref. [54].

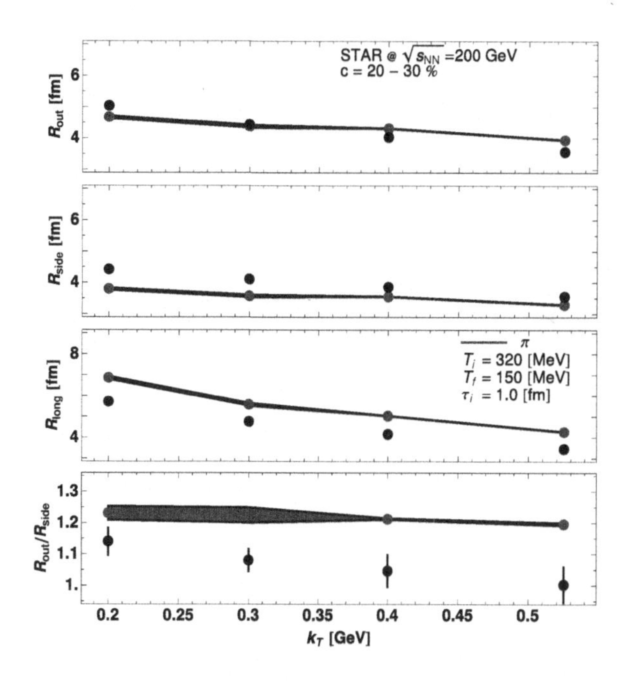

Fig. 22.7 The pionic HBT radii plotted as functions of the average transverse momentum of the pair compared to the STAR data [48] at centrality 20–30%. As in the case of the central collisions, the calculation uses the two-particle method and includes the Coulomb effects. The values of the Cracow 2+1 model parameters [18] are the same as in Fig. 22.4.

22.2 3+1 hydrodynamic models

One of the first 3+1 hydrodynamic codes used to analyze the RHIC data was worked out by Hirano and Tsuda [25]. They assumed the first order phase transition with the latent heat 1.7 GeV/fm^3. The starting time of the hydrodynamics was 0.6 fm. A special feature of this approach was the incorporation of the partial chemical equilibrium in the hadronic stage. The model reproduced the pion spectra and proton v_2. On the other hand, the proton spectra were overpredicted and the pion v_2 was not well reproduced. The important progress was achieved by Nonaka and Bass [28] who used RQMD as the hadronic after-burner to model the freeze-out. The hydrodynamic model of Ref. [28] describes well all essential one-particle observables.

22.2.1 *Cracow 3+1 hydrodynamic model*

In this Section we discuss a generalization of the 2+1 model presented in Sec. 22.1.1 to the 3+1 version. This improvement has been done recently by Bozek and Wyskiel [55] and we refer to it as to the Cracow 3+1 model. By comparing the 2+1 and 3+1 models we may select the effects of the non-trivial expansion along the beam axis. We may also reveal various effects connected with the assumption of the

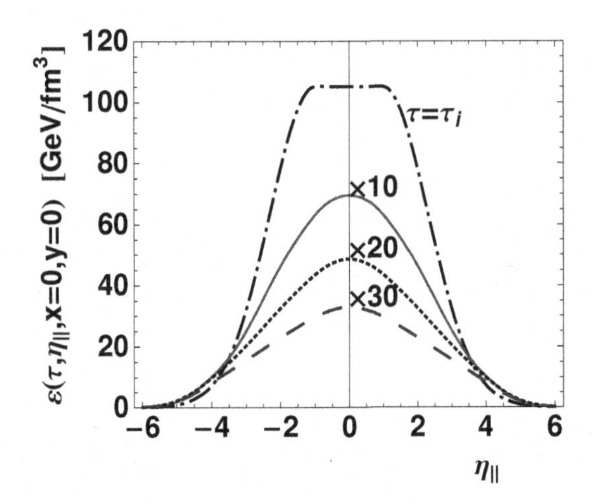

Fig. 22.8 Hydrodynamic expansion of matter in the Cracow 3+1 model [55]. The energy density of the system is plotted as a function of the spacetime rapidity $\eta_{\parallel}$ for different proper times, $\tau = 0.25$, 2, 4, 6 fm (the dashed-doted, solid, dotted, and dashed line, respectively).

boost-invariance used in 2+1 codes.

In Ref. [55] the same equation of state is used as in Ref. [18]. The initial conditions are taken in the form (15.10) with the parameters: $\Delta\eta_{\parallel} = 2$, $\delta\eta_{\parallel} = 1.3$, and $\kappa = 0.145$. The proportionality constant in Eq. (15.10) is chosen in such a way that the initial energy density at the center of the system at $\tau_{\rm i} = 0.25$ fm is equal to 107 GeV/fm^3. This corresponds to the initial central temperature of 510 MeV, that is well above the critical temperature. The initial longitudinal flow rapidity satisfies the condition $\vartheta_{\parallel} = \eta_{\parallel}$. The shift in the spacetime rapidity $\eta_{\parallel}^{\rm sh}$ is calculated for the appropriate centrality as a function of the transverse coordinates x and y. It will be shown below that this choice of the parameters leads to the good description of the transverse-momentum spectra, elliptic flow, and HBT radii of pions for different values of centrality and rapidity.

The spacetime-rapidity profiles of the energy density at the initial time ($\tau = \tau_{\rm i} = 0.25$ fm) and for later times ($\tau = 2$, 4, and 6 fm) are shown in Fig. 22.8. Due to the strong longitudinal acceleration at the edges of the system, the initial distribution with a plateau is changed to a Gaussian distribution that becomes wider with time. Such a change reminds us of the time evolution expected in the Landau model, although there exist important differences: the initial plateau in the Landau model is defined for a constant laboratory time, $t = t_{\rm i}$, and its width is narrower, since it is determined by the Lorentz contraction factor of the colliding nuclei. Moreover, there is no initial longitudinal flow in the Landau model, i.e., $\vartheta_{\parallel} = 0$ at $t = t_{\rm i}$.

The presence of the longitudinal acceleration makes the longitudinal fluid rapid-

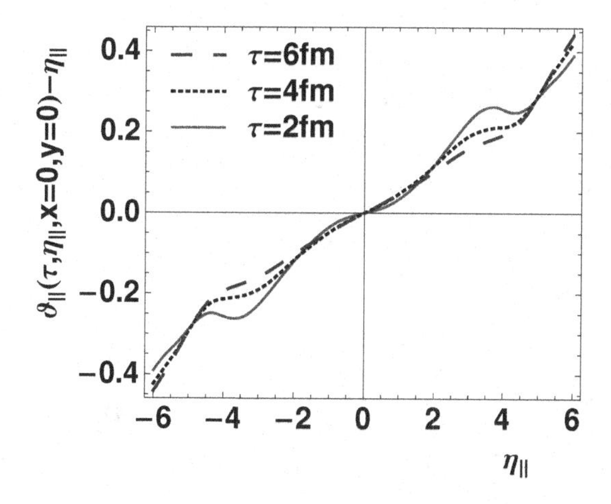

Fig. 22.9 The difference between the longitudinal fluid rapidity and the spacetime rapidity $\vartheta_{\parallel} - \eta_{\parallel}$ obtained in the Cracow 3+1 model [55] for several values of the proper time, $\tau = 2$, 4, 6 fm (the solid, dotted, and dashed line, respectively).

ity larger than the spacetime rapidity. This effect is illustrated in Fig. 22.9. It was shown in Ref. [56] that the longitudinal acceleration depends on the equation of state. It depends also on the viscous effects [57]. For the perfect fluid with a semi-hard equation of state, as considered in this Section, the longitudinal acceleration is quite substantial, making $\vartheta_{\parallel}$ significantly larger than $\eta_{\parallel}$.

To study further effects of the longitudinal acceleration, a comparison of the

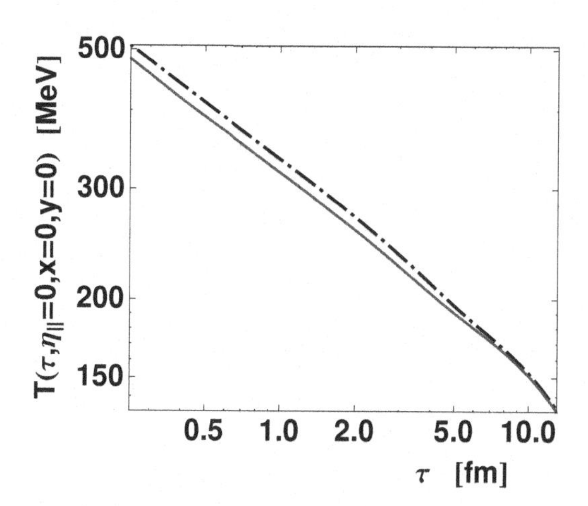

Fig. 22.10 The central temperature of the fireball shown as a function of the proper time for the Cracow 3+1 model [55] (dashed-dotted line) and the reference 2+1 evolution (solid line) ($b = 2.1$ fm).

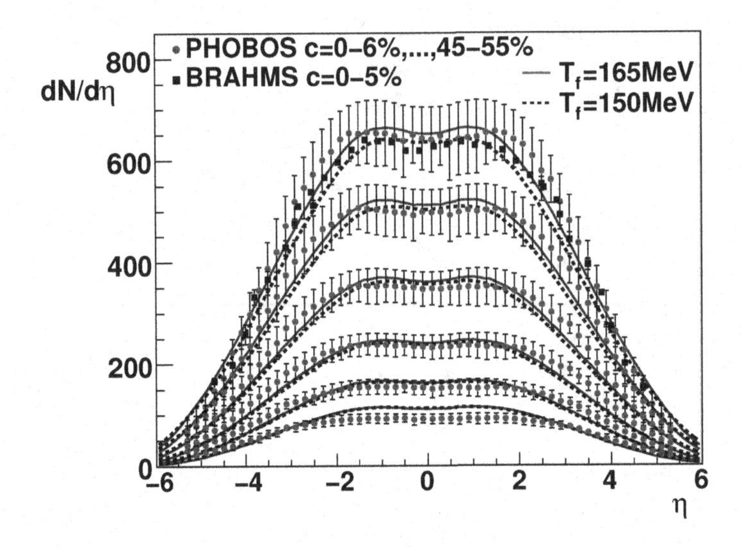

Fig. 22.11 Pseudorapidity distributions of charged particles for centrality classes 0–6%, 6–15%, 15–25%, 25–35%, 35–45%, and 5–55% calculated in the Cracow 3+1 model [55] for the freeze-out temperatures $T_{\rm f}$ = 165 and 150 MeV (solid and dashed lines respectively) compared to PHOBOS data (dots) [58]. The squares represent the BRAHMS data for centrality 0–5% [59].

3+1 and 2+1 hydrodynamic models has been done in [55]. The 2+1 hydrodynamic evolution is initialized at $\tau = \tau_{\rm i} = 0.25$ fm with the profile (15.10) without the $\eta_{\parallel}$ dependence, i.e., with H = const. The initial energy density that leads to the correct description of the data is 86 GeV/fm^3 ($T_{\rm i}$ = 485 MeV). The lower initial energy density required to describe the data in the 2+1 case may be explained by the additional cooling present in the non-boost invariant approach. Interestingly, the lifetimes of the 2+1 and 3+1 systems are very similar. This is again understood by the overall faster cooling of the initially hotter 3+1 system. The time dependence of the central temperature in the two cases is shown in Fig. 22.10.

In the 2+1 case which serves as a reference, the temperature at τ = 1 fm is about 320 MeV. This is consistent with the initial value used in the Cracow 2+1 model at $\tau = \tau_{\rm i}$ = 1 fm. However, when looking at details, an earlier starting point of the reference 2+1 hydrodynamic evolution makes it different from the Cracow 2+1 model. In the reference 2+1 case the transverse flow at τ = 1 fm is already developed, while in the 2+1 Cracow model the transverse flow at τ = 1 fm is exactly zero [2].

In Fig. 22.11 the pseudorapidity distributions obtained from the 3+1 hydrodynamic code are shown and compared to the experimental PHOBOS [58] and

[2]This discussion illustrates a possibility that a non-zero transverse flow in the initial conditions may be included by an earlier start of hydrodynamics.

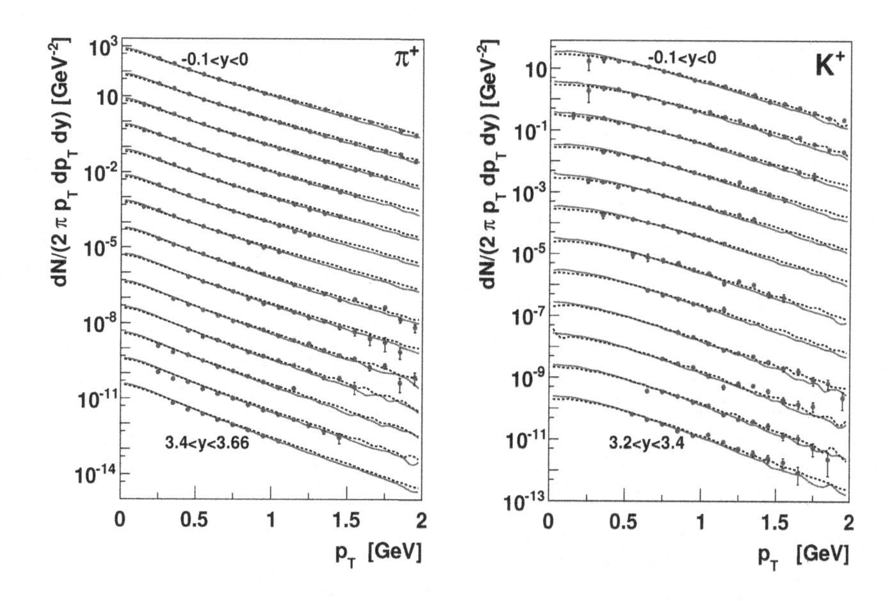

Fig. 22.12 Transverse momentum spectra of π^+ (left) and K^+ (right) calculated in the Cracow 3+1 model [55] for different rapidity windows and the centrality class 0–5% (results for different rapidity bins are scaled down by powers of 1/10). The dots represent the data of the BRAHMS Collab. [60].

BRAHMS [59] data. The Cracow 3+1 hydrodynamic model can successfully reproduce the data for different centralities. The results are shown for two thermal freeze-out temperatures $T_{\rm f} = 165$ and 150 MeV. The first one corresponds to the estimate of the chemical freeze-out temperature in Au+Au collisions [50]. The second is the same as that used in the 2+1 Cracow model. With the freeze-out temperature decreasing from $T_{\rm f} = 165$ MeV to $T_{\rm f} = 150$ MeV, the particle multiplicity drops down, however, this effect is small. We stress, that the centrality dependence predicted by the model comes solely from the geometrical scaling assumed in Eq. (15.10), other parameters remain unchanged.

In Fig. 22.12 the transverse-momentum spectra of pions and kaons are shown. The 3+1 Cracow model calculations are compared to the experimental data collected at different rapidities in the centrality class 0–5%. In the whole rapidity range, where identified particle spectra are available, one finds a very good agreement between the results of the model results and the BRAHMS data [60]. The best agreement is achieved for the freeze-out temperature of 150 MeV. The very good agreement between the data and the hydrodynamic model indicates that the matter created in central collisions behaves as a thermally equilibrated medium in the wide rapidity range, $-3.5 < y < 3.5$, despite the fact that the system under study is not boost-invariant.

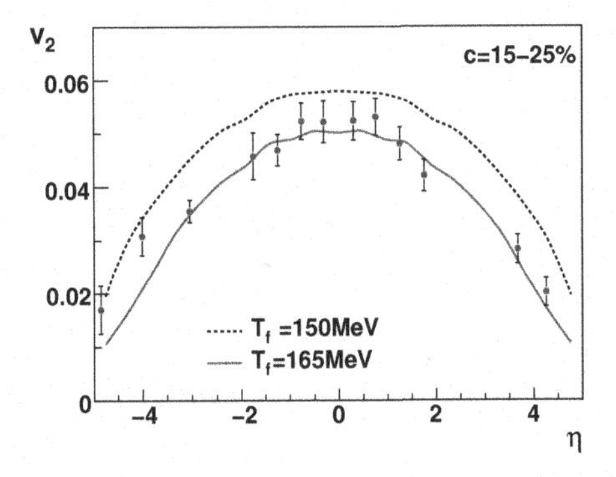

Fig. 22.13 Pseudorapidity dependence of the elliptic flow coefficient v_2 of charged particles obtained in the Cracow 3+1 model [55] for the centrality class 15–25% and two freeze-out temperatures $T_{\rm f} = 150$ and 165 MeV (dashed and solid lines respectively). The PHOBOS data are denoted by dots [61].

The measured elliptic flow coefficient of charged particles exhibits a clear pseudorapidity dependence [61]. To reproduce this shape of the v_2 in the hydrodynamic calculations, the initial conditions in energy density with a relatively narrow plateau in $\eta_{\parallel}$ must be chosen, see Eq. (15.10). Such initial conditions, combined with a hard equation of state and an early initial time of the evolution result in a disappearance of the Bjorken plateau [3] in the final hadron $v_2(\eta_{\parallel})$, see Fig. 22.13. The elliptic flow as a function of $p_{\perp}$ for identified particles is shown in Fig. 22.14. The discussed 3+1 hydrodynamic calculations describe the elliptic flow for mesons with $p_{\perp} < 1.5$ GeV, but to reproduce the saturation of v_2 for large $p_{\perp}$ dissipative or viscosity effects must be invoked. For the considered freeze-out temperatures, the elliptic flow of baryons is overpredicted .

The results of the 3+1 hydrodynamic code describing the HBT radii are presented in Fig. 22.15. We notice that the 3+1 and 2+1 calculations that describe well the transverse momentum spectra give also very similar HBT radii. Similarity in the spectra means that the transverse collective flow is the same. Together with the similarity of the freeze-out hypersurfaces, the similarity of the flow explains why the HBT radii from the 3+1 and 2+1 evolutions come out so close. The discrepancies between the 3+1 calculations and the data are smaller than 10%. The ratio $R_{\rm out}/R_{\rm side}$ from the model comes out close to the data as well.

22.3 RHIC HBT puzzle

In the plots showing the transverse-momentum dependence of v_2, see Figs. 22.6 and 22.14, we can see that the theoretical values of the proton ellip-

[3]Perhaps an onset of the Bjorken plateau may be seen in the range $-1 < y < 1$, see Fig. 22.13.

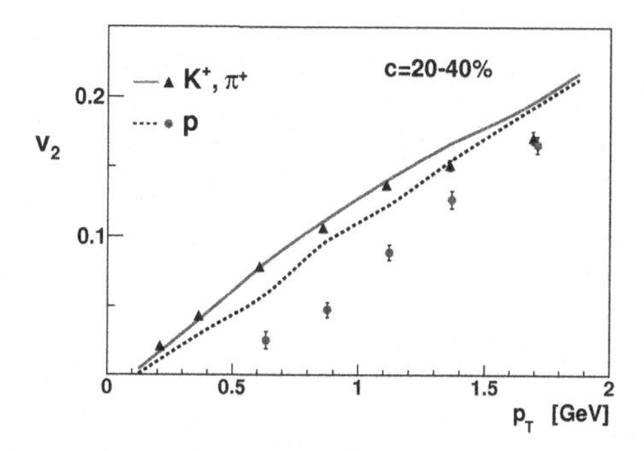

Fig. 22.14 Transverse-momentum dependence of the elliptic flow coefficient obtained in the Cracow 3+1 model [55] for protons (dotted line) and for π^+ and K^+ (solid line). The calculations are performed for $T_f = 150$ MeV. The PHENIX data (dots for protons and triangles for pions and kaons) are from Ref. [53].

tic flow are larger than the experimental values — the theoretical splitting of the pion and proton v_2 is not sufficiently large, as the data suggest. This behavior may be improved if the hydrodynamic evolution lasts longer, which also means that the freeze-out temperature is lower. Unfortunately, in this case the improvement of the results describing v_2 goes together with a deterioration of the good description of the HBT data.

The notorious difficulty in obtaining a simultaneous correct description of the elliptic flow and the HBT radii for various hadronic species in a single hydrodynamic model received the name of the *RHIC HBT puzzle*. Recently, it has been suggested by Pratt [62, 63] that this puzzle may be solved if several shortcomings of the original hydrodynamic models are removed at the same time. According to Refs. [62, 63] the necessary improvements include: **i)** the inclusion of the pre-equilibrium transverse flow (alternatively, a very early start of hydrodynamics), **ii)** the use of the realistic equation of state without the soft point where the sound velocity becomes essentially zero, **iii)** the inclusion of the shear viscosity, and finally **iv)** the improved calculation of the correlation functions. Unfortunately, at the moment no hydrodynamic calculation exists which explicitly shows that the inclusion of the discussed effects allows for the consistent description of the HBT radii and the elliptic flow. The approach of Refs. [62, 63] is restricted to the azimuthally symmetric situations and in the present form indicates that the effects **i)** – **iv)** help to reproduce the HBT radii together with the transverse-momentum spectra.

Perhaps, it is worth to make a comment about the role of the equation of state in the femtoscopic analyses, the point **ii)** above. Before RHIC started its operation, many hydrodynamic calculations were suggesting the appearance of long living fireballs in such high-energy collisions and, consequently, predicted $R_{\rm out}/R_{\rm side} \gg 1$. The arguments were based on the use of the first order phase transition with a distinct soft point [3, 64–67]. Since the RHIC data indicate $R_{\rm out}/R_{\rm side} \sim 1$, they

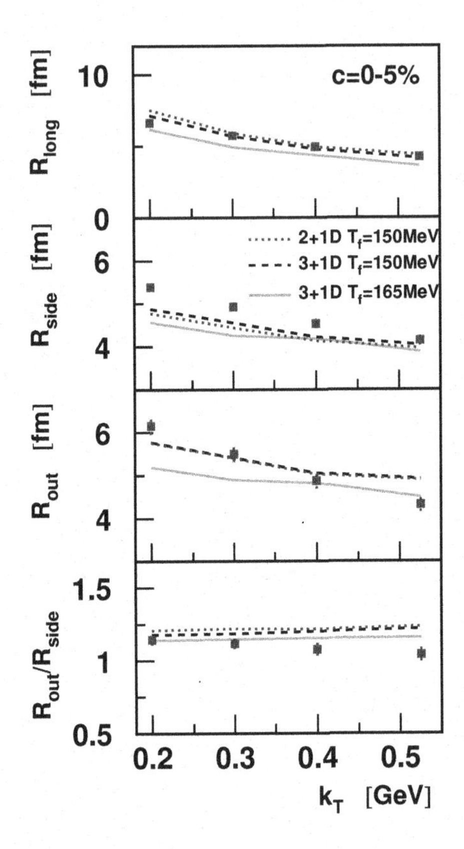

Fig. 22.15 HBT radii for Au+Au collisions at the centrality 0–5%. The $3+1$ calculations [55] with $T_{\rm f} = 165$ MeV (solid line) and $T_{\rm f} = 150$ MeV (dashed line), the reference $2+1$ calculation with $T_{\rm f} = 150$ MeV (dotted line) and STAR data [48] (squares) are shown.

indirectly exclude the first order phase transition. By the way, this is in agreement with the lattice simulations of QCD at zero baryon chemical potential.

22.3.1 *Cracow 2+1 model with Gaussian initial conditions*

The problems with the consistent description of one- and two-particle observables in the hydrodynamic models suggest that certain modifications of the standard hydrodynamic picture might be required. Clearly, the viscosity effects play an

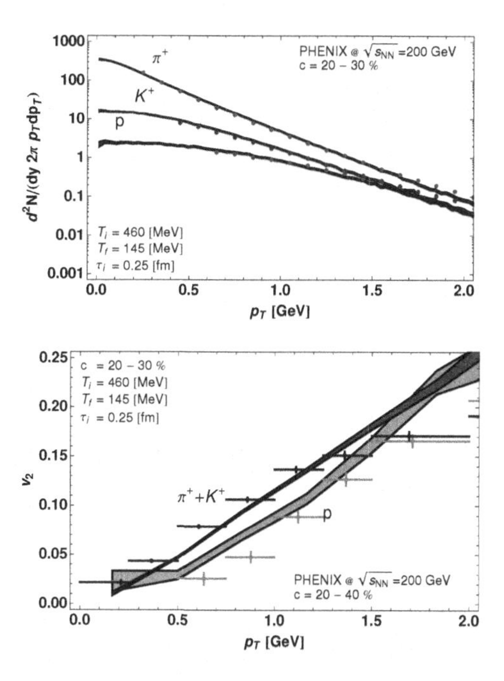

Fig. 22.16 The transverse-momentum spectra of pions, kaons and protons for the centrality class 20–30% (top) and the elliptic flow coefficient v_2 for the centrality class 20–40% (bottom). The Cracow 2+1 model with the Gaussian initial conditions [19] are plotted as functions of the transverse momentum and compared to the data from [47, 53]. The model results are obtained with the thermalization time $\tau_{\rm i} = 0.25$ fm, the initial central temperature $T_{\rm i} = 460$ MeV, and with the final temperature $T_{\rm f} = 145$ MeV. The width parameters used in the definition of the initial conditions: $a = 2.00$ fm and $b = 2.59$ fm.

important role and they are intensively studied at the moment.

Another possible way to modify the standard framework is connected with the use of non-standard initial conditions. This route has been followed in Ref. [19], where a simple Gaussian parameterization of the initial energy density profile in the transverse plane at $\tau = \tau_{\rm i} = 0.25$ fm is used,

$$\varepsilon(\mathbf{x}_\perp) = \varepsilon_{\rm i} \exp\left(-\frac{x^2}{2a^2} - \frac{y^2}{2b^2}\right). \tag{22.1}$$

The values of the width parameters a and b depend on the centrality class and are obtained by matching to the results for $\langle x^2 \rangle$ and $\langle y^2 \rangle$ from the Glauber Monte-Carlo model `GLISSANDO` [52]. This procedure implements the eccentricity fluctuations of the system in the Glauber approach [68, 69]. The expression $(a^2 + b^2)/2$ describes the overall transverse size of the system, whereas $(b^2 - a^2)/(b^2 + a^2)$ parameterizes its eccentricity.

The approach proposed in [19] follows closely the framework used in the 2+1 Cracow model. In particular, the same equation of state is used and the freeze-out is modeled with THERMINATOR. Thus, we shall refer to the model proposed in [19]

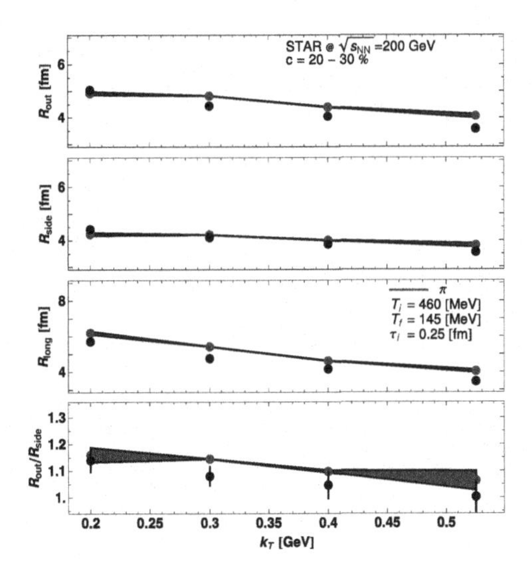

Fig. 22.17 The pion HBT radii $R_{\rm side}$, $R_{\rm out}$, $R_{\rm long}$, and the ratio $R_{\rm out}/R_{\rm side}$ for the centrality class 20–30%. The results of the Cracow 2+1 model with the Gaussian initial conditions [19] are compared to the data from [48]. Model parameters are the same as in Fig. 22.16.

as to the Cracow 2+1 model with Gaussian initial conditions. The main physical result connected with the use of the Gaussian profile (22.1), as compared to the Glauber initial conditions, is that it leads to a faster development of the transverse flow. This effect is crucial for the improved model description of the HBT radii and v_2.

In the upper part of Fig. 22.16 we show the transverse-momentum spectra of pions, kaons and protons for the centrality class 20–30%. The results of the Cracow 2+1 model with the Gaussian initial conditions are represented by the lines, while the PHENIX data [47] are indicated by the points. The lower part of Fig. 22.16 shows the model results and the PHENIX data [53] for v_2. The model parametrs are the following: $\tau_{\rm i} = 0.25$ fm, $T_{\rm i} = 460$ MeV, $T_{\rm f} = 145$ MeV, $a = 2.00$ fm, and $b = 2.59$ fm.

The change of the initial conditions leads to a better description of both the proton v_2 and the HBT radii which are shown in Fig. 22.17. The splitting between the slopes of the pion and proton v_2 becomes larger and the ratio $R_{\rm out}/R_{\rm side}$ agrees with the data. The remaining discrepancy between the theoretical and experimental proton v_2 may be attributed to final state interactions which are neglected in the Cracow model.

The example of the modified initial conditions discussed in this Section indicates that a significant improvement in the theoretical description of the data may be achieved if we abandon the standard picture of the initial conditions based on the optical limit of the Glauber model. On the other hand, we do not have any microscopic justification of the Gaussian initial conditions, hence, the RHIC HBT puzzle remains a debatable point.

22.4 Outlook at LHC

Within a 2+1 hydrodynamic approach applied to the midrapidity region, we expect that the increase of the initial beam energy from $\sqrt{s_{NN}} = 200$ GeV at RHIC to $\sqrt{s_{NN}} = 5.5$ TeV at LHC results essentially in a higher initial temperature $T_{\rm i}$. Other, less significant changes include the use of a larger value of the nucleon-nucleon cross section in the definition of the initial conditions and the application of different values of the chemical potentials at freeze-out.

Having this general idea in mind, the Cracow 2+1 model has been used to make predictions for the collisions at the LHC energy. Three different values for $T_{\rm i}$ are studied in [18], which are higher than those used at RHIC, namely, $T_{\rm i} = 400$, 450, and 500 MeV. Of course, it is not clear which value of $T_{\rm i}$ may be realized at LHC. Simple extrapolations suggest an increase of the multiplicity by a factor of 2 compared to the highest RHIC energies [70], which would favor $T_{\rm i}$ around 400 MeV. However, to study a broader range of possibilities, much higher temperatures have been also included [4].

In Tables 22.1 and 22.2 we list the results obtained in [18]. The following quantities are shown: the total π^+ multiplicity dN/dy at y = 0, the inverse-slope parameter for pions λ, the pion+kaon elliptic flow v_2 at $p_\perp = 1$ GeV, and the three HBT radii calculated at the pion pair momentum $k_\perp = 300$ MeV. The inverse-slope parameter given in Tables 22.1 and 22.2 is obtained from the formula given in Sec. 2.2.1.

The tables show the expected behavior — as the initial temperature increases, the multiplicity grows. This is so because a larger initial entropy causes a larger size of the system at freeze-out. The following approximate parameterizations work very well for the multiplicity of positive pions at LHC for the two considered centrality cases:

$$\begin{aligned} \frac{dN}{dy} &= 12000\,(T_{\rm i}/{\rm GeV})^{3.4}, \ (b = 2.4 \text{ fm}) \\ \frac{dN}{dy} &= 6000\,(T_{\rm i}/{\rm GeV})^{3.4}, \quad (b = 7.6 \text{ fm}). \end{aligned} \tag{22.2}$$

The power 3.4 works remarkably well. This behavior reflects the dependence of the initial entropy on T as shown in 6.3 of Sec. 6.6, where for the relevant temperature range of 300–500 MeV we have the approximate scaling $\sigma/T^3 \sim T^{0.4}$. Similarly,

[4] A larger value of the nucleon-nucleon cross section in the definition of the initial conditions ($\sigma_{\rm in} = 63$ mb, $\kappa = 0.2$) and of the nuclear profile ($A = 208$) imply that for LHC one may use $b = 2.4$ fm for $c = 2.5\%$ and $b = 7.6$ fm for $c = 25\%$. On the other hand, the chemical potentials used in the hadronization made by THERMINATOR are: $\mu_{\rm B} = 0.8$ MeV and $\mu_{\rm s} = \mu_{\rm I_3} = 0$ [71].

Table 22.1 Central collisions at RHIC corresponding to the results from Sec. 22.1.1 (the second row) and the predictions for central collisions at LHC (the three lower rows): A set of our results obtained for four different values of the initial temperature: T_i = 320, 400, 450, and 500 MeV. The columns contain the following information: dN/dy – the total pion multiplicity (positive pions only), λ – the inverse-slope parameter for positive pions at $p_\perp$ = 1 GeV, $R_{side}, R_{out}, R_{long}$ – the three HBT radii calculated at the average momentum $k_\perp$ = 300 MeV.

T_i [MeV]	$\frac{dN}{dy}$	λ [MeV]	R_{side} [fm]	R_{out} [fm]	R_{long} [fm]
320	274	237	4.2	5.4	6.5
400	543	288	5.2	5.9	7.7
450	802	320	5.6	6.1	8.2
500	1136	352	6.0	6.2	8.8

Table 22.2 Non-central collisions at RHIC corresponding to the results from Sec. 22.1.1 (the second row) and the predictions for non-central collisions at LHC (the three lower rows): The same quantities shown as in Table 22.1 with the additional information on the pion elliptic flow v_2 at $p_\perp$ = 1 GeV.

T_i [MeV]	$\frac{dN}{dy}$	λ [MeV]	$v_2^{\pi+K}$	R_{side} [fm]	R_{out} [fm]	R_{long} [fm]
320	152	244	0.14	3.6	4.5	5.6
400	272	299	0.16	4.2	4.5	6.2
450	401	331	0.16	4.6	4.6	6.7
500	569	363	0.15	5.2	4.9	7.4

for the slopes in the studied domain the authors of [18] find

$$\begin{aligned} \lambda(1 \text{ GeV}) &= 0.639\, T_i + 0.033 \text{ GeV}, \ (b = 2.4 \text{ fm}) \\ \lambda(1 \text{ GeV}) &= 0.661\, T_i + 0.033 \text{ GeV}, \ (b = 7.6 \text{ fm}). \end{aligned} \tag{22.3}$$

The HBT radii increase rather moderately with T_i, as can be seen from Tables 22.1 and 22.2. The ratio R_{out}/R_{side} decreases by about 10% which is an effect of the larger transverse flow caused by the larger initial temperature.

22.4.1 *Saturation of elliptic flow?*

The results shown in Table 22.2 indicate the saturation of the elliptic flow of light particles with increasing initial temperature (energy density) for a given initial space asymmetry. This observation is consistent with the findings of Kestin and Heinz discussed in Ref. [70]. Exactly this type of behavior suggests the existence of the hydrodynamic limit for v_2, which has been reached by the RHIC data. One of the most intriguing first measurements done at LHC will be to check if this limit still holds for higher energy collisions. If the measured LHC v_2 is significantly larger than the RHIC v_2, this means that the hydrodynamic picture should be substantially modified or even abandoned. In fact, Busza has showed [72] that simple extrapolations of the v_2 measured at RHIC and lower energies indicate that the hydrodynamic limit may be indeed surpassed. At the moment we may only expect the fascinating confrontation between theory and experiment.

Bibliography to Chapter 22

[1] A. Dumitru and D. H. Rischke, "Collective dynamics in highly relativistic heavy-ion collisions," *Phys. Rev.* **C59** (1999) 354–363.
[2] P. F. Kolb, J. Sollfrank, and U. W. Heinz, "Elliptic and hexadecupole flow from AGS to LHC energies," *Phys. Lett.* **B459** (1999) 667–673.
[3] D. Teaney and E. V. Shuryak, "An unusual space-time evolution for heavy ion collisions at high energies due to the QCD phase transition," *Phys. Rev. Lett.* **83** (1999) 4951–4954.
[4] P. F. Kolb, J. Sollfrank, and U. W. Heinz, "Anisotropic transverse flow and the quark-hadron phase transition," *Phys. Rev.* **C62** (2000) 054909.
[5] D. Teaney, J. Lauret, and E. V. Shuryak, "Flow at the SPS and RHIC as a quark gluon plasma signature," *Phys. Rev. Lett.* **86** (2001) 4783–4786.
[6] P. F. Kolb, P. Huovinen, U. W. Heinz, and H. Heiselberg, "Elliptic flow at SPS and RHIC: From kinetic transport to hydrodynamics," *Phys. Lett.* **B500** (2001) 232–240.
[7] P. Huovinen, P. F. Kolb, U. W. Heinz, P. V. Ruuskanen, and S. A. Voloshin, "Radial and elliptic flow at RHIC: further predictions," *Phys. Lett.* **B503** (2001) 58–64.
[8] P. F. Kolb, U. W. Heinz, P. Huovinen, K. J. Eskola, and K. Tuominen, "Centrality dependence of multiplicity, transverse energy, and elliptic flow from hydrodynamics," *Nucl. Phys.* **A696** (2001) 197–215.
[9] D. Teaney, J. Lauret, and E. V. Shuryak, "Hydro + cascade, flow, the equation of state, predictions and data," *Nucl. Phys.* **A698** (2002) 479–482.
[10] D. Teaney, J. Lauret, and E. V. Shuryak, "A hydrodynamic description of heavy ion collisions at the SPS and RHIC," `nucl-th/0110037`.
[11] U. W. Heinz and P. F. Kolb, "Early thermalization at RHIC," *Nucl. Phys.* **A702** (2002) 269–280.
[12] U. W. Heinz and P. F. Kolb, "Emission angle dependent pion interferometry at RHIC and beyond," *Phys. Lett.* **B542** (2002) 216–222.
[13] P. F. Kolb and U. W. Heinz, "Emission angle dependent HBT at RHIC and beyond," *Nucl. Phys.* **A715** (2003) 653–656.
[14] P. F. Kolb and R. Rapp, "Transverse flow and hadro-chemistry in Au+Au collisions

at $\sqrt{s_{\rm NN}} = 200$ GeV," *Phys. Rev.* **C67** (2003) 044903.
[15] P. F. Kolb, "Expansion rates at RHIC," *Heavy Ion Phys.* **21** (2004) 243–248.
[16] P. F. Kolb and U. Heinz, "Hydrodynamic description of ultrarelativistic heavy-ion collisions," nucl-th/0305084.
[17] M. Chojnacki, W. Florkowski, and T. Csorgo, "On the formation of Hubble flow in Little Bangs," *Phys. Rev.* **C71** (2005) 044902.
[18] M. Chojnacki, W. Florkowski, W. Broniowski, and A. Kisiel, "Soft heavy-ion physics from hydrodynamics with statistical hadronization: Predictions for collisions at $\sqrt{s_{\rm NN}} = 5.5$ TeV," *Phys. Rev.* **C78** (2008) 014905.
[19] W. Broniowski, M. Chojnacki, W. Florkowski, and A. Kisiel, "Uniform description of soft observables in heavy-oon collisions at $\sqrt{s_{\rm NN}} = 200$ GeV," *Phys. Rev. Lett.* **101** (2008) 022301.
[20] A. Kisiel, W. Broniowski, M. Chojnacki, and W. Florkowski, "Azimuthally-sensitive femtoscopy from RHIC to LHC in hydrodynamics with statistical hadronization," *Phys. Rev.* **C79** (2009) 014902.
[21] C. Nonaka, E. Honda, and S. Muroya, "(3+1)-dimensional relativistic hydrodynamical expansion of hot and dense matter in ultra-relativistic nuclear collision," *Eur. Phys. J.* **C17** (2000) 663–673.
[22] T. Hirano, "Is early thermalization achieved only near midrapidity in Au+Au collisions at $\sqrt{s_{\rm NN}} = 130$ GeV?," *Phys. Rev.* **C65** (2002) 011901.
[23] T. Hirano, K. Morita, S. Muroya, and C. Nonaka, "Hydrodynamical analysis of hadronic spectra in the 130 GeV/nucleon Au+Au collisions," *Phys. Rev.* **C65** (2002) 061902.
[24] K. Morita, S. Muroya, C. Nonaka, and T. Hirano, "Comparison of space-time evolutions of hot/dense matter in $\sqrt{s_{\rm NN}} = 17$ GeV and 130 GeV relativistic heavy ion collisions based on a hydrodynamical model," *Phys. Rev.* **C66** (2002) 054904.
[25] T. Hirano and K. Tsuda, "Collective flow and two pion correlations from a relativistic hydrodynamic model with early chemical freeze out," *Phys. Rev.* **C66** (2002) 054905.
[26] C. Nonaka and M. Asakawa, "Hydrodynamical evolution near the QCD critical end point," *Phys. Rev.* **C71** (2005) 044904.
[27] T. Hirano, U. W. Heinz, D. Kharzeev, R. Lacey, and Y. Nara, "Hadronic dissipative effects on elliptic flow in ultrarelativistic heavy-ion collisions," *Phys. Lett.* **B636** (2006) 299–304.
[28] C. Nonaka and S. A. Bass, "Space-time evolution of bulk QCD matter," *Phys. Rev.* **C75** (2007) 014902.
[29] U. W. Heinz, H. Song, and A. K. Chaudhuri, "Dissipative hydrodynamics for viscous relativistic fluids," *Phys. Rev.* **C73** (2006) 034904.
[30] A. K. Chaudhuri, "Dissipative hydrodynamics in 2+1 dimension," *Phys. Rev.* **C74** (2006) 044904.
[31] R. Baier, P. Romatschke, and U. A. Wiedemann, "Dissipative hydrodynamics and heavy ion collisions," *Phys. Rev.* **C73** (2006) 064903.
[32] R. Baier and P. Romatschke, "Causal viscous hydrodynamics for central heavy-ion collisions," *Eur. Phys. J.* **C51** (2007) 677–687.
[33] P. Romatschke, "Causal viscous hydrodynamics for central heavy-ion collisions. II: Meson spectra and HBT radii," *Eur. Phys. J.* **C52** (2007) 203–209.
[34] A. K. Chaudhuri, "Causal dissipative hydrodynamics for QGP fluid in 2+1 dimensions," 0704.0134.
[35] P. Romatschke and U. Romatschke, "Viscosity information from relativistic nuclear collisions: How perfect is the fluid observed at RHIC?," *Phys. Rev. Lett.* **99** (2007)

172301.
[36] H. Song and U. W. Heinz, "Causal viscous hydrodynamics in 2+1 dimensions for relativistic heavy-ion collisions," *Phys. Rev.* **C77** (2008) 064901.
[37] H. Song and U. W. Heinz, "Suppression of elliptic flow in a minimally viscous quark- gluon plasma," *Phys. Lett.* **B658** (2008) 279–283.
[38] M. Luzum and P. Romatschke, "Conformal relativistic viscous hydrodynamics: Applications to RHIC results at $\sqrt{s_{NN}} = 200$ GeV," *Phys. Rev.* **C78** (2008) 034915.
[39] A. K. Chaudhuri, "Viscous fluid dynamics in Au+Au collisions at RHIC," `arXiv:0801.3180`.
[40] H. Song and U. W. Heinz, "Multiplicity scaling in ideal and viscous hydrodynamics," *Phys. Rev.* **C78** (2008) 024902.
[41] G. S. Denicol, T. Kodama, T. Koide, and P. Mota, "Shock propagation and stability in causal dissipative hydrodynamics," *Phys. Rev.* **C78** (2008) 034901.
[42] G. S. Denicol, T. Kodama, T. Koide, and P. Mota, "Stability and Causality in relativistic dissipative hydrodynamics," *J. Phys.* **G35** (2008) 115102.
[43] G. S. Denicol, T. Kodama, T. Koide, and P. Mota, "Non-linearity induced by finite size of fluid cell in causal dissipative hydrodynamics," `arXiv:0808.3170`.
[44] U. W. Heinz and H. Song, "Causal relativistic hydrodynamics for viscous fluids," *J. Phys.* **G35** (2008) 104126.
[45] H. Song and U. W. Heinz, "Extracting the QGP viscosity from RHIC data – a status report from viscous hydrodynamics," `arXiv:0812.4274`.
[46] **PHENIX** Collaboration, K. Adcox *et al.*, "Formation of dense partonic matter in relativistic nucleus nucleus collisions at RHIC: Experimental evaluation by the PHENIX collaboration," *Nucl. Phys.* **A757** (2005) 184–283.
[47] **PHENIX** Collaboration, S. S. Adler *et al.*, "Identified charged particle spectra and yields in Au+Au collisions at $\sqrt{s_{NN}} = 200$ GeV," *Phys. Rev.* **C69** (2004) 034909.
[48] **STAR** Collaboration, J. Adams *et al.*, "Pion interferometry in Au+Au collisions at $\sqrt{s_{NN}} = 200$ GeV," *Phys. Rev.* **C71** (2005) 044906.
[49] A. Kisiel, T. Taluc, W. Broniowski, and W. Florkowski, "THERMINATOR: Thermal heavy-ion generator," *Comput. Phys. Commun.* **174** (2006) 669–687.
[50] W. Florkowski, W. Broniowski, and M. Michalec, "Thermal analysis of particle ratios and p_T-spectra at RHIC," *Acta Phys. Polon.* **B33** (2002) 761–769.
[51] A. Baran, W. Broniowski, and W. Florkowski, "Description of the particle ratios and transverse-momentum spectra for various centralities at rhic in a single-freeze-out model," *Acta Phys. Polon.* **B35** (2004) 779–798.
[52] W. Broniowski, M. Rybczynski, and P. Bozek, "GLISSANDO: GLauber Initial-State Simulation AND mOre.," *Comput. Phys. Commun.* **180** (2009) 69–83.
[53] **PHENIX** Collaboration, S. S. Adler *et al.*, "Elliptic flow of identified hadrons in Au+Au collisions at $\sqrt{s_{NN}} = 200$ GeV," *Phys. Rev. Lett.* **91** (2003) 182301.
[54] A. Kisiel, W. Florkowski, and W. Broniowski, "Femtoscopy in hydro-inspired models with resonances," *Phys. Rev.* **C73** (2006) 064902.
[55] P. Bozek and I. Wyskiel, "Rapid hydrodynamic expansion in relativistic heavy-ion collisions," *Phys. Rev.* **C79** (2009) 044916.
[56] L. M. Satarov, A. V. Merdeev, I. N. Mishustin, and H. Stoecker, "Longitudinal fluid dynamics for ultrarelativistic heavy- ion collisions," *Phys. Rev.* **C75** (2007) 024903.
[57] P. Bozek, "Viscous evolution of the rapidity distribution of matter created in relativistic heavy-ion collisions," *Phys. Rev.* **C77** (2008) 034911.
[58] B. B. Back *et al.*, "The significance of the fragmentation region in ultrarelativistic heavy ion collisions," *Phys. Rev. Lett.* **91** (2003) 052303.
[59] **BRAHMS** Collaboration, I. G. Bearden *et al.*, "Pseudorapidity distributions of

charged particles from Au+Au collisions at the maximum RHIC energy," *Phys. Rev. Lett.* **88** (2002) 202301.

[60] **BRAHMS** Collaboration, I. G. Bearden *et al.*, "Charged meson rapidity distributions in central Au+Au collisions at $\sqrt{s_{\rm NN}} = 200$ GeV," *Phys. Rev. Lett.* **94** (2005) 162301.

[61] **PHOBOS** Collaboration, B. B. Back *et al.*, "Centrality and pseudorapidity dependence of elliptic flow for charged hadrons in Au+Au collisions at $\sqrt{s_{\rm NN}} = 200$ GeV," *Phys. Rev.* **C72** (2005) 051901.

[62] S. Pratt, "Resolving the HBT Puzzle in Relativistic Heavy Ion Collision," *Phys. Rev. Lett.* **102** (2009) 232301.

[63] S. Pratt, "The Long Slow Death of the HBT Puzzle, Proceedings for QM 2009," `arXiv:0907.1094`.

[64] C. M. Hung and E. V. Shuryak, "Hydrodynamics near the QCD phase transition: Looking for the longest lived fireball," *Phys. Rev. Lett.* **75** (1995) 4003–4006.

[65] D. H. Rischke, Y. Pursun, J. A. Maruhn, H. Stoecker, and W. Greiner, "The phase transition to the quark-gluon plasma and its effects on hydrodynamic flow," *Heavy Ion Phys.* **1** (1995) 309.

[66] D. H. Rischke and M. Gyulassy, "The maximum lifetime of the quark-gluon plasma," *Nucl. Phys.* **A597** (1996) 701–726.

[67] D. H. Rischke and M. Gyulassy, "The time-delay signature of quark-gluon plasma formation in relativistic nuclear collisions," *Nucl. Phys.* **A608** (1996) 479–512, `nucl-th/9606039`.

[68] R. Andrade, F. Grassi, Y. Hama, T. Kodama, and J. Socolowski, O., "On the necessity to include event-by-event fluctuations in experimental evaluation of elliptical flow," *Phys. Rev. Lett.* **97** (2006) 202302.

[69] Y. Hama *et al.*, "NeXSPheRIO results on elliptic-flow fluctuations at RHIC," *Phys. Atom. Nucl.* **71** (2008) 1558–1564.

[70] N. Armesto *et al.*, "Heavy Ion Collisions at the LHC - Last Call for Predictions," *J. Phys.* **G35** (2008) 054001.

[71] A. Andronic, P. Braun-Munzinger, and J. Stachel, "Hadron production in central nucleus nucleus collisions at chemical freeze-out," *Nucl. Phys.* **A772** (2006) 167–199.

[72] W. Busza, "Trends in multiparticle production and some 'predictions' for pp and PbPb collisions at LHC," *J. Phys.* **G35** (2008) 044040.

Chapter 23

Exercises to PART VI

Exercise 23.1. *Non-relativistic phase space density in the Fermi model.*
Neglecting the spin properties of the particles and disregarding the momentum conservation find that the phase space density of n non-relativistic particles with masses $m_1, ..., m_n$ (i.e., the analog of Eq. (19.2) for n particles, without the delta function expressing the momentum conservation) is given by the expression

$$\frac{dQ_n(T_{\rm kin})}{dT_{\rm kin}} = (2\pi)^{3n/2}(m_1...m_n)^{3/2}\frac{(T_{\rm kin})^{3n/2-1}}{\Gamma(3n/2)}. \tag{23.1}$$

Here $T_{\rm kin}$ is the total classical kinetic energy. *Hint*: Consider the identity

$$\begin{aligned}\left(\frac{\sqrt{\pi}}{2}\right)^n &= \int_{-\infty}^{+\infty} dx_1 x_1^2 ... \int_{-\infty}^{+\infty} dx_n x_n^2 \exp\left(-\sum_{i=1}^{n} x_i^2\right) \\ &= \int_0^{+\infty} dr \int_{-\infty}^{+\infty} dx_1 x_1^2 ... \int_{-\infty}^{+\infty} dx_n x_n^2 \exp(-r)\, \delta\left(r - \sum_{i=1}^{n} x_i^2\right)\end{aligned} \tag{23.2}$$

and show that

$$\int_{-\infty}^{+\infty} dx_1 x_1^2 ... \int_{-\infty}^{+\infty} dx_n x_n^2\, \delta\left(T_{\rm kin} - \sum_{i=1}^{n} x_i^2\right) = \left(\frac{\sqrt{\pi}}{2}\right)^n \frac{(T_{\rm kin})^{3n/2-1}}{\Gamma(3n/2)}. \tag{23.3}$$

The energy density of the phase space $Q_n(T_{\rm kin})$ may be easily expressed by the integral above, which leads directly to Eq. (23.1).

Exercise 23.2. *Ultra-relativistic phase space density in the Fermi model.*
Again neglecting the spin properties of the particles and disregarding the momentum conservation find that the phase space density of n ultra relativistic particles (i.e., the analog of Eq. (19.2) for n massless particles) is given by the expression

$$\frac{dQ_n(W)}{dW} = 2^{3n}\pi^n \frac{W^{3n-1}}{\Gamma(3n)}. \tag{23.4}$$

Here W is the total energy of n particles including the rest energy. *Hint*: In this case consider the identity

$$1 = \int_0^{+\infty} dx_1 x_1^5 ... \int_0^{+\infty} dx_n x_n^5 \exp\left(-\sum_{i=1}^n x_i^2\right) \tag{23.5}$$
$$= \int_0^{+\infty} dr \int_0^{+\infty} dx_1 x_1^5 ... \int_0^{+\infty} dx_n x_n^5 \exp(-r)\, \delta\left(r - \sum_{i=1}^n x_i^2\right)$$

and show that

$$\int_0^{+\infty} dx_1 x_1^5 ... \int_0^{+\infty} dx_n x_n^5\, \delta\left(W - \sum_{i=1}^n x_i^2\right) = \frac{W^{3n-1}}{\Gamma(3n)}. \tag{23.6}$$

Equation (23.6) may be used to derive Eq. (23.4).

Exercise 23.3. *Phase space density for a mixture of classical and ultra-relativistic particles in the Fermi model.*
Generalize the two previous cases to the situation where the final state consists of s classical particles with mass m_N and n ultra-relativistic (massless) particles. The initial total energy is W and the final kinetic energy is $W - sm_N$. Show that in this case the phase space density is given by the formula

$$\frac{dQ_{s+n}(W - sm_N)}{dW} = 2^{3n+3s/2} \pi^{n+3s/2} m_N^{3s/2} \frac{(W - sm_N)^{3n+3s/2-1}}{\Gamma(3n + 3s/2)}. \tag{23.7}$$

Exercise 23.4. *Khalatnikov's equation.*
i) Calculate the derivatives of the Khalatnikov function:

$$\frac{\partial \chi}{\partial Y} = -\frac{l T_i}{c_s} e^Y \left[\int_{c_s\vartheta}^{-Y} e^{(1+b_s)Y'} I_0(b_s X')\, dY' - e^{-(1+b_s)Y} I_0(b_s X) \right], \tag{23.8}$$

$$\frac{\partial^2 \chi}{\partial Y^2} = \frac{\partial \chi}{\partial Y} - \frac{l T_i}{c_s} b_s\, e^{-b_s Y} \left[I_0(b_s X) - I_1(b_s X) \frac{Y}{X} \right], \tag{23.9}$$

$$\frac{\partial \chi}{\partial \vartheta} = \frac{l T_i}{c_s} e^Y \left[c_s e^{(1+b_s)c_s\vartheta} + b_s c_s^2 \vartheta \int_{c_s\vartheta}^{-Y} e^{(1+b_s)Y'} \frac{I_1(b_s X')}{X'} dY' \right], \tag{23.10}$$

$$\frac{\partial^2 \chi}{\partial \vartheta^2} = \frac{l T_i}{c_s} e^Y \left[\left(c_s^2 (1+b_s) - \frac{1}{2} b_s^2 c_s^3 \vartheta \right) e^{(1+b_s)c_s\vartheta} + b_s c_s^2 \right. \tag{23.11}$$

$$\left. \times \int_{c_s\vartheta}^{-Y} e^{(1+b_s)Y'} \left(\frac{Y'^2}{X'^3} I_1(b_s X') - \frac{b_s c_s^2 \vartheta^2}{2X'^2} \left[I_0(b_s X') + I_2(b_s X') \right] \right) dY' \right].$$

In Eqs. (23.8) – (23.10) we have introduced the short-hand notation

$$b_s = \frac{1-c_s^2}{2c_s^2}, \quad Y = \ln(T/T_i),$$
$$X' = \sqrt{Y'^2 - c_s^2\vartheta^2}, \quad X = \sqrt{Y^2 - c_s^2\vartheta^2}. \tag{23.12}$$

The functions $I_0(x)$ and $I_1(x)$ are the modified Bessel functions of the first kind, $dI_0(x)/dx = I_1(x)$ and $dI_1(x)/dx = [I_0(x) + I_2(x)]/2$.

ii) Prove that χ satisfies the Khalatnikov equation (20.15). *Hint:* Substitute the derivatives (23.8), (23.9), and (23.11) into Eq. (20.15). Replace the Bessel functions $I_0(X)$ and $I_1(X)$ in the formulas (23.8) and (23.9) according to the rules

$$-e^{-b_s Y} I_1(b_s X) \frac{Y}{X} = e^Y \left[e^{(1+b_s)Y'} I_1(b_s X') \frac{Y'}{X'} \bigg|_{c_s\vartheta}^{-Y} + \frac{b_s c_s \vartheta}{2} e^{(1+b_s)c_s\vartheta} \right],$$

$$e^{-b_s Y} I_0(b_s X) = e^Y \left[e^{(1+b_s)Y'} I_0(b_s X') \bigg|_{c_s\vartheta}^{-Y} + e^{(1+b_s)c_s\vartheta} \right],$$

where we use the notation

$$f(Y')|_{c_s\vartheta}^{-Y} \quad = \quad f(Y' = -Y) - f(Y' = c_s\vartheta). \tag{23.13}$$

The terms proportional to $e^{(1+b_s)c_s\vartheta}$ cancel. Using the rule of the integration by parts,

$$I_n(X') f(Y')|_{c_s\vartheta}^{-Y} = \int_{c_s\vartheta}^{-Y} \frac{dI_n(X')}{dY'} f(Y') dY' + \int_{c_s\vartheta}^{-Y} I_n(X') \frac{df(Y')}{dY'} dY',$$

we rewrite all the terms as the integral over Y'. With the help of the recurrence relations for the Bessel functions we show that the integrand is zero, hence χ fulfills the Khalatnikov equation.

Exercise 23.5. *Bjorken model.*
Calculate the four-acceleration of the Bjorken flow.

PART VII

FREEZE-OUT MODELS

Chapter 24

Thermal Models

24.1 Thermal versus hydro-inspired models

In the following Part we introduce a large class of models describing hadron production in heavy-ion collisions. These models are very often used to analyze and interpret the heavy-ion soft hadronic data. Their common feature is that they all use simple thermodynamic and hydrodynamic concepts. This is reflected in their names — they are popularly known as *thermal models*, *statistical models*, or *hydro-inspired models*. Since all these models describe the properties of matter at freeze-out (chemical and/or thermal) we shall refer to them generally as to the *freeze-out* models.

The names *thermal models* and *statistical models* are used for thermodynamic approaches which try to explain the relative abundances of hadrons, i.e., the *ratios of hadron multiplicities*. This analysis determines the values of thermodynamic parameters at the chemical freeze-out, see Sec. 4.5. On the other hand, the *hydro-inspired models* are used to analyze the hadronic *transverse-momentum spectra and correlations* [1]. In this case, the name emphasizes the fact that such models do not include the full hydrodynamic evolution but use various hydrodynamics-motivated assumptions about the state of matter at the thermal freeze-out, see Sec. 4.4.

Using a few thermodynamic parameters, the thermal models usually do very good job in describing the relative yields of many hadronic species at the chemical freeze-out. For example, within the grand canonical version of the thermal model many successful fits were obtained with only two independent parameters, $T_{\rm chem}$ and $\mu^B_{\rm chem}$, which are the values of the temperature and baryon chemical potential at the chemical freeze-out.

The input for the hydro-inspired models is more complicated as it involves the ansatz describing the thermal freeze-out hypersurface. Its particular shape in the Minkowski space should be precisely defined. In this case, the real hydrodynamic calculations may be useful, as they give suggestions about the forms of the thermal

[1]Of course the classification introduced in this Chapter is not unique. It is only an attempt to characterize the main streams of the activities carried out in this wide subfield of heavy-ion physics.

freeze-out hypersurface. In addition, the hydrodynamic calculations may give hints about the values of the hydrodynamic flow at the thermal freeze-out. Nevertheless, *the common practice in hydro-inspired models is that the conditions at the thermal freeze-out are simply assumed* and only later confirmed/falsified by the comparison of the model results with the experimental data.

The perfect methodology would be to match the freeze-out hypersurfaces that have been postulated and successfully verified in the framework of the hydro-inspired models with the freeze-out hypersurfaces obtained from the hydrodynamic calculations (one can easily notice that the freeze-out hypersurfaces are the "input" for hydro-inspired models and the "output" of hydrodynamic calculations). Unfortunately, this kind of work is practically never done. This is so because, firstly, different groups work in these two different subfields of heavy-ion physics, secondly, the freeze-out hypersurfaces assumed in the hydro-inspired models are usually quite different from the hypersurfaces obtained from the hydrodynamic calculations — at first sight the successful matching seems very hard or even impossible to achieve. Therefore, the hydro-inspired models form a kind of independent subfield of heavy-ion physics, where the emphasis is put on the successful fitting of various experimental data. This situation does not exclude the case, however, that stronger and interesting links between the two approaches may be found in the near future.

The thermal models treat the matter formed at the chemical freeze-out as the multi-component hadronic gas. One may check that the evident success of this approach is related to the fact that practically all known resonances are included in these analyses. On the other hand, the hydro-inspired models usually ignore the effects of the resonance decays or include only few species (the most important, light resonances such as ρ or Δ). This attitude is partially explained by the fact that the thermal freeze-out temperature $T_{\rm therm}$ is smaller than the chemical freeze-out temperature $T_{\rm chem}$, and the role of heavy resonances at $T_{\rm therm}$ is diminished.

It is interesting to mention that there are approaches where the chemical and thermal freeze-out happen at the same time. In the *single-freeze-out model* one describes the spectra and predicts their absolute normalization [2]. Thus, the integrated spectra give the yields of hadrons. Within this approach one can describe many aspects of the RHIC data, provided the contributions from the resonance decays are consistently included in all the calculations [3].

In the remaining part of this Chapter we present the main assumptions of the thermal models. We show how they are used to analyze the yields of hadronic abundances at the chemical freeze-out. In the next Chapter we present various

[2]It is a common practice that the theoretical modeling of the spectra includes free parameters for the normalization of each hadron spectrum. The stress is put on the correct reproduction of the shapes of the spectra only. Of course, this way of fitting cannot deliver any important information about the ratios of hadronic multiplicities.

[3]This requirement should not be surprising — as we have learned in Secs. 6.4 and 6.5, the inclusion of resonances accounts for the interactions of stable hadrons such as pions, kaons, and protons.

kinds of the hydro-inspired models, including the very popular blast-wave model.

24.2 Static fireball

The simplest form of the freeze-out hypersurface is that defined by the condition of constant laboratory time, $t = \text{const}$. In this case, Eq. (15.19) leads to the trivial result

$$d\Sigma^\mu = (dV, 0, 0, 0) = (d^3x, 0, 0, 0) \tag{24.1}$$

and Eq. (15.18) yields

$$N = \int d^3p \int dV f(x, p). \tag{24.2}$$

If $f(x, p)$ is the equilibrium distribution and the hydrodynamic expansion of the fluid is negligible, $u^\mu(x) \approx (1, 0, 0, 0)$, then the hadron multiplicities may be obtained from the standard thermodynamic expression

$$N_i = g_i V \int \frac{d^3p}{(2\pi)^3} \left[\exp\left(\frac{E_i(p) - \mu_i}{T}\right) - \epsilon\right]^{-1}. \tag{24.3}$$

In general, for strongly interacting matter the chemical potential μ_i becomes a linear combination of the baryon chemical potential, the strange chemical potential, and the isospin chemical potential

$$\mu_i = \mu_B B_i + \mu_S S_i + \mu_I I_i. \tag{24.4}$$

Here B_i, S_i, and I_i are the baryon number, strangeness, and the third component of isospin of the ith hadron, respectively. The use of the chemical potentials μ_B, μ_S and μ_I, allows us to satisfy the appropriate conservation laws. In particular, if the matter at freeze-out originates from the initial nuclear matter of the two colliding nuclei, the total strangeness of the system at freeze-out must be zero

$$\sum_i S_i N_i = 0, \tag{24.5}$$

and the ratio of the electric charge to the baryon number in the hadronic fireball at freeze-out is the same as in the colliding nuclei

$$\frac{\sum_i Q_i N_i}{\sum_i B_i N_i} = \frac{Z}{A}. \tag{24.6}$$

Here Q_i is the electric charge of the ith hadron. According to the Gell-Mann – Nishijima formula we have

$$Q_i = I_i + \frac{B_i + S_i}{2}. \tag{24.7}$$

Equations (24.5) and (24.6) indicate that not all thermodynamic quantities are independent. The most popular choice is to use Eqs. (24.5) and (24.6) to eliminate μ_S and μ_I. In this way, the produced matter may be characterized by two intensive parameters: temperature and baryon chemical potential.

24.3 Thermal analysis of ratios of hadron multiplicities

Equation (24.3) is the basis of a very popular thermal analysis of the hadron yields [4]. In this approach, the measured multiplicities of different hadrons are used to construct all independent ratios of the multiplicities, $R_k^{\rm exp}$, and such experimental ratios are confronted with the model results, $R_k^{\rm model}$, obtained with the help of Eq. (24.3). In this procedure we look for the optimal values of the temperature and the baryon chemical potential, i.e., for the values which lead to the best agreement of the model thermal ratios with the measured ratios. Most often, this is achieved by minimizing the expression

$$\chi^2 = \sum_{k=1}^{n} \frac{\left(R_k^{\rm exp} - R_k^{\rm model}\right)^2}{\sigma_k^2}, \tag{24.8}$$

where $R_k^{\rm exp}$ is the kth measured ratio, σ_k is the corresponding error, and $R_k^{\rm model}$ is the theoretical result for the same ratio. The total number of different ratios included in the analysis is denoted here by n. If m multiplicities are measured, $n = m(m-1)/2$.

Introducing the short-hand notation $\alpha_1 = T_{\rm chem}$ and $\alpha_2 = \mu_{\rm chem}^B$ we may write

$$\chi^2(\alpha) = \chi^2(\alpha_{\rm min}) + (\alpha - \alpha_{\rm min})^T V^{-1} (\alpha - \alpha_{\rm min}) + \cdots, \tag{24.9}$$

where $\alpha = (\alpha_1, \alpha_2)$, $\alpha_{\rm min} = (\alpha_{1\,\rm min}, \alpha_{2\,\rm min})$ is the optimal pair of the parameters, and V is the variance matrix of the parameters α. We note that $\chi^2(\alpha)$ has a minimum at $\alpha = \alpha_{\rm min}$, so the first derivatives of $\chi^2(\alpha)$ vanish at this point. If $F(\alpha)$ is a function of the fitted parameters α, the variance of F is given by

$$(\Delta F)^2 = \sum_{mn} \frac{\partial F}{\partial \alpha_m} \frac{\partial F}{\partial \alpha_n} V_{mn}. \tag{24.10}$$

In the special cases where $F = \alpha_1$ and $F = \alpha_2$, Eq. (24.10) can be used to make an estimate of the errors of the fitted temperature and baryon chemical potential

$$(\Delta T_{\rm chem})^2 = V_{11}, \qquad \left(\Delta \mu_{\rm chem}^B\right)^2 = V_{22}. \tag{24.11}$$

[4]Equation (24.3) may be also used to calculate the transverse-mass and rapidity distributions, see Ex. 27.1. However, in this case one finds difficulties to reproduce the experimental data. This is understood as a consequence of neglecting the expansion of matter.

24.4 Inclusion of expansion

The advantage of the studies of the ratios of hadron multiplicities follows from the fact that most of the effects connected with the hydrodynamic flow cancel in this kind of the analysis [1]. For example, let us assume that our system at freeze-out is not a single static fireball but it is formed by a group of different fireballs moving with different four-velocities $u^\mu(x)$. If all these fireballs undergo the freeze-out process at the same thermodynamic conditions (established in their local rest frames), the ratios measured in the full phase space are identical with those obtained for a single static fireball characterized by the same temperature and baryon chemical potential. This reasoning may be cast in a more quantitative form if one uses the Cooper-Frye formula. Equation (15.18) may be rewritten in terms of the particle number current

$$N_i = \int d\Sigma_\mu(x)\, n_i^\mu(x), \tag{24.12}$$

where

$$\begin{aligned} n_i^\mu(x) &= 2\int d^4p\, \theta\left(p^0\right) \delta\left(p^2 - m_i^2\right) p^\mu \;\; f_i\left(p\cdot u\right) \\ &= 2\int d^4p\, \theta\left(p^0\right) \delta\left(p^2 - m_i^2\right) p^\mu \, \frac{g_i}{(2\pi)^3}\left[\exp\left(\frac{p\cdot u - \mu_i}{T}\right) - \epsilon\right]^{-1}. \end{aligned} \tag{24.13}$$

Equation (24.13) is written in a manifestly Lorentz-covariant way. In local thermal equilibrium the number current is proportional to the four-velocity,

$$n_i^\mu(x) = n_i(x)\; u^\mu(x), \tag{24.14}$$

and

$$\begin{aligned} n_i(x) &= u_\mu(x)\;\; n_i^\mu(x) \\ &= 2\int d^4p\, \theta\left(p^0\right) \delta\left(p^2 - m_i^2\right)\, p^\mu u_\mu\; f_i\left(p\cdot u\right) \\ &= \int d^3p\; f_i\left(E_p\right) = n_i\left(T(x), \mu^i(x)\right). \end{aligned} \tag{24.15}$$

Thus, $n_i(x)$ denotes the equilibrium particle density at the temperature $T(x)$ and the chemical potential $\mu_i(x)$. The total particle yield of the species i is therefore

$$N_i = \int d\Sigma_\mu\, n_i\left(T(x), \mu_i(x)\right)\, u^\mu(x). \tag{24.16}$$

We observe that the particle ratios do not depend on a particular shape of the freeze-out surface as long as the local thermodynamic parameters are independent of x. In this case we have

$$\frac{N_i}{N_j} = \frac{n_i(T,\mu_i)\int d\Sigma_\mu\, u^\mu(x)}{n_j(T,\mu_j)\int d\Sigma_\mu\, u^\mu(x)} = \frac{n_i(T,\mu_i)}{n_j(T,\mu_j)}, \tag{24.17}$$

so the ratios are the same as those in a static fireball.

The argumentation presented above sheds light on the celebrated discussion of the problem if it is better to consider the 4π ratios or the midrapidity ratios. The 4π ratios are the ratios of the multiplicities measured in the full solid angle (full phase space), while the midrapidity ratios are the ratios of the multiplicities measured in a certain rapidity interval around $\mathrm{y} = 0$, for example, for $|\mathrm{y}| < 1$. If the 4π ratios are considered, it is hard to imagine that the freeze-out conditions are the same for all spacetime points belonging to the freeze-out hypersurface. In particular, the freeze-out baryon-rich conditions in the fragmentation regions are different from the baryon-poor conditions characterizing the midrapidity region. In this case, the thermal analysis of the ratios gives us only the estimates of the averaged thermal parameters. On the other hand, if the system is boost-invariant, then we have

$$\left.\frac{dN_i/dy}{dN_j/dy}\right|_{\mathrm{y}=0} = \frac{N_i}{N_j} = \frac{n_i}{n_j}. \tag{24.18}$$

Here, the first equality follows trivially from the assumed boost invariance, while the second one reflects the factorization of the volume of the system. Hence, the midrapidity ratios may be used to fit the thermal parameters of the boost-invariant model. We conclude with the following remarks: In ultra-relativistic heavy-ion collisions (top RHIC energies), where the central region is approximately boost-invariant, the use of the midrapidity ratios is preferable. At lower energies, the use of the 4π ratios as well as of the midrapidity ratios suffers from certain limitations. In this case, the thermal analysis gives information about the averaged thermal parameters only.

24.5 Examples of data analysis

As an example of the thermal analysis introduced in Sec. 24.3, in Table 24.1 we present the results of Ref. [2]. In this work, the particle ratios measured in the midrapidity region of central Au+Au collisions at $\sqrt{s_{\mathrm{NN}}} = 200$ GeV have been studied. As always in such studies, the particle multiplicities are denoted by the symbols of the appropriate particles. We observe a very good agreement between the data and the model results. Having in mind the simplicity of the thermal approach and complexity of heavy-ion collisions, such a good agreement is indeed remarkable.

Similarly to other thermal-model studies, the optimal value of the temperature found in [2], $T_{\mathrm{chem}} = 165$ MeV, turns out to be very close to the temperature of the crossover phase transition in QCD at $\mu_B = 0$, $T_{\mathrm{c}} \sim 170$ MeV, see Sec. 5.4.1. The closeness of these two temperatures suggests a connection between the phase transition and chemical equilibration. Several possible explanations of this connection have been put forward. For example, in Ref. [8] it is argued that multi-particle collisions are strongly enhanced in the phase transition region. Then, a rapid fall-off of the scattering rates takes place, such that the experimentally determined chemical

Table 24.1 Comparison of the thermal model predictions with the RHIC data describing the particle ratios measured in the central Au+Au collisions at $\sqrt{s_{\rm NN}} = 200$ GeV.

	thermal model [2]	experiment
fitted thermal parameters		
T [MeV]	165.6±4.5	
μ_B [MeV]	28.5±3.7	
μ_S [MeV]	6.9	
μ_I [MeV]	-0.9	
ratios used in thermal analysis		
π^-/π^+	1.009 ± 0.003	$1.025 \pm 0.006 \pm 0.018$ [3] (0–12%) $1.02 \pm 0.02 \pm 0.10$ [4] (0–5%)
K^-/K^+	0.939 ± 0.008	$0.95 \pm 0.03 \pm 0.03$ [3] (0–12%) $0.92 \pm 0.03 \pm 0.10$ [4] (0–5%)
$\overline{p}/p$	0.74 ± 0.04	$0.73 \pm 0.02 \pm 0.03$ [3] (0–12%) $0.70 \pm 0.04 \pm 0.10$ [4] (0–5%) 0.78 ± 0.05 [5] (0–5%)
$\overline{p}/\pi^-$	0.104 ± 0.010	0.083 ± 0.015 [6] (0–5%)
K^-/π^-	0.174 ± 0.001	0.156 ± 0.020 [6] (0–5%)
$\Omega/h^- \times 10^3$	0.990 ± 0.120	$0.887 \pm 0.111 \pm 0.133$ [7] (0–10%)
$\overline{\Omega}/h^- \times 10^3$	0.900 ± 0.124	$0.935 \pm 0.105 \pm 0.140$ [7] (0–10%)

freeze-out temperature is a good measure of the phase transition temperature. In Ref. [9] it is suggested that the very hadronization process leads to the equilibrium distributions of hadrons. Yet another interesting explanation of the relation between the phase transition and chemical equilibration is proposed in [10, 11]. Since color confinement does not allow colored constituents to exist in the physical vacuum, it creates a situation similar to the gravitational confinement provided by black holes. Quarks and gluons can cross the barrier to the physical vacuum only by quantum tunneling, i.e., through the QCD counterpart of Hawking radiation by black holes. This leads to the thermal hadron production in the form of Hawking-Unruh radiation. We note, that this mechanism has attractive features, since it explains the fact that similar values of the temperature are obtained in the analysis of simpler systems, including e^+e^- annihilation, where no rescattering of the produced particles is expected — see also our discussion of the early thermalization problem in Sec. 4.2.

Another important outcome of the thermal analysis of the heavy-ion data obtained at RHIC top energies is the low value of the baryon chemical potential, $\mu^B_{\rm chem} \sim 40$ MeV. This is a consequence of the high $\bar{p}/p$ ratio measured in the central region. The corresponding low value of the ratio $\mu^B_{\rm chem}/T_{\rm chem}$ suggests that the thermodynamic properties of the central region are determined mainly by the

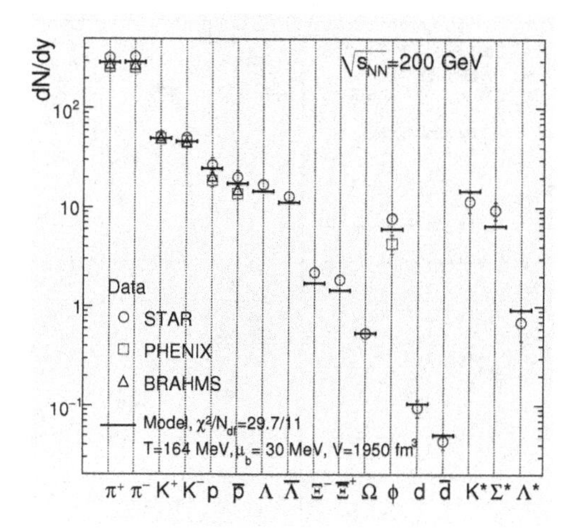

Fig. 24.1 Comparison of the thermal model predictions and the RHIC data for Au+Au central collisions at $\sqrt{s_{\rm NN}}$ = 200 GeV. Reprinted from [12] with permission from Acta Phys. Pol. **B**.

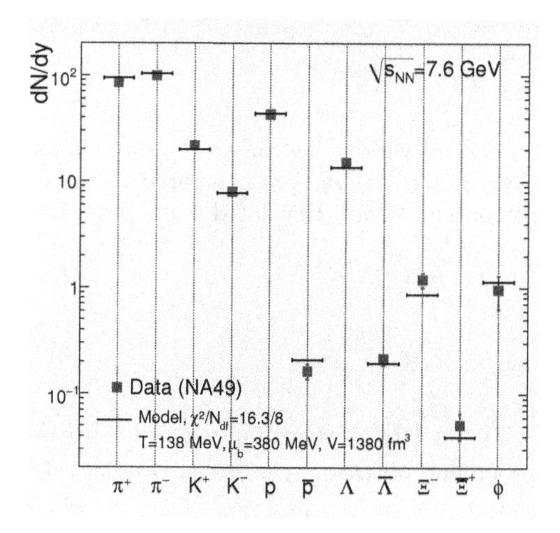

Fig. 24.2 Comparison of the thermal model predictions and the SPS data for Pb+Pb central collisions at $\sqrt{s_{\rm NN}}$ = 7.6 GeV. Reprinted from [12] with permission from Acta Phys. Pol. **B**.

values of the temperature. This is the reason why the baryon chemical potential was neglected in the hydrodynamic 2+1 calculations presented in Sec. 22.1.1.

The results of more recent and complex studies of the particle ratios, done in the framework of the grand canonical ensemble by Andronic, Braun-Munzinger, and Stachel, have been presented in [12], see also [13]. The results describing central

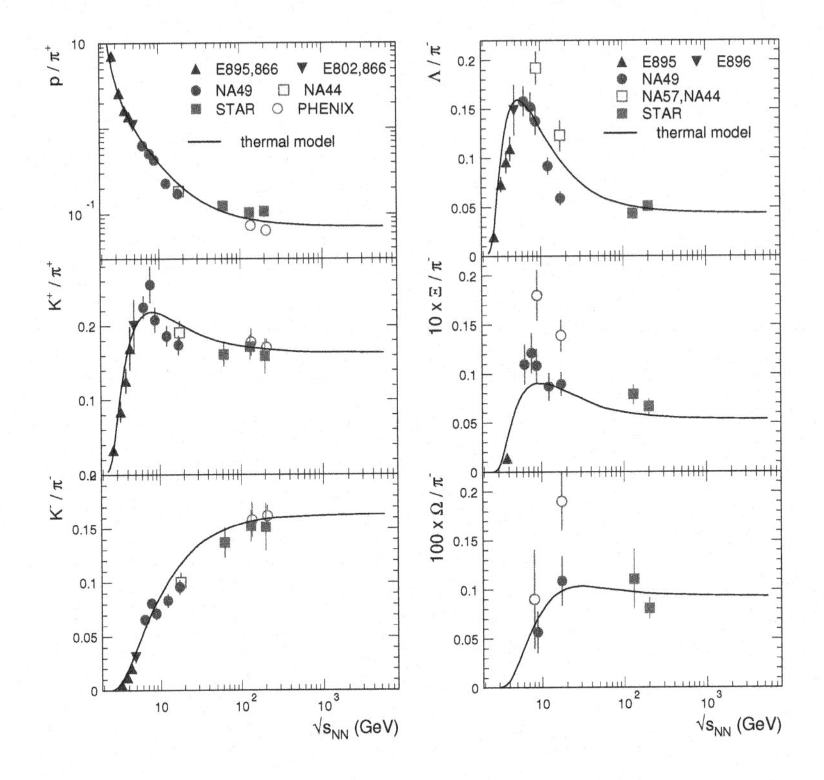

Fig. 24.3 Energy dependence of various hadron yields normalized to the pion yield. The thermal-model results by Andronic et al. are compared to the experimental results obtained for central heavy-ion collisions. Reprinted from [12] with permission from Acta Phys. Pol. **B**.

Au+Au collisions at the top RHIC beam energy are reproduced in Fig. 24.1. We observe very good agreement between the model predictions and the RHIC data. The values of the optimal thermal parameters, $T_{\rm chem} = 164$ MeV and $\mu^B_{\rm chem} = 40$, coincide with the values found in [2]. The thermal-model results describing Pb+Pb collisions at the beam energy of 30 A GeV ($\sqrt{s_{\rm NN}} = 7.6$ GeV) are shown and compared to the NA49 data in Fig. 24.2. Again, the very good agreement is found. The values of the thermal parameters are different from those found at the higher beam energy. The value of the temperature is slightly reduced, while the value of the baryon chemical potential is about ten times larger, $T_{\rm chem} = 138$ MeV and $\mu^B_{\rm chem} = 380$ MeV. Clearly, a higher value of the baryon chemical potential indicates larger stopping power of the heavy ions colliding at lower beam energies, see our discussion in Sec. 2.6.

The empirical parameterization of the energy dependence of $T_{\rm chem}$ and $\mu^B_{\rm chem}$,

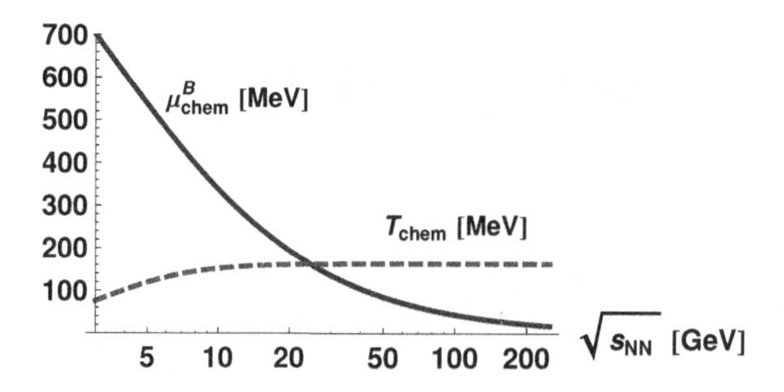

Fig. 24.4 Beam-energy dependence of the optimal chemical-freeze-out parameters for central heavy-ion collisions as found in Ref. [12] and given by Eqs. (24.19) and (24.20). Interestingly, the chemical-freeze-out temperature saturates very fast at the value 164 MeV that is close to the critical temperature in QCD at zero baryon chemical potential.

that has been worked out in Ref. [12], has the following form:

$$T_{\rm chem} = \frac{164\,{\rm MeV}}{1+\exp(2.60 - \ln(\sqrt{s_{\rm NN}[{\rm GeV}]})/0.45)} \tag{24.19}$$

and

$$\mu^B_{\rm chem} = \frac{1303\,{\rm MeV}}{1+0.286\sqrt{s_{\rm NN}[{\rm GeV}]}}. \tag{24.20}$$

The thermal-model results obtained in [12] for different beam energies (AGS, SPS, and RHIC) are shown in Fig. 24.3 and compared to the data obtained in central heavy-ion collisions. The graphical representation of Eqs. (24.19) and (24.20) is given in Fig. 24.4. With the increasing beam energy one can see a fast saturation of the temperature and a monotonous decrease of the baryon chemical potential.

24.6 Gaździcki's horn

A characteristic feature in the energy dependence of various experimental particle ratios shown in Fig. 24.3 is the sharp maximum seen in the K^+/π^+ behavior at $\sqrt{s_{\rm NN}} \sim 10$ GeV. Very interestingly, such a structure was predicted within a simple model of quark-gluon plasma formation by Gaździcki and Gorenstein in 1998 [14]. Their prediction was confirmed later by the NA49 Collaboration [15]. Moreover, the very recent STAR measurements [16], done at the lowered RHIC energy $\sqrt{s_{\rm NN}} = 9.2$ GeV, show consistency with the NA49 results.

Nowadays, one refers to the sharp peak in the K^+/π^+ ratio as to "Gaździcki's horn". The existence of this sharp structure is very difficult to explain in the thermal calculations and hadronic cascade models. Therefore, many physicists consider the horn structure as an experimental evidence for the onset of deconfinement and QGP formation.

24.7 Chemical non-equilibrium

In many thermal approaches to heavy-ion physics one considers the situations without full chemical equilibrium [5]. In such cases, one calculates the hadron densities using the following formula [17]

$$n_i = g_i \int \frac{d^3p}{(2\pi)^3} \frac{1}{\Upsilon_i^{-1} \exp(\sqrt{p^2+m_i^2}/T) - \epsilon},$$
$$= \frac{g_i}{2\pi^2}\epsilon \sum_{\kappa=1}^{\infty} \epsilon^{\kappa} \Upsilon_i^{\kappa} \frac{T\, m_i^2}{\kappa} K_2\left(\frac{\kappa m_i}{T}\right). \tag{24.21}$$

Here $\epsilon = +1$ corresponds to bosons, while $\epsilon = -1$ corresponds to fermions, compare Table 8.1. The quantity Υ_i is the fugacity factor. It is defined through the parameters $\lambda_{I_i}, \lambda_q, \lambda_s, \lambda_c$ (expressing, respectively, the isospin, light, strange and charm *quark fugacity factors*), and $\gamma_q, \gamma_s, \gamma_c$ (expressing the light, strange and charm *quark phase space occupancies*), namely

$$\Upsilon_i = \lambda_{I_i} \left(\lambda_q\gamma_q\right)^{N_q^i} \left(\lambda_s\gamma_s\right)^{N_s^i} \left(\lambda_c\gamma_c\right)^{N_c^i} \left(\lambda_{\bar{q}}\gamma_{\bar{q}}\right)^{N_{\bar{q}}^i} \left(\lambda_{\bar{s}}\gamma_{\bar{s}}\right)^{N_{\bar{s}}^i} \left(\lambda_{\bar{c}}\gamma_{\bar{c}}\right)^{N_{\bar{c}}^i}, \tag{24.22}$$

where

$$\lambda_q = \lambda_{\bar{q}}^{-1}, \qquad \lambda_s = \lambda_{\bar{s}}^{-1}, \qquad \lambda_c = \lambda_{\bar{c}}^{-1}, \tag{24.23}$$

and

$$\gamma_q = \gamma_{\bar{q}}, \qquad \gamma_s = \gamma_{\bar{s}}, \qquad \gamma_c = \gamma_{\bar{c}}. \tag{24.24}$$

Here, N_q^i, N_s^i and N_c^i are the numbers of light (u, d), strange (s) and charm (c) quarks in the ith hadron, and $N_{\bar{q}}^i$, $N_{\bar{s}}^i$ and $N_{\bar{c}}^i$ are the numbers of the corresponding antiquarks in the same hadron.

In the case of the models assuming chemical equilibrium ($\gamma_q = \gamma_s = 1$) and absence of charm ($N_c^i = N_{\bar{c}}^i = 0$) one has

$$\Upsilon_i^{\rm eq} = \exp\left(\frac{\mu_B B_i + \mu_S S_i + \mu_I I_i}{T}\right), \tag{24.25}$$

where B_i, S_i, and I^i are the baryon number, strangeness, and the third component of the isospin of the ith particle, and μ's are the corresponding chemical potentials. In this case, the two formulations are related by equations:

$$\lambda_q^{\rm eq} = e^{\mu_B/3T}, \tag{24.26}$$
$$\lambda_s^{\rm eq} = e^{(-3\mu_s+\mu_B)/3T}, \tag{24.27}$$
$$\lambda_{I_i}^{\rm eq} = \lambda_{I_i} = e^{I_i\mu_I/T}. \tag{24.28}$$

[5]The full chemical equilibrium corresponds to the case where the grand canonical ensemble may be applied, with the chemical potentials accounting for the conservation of the baryon number, strangeness, and isospin.

Thermal studies of the particle ratios based on Eq. (24.21) have been performed by Rafelski, Letessier, and Torrieri [18–20]. The analysis of the RHIC PHENIX data for Au+Au central collisions at $\sqrt{s_{\rm NN}}$ = 200 GeV gives: $T = 141.1$ MeV, $\mu_B = 25.7$ MeV, $\mu_S = 5.6$ MeV, $\lambda_I = \exp(\mu_I/T) = 0.9967$, $\gamma_s = 2.430$, and $\gamma_q = 1.613$ [20]. One can notice that this fit favors different from unity values of γ's and gives a lower value of the freeze-out temperature than the chemical equilibrium fits discussed earlier.

The situation where the same set of data may be described by two essentially different physical scenarios (chemical equilibrium vs. chemical non-equilibrium) is not comfortable. Clearly, further studies involving other observables are necessary to distinguish between the two cases. At the moment most physicists working on thermal models assume chemical equilibrium or include the possibility that $\gamma_s \neq 1$ but keeping $\gamma_q = 1$.

As a matter of fact, the central heavy-ion collisions from AGS to RHIC have been analyzed in the model with $\gamma_s \neq 1$ by Becattini, Manninen, and Gaździcki [21]. The results of this work indicate that $\gamma_s \sim 0.65$ in Au+Au collisions at AGS. With the increased beam energy γ_s grows and reaches unity in Au+Au collisions at RHIC. This observation suggests that the application of the equilibrium model at lower energies, and also in peripheral RHIC collisions, as shown in [22], might be inadequate. However, the discrepancy between the results of the equilibrium and non-equilibrium models may be attributed to the differences in the data selection. The analysis [12] uses the midrapidity ratios, while [21] uses the 4π ratios. Unfortunately, for the systems which are not boost-invariant and have locally different freeze-out conditions neither of the two approaches is unquestionably suitable, see our comments in the end of Sec. 24.4. Perhaps, one way to solve this problem is to use the rapidity dependent thermodynamic variables, as proposed in Refs. [23, 24].

24.8 Strangeness enhancement

In 1982 Rafelski and Müller proposed that the enhancement of strangeness production in heavy-ion collisions may be interpreted as a signature of the quark-gluon plasma production [25]. Their arguments used the idea that the threshold for strangeness production in the plasma is significantly lower than in the hadron gas (the mass of two strange quarks rather than the mass of two kaons) and the abundant strangeness production may be observed provided it survives the hadronization process.

The enhanced relative production of strange quarks with respect to light quarks may be studied with the help of the Wróblewski ratio

$$\lambda_s = \frac{2\langle s\bar{s}\rangle}{\langle u\bar{u}\rangle + \langle d\bar{d}\rangle}. \tag{24.29}$$

The increase of this factor when going from elementary to heavy-ion collisions has been indeed observed [26]. Moreover, the hyperons show a hierarchical enhancement in central Pb+Pb collisions with respect to peripheral Pb+Pb and p-Pb collisions at the top SPS energy [27]. The important question remains, however, if the observed enhancement may be attributed to the originally proposed mechanism. Perhaps other production mechanisms, such as the hadronic rescattering in the final stage, may be responsible for this effect. In this respect, the thermal analyses of the ratios of strange to non-strange particles are very useful. In particular, the observed equilibration of strangeness in central Au+Au collisions at the top RHIC energy, combined with the extracted short timescales of the collision process, suggests a partonic mechanism of strangeness production, which is in agreement with the original idea by Rafelski and Müller.

Bibliography to Chapter 24

[1] U. W. Heinz, "Strange messages: Chemical and thermal freeze-out in nuclear collisions," *J. Phys.* **G25** (1999) 263–274.

[2] A. Baran, W. Broniowski, and W. Florkowski, "Description of the particle ratios and transverse-momentum spectra for various centralities at RHIC in a single-freeze-out model," *Acta Phys. Polon.* **B35** (2004) 779–798.

[3] **PHOBOS** Collaboration, B. Wosiek *et al.*, "Identified particles in Au+Au collisions at $\sqrt{s_{\rm NN}}$ = 200 GeV," *Nucl. Phys.* **A715** (2003) 510–513.

[4] **PHENIX** Collaboration, T. Chujo, "Results on identified hadrons from the PHENIX experiment at RHIC," *Nucl. Phys.* **A715** (2003) 151–160.

[5] **STAR** Collaboration, G. Van Buren, "Soft physics in STAR," *Nucl. Phys.* **A715** (2003) 129–139.

[6] **STAR** Collaboration, O. Y. Barannikova and F. Wang, "Midrapidity π^{+-}, K^{+-}, and $\bar{p}$ spectra and particle ratios from STAR," *Nucl. Phys.* **A715** (2003) 458–461.

[7] **STAR** Collaboration, C. Suire, "Ω^- and $\bar{\Omega}^+$ production in Au+Au collisions at $\sqrt{s_{\rm NN}}$ = 130 GeV and 200 GeV," *Nucl. Phys.* **A715** (2003) 470–4735.

[8] P. Braun-Munzinger, J. Stachel, and C. Wetterich, "Chemical freeze-out and the QCD phase transition temperature," *Phys. Lett.* **B596** (2004) 61–69.

[9] U. W. Heinz, "The little bang: Searching for quark-gluon matter in relativistic heavy-ion collisions," *Nucl. Phys.* **A685** (2001) 414–431.

[10] D. Kharzeev and K. Tuchin, "From color glass condensate to quark-gluon plasma through the event horizon," *Nucl. Phys.* **A753** (2005) 316–334.

[11] P. Castorina, D. Kharzeev, and H. Satz, "Thermal hadronization and Hawking-Unruh radiation in QCD," *Eur. Phys. J.* **C52** (2007) 187–201.

[12] A. Andronic, P. Braun-Munzinger, and J. Stachel, "Thermal hadron production in relativistic nuclear collisions," *Acta Phys. Polon.* **B40** (2009) 1005–1012.

[13] A. Andronic, P. Braun-Munzinger, and J. Stachel, "Hadron production in central nucleus nucleus collisions at chemical freeze-out," *Nucl. Phys.* **A772** (2006) 167–199.

[14] M. Gazdzicki and M. I. Gorenstein, "On the early stage of nucleus nucleus collisions," *Acta Phys. Polon.* **B30** (1999) 2705.

[15] **NA49** Collaboration, C. Alt *et al.*, "Pion and kaon production in central Pb+Pb collisions at 20A and 30A GeV: Evidence for the onset of deconfinement," *Phys. Rev.* **C77** (2008) 024903.

[16] **STAR** Collaboration, D. Das, "Recent results from STAR experiment in Au+Au

collisions at $\sqrt{s_{NN}} = 9.2$ GeV," arXiv:0906.0630.

[17] G. Torrieri *et al.*, "Share: Statistical hadronization with resonances," *Comput. Phys. Commun.* **167** (2005) 229–251.

[18] J. Rafelski, J. Letessier, and G. Torrieri, "Strange hadrons and their resonances: A diagnostic tool of QGP freeze-out dynamics," *Phys. Rev.* **C64** (2001) 054907.

[19] J. Rafelski and J. Letessier, "Strangeness and statistical hadronization: How to study quark-gluon plasma," *Acta Phys. Polon.* **B34** (2003) 5791–5824.

[20] J. Rafelski, J. Letessier, and G. Torrieri, "Centrality dependence of bulk fireball properties at RHIC," *Phys. Rev.* **C72** (2005) 024905.

[21] F. Becattini, J. Manninen, and M. Gazdzicki, "Energy and system size dependence of chemical freeze-out in relativistic nuclear collisions," *Phys. Rev.* **C73** (2006) 044905.

[22] J. Manninen and F. Becattini, "Chemical freeze-out in ultra-relativistic heavy ion collisions at $\sqrt{s_{NN}} = 130$ and 200 GeV," *Phys. Rev.* **C78** (2008) 054901.

[23] B. Biedron and W. Broniowski, "Rapidity-dependent spectra from a single-freeze-out model of relativistic heavy-ion collisions," *Phys. Rev.* **C75** (2007) 054905.

[24] F. Becattini and J. Cleymans, "Chemical equilibrium in heavy ion collisions: Rapidity dependence," *J. Phys.* **G34** (2007) S959–964.

[25] J. Rafelski and B. Muller, "Strangeness production in the quark-gluon plasma," *Phys. Rev. Lett.* **48** (1982) 1066.

[26] F. Becattini, M. Gazdzicki, A. Keranen, J. Manninen, and R. Stock, "Study of chemical equilibrium in nucleus nucleus collisions at AGS and SPS energies," *Phys. Rev.* **C69** (2004) 024905.

[27] E. Andersen *et al.*, "Enhancement of central Lambda, Xi and Omega yields in Pb Pb collisions at 158-A-GeV/c," *Phys. Lett.* **B433** (1998) 209–216.

Chapter 25

Hydro-Inspired Models

In the previous Chapter we discussed thermal models which are used to study the ratios of hadronic abundances. They are applied to determine the properties of matter at the chemical freeze-out. Now, we turn to the discussion of the hydro-inspired models. These models serve to analyze the momentum distributions and give us information about the properties of matter at the thermal freeze-out.

25.1 Blast-wave model of Siemens and Rasmussen

In 1979 Siemens and Rasmussen introduced a model of hadron production in Ne+NaF reactions at the beam energy of 800 MeV per nucleon [1]. The main physical picture adopted in the model was that the very fast hydrodynamic expansion of matter leads to a sudden decoupling of hadrons and freezing of their momentum distributions. In this kind of a framework, the hadron momentum distributions continue to have thermal character (although modified by earlier collective expansion effects) until the observation point. In the original article one may read: "central collisions of heavy nuclei at kinetic energies of a few hundred MeV per nucleon produce fireballs of hot, dense nuclear matter; such fireballs explode, producing blast waves of nucleons and pions". In this way, with Ref. [1], the idea of the blast waves of hadrons and the blast-wave model itself appeared in the field of heavy-ion collisions. In this Section we briefly discuss the original version of the blast-wave model. Our presentation follows Refs. [2, 3].

Even though the model of Siemens and Rasmussen was inspired by an earlier hydrodynamic calculation by Bondorf, Garpman, and Zimanyi [4], the results presented in Ref. [1] did not follow from the solutions of the hydrodynamic equations — they were only consequences of the specific assumptions about the freeze-out conditions. This is a characteristic situation for the hydro-inspired models.

The most important feature of the model was the spherically symmetric expansion of the shells of matter with constant radial velocity. With an additional assumption about the times when such shells decay into freely streaming hadrons (this point will be discussed in a greater detail below) Siemens and Rasmussen obtained the formula for the momentum distribution of the emitted hadrons [1]

$$\frac{dN}{d^3p} = Z \exp\left(-\frac{\gamma E_p}{T}\right) \left[\left(1 + \frac{T}{\gamma E_p}\right) \frac{\mathrm{sinh} a}{a} - \frac{T}{\gamma E_p} \mathrm{cosh} a\right]. \tag{25.1}$$

In Eq. (25.1) Z is the normalization factor, $E_p = \sqrt{m^2 + \mathbf{p}^2}$ denotes the hadron energy, T is the temperature of the fireball (the same for all fluid shells), and $\gamma = (1 - v^2)^{-1/2}$ is the Lorentz gamma factor with v denoting the radial collective velocity (radial flow). A dimensionless parameter a is defined by the equation

$$a = \frac{\gamma v p}{T}. \tag{25.2}$$

Small values of v (and a) correspond to small expansion rate and, as expected, a simple Boltzmann factor is obtained from Eq. (25.1) in the limit $v \to 0$,

$$\frac{dN}{d^3p} \to Z \exp\left(-\frac{E_p}{T}\right). \tag{25.3}$$

In this way we obtain the picture of a static fireball studied in Sec. 24.2.

The fitting of Eq. (25.1) to the data gave $T = 44$ MeV and $v = 0.373$. The interesting feature of the fit was that the value of the radial flow v turned out to be quite large. This suggested the strong collective behavior. In the original paper the authors wrote: "Monte Carlo studies suggest that Ne+NaF system is too small for multiple collisions to be very important, thus, this evidence for a blast feature may be an indication that pion exchange is enhanced, and the effective nucleon mean free path shortened in dense nuclear matter".

In the remaining part of this Section we show how Eq. (25.1) may be derived within the general framework introduced in Chap. 15. For spherically symmetric freeze-outs it is natural to introduce the following parameterization of the space-time points on the freeze-out hypersurface [2]

$$x^\mu = (t, x, y, z) = (t(\zeta), r(\zeta) \sin\theta \cos\phi, r(\zeta) \sin\theta \sin\phi, r(\zeta) \cos\theta). \tag{25.4}$$

The freeze-out hypersurface is completely defined if a curve, i.e., the mapping $\zeta \longrightarrow (t(\zeta), r(\zeta))$ in the $t - r$ space is given. This curve defines the (freeze-out) times when the hadrons in the shells of radius r decouple. The range of ζ may be always restricted to the interval: $0 \le \zeta \le 1$. The three coordinates: $\phi \in [0, 2\pi], \theta \in [0, \pi]$, and $\zeta \in [0, 1]$ play the role of the variables α, β, γ appearing in Eq. (15.19) [1]. Thus,

[1] Note that ζ in Eq. (25.4) does not coincide with the angle introduced in Eq. (15.22). However, it plays a similar role of the third variable required to parameterize the three-dimensional freeze-out hypersurface.

the element of the spherically symmetric hypersurface has the form

$$d\Sigma^\mu = (r'(\zeta), t'(\zeta)\sin\theta\cos\phi, t'(\zeta)\sin\theta\sin\phi, t'(\zeta)\cos\theta)\, r^2(\zeta)\sin\theta\, d\theta\, d\phi\, d\zeta, \tag{25.5}$$

where the prime denotes the derivatives taken with respect to ζ, see Ex. 27.3. In addition to the spherically symmetric hypersurface we introduce the spherically symmetric flow

$$u^\mu = \gamma(\zeta)\,(1, v(\zeta)\sin\theta\cos\phi, v(\zeta)\sin\theta\sin\phi, v(\zeta)\cos\theta)\,, \tag{25.6}$$

where $\gamma(\zeta)$ is the Lorentz factor,

$$\gamma(\zeta) = (1 - v^2(\zeta))^{-1/2}.$$

Similarly, the four-momentum of a hadron is parameterized as

$$p^\mu = (E_p, p\sin\theta_p\cos\phi_p, p\sin\theta_p\sin\phi_p, p\cos\theta_p)\,, \tag{25.7}$$

and we find the two useful expressions:

$$p\cdot u = [E_p - pv(\zeta)\cos\theta]\;\gamma(\zeta), \tag{25.8}$$

$$d\Sigma\cdot p = [E_p\, r'(\zeta) - p\, t'(\zeta)\cos\theta]\, r^2(\zeta)\sin\theta\, d\theta\, d\phi\, d\zeta. \tag{25.9}$$

We note that the spherical symmetry allows us to restrict our considerations to the special case $\theta_p = 0$.

In the case of the Boltzmann statistics, with the help of Eqs. (15.29), (25.8) and (25.9), we obtain the following form of the momentum distribution

$$E_p\frac{dN}{d^3p} = \int\limits_0^1 \frac{e^{-(E_p\gamma-\mu)/T}}{2\pi^2}\left[E_p\,\frac{\mathrm{sinh}a}{a}\,\frac{dr}{d\zeta} + T\,\frac{(\mathrm{sinh}a - a\mathrm{cosh}a)}{a\gamma v}\,\frac{dt}{d\zeta}\right] r^2(\zeta)d\zeta. \tag{25.10}$$

Here v, γ, r and t are functions of ζ, and the parameter a is defined by Eq. (25.2). The thermodynamic parameters T and μ may also depend on ζ. To proceed further we need to make certain assumptions about the ζ-dependence of these quantities. In particular, to obtain the model of Siemens and Rasmussen we assume that the thermodynamic parameters as well as the transverse velocity are constant

$$T = \text{const.},\quad \mu = \text{const.},\quad v = \text{const.}\quad (\gamma = \text{const.},\quad a = \text{const.}). \tag{25.11}$$

Moreover, we should assume that the freeze-out curve in the $t - r$ space satisfies the condition

$$dt = v\,dr,\quad t = t_0 + vr. \tag{25.12}$$

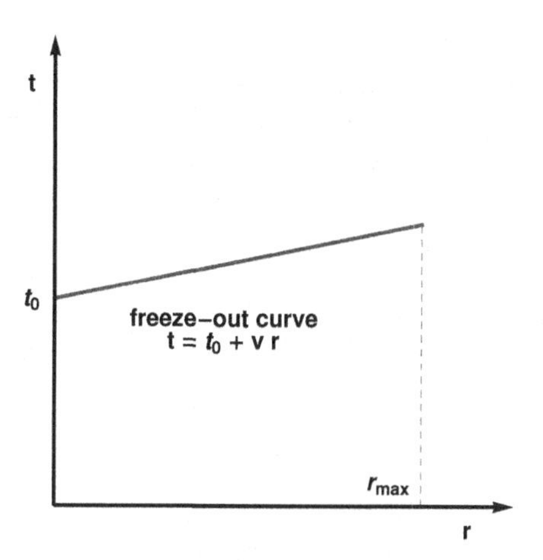

Fig. 25.1 The time-like freeze-out curve assumed in the blast-wave model of Siemens and Rasmussen is represented by the solid line, compare Eq. (25.12).

In this case we obtain the formula

$$\frac{dN}{d^3p} = \frac{e^{-(E_p\gamma-\mu)/T}}{2\pi^2}\left[\left(1+\frac{T}{\gamma E_p}\right)\frac{\sinh a}{a} - \frac{T}{\gamma E_p}\cosh a\right]\int\limits_0^1 r^2(\zeta)\frac{dr}{d\zeta}d\zeta. \qquad (25.13)$$

Equation (25.13) agrees with Eq. (25.1) if we make the following identifications

$$Z = \frac{V}{(2\pi)^3}\exp\left(\frac{\mu}{T}\right), \quad \int\limits_0^1 r^2(\zeta)\frac{dr}{d\zeta}d\zeta = \frac{r^3_{\rm max}}{3}, \quad V = \frac{4}{3}\pi r^3_{\rm max}. \qquad (25.14)$$

Note that the quantity $r_{\rm max}$ denotes the maximum value of the radius of the system.

An interesting and perhaps unexpected feature of the model proposed by Siemens and Rasmussen is the relation between the times and positions of the freeze-out points, see Eq. (25.12) illustrated in Fig. 25.1. Equation (25.12) indicates that the fluid elements, which are further away from the center, freeze-out later. Moreover, taking into account Eq. (25.12) in the formula (25.5) we find that the four-vector describing the hypersurface is parallel to the four-vector describing the flow, compare Eqs. (25.5) and (25.6) giving $d^3\Sigma^\mu \sim u^\mu$ in this case. As we shall see the same features are assumed in the single-freeze-out model of Refs. [5–7].

25.2 Blast-wave model of Schnedermann, Sollfrank, and Heinz

The model presented above is appropriate for the low-energy scattering processes where the two nuclei completely merge at the initial stage of the collision and further expansion of the system is to large extent isotropic. At higher energies such a picture is not valid anymore and, following the famous paper by Bjorken [8], the boost-invariant and cylindrically symmetric models have been introduced to describe the ultra-relativistic collisons [2].

The boost-invariance may be incorporated in the hydrodynamic equations, kinetic equations, and also in the modeling of the freeze-out process. In the latter case, the appropriate formalism was developed by Schnedermann, Sollfrank, and Heinz [9], see also Refs. [2, 10]. The boost-invariant, cylindrically symmetric hypersurface may be parametrized by the three coordinates: the azimuthal angle ϕ, the spacetime rapidity $\eta_\parallel$, and an additional quantity ζ, used to parametrize the freeze-out hypersurface in the $\tau - r$ plane,

$$x^\mu = \left(\tau(\zeta)\cosh\eta_\parallel, r(\zeta)\cos\phi, r(\zeta)\sin\phi, \tau(\zeta)\sinh\eta_\parallel\right). \quad (25.15)$$

Equation (25.15) may be treated as the special case of the parameterization (15.22), where the function d depends only on ζ, namely

$$\begin{aligned}\tau(\zeta) &= [\tau_{\rm i} + d(\zeta)\sin\zeta] \\ r(\zeta) &= d(\zeta)\cos\zeta.\end{aligned} \quad (25.16)$$

We note, however, that ζ appearing in Eq. (25.15) may be any variable used to define uniquely the freeze-out curve $(\tau(\zeta), r(\zeta))$. It does not have to be necessarily the angle introduced in Eq. (15.22). In particular, in this Chapter it is convenient to choose the range of ζ in such a way that $0 \leq \zeta \leq 1$ and $r = 0$ for $\zeta = 0$.

The volume element of the hypersurface defined by Eq. (25.15) is then

$$d\Sigma^\mu = \left(\frac{dr}{d\zeta}\cosh\eta_\parallel, \frac{d\tau}{d\zeta}\cos\phi, \frac{d\tau}{d\zeta}\sin\phi, \frac{dr}{d\zeta}\sinh\eta_\parallel\right) r(\zeta)\tau(\zeta)\, d\zeta\, d\eta_\parallel d\phi. \quad (25.17)$$

With the standard parameterization of the particle four-momentum,

$$p^\mu = (m_\perp\cosh{\rm y},\, p_\perp\cos\phi_p,\, p_\perp\sin\phi_p,\, m_\perp\sinh{\rm y}), \quad (25.18)$$

we find

$$p\cdot d\Sigma = \left[m_\perp\cosh({\rm y}-\eta_\parallel)\frac{dr}{d\zeta} - p_\perp\cos(\phi-\phi_p)\frac{d\tau}{d\zeta}\right] r(\zeta)\tau(\zeta)d\zeta d\eta_\parallel\, d\phi. \quad (25.19)$$

In addition we assume the boost-invariant form of the fluid four-velocity [3]

$$\begin{aligned}u^\mu &= \left(\cosh\vartheta_\perp(\zeta)\cosh\eta_\parallel, \sinh\vartheta_\perp(\zeta)\cos\phi, \sinh\vartheta_\perp(\zeta)\sin\phi, \cosh\vartheta_\perp(\zeta)\sinh\eta_\parallel\right) \\ &= \cosh\vartheta_\perp(\zeta)\cosh\eta_\parallel\left(1, \frac{\tanh\vartheta_\perp(\zeta)}{\cosh\eta_\parallel}\cos\phi, \frac{\tanh\vartheta_\perp(\zeta)}{\cosh\eta_\parallel}\sin\phi, \tanh\eta_\parallel\right).\end{aligned} \quad (25.20)$$

[2]The data delivered by the BRAHMS Collaboration indicate that the systems produced at RHIC may be treated as boost-invariant only in the limited rapidity range $-1 < {\rm y} < 1$, see Fig. 2.2. Moreover, the assumption about the cylindrical symmetry is valid only for the most central data.

[3]With the substitutions: $t = \tau\cosh\eta_\parallel$, $\bar{v}_x(\tau(\zeta), r(\zeta)) = v_\perp(\zeta)\cos\phi$, $\bar{v}_y(\tau(\zeta), r(\zeta)) = v_\perp(\zeta)\sin\phi$, $z = \tau\sinh\eta_\parallel$, and $v_\perp(\zeta) = \tanh\vartheta_\perp(\zeta)$, one may check that Eq. (25.20) is the special case of Eq. (2.51). The latter was introduced in Sec. 2.7 where general concepts related to boost-invariance were discussed.

The function $\vartheta_\perp(\zeta)$ describes the transverse rapidity of the fluid element in the plane $z = 0$. Its dependence on ζ reflects the fact that the transverse flow depends on the spacetime coordinates of the freeze-out points. As in the one-dimensional Bjorken model we have

$$v_z = \tanh \eta_\parallel = \frac{z}{t}. \tag{25.21}$$

On the other hand, the transverse components of the flow are

$$\begin{aligned} v_x &= \frac{\tanh \vartheta_\perp(\zeta)}{\cosh \eta_\parallel} \cos\phi, \\ v_y &= \frac{\tanh \vartheta_\perp(\zeta)}{\cosh \eta_\parallel} \sin\phi. \end{aligned} \tag{25.22}$$

We also find

$$p \cdot u = m_\perp \cosh\vartheta_\perp(\zeta) \cosh(\eta_\parallel - \mathrm{y}) - p_\perp \sinh\vartheta_\perp(\zeta) \cos(\phi - \phi_p). \tag{25.23}$$

25.2.1 *Expansion with constant transverse flow*

The boost-invariant and cylindrically symmetric freeze-out is completely defined if the three functions of the parameter ζ are given: $\tau(\zeta), r(\zeta)$ and $\vartheta_\perp(\zeta)$. In the following we shall present two versions of the blast-wave model which use different forms of those functions. In the first version of the model we shall assume that the transverse component of the fluid velocity is constant, i.e., for all values of ζ the function $\vartheta_\perp(\zeta)$ has the same value

$$v_\perp\left(\vartheta_\perp(\zeta), \eta_\parallel = 0\right) = \tanh \vartheta_\perp(\zeta) = \text{const.} \tag{25.24}$$

The condition of constant transverse flow is similar to the condition assumed in the blast-wave model of Siemens and Rasmussen.

25.2.1.1 *Momentum distribution function*

In this case, the Cooper-Frye formalism, Eqs. (15.34), (25.19) and (25.23), gives directly the one-particle distribution function

$$\begin{aligned} \frac{dN}{d\mathrm{y}d^2p_\perp} &= \frac{e^{\beta\mu}}{(2\pi)^3} \int\limits_0^{2\pi} d\phi \int\limits_{-\infty}^{\infty} d\eta_\parallel \int\limits_0^1 d\zeta \; r(\zeta)\tau(\zeta) \\ &\times \left[m_\perp \cosh(\eta_\parallel - \mathrm{y}) \frac{dr}{d\zeta} - p_\perp \cos(\phi - \phi_p) \frac{d\tau}{d\zeta} \right] \\ &\times \exp\left[-\beta m_\perp \cosh(\vartheta_\perp) \cosh(\eta_\parallel - \mathrm{y}) + \beta p_\perp \sinh(\vartheta_\perp) \cos(\phi - \phi_p) \right]. \end{aligned} \tag{25.25}$$

In accordance with our initial assumptions the momentum distribution (25.25) is independent of y and ϕ_p, i.e., our system is boost-invariant and cylindrically symmetric. Hence, in the following we may set $\mathrm{y} = 0$ and $\phi_p = 0$. Moreover, the integrals over $\eta_\parallel$ and ϕ are analytic and we easily find

$$
\begin{aligned}
\frac{dN}{dyd^2p_\perp} &= \frac{e^{\beta\mu}}{2\pi^2} \int_0^1 d\zeta \; r(\zeta)\tau(\zeta) \left\{ \frac{dr}{d\zeta} m_\perp K_1 \left[\beta m_\perp \cosh(\vartheta_\perp)\right] I_0 \left[\beta p_\perp \sinh(\vartheta_\perp)\right] \right. \\
&\quad \left. - \frac{d\tau}{d\zeta} p_\perp K_0 \left[\beta m_\perp \cosh(\vartheta_\perp)\right] I_1 \left[\beta p_\perp \sinh(\vartheta_\perp)\right] \right\} \\
&= \frac{e^{\beta\mu}}{2\pi^2} m_\perp K_1 \left[\frac{\beta m_\perp}{\sqrt{1-v_\perp^2}}\right] I_0 \left[\frac{\beta p_\perp v_\perp}{\sqrt{1-v_\perp^2}}\right] \int_0^1 d\zeta \; r(\zeta)\tau(\zeta) \frac{dr}{d\zeta} \\
&\quad - \frac{e^{\beta\mu}}{2\pi^2} p_\perp K_0 \left[\frac{\beta m_\perp}{\sqrt{1-v_\perp^2}}\right] I_1 \left[\frac{\beta p_\perp v_\perp}{\sqrt{1-v_\perp^2}}\right] \int_0^1 d\zeta \; r(\zeta)\tau(\zeta) \frac{d\tau}{d\zeta}.
\end{aligned}
\tag{25.26}
$$

In Eq. (25.26) we used the condition (25.24). An additional simplification appears, if we assume that τ is constant at freeze-out, see Fig. 25.2,

$$
\tau(\zeta) = \tau_{\mathrm{f}} = \text{const.} \tag{25.27}
$$

In this case one finds

$$
\frac{dN}{dyd^2p_\perp} = \frac{e^{\beta\mu}}{4\pi^2} \tau_{\mathrm{f}}\, r_{\max}^2 m_\perp K_1 \left[\frac{\beta m_\perp}{\sqrt{1-v_\perp^2}}\right] I_0 \left[\frac{\beta p_\perp v_\perp}{\sqrt{1-v_\perp^2}}\right]. \tag{25.28}
$$

Here we have also assumed that r is a monotonic function of ζ, and $r_{\max}$ is the radius of the firecylinder at freeze-out.

It is not an exaggeration to say that Eq. (25.28) is one of the most popular theoretical expressions used in the relativistic heavy-ion physics. It was employed in hundreds of papers to fit the transverse momentum spectra of hadrons. The fits based on (25.28) yield the values of the temperature and transverse flow at the thermal freeze-out. For example, the BRAHMS Collaboration, studying Au+Au collisions at RHIC, found the temperatures in the range $125\,\mathrm{MeV} > T > 115\,\mathrm{MeV}$ and the transverse flow in the range $0.7 < v_\perp < 0.8$ for different centralities corresponding to the range $60 < N_{\mathrm{part}} < 340$ [11]. The larger values of N_{part} correspond to smaller values of the temperature and larger values of the transverse flow, an effect understood by the longer lifetime of the fireball formed in the central collisions.

We stress that such fits usually treat the factor $e^{\beta\mu}\tau_{\mathrm{f}}\, r_{\max}^2/(4\pi^2)$ as an arbitrary normalization constant that is adjusted separately for each hadron species. For light particles, where $m_\perp \approx p_\perp$, and for large arguments of the Bessel functions, we may use the expansions (31.11) and (31.16) in Eq. (25.28) to find

$$
\frac{dN}{dyd^2p_\perp} \sim \exp\left[-\beta p_\perp \sqrt{\frac{1-v_\perp}{1+v_\perp}}\right]. \tag{25.29}
$$

This expression indicates directly that the transverse flow changes the slope of the transverse-momentum spectra, see Sec. 2.2.1. The change of the slope may be

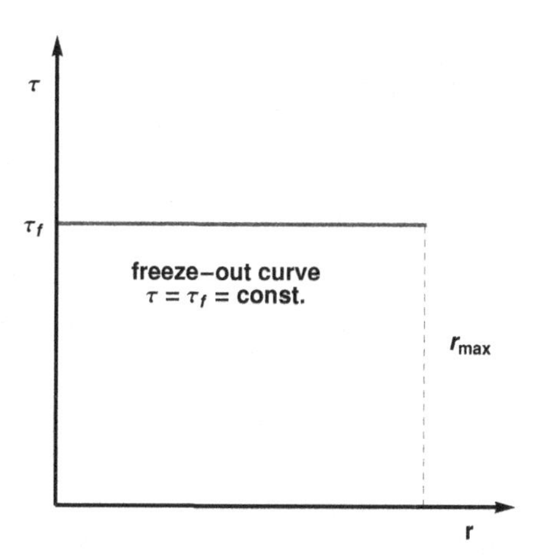

Fig. 25.2 The freeze-out curve assumed in the most popular boost-invariant version of the blast-wave model.

interpreted as the appearance of the effective temperature λ that is larger than the original "true" temperature $T_\mathrm{f} = 1/\beta$,

$$\lambda = T_\mathrm{f} \sqrt{\frac{1+v_\perp}{1-v_\perp}}. \tag{25.30}$$

Equation (25.30) may be interpreted as the blue shift of the observed hadronic spectrum, see Ex. 27.4.

25.2.1.2 *Side correlation function*

In this case, the parameterization (17.58) yields

$$\tilde{q}_\mathrm{side} \cdot x = -q_\mathrm{side}\, y = -q_\mathrm{side}\, r \sin\phi, \tag{25.31}$$

and the appropriate source term follows from Eqs. (15.35), and (17.57)

$$\begin{aligned} S(k_\perp, q_\mathrm{side}) &= \frac{e^{\beta\mu}}{2\pi^2} m_\perp K_1 \left[\frac{\beta m_\perp}{\sqrt{1-v_\perp^2}}\right] \int\limits_0^1 d\zeta\; r(\zeta)\tau(\zeta) \frac{dr}{d\zeta} \mathcal{I}_0 \left[\frac{\beta k_\perp v_\perp}{\sqrt{1-v_\perp^2}}, q_\mathrm{side}\, r(\zeta)\right] \\ &- \frac{e^{\beta\mu}}{2\pi^2} k_\perp K_0 \left[\frac{\beta m_\perp}{\sqrt{1-v_\perp^2}}\right] \int\limits_0^1 d\zeta\; r(\zeta)\tau(\zeta) \frac{d\tau}{d\zeta} \mathcal{I}_1 \left[\frac{\beta k_\perp v_\perp}{\sqrt{1-v_\perp^2}}, q_\mathrm{side}\, r(\zeta)\right]. \end{aligned} \tag{25.32}$$

Here we have defined two functions:

$$\mathcal{I}_0(a,b) = \frac{1}{2\pi} \int\limits_0^{2\pi} d\phi \exp(a\cos\phi - ib\sin\phi) \tag{25.33}$$

and

$$\mathcal{I}_1(a,b) = \frac{\partial}{\partial a}\mathcal{I}_0(a,b) = \frac{1}{2\pi}\int_0^{2\pi} d\phi\, \cos\phi\, \exp(a\cos\phi - ib\sin\phi). \tag{25.34}$$

For $b = 0$ these two functions are reduced to the modified Bessel functions of the first kind $I_0(a)$ and $I_1(a)$, see Ex. 27.5 and Sec. 31.3. If τ is constant at the freeze-out, the source function is simply

$$S(k_\perp, q_{\rm side}) = \frac{e^{\beta\mu}}{2\pi^2}\,\tau_{\rm f}\, m_\perp K_1\left[\frac{\beta m_\perp}{\sqrt{1-v_\perp^2}}\right]\int_0^{r_{\rm max}} dr\, r\, \mathcal{I}_0\left[\frac{\beta k_\perp v_\perp}{\sqrt{1-v_\perp^2}}, q_{\rm side}\, r\right]. \tag{25.35}$$

25.2.1.3 *Out correlation function*

In this case we find

$$\tilde{q}_{\rm out}\cdot x = \Delta E_{\rm out}\, t - q_{\rm out}\, x = \Delta E_{\rm out}\, \tau\cosh\eta_\parallel - q_{\rm out}\, r\cos\phi, \tag{25.36}$$

where the energy difference $\Delta E_{\rm out}$ is defined by Eq. (17.59). The corresponding source term is obtained from Eqs. (15.32) and (17.57)

$$\begin{aligned} S(k_\perp, q_{\rm out}) = {} & \frac{e^{\beta\mu}}{2\pi^2} m_\perp \int_0^1 d\zeta\; r(\zeta)\tau(\zeta)\frac{dr}{d\zeta} \\ & \times K_1\left[\frac{\beta m_\perp}{\sqrt{1-v_\perp^2}} - i\Delta E_{\rm out}\tau(\zeta)\right] I_0\left[\frac{\beta k_\perp v_\perp}{\sqrt{1-v_\perp^2}} - iq_{\rm out}\, r(\zeta)\right] \\ & - \frac{e^{\beta\mu}}{2\pi^2} k_\perp \int_0^1 d\zeta\; r(\zeta)\tau(\zeta)\frac{d\tau}{d\zeta} \\ & \times K_0\left[\frac{\beta m_\perp}{\sqrt{1-v_\perp^2}} - i\Delta E_{\rm out}\tau(\zeta)\right] I_1\left[\frac{\beta k_\perp v_\perp}{\sqrt{1-v_\perp^2}} - iq_{\rm out}\, r(\zeta)\right]. \end{aligned} \tag{25.37}$$

For $\tau = \tau_{\rm f}$ we find

$$\begin{aligned} S(k_\perp, q_{\rm out}) = {} & \frac{e^{\beta\mu}}{2\pi^2}\tau_{\rm f}\, m_\perp K_1\left[\frac{\beta m_\perp}{\sqrt{1-v_\perp^2}} - i\Delta E_{\rm out}\,\tau_{\rm f}\right] \\ & \times \int_0^{r_{\rm max}} dr\, r I_0\left[\frac{\beta k_\perp v_\perp}{\sqrt{1-v_\perp^2}} - iq_{\rm out}\, r\right]. \end{aligned} \tag{25.38}$$

25.2.1.4 *Long correlation function*

Similarly to the two previous cases we find

$$\tilde{q}_{\rm long}\cdot x = -q_{\rm long}\, z = -q_{\rm long}\, \tau\sinh\eta_\parallel \tag{25.39}$$

and

$$
\begin{aligned}
S(k_\perp, q_{\rm long}) = & \frac{e^{\beta\mu}}{2\pi^2} m_\perp\, I_0 \left[\frac{\beta k_\perp v_\perp}{\sqrt{1-v_\perp^2}} \right] \int\limits_0^1 d\zeta\; r(\zeta)\tau(\zeta) \frac{dr}{d\zeta} \\
& \times \mathcal{K}_1 \left[\frac{\beta m_\perp}{\sqrt{1-v_\perp^2}}, q_{\rm long}\, \tau(\zeta) \right] \\
& - \frac{e^{\beta\mu}}{2\pi^2} k_\perp\, I_1 \left[\frac{\beta k_\perp v_\perp}{\sqrt{1-v_\perp^2}} \right] \int\limits_0^1 d\zeta\; r(\zeta)\tau(\zeta) \frac{d\tau}{d\zeta} \\
& \times \mathcal{K}_0 \left[\frac{\beta m_\perp}{\sqrt{1-v_\perp^2}}, q_{\rm long}\, \tau(\zeta) \right],
\end{aligned}
\tag{25.40}
$$

where:

$$
\mathcal{K}_0(a,b) = \frac{1}{2} \int\limits_{-\infty}^{\infty} d\alpha \exp(-a \cosh\alpha - ib \sinh\alpha), \tag{25.41}
$$

and

$$
\mathcal{K}_1(a,b) = -\frac{\partial}{\partial a} \mathcal{K}_0(a,b) = \frac{1}{2} \int\limits_{-\infty}^{\infty} d\alpha \cosh\alpha \exp(-a \cosh\alpha - ib \sinh\alpha). \tag{25.42}
$$

For $b = 0$ the functions $\mathcal{K}_0(a,b)$ and $\mathcal{K}_1(a,b)$ are reduced to the modified Bessel functions of the second kind $K_0(a)$ and $K_1(a)$, see Sec. 31.2. In the case $\tau = \tau_{\rm f}$ one finds

$$
S(k_\perp, q_{\rm long}) = \frac{e^{\beta\mu}}{2\pi^2} \tau_{\rm f}\, m_\perp\, I_0 \left[\frac{\beta k_\perp v_\perp}{\sqrt{1-v_\perp^2}} \right] \int\limits_0^{r_{\rm max}} dr\, r\, \mathcal{K}_1 \left[\frac{\beta m_\perp}{\sqrt{1-v_\perp^2}}, q_{\rm long}\, \tau_{\rm f} \right]. \tag{25.43}
$$

25.2.2 *Hubble-like expansion*

In the second version of the boost-invariant and cylindrically symmetric model of freeze-out we assume that the hypersurface is defined by the condition

$$
x^\mu x_\mu = t^2 - x^2 - y^2 - z^2 = \tau^2(\zeta) - r^2(\zeta) = \tau_{3\rm f}^2 = \text{const.}, \tag{25.44}
$$

see Fig. 25.3, and the fluid four-velocity is proportional to the space-time position

$$
u^\mu = \gamma(1, \mathbf{v}) = \frac{x^\mu}{\tau} = \frac{t}{\tau} \left(1, \frac{\mathbf{x}}{t} \right). \tag{25.45}
$$

The index 3 in the symbol $\tau_{3\rm f}$ denotes that we assume now that the freeze-out takes place on the three-dimensional hypersurface of constant proper time. Equations (25.44) and (25.45) imply that the parameterization of the freeze-out hypersurface

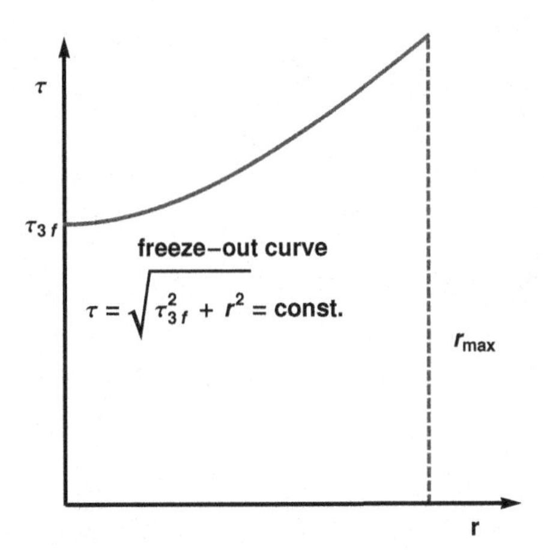

Fig. 25.3 The freeze-out curve assumed in the the blast-wave model with the Hubble flow profile.

(25.15) and the parameterization of the four-velocity field (25.20) are connected, and the functions $\tau(\zeta)$ and $r(\zeta)$ may be expressed as the function of $\vartheta_\perp$. To check this property we first rewrite Eq. (25.22) using Eq. (25.45),

$$\frac{x}{t} = \frac{\tanh \vartheta_\perp(\zeta)}{\cosh \eta_\parallel} \cos \phi, \quad \frac{y}{t} = \frac{\tanh \vartheta_\perp(\zeta)}{\cosh \eta_\parallel} \sin \phi. \tag{25.46}$$

In the next step Eqs. (25.46) and (25.44) yield

$$\begin{aligned} r &= \tau \tanh\vartheta_\perp = \tau_{3f} \ \sinh \vartheta_\perp, \\ \tau &= \sqrt{\tau_{3f}^2 + r^2} = \tau_{3f} \ \cosh \vartheta_\perp. \end{aligned} \tag{25.47}$$

The calculation of the volume element of the freeze-out hypersurface shows that it is proportional to u^μ, as in the original Siemens and Rasmussen model,

$$d\Sigma^\mu = u^\mu \tau_{3f}^3 \cosh(\vartheta_\perp)\sinh(\vartheta_\perp)\, d\vartheta_\perp\, d\eta_\parallel\, d\phi = u^\mu \tau_{3f}\, r\, dr\, d\eta_\parallel\, d\phi. \tag{25.48}$$

25.2.2.1 *Momentum distribution function*

As in the blast-wave model, the one-particle distribution functions are obtained from the Cooper-Frye formalism

$$\begin{aligned} \frac{dN}{dyd^2p_\perp} = \tau_{3f}^3 \frac{e^{\beta\mu}}{(2\pi)^3} \int_0^{2\pi} d\phi \int_{-\infty}^{\infty} d\eta_\parallel \int_0^{\vartheta_\perp^{\max}} & d\vartheta_\perp \cosh(\vartheta_\perp)\sinh(\vartheta_\perp) \\ \times \left[m_\perp \cosh(\vartheta_\perp)\cosh(\eta_\parallel - \mathrm{y}) - p_\perp \sinh(\vartheta_\perp)\cos(\phi - \phi_p)\right] & \\ \times \exp\left[-\beta m_\perp \cosh(\vartheta_\perp)\cosh(\eta_\parallel - \mathrm{y}) + \beta p_\perp \sinh(\vartheta_\perp)\cos(\phi - \phi_p)\right]. & \end{aligned} \tag{25.49}$$

A difference between the formula (25.25) used in the blast-wave model and Eq. (25.49) is that $\vartheta_\perp$ is not fixed now, and we have to integrate over the velocity field. Nevertheless, the integrals over $\eta_{\parallel}$ and ϕ are again analytic

$$\begin{aligned}
\frac{dN}{dyd^2p_\perp} &= \frac{e^{\beta\mu}}{(2\pi)^3}\int_0^{2\pi} d\phi \int_{-\infty}^{\infty} d\eta_{\parallel} \int_0^{r_{\max}} dr\, r \left[m_\perp \sqrt{\tau_{3f}^2 + r^2}\cosh(\eta_{\parallel} - \mathrm{y}) - p_\perp r\cos(\phi - \phi_p)\right] \\
&\quad \times \exp\left[-\beta m_\perp \sqrt{1 + \frac{r^2}{\tau_{3f}^2}}\cosh(\eta_{\parallel} - \mathrm{y}) + \frac{\beta p_\perp r}{\tau_{3f}}\cos(\phi - \phi_p)\right] \\
&= \frac{e^{\beta\mu}}{2\pi^2}\int_0^{r_{\max}} dr\, r \left\{ m_\perp \sqrt{\tau_{3f}^2 + r^2} K_1\left[\beta m_\perp \sqrt{1 + \frac{r^2}{\tau_{3f}^2}}\right] I_0\left[\frac{\beta p_\perp r}{\tau_{3f}}\right]\right. \\
&\quad \left. - p_\perp r K_0\left[\beta m_\perp \sqrt{1 + \frac{r^2}{\tau_{3f}^2}}\right] I_1\left[\frac{\beta p_\perp r}{\tau_{3f}}\right]\right\} \qquad (25.50)
\end{aligned}$$

25.2.2.2 *Side, out, and long correlation functions*

In this case, similarly to the blast-wave model we find the following expressions for the source functions:

$$\begin{aligned}
S(k_\perp, q_{\text{side}}) = \frac{e^{\beta\mu}}{2\pi^2}\int_0^{r_{\max}} dr\, r &\left\{ m_\perp \sqrt{\tau_{3f}^2 + r^2} \right. \qquad (25.51) \\
&\times K_1\left[\beta m_\perp \sqrt{1 + \frac{r^2}{\tau_{3f}^2}}\right] \mathcal{I}_0\left[\beta k_\perp \frac{r}{\tau_{3f}}, q_{\text{side}}\, r(\zeta)\right] \\
&\left. - k_\perp r K_0\left[\beta m_\perp \sqrt{1 + \frac{r^2}{\tau_{3f}^2}}\right] \mathcal{I}_1\left[\beta k_\perp \frac{r}{\tau_{3f}}, q_{\text{side}}\, r(\zeta)\right]\right\},
\end{aligned}$$

$$\begin{aligned}
S(k_\perp, q_{\text{out}}) &= \frac{e^{\beta\mu}}{2\pi^2}\int_0^{r_{\max}} dr\, r \left\{ m_\perp \sqrt{\tau_{3f}^2 + r^2} \right. \qquad (25.52) \\
&\times K_1\left[\beta m_\perp \sqrt{1 + \frac{r^2}{\tau_{3f}^2}} - i\Delta E_{\text{out}}\sqrt{\tau_{3f}^2 + r^2}\right] I_0\left[\frac{\beta k_\perp r}{\tau_{3f}} - i q_{\text{out}}\, r\right] \\
&\left. - k_\perp r K_0\left[\beta m_\perp \sqrt{1 + \frac{r^2}{\tau_{3f}^2}} - i\Delta E_{\text{out}}\sqrt{\tau_{3f}^2 + r^2}\right] I_1\left[\frac{\beta k_\perp r}{\tau_{3f}} - i q_{\text{out}}\, r\right]\right\},
\end{aligned}$$

$$S(k_\perp, q_{\rm long}) = \frac{e^{\beta\mu}}{2\pi^2} \int\limits_0^{r_{\rm max}} dr\, r \left\{ m_\perp \sqrt{\tau_{\rm 3f}^2 + r^2} \right. \tag{25.53}$$

$$\times \mathcal{K}_1 \left[\beta m_\perp \sqrt{1 + \frac{r^2}{\tau_{\rm 3f}^2}}, q_{\rm long} \sqrt{\tau_{\rm 3f}^2 + r^2} \right] I_0 \left[\frac{\beta k_\perp r}{\tau_{\rm 3f}} \right]$$

$$\left. - k_\perp r\, \mathcal{K}_0 \left[\beta m_\perp \sqrt{1 + \frac{r^2}{\tau_{\rm 3f}^2}}, q_{\rm long} \sqrt{\tau_{\rm 3f}^2 + r^2} \right] I_1 \left[\frac{\beta k_\perp r}{\tau_{\rm 3f}} \right] \right\}.$$

In Eqs. (25.51)–(25.53) $m_\perp = \sqrt{m_\pi^2 + k_\perp^2}$.

25.3 Buda-Lund model

The initial version of the Buda-Lund model was formulated to describe cylindrically symmetric systems with no constraint of boost-invariance [12, 13]. Later the model was developed to describe ellipsoidally symmetric systems [14]. In contrast to the blast-wave models discussed before, the Buda-Lund model cannot be treated as the special case of the Cooper-Frye formula. In this approach a special ansatz for the emission function is adopted, which allows for the production of particles from the whole spacetime volume of the source. The standard emission function of the model has the form

$$S_{\rm core}(x,p) = \frac{g}{(2\pi)^3} \frac{m_\perp \cosh(\eta_\parallel - {\rm y})}{\exp\left(\frac{u^\mu(x)p_\mu}{T(x)} - \frac{\mu(x)}{T(x)}\right) \pm 1} \frac{1}{(2\pi\Delta\tau^2)^{1/2}} \exp\left[-\frac{(\tau-\tau_0)^2}{2\Delta\tau^2}\right], \tag{25.54}$$

where the temperature and chemical potential depend on the position coordinates

$$\frac{\mu(x)}{T(x)} = \frac{\mu_0}{T_0} - \frac{r^2}{2R_G^2} - \frac{\eta_\parallel^2}{2\Delta\eta^2}, \tag{25.55}$$

$$\frac{1}{T(x)} = \frac{1}{T_0}\left(1 + \frac{T_0 - T_s}{T_s}\frac{r^2}{2R_G^2}\right)\left(1 + \frac{T_0 - T_{\rm e}}{T_{\rm e}}\frac{(\tau-\tau_0)^2}{2\Delta\tau^2}\right). \tag{25.56}$$

In Eq. (25.56), the quantity T_0 is the temperature in the center at the mean freeze-out time, $T_{\rm s}$ is the temperature on the surface at the mean freeze-out time, and $T_{\rm e}$ is the temperature in the center at the end of particle emission. The flow pattern assumed in the Buda-Lund model has a Hubble-like structure, with the velocity of the fluid element proportional to its distance from the center. Such patterns appear in the analytic [15] and numerical [16] solutions of the equations of the relativistic hydrodynamics. The emission function (25.54) is assumed to include the effects of short-lived resonances and is called the core function. The value of the central chemical potential μ_0 is determined from the data for each particle separately, from the absolute normalization as well as the shape of the single particle spectra. For physical interpretation of other model parameters we refer to the original papers [12, 13].

The Buda-Lund model gives a very good description of the $p_\perp$-spectra, v_2 and HBT radii in the full rapidity range. The data fits indicate a freeze-out temperature of 100–120 MeV in Au+Au collisions. However, about 1/8th of the particles is found to be emitted from a very hot center with $T \sim 200$ MeV. A comparison of this result with the lattice QCD results for the critical temperature was considered as an indication of quark deconfinement.

25.4 Seattle model

The Seattle model formulated in Refs. [17, 18] describes cylindrically symmetric systems. No assumption about the boost-invariance is made. The main characteristic feature of the model is inclusion of the effects of the final-state interaction of the outgoing pions. Such effects are taken into account by the use of the distorted wave functions $\Psi_{\mathbf{p}}^{(-)}(x)$. The formalism, following Ref. [19], is based on the generalized emission function, which may be used to get the one-particle and two-particle distributions in the momentum space

$$S(x,p,q) = \int d^4K' S_0(x,K') \int \frac{d^4x'}{(2\pi)^4}\, e^{-iK'\cdot x'} \Psi_{\mathbf{p_1}}^{(-)}(x+x'/2)\Psi_{\mathbf{p_2}}^{(-)*}(x-x'/2), \tag{25.57}$$

$$p = (p_1+p_2)/2, \quad q = p_1 - p_2. \tag{25.58}$$

The quantities p_1 and p_2 are the pion four-momenta. In the special case, where the distorted wave functions are replaced by the plane waves, the standard formulation is recovered with the emission function reduced to S_0,

$$\Psi_{\mathbf{p}}^{(-)}(x) \to e^{ip\cdot x}, \quad S(x,p,q) \to S_0(x,p)e^{iq\cdot x}, \quad S(x,p,0) \to S_0(x,p). \tag{25.59}$$

The function S_0 resembles the parameterization used in the Buda-Lund model, however, in this case the thermodynamic parameters are constant,

$$S_0(x,p) = \frac{m_\perp \cosh\eta_\parallel}{\sqrt{2\pi(\Delta\tau)^2}} \exp\left[-\frac{(\tau-\tau_0)^2}{2\Delta\tau^2} - \frac{\eta_\parallel^2}{2\Delta\eta^2}\right] \frac{\Omega(r)}{(2\pi)^3} \frac{1}{\exp(\frac{p\cdot u-\mu_\pi}{T})-1}. \tag{25.60}$$

The distribution of matter is characterized by the function

$$\Omega(r) = \frac{1}{\left[\exp\left(\frac{r-R_{WS}}{a_{WS}}\right)+1\right]^2}, \quad r = \sqrt{x^2+y^2}. \tag{25.61}$$

The optimal values of the parameters obtained from the fit to the STAR data [20] describing the Au+Au collisions at the beam energy $\sqrt{s_{\rm NN}} = 200$ GeV are the following [18]: $T = 215$ MeV, $\mu_\pi = 123$ MeV, $\tau_0 = 8$ fm, $\Delta\tau = 2.7$ fm, $R_{WS} = 12$ fm, $a_{WS} = 0.8$ fm, $\Delta\eta = 1$, $\eta_f = 1.5$, $w_0 = 0.142 \pm 0.046$ fm^{-2}, and $w_2 = 0.582 \pm 0.014 + i(0.123 \pm 0.002)$. The parameter η_f defines the magnitude of the transverse flow [17], whereas the parameters w_0 and w_2 define the optical potential which modifies the outgoing pion wave functions. The strength of the attraction inside the medium is greater than m_π^2, hence a pion behaves to large extent as a massless particle. This behavior may be related to the phenomena discussed in Refs. [21, 22].

The Seattle model describes successfully the pion transverse-momentum spectra and the HBT radii in the region $p_\perp > 100$ MeV. On the other hand, the model predicts non-monotonic structures in the transverse-momentum region below 70 MeV. Such predictions were confronted with the PHOBOS data in Ref. [23] and certain discrepancies between the data and the model were found. Very likely the non-monotonic structures are due to the sudden change of the optical potential in the transverse direction and a better agreement with the data may be achieved by the modification of this dependence.

Bibliography to Chapter 25

[1] P. J. Siemens and J. O. Rasmussen, "Evidence for a blast wave from compressed nuclear matter," *Phys. Rev. Lett.* **42** (1979) 880–887.

[2] D. H. Rischke and M. Gyulassy, "The time-delay signature of quark-gluon plasma formation in relativistic nuclear collisions," *Nucl. Phys.* **A608** (1996) 479–512.

[3] W. Florkowski and W. Broniowski, "Hydro-inspired parameterizations of freeze-out in relativistic heavy-ion collisions," *Acta Phys. Polon.* **B35** (2004) 2895–2910.

[4] J. P. Bondorf, S. I. A. Garpman, and J. Zimanyi, "A simple analytic hydrodynamic model for expanding fireballs," *Nucl. Phys.* **A296** (1978) 320–332.

[5] W. Broniowski and W. Florkowski, "Explanation of the RHIC p_T-spectra in a thermal model with expansion," *Phys. Rev. Lett.* **87** (2001) 272302.

[6] W. Broniowski and W. Florkowski, "Strange particle production at RHIC in a single-freeze-out model," *Phys. Rev.* **C65** (2002) 064905.

[7] W. Broniowski, A. Baran, and W. Florkowski, "Thermal approach to RHIC," *Acta Phys. Polon.* **B33** (2002) 4235–4258.

[8] J. D. Bjorken, "Highly relativistic nucleus-nucleus collisions: The central rapidity region," *Phys. Rev.* **D27** (1983) 140–151.

[9] E. Schnedermann, J. Sollfrank, and U. W. Heinz, "Thermal phenomenology of hadrons from 200-A/GeV S+S collisions," *Phys. Rev.* **C48** (1993) 2462–2475.

[10] P. V. Ruuskanen, "Transverse hydrodynamics with a first order phase transition in very high-energy nuclear collisions," *Acta Phys. Polon.* **B18** (1987) 551–590.

[11] **BRAHMS** Collaboration, I. Arsene *et al.*, "Quark gluon plasma and color glass condensate at RHIC? The perspective from the BRAHMS experiment," *Nucl. Phys.* **A757** (2005) 1–27.

[12] T. Csorgo and B. Lorstad, "Bose-Einstein correlations for three-dimensionally expanding, cylindrically symmetric, finite systems," *Phys. Rev.* **C54** (1996)

1390–1403.

[13] T. Csorgo, "Particle interferometry from 40-MeV to 40-TeV," *Heavy Ion Phys.* **15** (2002) 1–80.

[14] M. Csanad, T. Csorgo, and B. Lorstad, "Buda-Lund hydro model for ellipsoidally symmetric fireballs and the elliptic flow at RHIC," *Nucl. Phys.* **A742** (2004) 80–94.

[15] T. Csorgo, F. Grassi, Y. Hama, and T. Kodama, "Simple solutions of relativistic hydrodynamics for longitudinally and cylindrically expanding systems," *Phys. Lett.* **B565** (2003) 107–115.

[16] M. Chojnacki, W. Florkowski, and T. Csorgo, "On the formation of Hubble flow in Little Bangs," *Phys. Rev.* **C71** (2005) 044902.

[17] J. G. Cramer, G. A. Miller, J. M. S. Wu, and J.-H. Yoon, "Quantum opacity, the RHIC HBT puzzle, and the chiral phase transition," *Phys. Rev. Lett.* **94** (2005) 102302.

[18] G. A. Miller and J. G. Cramer, "Polishing the Lens: I Pionic Final State Interactions and HBT Correlations- Distorted Wave Emission Function (DWEF) Formalism and Examples," *J. Phys.* **G34** (2007) 703–740.

[19] M. Gyulassy, S. K. Kauffmann, and L. W. Wilson, "Pion interferometry of nuclear collisions. I. Theory," *Phys. Rev.* **C20** (1979) 2267–2292.

[20] **STAR** Collaboration, J. Adams *et al.*, "Identified particle distributions in pp and Au+Au collisions at $\sqrt{s_{\rm NN}} = 200$ GeV," *Phys. Rev. Lett.* **92** (2004) 112301.

[21] T. D. Cohen and W. Broniowski, "Vanishing condensates and anomalously light Goldstone modes in medium," *Phys. Lett.* **B342** (1995) 25–31.

[22] D. T. Son and M. A. Stephanov, "Pion propagation near the QCD chiral phase transition," *Phys. Rev. Lett.* **88** (2002) 202302.

[23] **PHOBOS** Collaboration, B. B. Back *et al.*, "Comment on "Quantum opacity, the RHIC HBT puzzle, and the chiral phase transition"," `nucl-ex/0506008`.

Chapter 26

Decays of Resonances

The momentum spectra of emitted hadrons, when considered in the rest frame of the fluid element at the thermal freeze-out, contain two contributions. The first one, commonly called the *primordial contribution*, is purely thermal and described by the equilibrium distribution functions characterized by the freeze-out thermodynamic parameters. The second one comes from heavier resonances, whose sequential decays populate lower energy states. Since a substantial part of the produced light particles comes from resonance decays, one may expect that the measured spectra are significantly changed by this effect. Indeed, it has been already known for a long time that the resonance decays modify the low-$p_\perp$ spectrum [1–5].

In the natural way, the measured contributions are modified additionally by the effects connected with the expansion of matter. Thus, the observed spectra are usually interpreted as the primordial distributions modified by the resonance decays and expansion. Any realistic modeling of the hadronic spectra should include these two phenomena. We sketch the method of dealing with two- and three-body decays. We also show how the sequential decays may be treated in the framework of the hydro-inspired models. For this purpose several iterative procedures are worked out. In the end of this Chapter we introduce the single-freeze-out model where the chemical and thermal freeze-outs are identified. This is possible if all resonance decays are included in the calculations of the spectra.

26.1 Two-body decays

We assume that the momentum distribution function of a hadronic resonance in the local rest frame of the fluid element is described by the function $f(\mathbf{k})$. Since we average over all possible polarization states, the momentum distribution of the emitted particles in the resonance's rest frame is isotropic. Thus, in the case of two-body decays, the spectrum of the emitted particle (denoted by the index 1) is obtained from the expression [6, 7]

$$f_1(\mathbf{q}) = b_R \frac{2J_R+1}{2J_1+1} \int d^3k \; f(\mathbf{k}) \int \frac{d^3p}{4\pi p^{*2}} \delta\left(|\mathbf{p}| - p^*\right) \delta^{(3)}\left(\hat{L}_k \mathbf{p} - \mathbf{q}\right), \qquad (26.1)$$

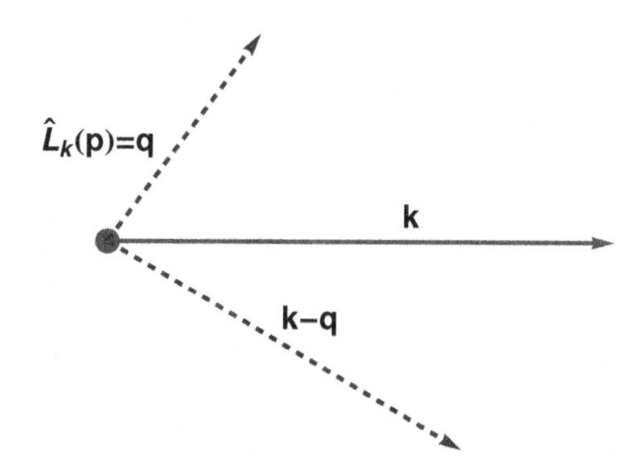

Fig. 26.1 A decay of the resonance with the three momentum $\mathbf{k}$ produces two "daughter" particles with momenta $\mathbf{q}$ and $\mathbf{k} - \mathbf{q}$.

where $\mathbf{p}$ is the momentum of the emitted particle in the rest frame of the resonance, and $\hat{L}_k$ is the Lorentz transformation to the rest frame of the fluid element, see Fig. 26.1 and Ex. 27.6,

$$\hat{L}_k \mathbf{p} = \mathbf{p} + [(\gamma_k - 1)\ v_k^{-2}\, \mathbf{v}_k \cdot \mathbf{p} + \gamma_k\ E^*]\mathbf{v}_k = \mathbf{q}. \tag{26.2}$$

Here $\mathbf{k}$ is the momentum of the resonance in the fluid local rest frame, $\mathbf{v}_k$ is the corresponding velocity,

$$\mathbf{v}_k = \frac{\mathbf{k}}{\sqrt{k^2 + m_R^2}}, \tag{26.3}$$

and γ_k is the Lorentz factor,

$$\gamma_k = \left(1 - v_k^2\right)^{-1/2}. \tag{26.4}$$

In Eq. (26.1) we also introduced the notation for the center-of-mass quantities $p^* = |\mathbf{p}|$ and $E^* = \sqrt{p_*^2 + m_1^2}$. The standard formula gives

$$p^* = \frac{\left[(m_R^2 - (m_1 + m_2)^2)(m_R^2 - (m_1 - m_2)^2)\right]^{1/2}}{2m_R}, \tag{26.5}$$

where m_R is the mass of the resonance, whereas m_1 and m_2 are the masses of the emitted particles, compare Ex. 10.8. The quantity b_R is the branching ratio for the considered channel, and J_R and J_1 are the spins of the resonance and particle 1, respectively. The physical interpretation of Eq. (26.1) is the following: the isotropic distribution of particle 1 in the resonance rest frame, $\delta\left(|\mathbf{p}| - p^*\right)/(4\pi p^{*2})$, is boosted to the fluid rest frame, and finally folded with the resonance distribution $f(\mathbf{k})$.

In order to write Eq. (26.1) in a more compact form, we change the variables,

$$\mathbf{p}' = \hat{L}_k \mathbf{p}, \qquad d^3p = \frac{E_p}{E_{p'}}\, d^3p' = \frac{E^*}{E_q} d^3p'. \tag{26.6}$$

Then, the integration over p' in (26.1) becomes trivial and we find

$$f_1(\mathbf{q}) = b_R \frac{2J_R+1}{2J_1+1} \frac{1}{4\pi p^{*2}} \frac{E^*}{E_q} \int d^3k \, f(\mathbf{k}) \, \delta\left(\left|\hat{L}_k^{-1}\mathbf{q}\right| - p^*\right). \tag{26.7}$$

Since the quantity $\hat{L}_k^{-1}\mathbf{q}$ is the momentum of the emitted particle in the reference frame connected with the resonance, we may rewrite Eq. (26.7) in the explicitly Lorentz-covariant way

$$E_q \, f_1(\mathbf{q}) = b_R \frac{2J_R+1}{2J_1+1} \frac{1}{4\pi p^*} \int \frac{d^3k}{E_k} E_k \, f(\mathbf{k}) \, \delta\left(\frac{k^\mu \cdot q_\mu}{m_R} - E^*\right). \tag{26.8}$$

Here k^μ and q^μ are the four-momenta of the resonance and the emitted particle, respectively. Equation (26.8) is a transformation rule which generates the spectrum of the emitted particles from the spectrum of the resonances. It can be reduced to a one-dimensional integral if the spectrum of the resonances is isotropic. In this case f depends only on the magnitude of the three-momentum, and the integration over the polar and azimuthal angles in (26.8) can be done analytically. In this way we obtain

$$f_1(q) = b_R \frac{2J_R+1}{2J_1+1} \frac{m_R}{2E_q p^* q} \int_{k_-(q)}^{k_+(q)} dk \, k \, f(k), \tag{26.9}$$

where the limits of the integration are

$$k_\pm(q) = \frac{m_R \left|E^* q \pm p^* E_q\right|}{m_1^2}. \tag{26.10}$$

Equation (26.9) is a relativistic generalization of the formula derived in Ref. [4]. Its simplicity turns out to be especially important in the numerical treatment of the sequential decays.

Denoting the volume of the fluid element in its rest frame by $\Delta^3 x$ we may write

$$f(\mathbf{k}) = \frac{\Delta N}{\Delta^3 k} = \Delta^3 x \frac{\Delta N}{\Delta^3 x \Delta^3 k} = \Delta^3 x f(x,k) \tag{26.11}$$

and

$$E_k f(\mathbf{k}) = E_k \, \Delta^3 x \frac{\Delta N}{\Delta^3 x \Delta^3 k} = E_k \, \Delta^3 x f(x,k) = k^\mu \Delta\Sigma_\mu(x) f(x,p). \tag{26.12}$$

Here $f(x,p)$ is the familiar phase-space distribution function, and

$$\Delta\Sigma_\mu(x) = (\Delta^3 x, 0, 0, 0)$$

is the four-vector describing the volume element $\Delta^3 x$. In the analogous way, we introduce the phase-space distribution of the emitted particles, $f_1(x,q)$, and obtain

$$q^\mu \Delta\Sigma_\mu(x)\, f_1\,(x,q) = b_R\, \frac{2J_R+1}{2J_1+1}\frac{1}{4\pi p^*} \int \frac{d^3k}{E_k}\, k^\mu \Delta\Sigma_\mu(x)\, f\,(x,k)\, \delta\left(\frac{k^\mu \cdot q_\mu}{m_R} - E^*\right). \tag{26.13}$$

For brevity of the notation we introduce the Lorentz-invariant function [8]

$$B(k,q) = b_R\, \frac{2J_R+1}{2J_1+1}\frac{1}{4\pi p^*}\delta\left(\frac{k^\mu \cdot q_\mu}{m_R} - E^*\right). \tag{26.14}$$

In this way, Eq. (26.13) may be cast in the form

$$q^\mu \Delta\Sigma_\mu(x)\, f_1\,(x,q) = \int \frac{d^3k}{E_k}\, B(k,q)\, k^\mu \Delta\Sigma_\mu(x)\, f\,(x,k)\,. \tag{26.15}$$

We note that Eq. (26.15) has a manifestly covariant form, so it may be easily used in different reference frames. The restriction exists, however, that the four-vector describing the fluid element is a time-like vector. In the special case, where the resonance distribution function $f(x,k)$ is given by the (local) equilibrium distribution function and the four-vector $\Sigma^\mu(x)$ is parallel to the four-velocity $u^\mu(x)$, the Lorentz structure of Eq. (26.15) implies that f_1 is a function of the product $q \cdot u$. Hence, all the calculations can be done in the reference frame where $u^\mu = (1,0,0,0)$. This procedure leads (after all necessary substitutions) directly to Eq. (26.9).

Another way of dealing with Eq. (26.8) is to replace the momentum distribution function $f(\mathbf{k})$ by the emission function $S(x,\mathbf{k})$. In this case we consider the particles in the infinitesimal neighborhood of the space-time point x, where

$$E_k f(\mathbf{k}) = E_k\, \Delta^4 x \frac{\Delta N}{\Delta^4 x \Delta^3 k} = \Delta^4 x\, S(x,\mathbf{k}). \tag{26.16}$$

Substituting of Eq. (26.16) into Eq. (26.8) yields

$$S_1\,(x,\mathbf{q}) = \int \frac{d^3k}{E_k}\, B(k,q)\, S\,(x,\mathbf{k})\,. \tag{26.17}$$

26.2 Three-body decays

In the case of three-body decays we can follow the same steps as above, with the extra modification connected with the fact that different values of p^* are possible now. The distribution of the allowed values of p^* may be obtained from the formula

$$g(p_1) = \frac{p_1^2}{E_{p_1}} \int d\Omega_1 \int \frac{d^3p_2}{E_{p_2}} \int \frac{d^3p_3}{E_{p_3}}$$
$$\times \delta(m_R - E_{p_1} - E_{p_2} - E_{p_3})\, \delta(\mathbf{p}_1 + \mathbf{p}_2 + \mathbf{p}_3)\, |M_{R\to P_1+P_2+P_3}|^2, \tag{26.18}$$

where $M_{R\to P_1+P_2+P_3}$ is the matrix element describing the decay and $d\Omega_1$ is the element of the solid angle. Here the decay is considered in the resonance rest frame, so we set $p^* = p_1$.

26.3 Sequential decays: spectra

In this Section we analyze sequential (chain) decays of hadronic resonances at freeze-out. The effects connected with such decays are very important, since the measured hadron multiplicities and spectra receive large contributions from the decaying heavy resonances. In the thermal approach, the proper inclusion of the resonances is crucial for obtaining a good agreement with the experimental data describing ratios of hadron multiplicities and hadron spectra.

We consider a chain of decays which are initialized by the decoupling of a heavy resonance, denoted by N, from the rest of hadronic matter at the spacetime point x_N and its decay at the spacetime point x_{N-1}. At this point the next hadron is created, denoted by $N-1$, which decays later at the spacetime point x_{N-2}. Such a process continues until a stable particle is created, which travels freely to the detector. We identify the last particle with a pion and generalize Eq. (26.17) to include the finite time and distance between individual decays, see Fig. 26.2.

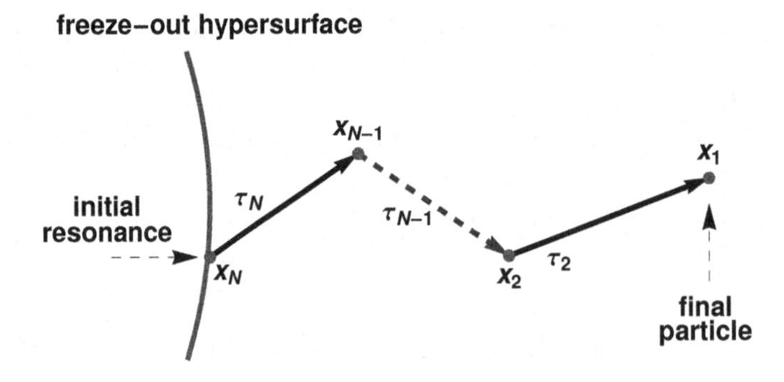

Fig. 26.2 A sequence of the resonance decays. The initial resonance decouples on the freeze-out hypersurface at the space-time point x_N, and decays after the average time $\tau_N \approx 1/\Gamma_N$. One of the decay products is formed at the point x_{N-1}. It decays again after the time τ_{N-1}. At the end of the cascade a stable particle is formed (e.g., pion, kaon, or proton), which is directly observed in the experiment.

The decay time is described by the exponential distribution expressed by the decay rate Γ, see Eqs. (6.1) and (6.2). For one particular sequence of the decays one obtains [8, 9]

$$
\begin{aligned}
S_\pi(x_1, p_1) = & \int \frac{d^3 p_2}{E_{p_2}} B(p_2, p_1) \int d\tau_2 \Gamma_2 e^{-\Gamma_2 \tau_2} \\
& \times \int d^4 x_2 \delta^{(4)} \left(x_2 + \frac{p_2 \tau_2}{m_2} - x_1 \right) \\
& \dots \\
& \times \int \frac{d^3 p_{N-1}}{E_{p_{N-1}}} B(p_{N-1}, p_{N-2}) \int d\tau_{N-1} \Gamma_{N-1} e^{-\Gamma_{N-1} \tau_{N-1}} \\
& \times \int d^4 x_{N-1} \delta^{(4)} \left(x_{N-1} + \frac{p_{N-1} \tau_{N-1}}{m_{N-1}} - x_{N-2} \right) \\
& \times \int \frac{d^3 p_N}{E_{p_N}} B(p_N, p_{N-1}) \int d\tau_N \Gamma_N e^{-\Gamma_N \tau_N} \\
& \times \int d\Sigma_\mu(x_N)\, p_N^\mu \, \delta^{(4)} \left(x_N + \frac{p_N \tau_N}{m_N} - x_{N-1} \right) f_N \left[p_N \cdot u(x_N) \right].
\end{aligned}
\tag{26.19}
$$

Integration over all space-time positions gives us the momentum distribution of pions

$$
\begin{aligned}
E_{p_1} \frac{dN_\pi}{d^3 p_1} &= \int d^4 x_1\, S_\pi(x_1, p_1) \\
&= \int \frac{d^3 p_2}{E_{p_2}} B(p_2, p_1) \int \frac{d^3 p_3}{E_{p_3}} B(p_3, p_2) \dots \int \frac{d^3 p_{N-1}}{E_{p_{N-1}}} B(p_{N-1}, p_{N-2}) \\
&\quad \times \int \frac{d^3 p_N}{E_{p_N}} B(p_N, p_{N-1}) \int d\Sigma_\mu(x_N)\, p_N^\mu \, f_N \left[p_N \cdot u(x_N) \right]
\end{aligned}
\tag{26.20}
$$

The expressions presented above are significantly simplified in the case where the element of the freeze-out hypersurface is proportional to the four-velocity,

$$
d\Sigma_\mu(x_N) = d\Sigma(x_N)\, u_\mu(x_N). \tag{26.21}
$$

Using this assumption one finds

$$
\begin{aligned}
E_{p_1}\frac{dN_\pi}{d^3p_1} &= \int d\Sigma\left(x_N\right)\int\frac{d^3p_2}{E_{p_2}}B\left(p_2,p_1\right)...\int\frac{d^3p_{N-1}}{E_{p_{N-1}}}B\left(p_{N-1},p_{N-2}\right)\\
&\times\int\frac{d^3p_N}{E_{p_N}}B\left(p_N,p_{N-1}\right)p_N\cdot u\left(x_N\right)\,f_N\left[p_N\cdot u\left(x_N\right)\right]\\
&= \int d\Sigma\left(x_N\right)p_1\cdot u\left(x_N\right)\,f_1\left[p_1\cdot u\left(x_N\right)\right].
\end{aligned}
\tag{26.22}
$$

Here we introduced the following transformation

$$
p_{l-1}\cdot u\left(x_N\right)\,f_{l-1}\left[p_{l-1}\cdot u\left(x_N\right)\right] = \int\frac{d^3p_l}{E_{p_l}}B\left(p_l,p_{l-1}\right)p_l\cdot u\left(x_N\right)\,f_l\left[p_l\cdot u\left(x_N\right)\right].
\tag{26.23}
$$

In order to calculate the pion spectrum (26.22), the transformation (26.23) should be performed iteratively starting with $l = N$ and ending at $l = 2$. All calculations can be done in the local rest frame where $u^\mu(x_N) = (1,0,0,0)$.

For arbitrary freeze-out surfaces we may proceed in a different way and do the integration over the freeze-out hypersurface first. In this case we have

$$
\begin{aligned}
E_{p_1}\frac{dN_\pi}{d^3p_1} &= \int\frac{d^3p_2}{E_{p_2}}B\left(p_2,p_1\right)\int\frac{d^3p_3}{E_{p_3}}B\left(p_3,p_2\right)...\\
&\times\int\frac{d^3p_{N-1}}{E_{p_{N-1}}}B\left(p_{N-1},p_{N-2}\right)\int\frac{d^3p_N}{E_{p_N}}B\left(p_N,p_{N-1}\right)\,F_N\left(p_N\right),
\end{aligned}
\tag{26.24}
$$

where

$$
F_N\left(p_N\right) = \int d\Sigma_\mu\left(x_N\right)\,p_N^\mu\,f_N\left[p_N\cdot u\left(x_N\right)\right]
\tag{26.25}
$$

Introducing a sequence of the transformation rules (for $l = N, N-1, ..., 2$)

$$
F_{l-1}(p_{l-1}) = \int\frac{d^3p_l}{E_{p_l}}B\left(p_l,p_{l-1}\right)\,F_l\left(p_l\right)
\tag{26.26}
$$

we find

$$
E_{p_1}\frac{dN_\pi}{d^3p_1} = F_1(p_1).
\tag{26.27}
$$

26.3.1 *Boost-invariant, cylindrically symmetric case*

In general, the functions $F(p)$ may depend on all three components of the three-momentum $\mathbf{p}$. This makes the practical application of the transformation (26.26) much more complicated than the use of Eq. (26.23). In many situations, however, the symmetry of the freeze-out surface reduces the number of the independent variables. This is so for the boost-invariant, cylindrically symmetric freeze-outs, where only the magnitude of the transverse momentum enters and $F_i(p_i) = f_i(p_{i\perp})$. In this case the transformation (26.26) becomes [1]

$$
\begin{aligned}
g(q_\perp) &= \int \frac{d^3k}{E_k} B(k,q)\, f(k_\perp) \qquad (26.28)\\
&= \frac{b_R}{4\pi p^*} \int \frac{d^3k}{E_k} \delta\left(\frac{k\cdot q}{m_R} - E^*\right) f(k_\perp)\\
&= \frac{b_R m_R}{4\pi p^*} \int dy\; k_\perp dk_\perp\; d\phi\; \delta\left(m_{1\perp} m_{R\perp} \cosh y - k_\perp q_\perp \cos\phi - m_R E^*\right) f(k_\perp).
\end{aligned}
$$

The integral over the rapidity y can be easily done, and we obtain

$$
g(q_\perp) = \frac{b_R m_R}{\pi p^* m_{1\perp}} \int dk_\perp \frac{k_\perp}{m_{R\perp}} f(k_\perp) \int_0^\pi d\phi \frac{\theta(x-1)}{\sqrt{(x-1)(x+1)}}, \qquad (26.29)
$$

where x is a function of ϕ, namely

$$
x(\phi) = \frac{m_R E^* + k_\perp q_\perp \cos\phi}{m_{1\perp} m_{R\perp}}. \qquad (26.30)
$$

We note that the factor $1/4$ disappeared in the transition from (26.28) to (26.29), because there are two opposite values of rapidity for which the argument of the Dirac function vanishes and we restricted the integration contour to the range $(0,\pi)$. The step function $\theta(x-1)$ in Eq. (26.29) reflects the condition $\cosh y \geq 1$. The integral over the angle ϕ in (26.29) can be rewritten in the form

$$
\begin{aligned}
\int_0^\pi d\phi \frac{\theta(x-1)}{\sqrt{(x-1)(x+1)}} &= \theta(x(0)-1)\theta(1-x(\pi)) \int_0^{\phi_0} d\phi \frac{1}{\sqrt{(x-1)(x+1)}}\\
&+ \theta(x(\pi)-1) \int_0^\pi d\phi \frac{1}{\sqrt{(x-1)(x+1)}}, \qquad (26.31)
\end{aligned}
$$

where

$$
\phi_0 = \arccos\left(\frac{m_{1\perp} m_{R\perp} - m_R E^*}{k_\perp q_\perp}\right). \qquad (26.32)
$$

The integrals on the right-hand-side of Eq. (26.31) are analytic (they can be expressed through the elliptic functions), so the transformation (26.29) involves only a single numerical integration over the transverse momenta.

[1] In order to reduce the number of indices we use the notation $F_l = f, p_l = k$, and $p_\perp^l = k_\perp$, for the "parent" particle, and $F_{l-1} = g, p_{l-1} = q$, and $p_\perp^{l-1} = q_\perp$, for the "daughter" particle. Moreover, in the calculations we omitted the trivial factor $(J_R+1)/(J_1+1)$, Since the considered system is boost-invariant and cylindrically symmetric the quantity y is the relative rapidity, while ϕ is the difference between the azimuthal angles of the momenta.

26.3.2 *Boost-invariant, cylindrically asymmetric case*

The cylindrically asymmetric case is important in the context of the studies of the strong elliptic flow observed at RHIC. From the technical point of view, in this case we may repeat the procedure described in the previous Section, however, the functions f and g depend now not only on the transverse momenta $q_\perp$ and $k_\perp$ but also on the azimuthal angles ϕ_q and ϕ_k, respectively. The straightforward generalization of Eq. (26.29) is [10, 11]

$$g(q_\perp, \phi_q) = \frac{b_R m_R}{2\pi p^* m_{1\perp}} \int dk_\perp \frac{k_\perp}{m_{R\perp}} \int_{-\pi}^{\pi} d\phi_k f(k_\perp, \phi_k) \frac{\theta(x-1)}{\sqrt{(x-1)(x+1)}} \tag{26.33}$$

where x is a function of $\phi_k - \phi_q$, namely

$$x(\phi_k - \phi_q) = \frac{m_R E^* + k_\perp q_\perp \cos(\phi_k - \phi_q)}{m_{1\perp} m_{R\perp}}. \tag{26.34}$$

Changing the integration variable, $\phi_k \to \phi_k + \phi_q$, and using the fact that we deal with periodic functions, we rewrite Eq. (26.33) in a compact form

$$g(q_\perp, \phi_q) = \int dk_\perp k_\perp \int_{-\pi}^{\pi} d\phi_k f(k_\perp, \phi_k + \phi_q) J(k_\perp, q_\perp, \phi_k) \tag{26.35}$$

where

$$J(k_\perp, q_\perp, \phi_k) = \frac{b_R m_R}{2\pi p^* m_{1\perp} m_{R\perp}} \frac{\theta(x-1)}{\sqrt{(x-1)(x+1)}} \tag{26.36}$$

and x is given by Eq. (26.34) with $\phi_q = 0$. The periodicity of the functions $g(q_\perp, \phi_q)$ and $f(k_\perp, \phi_k)$ allows us to use the Fourier expansion

$$\begin{aligned} g(q_\perp, \phi_q) &= \sum_n \left(\cos n\phi_q \, g_n(q_\perp) + \sin n\phi_q \, g_n^-(q_\perp)\right), \\ f(k_\perp, \phi_k + \phi_q) &= \sum_n \left(\cos n(\phi_k + \phi_q) \, f_n(k_\perp) + \sin n(\phi_k + \phi_q) \, f_n^-(k_\perp)\right). \end{aligned} \tag{26.37}$$

Substituting Eq. (26.37) into Eq. (26.35) and using the orthogonality properties of the trigonometric functions one finds

$$\begin{aligned} g_m(q_\perp) &= \int k_\perp dk_\perp \int_{-\pi}^{\pi} d\phi_k f_m(k_\perp) J(k_\perp, q_\perp, \phi_k) \cos m\phi_k \\ g_m^-(q_\perp) &= \int k_\perp dk_\perp \int_{-\pi}^{\pi} d\phi_k f_m^-(k_\perp) J(k_\perp, q_\perp, \phi_k) \sin m\phi_k. \end{aligned} \tag{26.38}$$

In the derivation of this formula we used the property that J depends on $\cos\phi_k$. In this way one finds that: **i)** the coefficients in the expansion (26.37) evolve in the independent way in the cascade and **ii)** the coefficients $g_m^-(q_\perp)$ vanish. Thus, the most important contributions come from the first coefficients $g_m(q_\perp)$,

$$\begin{aligned} g_0(q_\perp) &= \int k_\perp dk_\perp \int_{-\pi}^{\pi} d\phi_k f_0(k_\perp) J(k_\perp, q_\perp, \phi_k), \\ g_2(q_\perp) &= \int k_\perp dk_\perp \int_{-\pi}^{\pi} d\phi_k f_2(k_\perp) J(k_\perp, q_\perp, \phi_k) \cos 2\phi_k, \\ g_4(q_\perp) &= \int k_\perp dk_\perp \int_{-\pi}^{\pi} d\phi_k f_4(k_\perp) J(k_\perp, q_\perp, \phi_k) \cos 4\phi_k. \end{aligned} \tag{26.39}$$

Here we assumed that initially only the even coefficients in the expansion of f are present.

26.4 Sequential decays: correlations

The Fourier transform of the pion emission function can be done step by step for all particles in a cascade. The first integration over the pion space-time coordinates gives [8]

$$\begin{aligned} &\int d^4x_1 e^{i\,q\cdot x_1}\, S_\pi(x_1, p_1) \\ &= \int d^4x_1 e^{i\,q\cdot x_1} \int \frac{d^3p_2}{E_{p_2}} B(p_2, p_1) \int d\tau_2 \Gamma_2 e^{-\Gamma_2\tau_2} \int d^4x_2\, \delta^{(4)}\left(x_2 + \frac{p_2\tau_2}{m_2} - x_1\right) \ldots \\ &= \int \frac{d^3p_2}{E_{p_2}} B(p_2, p_1) \int d\tau_2 \Gamma_2 e^{-\left(\Gamma_2 - i\frac{q\cdot p_2}{m_2}\right)\tau_2} \int d^4x_2\; e^{q\cdot x_2} \ldots \\ &= \int \frac{d^3p_2}{E_{p_2}} B(p_2, p_1) \frac{1}{1 - i\frac{q\cdot p_2}{m_2\Gamma_2}} \int d^4x_2\; e^{iq\cdot x_2} \ldots \end{aligned} \tag{26.40}$$

Repeating the same procedure for all resonances in the cascade gives

$$\int d^4x_1 e^{i\,q\cdot x_1}\, S_\pi(x_1, p_1) \tag{26.41}$$

$$\begin{aligned} &= \int \frac{d^3p_2}{E_{p_2}} B(p_2, p_1) \frac{1}{1 - i\frac{q\cdot p_2}{m_2\Gamma_2}} \int \frac{d^3p_3}{E_{p_3}} B(p_3, p_2) \frac{1}{1 - i\frac{q\cdot p_3}{m_3\Gamma_3}} \cdots \\ &\times \int \frac{d^3p_N}{E_{p_N}} B(p_N, p_{N-1}) \frac{1}{1 - i\frac{q\cdot p_N}{m_N\Gamma_N}} \int d\Sigma_\mu(x_N)\, p_N^\mu\, e^{i\,q\cdot x_N} f_N\left[p_N \cdot u(x_N)\right] \end{aligned}$$

Now we define

$$F_N(p_N, q) = \int d\Sigma_\mu(x_N)\, p_N^\mu\, e^{i\,q\cdot x_N} f_N\left[p_N \cdot u(x_N)\right], \tag{26.42}$$

and

$$F_{l-1}(p_{l-1}, q) = \int \frac{d^3p_l}{E_{p_l}} B(p_l, p_{l-1}) \frac{1}{1 - i\frac{q\cdot p_l}{m_l\Gamma_l}} F_l(p_l, q), \tag{26.43}$$

where the index l runs from $l = N$ to $l = 2$. With this notation, our initial expression (26.40) can be written as

$$\int d^4x_1 e^{i\,q\cdot x_1}\, S_\pi(x_1, p_1) = F_1(p_1, q). \tag{26.44}$$

26.5 Single-freeze-out model

We have distinguished thermal models from hydro-inspired models by the specification of the range of their possible applications; the thermal models are used to analyze the ratios of hadron multiplicities, while the hydro-inspired models are used to study the hadron momentum distributions. Consequently, the thermal models give us information about the chemical freeze-out, i.e., the stage where the hadron abundances are fixed, whereas the hydro-inspired models give us information about the thermal freeze-out, i.e., the stage where the hadron momentum distributions are fixed.

In our previous discussion we have stressed the fact that the thermal-model calculations include the contributions from resonance decays. On the other hand, the hydro-inspired models usually neglect such contributions. One of the reasons for this situation is a trivial fact — the inclusion of many resonances in the calculations of the spectra is much more complicated and involved than their inclusion in the calculations of the yields. Another reason is that the thermal freeze-out happens at a lower temperature, as compared to the chemical freeze-out, and the contributions from the resonance decays are expected to be smaller at T_{therm}. One has to remember, however, that although the massive states are suppressed by the Boltzmann factor, their number increases exponentially, as described by the Hagedorn mass spectrum discussed in Sec. 6.3. In this situation one usually has to include many heavy resonances until the physical results saturate [2].

In Ref. [7] a hydro-inspired model was proposed where the thermal-freeze-out temperature T_{therm} was set equal to the chemical-freeze-out temperature $T_{\text{chem}} = 165$ MeV. The use of such a high temperature guarantees that the ratios of the hadron multiplicities are correctly reproduced (this we know already from the analysis of the particle ratios described in Sec. 24.5). The space-time structure of the freeze-out hypersurface as well as the hydrodynamic flow profile used in [7] are defined by the expressions given in Sec. 25.2.2. A novel feature of [7] was that

[2] For example, in the analysis of the particle ratios measured at the RHIC top energies, the hadrons with masses up to 1.6 GeV should be included [12].

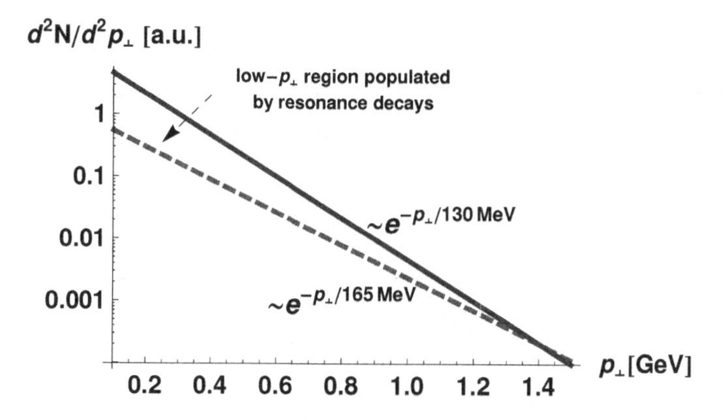

Fig. 26.3 Schematic illustration of the effect of sequential hadronic decays on the final transverse-momentum spectrum of pions. The dashed line shows the spectrum of the primordial pions characterized by the temperature $T_{\rm chem} = 165$ MeV, the solid line shows the spectrum of primordial and secondary pions that may be characterized by the effective temperature (inverse slope) $\lambda = 130$ MeV.

the contributions from all known hadronic resonances were included not only in the calculations of the hadron multiplicities but also in the calculation of the hadronic transverse-momentum spectra. In the numerical calculations the sequential decays were included with the help of the iterative methods presented earlier in this Chapter. The main physical effect of the resonance decays is that the spectrum of the final particles looks cooler. The particles originating from the resonance decays have little phase space and populate the low-$p_\perp$ region, see Fig. 26.3. Because of this behavior, the final spectra are steeper, which corresponds to an effective temperature that is smaller than $T_{\rm chem}$ (a smaller inverse slope of the measured spectra). For $T_{\rm chem} = 165$ MeV this effect is quite substantial, leading to the final effective temperature of about 130 MeV. Remarkably, such temperatures appear in the studies of the thermal freeze-out, and, indeed, the model introduced in [7] was successfully used to fit the hadron spectra, see Fig. 26.4. In this way an identification of the two freeze-outs had been accomplished.

Further studies showed that the single-freeze-out model describes well the HBT radii [11, 16], the elliptic flow [10, 11], and the balance functions [17]. On the other hand the model has a problem with the correct reproduction of the yields of short-living resonances [3] [18]. This indicates that the two freeze-out do not exactly coincide. Anyway, the observations made in [7] exclude the concepts of long living hadronic fireballs. This is in accordance with the proposed solutions of the HBT puzzle which also indicate fast expansion and breakup of matter produced in relativistic heavy-ion collisions.

[3] The correct description of the yields of strongly decaying resonances is a problem also for the standard thermal models.

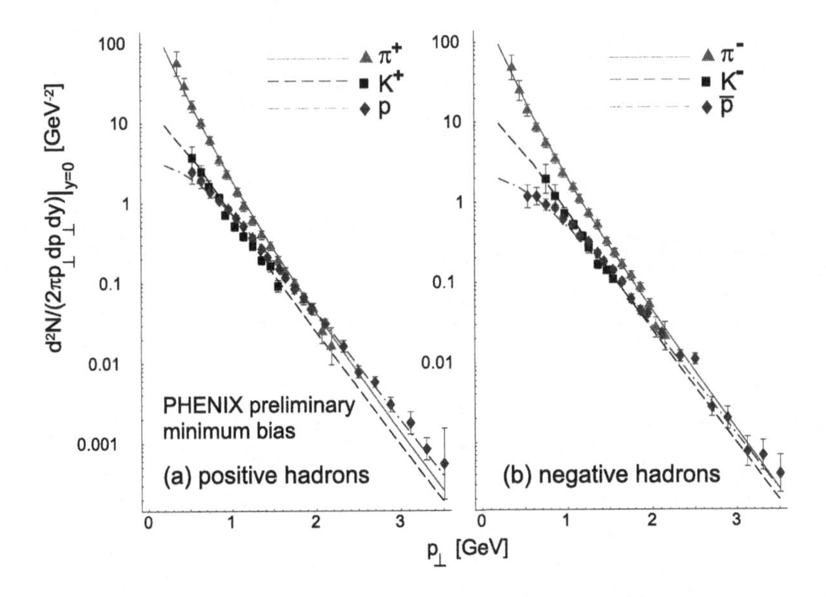

Fig. 26.4 The $p_\perp$-spectra of pions (solid line), kaons (dashed line) and protons or antiprotons (dashed-dotted line), as evaluated from the single-freeze-out model [7], compared to the PHENIX preliminary data [13]. Later official PHENIX data of Ref. [14] agree with the data used here. Feeding from the weak decays is included. Reprinted from [15] with permission from Acta Phys. Pol. **B**.

26.6 Open source computer codes for thermal and hydro-inspired models

Recently, several open source codes implementing the thermal and hydro-inspired models have been constructed. Their aim is to form a common platform for the interpretation of the data within the statistical approach and to tighten the link between the data and theory.

The SHARE package (Statistical HAdronization with REsonances), Torrieri et al. [19, 20], is a set of Fortran programs which perform various tasks. With the input of the thermal parameters, such as temperature and chemical potentials, the ratios of the hadron abundances may be calculated. More importantly, with the given experimental input of the hadron abundances the optimal thermal parameters are obtained with the help of the fitting procedure which delivers not only the minimum values of χ^2 but also the χ^2 contours and profiles. SHARE allows also for a selection of the hadronization model, i.e., the conditions for chemical equilibrium or non-equlibrium may be chosen. The weak decays in SHARE receive a special treatment; the particle acceptance factors are introduced for both decaying particles. SHARE has also the option to include the finite widths of the resonances in the calculation of the yields. For the modest task of using SHARE to find the ratios of hadron abundances which correspond to a given set of the thermal parameters, the special web-calculator SHAREONLINE is available on the web page [21]. SHARE-ONLINE is a very convenient tool for doing fast estimates within a thermal model. The recently developed second version of SHARE [20] provides the new feature allowing the studies of statistical fluctuations of the particle yields. The standard SHARE package contains also a program written in Mathematica, MathSHARE, which allows for the calculation of the transverse-momentum spectra [21].

The THERMUS package, Wheaton and Cleymans [22], is composed of C++ classes and functions developed for the incorporation into the ROOT environment [23]. THERMUS allows for the choice of the statistical ensemble. One may choose the grand-canonical ensemble (where the conservation laws for the energy and quantum numbers are enforced on the average through the temperature and chemical potentials), the canonical ensemble (where the conservation laws of quantum and particle numbers such as baryon number, strangeness, and charge are exactly enforced), and the mixed-canonical ensemble (where the strangeness in the system is fixed exactly, while the baryon number and charge are treated within the grand-canonical approach). In this respect THERMUS offers complementary features to those offered by SHARE. A distinctive feature of THERMUS is the possibility of inclusion of the Van der Waals corrections (excluded volume corrections), which become important when the particle densities predicted by the statistical models become large. Similarly to SHARE, THERMUS includes the finite widths of the resonances with the help of the Breit-Wigner formula.

The last statistical code we discuss here is THERMINATOR (THERMal hevy-IoN generATOR), Kisiel et al. [24]. THERMINATOR uses the same universal input for particle properties as SHARE, however, similarly to THERMUS, it is written in C++ and complies to the standards of the ROOT environment [23]. The main difference between THERMINATOR and the two previously discussed models is that THERMINATOR is a Monte-Carlo generator. At the beginning of each simulated event THERMINATOR generates all stable particles and resonances on the chosen freeze-out hypersurface. The complete chemical equilibrium is assumed in this approach and the local phase-space densities are given by the statistical Fermi-Dirac or Bose-Einstein distributions. In the next stage of the simulation of an event, the particles are allowed to move freely and decay. The contributions from the decays of resonances feed the yields of stable particles in the substantial way. The space-time points where the particles are born and their momenta are memorized, so the model allows for the calculation of different hadron characteristics, in particular one may study hadron spectra and correlations. THERMINATOR, as a Monte-Carlo program, is especially well suited for the inclusion of various experimental cuts in the theoretical calculations.

Bibliography to Chapter 26

[1] J. Sollfrank, P. Koch, and U. W. Heinz, "The influence of resonance decays on the p_T spectra from heavy ion collisions," *Phys. Lett.* **B252** (1990) 256–264.
[2] J. Sollfrank, P. Koch, and U. W. Heinz, "Is there a low p_T "anomaly" in the pion momentum spectra from relativistic nuclear collisions?," *Z. Phys.* **C52** (1991) 593–610.
[3] G. E. Brown, J. Stachel, and G. M. Welke, "Pions from resonance decay in Brookhaven relativistic heavy ion collisions," *Phys. Lett.* **B253** (1991) 19–22.
[4] W. Weinhold, B. Friman, and W. Norenberg, "Thermodynamics of Δ resonances," *Phys. Lett.* **B433** (1998) 236–242.
[5] W. Weinhold, B. L. Friman, and W. Noerenberg, "Thermodynamics of an interacting πN system," *Acta Phys. Polon.* **B27** (1996) 3249–3253.
[6] W. Florkowski, W. Broniowski, and M. Michalec, "Thermal analysis of particle ratios and p_T spectra at RHIC," *Acta Phys. Polon.* **B33** (2002) 761–769.
[7] W. Broniowski and W. Florkowski, "Explanation of the RHIC p_T-spectra in a thermal model with expansion," *Phys. Rev. Lett.* **87** (2001) 272302.
[8] J. Bolz, U. Ornik, M. Plumer, B. R. Schlei, and R. M. Weiner, "Resonance decays and partial coherence in Bose-Einstein correlations," *Phys. Rev.* **D47** (1993) 3860–3870.
[9] W. Broniowski and W. Florkowski, "Strange particle production at RHIC in a single-freeze-out model," *Phys. Rev.* **C65** (2002) 064905.
[10] A. Baran, "Description of the azimuthal asymmetry in relativistic heavy-ion collisions in the framework of the thermal model of particle production," PhD Thesis, Institute of Nuclear Physics, Cracow, 2004.
[11] W. Broniowski, A. Baran, and W. Florkowski, "Thermal model at RHIC. II: Elliptic flow and HBT radii," *AIP Conf. Proc.* **660** (2003) 185–195.
[12] M. Michalec, "Thermal description of particle production in ultra- relativistic heavy-ion collisions," `nucl-th/0112044`.
[13] **PHENIX** Collaboration, J. Velkovska, "Pt Distributions of Identified Charged Hadrons Measured with the PHENIX Experiment at RHIC," *Nucl. Phys.* **A698** (2002) 507–510.
[14] **PHENIX** Collaboration, K. Adcox *et al.*, "Centrality dependence of pi+-, K+-, p and anti-p production from $\sqrt{s_{\rm NN}} = 130$ GeV Au + Au collisions at RHIC," *Phys. Rev. Lett.* **88** (2002) 242301.
[15] W. Broniowski, A. Baran, and W. Florkowski, "Thermal approach to RHIC," *Acta Phys. Polon.* **B33** (2002) 4235–4258.
[16] A. Kisiel, W. Florkowski, and W. Broniowski, "Femtoscopy in hydro-inspired models with resonances," *Phys. Rev.* **C73** (2006) 064902.
[17] P. Bozek, W. Broniowski, and W. Florkowski, "Balance functions in a thermal model with resonances," *Acta Phys. Hung.* **A22** (2005) 149–157.
[18] W. Broniowski, W. Florkowski, and B. Hiller, "Thermal analysis of production of resonances in relativistic heavy-ion collisions," *Phys. Rev.* **C68** (2003) 034911.
[19] G. Torrieri *et al.*, "Share: Statistical hadronization with resonances," *Comput. Phys. Commun.* **167** (2005) 229–251.
[20] G. Torrieri, S. Jeon, J. Letessier, and J. Rafelski, "SHAREv2: Fluctuations and a comprehensive treatment of decay feed-down," *Comput. Phys. Commun.* **175** (2006) 635–649.
[21] SHAREONLINE `http://www.physics.arizona.edu/ gtshare/SHARE/sharev1.html`.

[22] S. Wheaton and J. Cleymans, "THERMUS: A thermal model package for ROOT," *Comput. Phys. Commun.* **180** (2009) 84–106.
[23] R. Brun and F. Rademakers, "Root: An object oriented data analysis framework," *Nucl. Instrum. Meth.* **A389** (1997) 81–86. (http://root.cern.ch).
[24] A. Kisiel, T. Taluc, W. Broniowski, and W. Florkowski, "THERMINATOR: Thermal heavy-ion generator," *Comput. Phys. Commun.* **174** (2006) 669–687.

Chapter 27

Exercises to PART VII

Exercise 27.1. *Static fireball.*
Consider Eq. (24.3) in the Boltzmann limit where

$$N = gV \int \frac{d^3p}{(2\pi)^3} \exp\left(-\frac{E_p - \mu}{T}\right). \tag{27.1}$$

Express the energy and longitudinal momentum in terms of the transverse mass and rapidity. **i)** Integrate Eq. (27.1) over rapidity and show that the transverse-mass distribution is given by the formula ($\beta = 1/T$)

$$\frac{dN}{m_\perp dm_\perp} = \frac{gVe^{\beta\mu}}{2\pi^2} m_\perp K_1(\beta m_\perp). \tag{27.2}$$

In the next step expand the Bessel function K_2 for large arguments and find that Eq. (27.2) for $\beta m_\perp \gg 1$ is reduced to the expression

$$\frac{dN}{m_\perp dm_\perp} = \frac{gVe^{\beta\mu}}{(2\pi)^{3/2}} \left(\frac{m_\perp}{\beta}\right)^{1/2} e^{-\beta m_\perp}. \tag{27.3}$$

This example teaches us that the non-trivial power of the transverse mass should be included in the calculations of the transverse-mass spectra in the static-fireball model. **ii)** Integrate Eq. (27.1) over the transverse mass first and find the rapidity distribution

$$\frac{dN}{dy} = \frac{gVe^{\beta\mu}T^3}{(2\pi)^2} \frac{e^{-\beta m \cosh \mathrm{y}} \left(2 + \beta m \cosh \mathrm{y}(2 + \beta m \cosh \mathrm{y})\right)}{\cosh^2 \mathrm{y}}, \tag{27.4}$$

where m is the particle's mass. In the limit $m \to 0$ this expression is reduced to

$$\frac{dN}{dy} = \frac{gVe^{\beta\mu}T^3}{2\pi^2 \cosh^2 \mathrm{y}}. \tag{27.5}$$

iii) By integration of the last equation over rapidity show that the density of massless classical particles is given by the formula

$$n = \frac{gVe^{\beta\mu}T^3}{\pi^2}. \tag{27.6}$$

Derive the last result directly from Eq. (27.1).

Exercise 27.2. *Ratios of hadron multiplicities/densities.*
Consider a hadron gas at the zero baryon chemical potential, $\mu_B = 0$, and three temperatures: $T = 100$, 130, and 165 MeV. For each value of the temperature calculate numerically the ratio of proton density to the positive pion density (p/π^+).

Exercise 27.3. *Spherically symmetric freeze-out hypersurface.*
Derive Eq. (25.5) that describes the element of the spherically symmetric hypersurface.

Exercise 27.4. *Inverse slope.*
Show that (25.30) may be interpreted as a formula describing the Doppler effect. *Hint:* Use the Lorentz transformation to show that the factor $\sqrt{(1+v_\perp)/(1-v_\perp)}$ describes the blue shift of the frequency of the source moving with the velocity $v_\perp$ towards the observer.

Exercise 27.5. *Functions $\mathcal{I}_n(a,b)$.*
Show that the functions $\mathcal{I}_0(a,b)$ and $\mathcal{I}_1(a,b)$ defined by Eqs. (25.33) and (25.34) may be alternatively defined by the expressions:

$$\mathcal{I}_0(a,b) = \frac{1}{\pi} \int\limits_0^\pi d\phi \cosh[a\cos\phi]\cos(b\sin\phi) \tag{27.7}$$

and

$$\mathcal{I}_1(a,b) = \frac{1}{\pi} \int\limits_0^\pi d\phi \, \cos\phi \, \sinh[a\cos\phi]\cos(b\sin\phi). \tag{27.8}$$

Exercise 27.6. *Lorentz boost.*
Derive Eq. (26.2). *Hint:* Consider the changes of the perpendicular and parallel components of $\mathbf{p}$ with respect to $\mathbf{v}_k$. Show that they transform according to the well known one-dimensional Lorentz transformation.

PART VIII

ELECTROMAGNETIC SIGNALS FROM HOT AND DENSE MATTER

Chapter 28

In-Medium Modifications of Hadron Properties

Since the QED coupling constant α is very small, photons and dileptons (e^+e^- or $\mu^+\mu^-$ pairs) produced at a certain stage of relativistic heavy-ion collisions move through the surrounding matter practically without any further interactions. Hence, they carry information directly from the point they were formed to detectors. This is not the case for hadrons, whose measured properties may be changed by rescattering effects in the late expansion phase.

Photons and dileptons are produced during the whole spacetime evolution of the collision process. First photons and lepton pairs are created during the initial stage when the two nuclei pass through each other, while the last photons and lepton pairs are formed at the point when the hadrons are decoupled and move freely to detectors. In fact, after decoupling of hadronic matter a huge number of photons and dileptons are produced by the decays of hadronic resonances.

The lepton pairs produced from the early initial hard collisions between the partons of the incoming nuclei are the best known part of the dilepton spectrum. These processes are called the Drell-Yan mechanism. Lepton pairs can be additionally produced through the semileptonic decays of charmed mesons produced in initial hard processes. However, the most interesting for us is the radiation from the hot and dense reaction volume created in central collisions, where the quark-gluon plasma and chiral restoration are expected. A source of such radiation might be the quark-antiquark annihilation in the quark-gluon plasma. Of course, our task is to identify the possible processes leading to the emission of photons from the reaction volume and to distinguish them from the background described by more conventional physics.

Dilepton spectra are usually divided into three invariant-mass regions. The low-mass region, $M < 1$ GeV, is dominated by hadronic interactions and hadronic decays after freeze-out. In the intermediate-mass region, 1 GeV $< M <$ 2.5 GeV, the contribution from the thermalized QGP might be seen. The high-mass region, $M \sim m_{J/\Psi}$, is important in connection with the suppressed J/Ψ production, proposed as a signal of the deconfinement phase transition [1].

The invariant mass spectrum of dileptons contains the peaks corresponding to the two-body decays of various mesons (for example: $\rho^0, \omega \to e^+e^-$). Such peaks appear above the continuous background of three-body Dalitz decays (for example: $\pi^0, \eta, \eta' \rightarrow e^+e^-\gamma, \omega \rightarrow \pi^0 e^+e^-$). *The positions and widths of the peaks reveal information about the hadron properties at the late stages of the collision process.* This issue is the main subject of this Chapter.

Dilepton production has been studied since the beginning of the SPS heavy-ion program by various experiments: HELIOS/NA34, NA38, CERES/NA45, NA50, and NA60. The first measurements were carried out by CERES, HELIOS, and NA38. The CERES Collaboration [2] measured dielectron spectra in the low-mass region, the HELIOS-3 Collaboration [3] measured dimuon spectra up to the J/Ψ region, and the NA38 Collaboration measured dimuon spectra in the intermediate- and high-mass region, with the particular emphasis on the J/Ψ suppression. The most detailed and comprehensive dilepton results at the SPS has been recently delivered by the NA60 Collaboration which measured dimuon production in In+In reactions [4]. At RHIC, the dielectron and photon spectra in Au+Au collisions at $\sqrt{s_{NN}} = 200$ GeV have been measured by the PHENIX Collaboration [5,6].

28.1 Low-mass dilepton enhancement

At the SPS the production of dileptons with low invariant masses was first measured by CERES [2] and HELIOS-3 [3] Collaborations. Both experiments reported an enhanced production in S-induced reactions. The enhanced production means, in this context, that the dilepton yield exceeds a simple extrapolation from more elementary pA collisions.

The low-mass e^+e^-–pair data of the CERES Collaboration from p+Be and p+Au collisions can be well explained by the expected decays of neutral mesons in the final state (in particular, they are well reproduced by the RQMD model [7], see also [8]). Since there is no precise data on neutral meson production in heavy-ion collisions, in Ref. [2] the CERES Collaboration makes the assumption that there are identical relative particle abundances in the final state of all collision systems. Applying this scaling law one obtains the expected neutral meson contribution to the dilepton spectrum in heavy-ion collisions.

For central S+Au collisions the measured spectrum presented in [2] agrees with the extrapolated contribution from neutral meson decays only for masses below 200 MeV (this region is dominated by π^0 Dalitz-decay). At higher masses the dilepton yield is strongly enhanced, especially in the region 300–500 MeV. Integrating over the mass region between 200 MeV and 1.5 GeV, the CERES Collaboration finds the dilepton yield enhanced by a factor of 5. An interesting feature of the measured spectrum is that it has a different shape from the spectrum extrapolated from neutral meson decays. In particular, it does not show the typical resonance structure of the $\rho, \omega \rightarrow e^+e^-$ decays around 780 MeV.

The more recent CERES results describing e^+e^- pairs from Pb+Au collisions [9,10] are shown in Fig. 28.1. They exhibit similar features to those discovered with

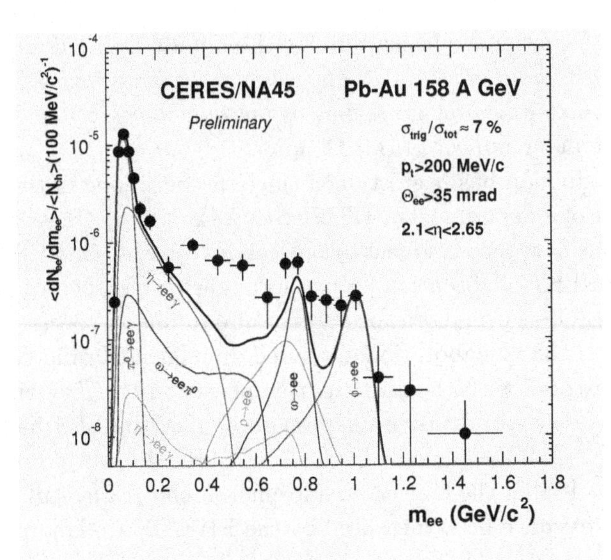

Fig. 28.1 Dielectron mass spectrum measured by the CERES Collaboration in Pb+Au collisions at the top SPS energy of 158 A GeV [9]. The plot describes also the "cocktail" contribution to the dielectron spectrum, i.e., the sum of the contributions from neutral meson decays in the final state. The dilepton production in the invariant-mass range between 200 and 600 MeV is strongly enhanced, compared with the theoretical predictions including the resonance decays in vacuum only. Reprinted from [9] with permission from Elsevier.

sulfur beams. Below 200 MeV the data agree well with the expected Dalitz decay contribution. Above 200 MeV the yield is strongly enhanced, especially around 500 MeV. The mass spectrum exhibits an enhancement of e^+e^- pairs in the mass range between 200 and 600 MeV by a factor of 2.8 [9]. We note that in this case the reference contribution from hadron decays in the final state, i.e., the hadronic "cocktail", has been evaluated with the help of the thermal model of Ref. [11], where the chemical freeze-out conditions for temperature and baryon chemical potential are obtained from a fit to hadronic observables at the SPS and AGS energies.

The muon pair measurements of the HELIOS-3 Collaboration provide another evidence for the enhancement. The data [3] show the enhanced production in S+W collisions (compared to p+W collisions). The spectral shape again deviates from the yield expected for hadron decays. Integrating from the dimuon threshold to 600 MeV one finds the enhancement factor 1.75 [1].

[1]It is worth to mention that the dilepton spectrum was also measured at LBL Bevelac by the DLS Collaboration [12, 13] in the heavy-ion collisions at incident energies around 1 GeV per nucleon (such energies are two orders of magnitude lower than at the SPS). Although the first published data based on the limited data set were consistent with the results obtained from transport model calculations, a later analysis including the full data set showed a considerable excess in the dilepton production [14]. The measurements at the similar beam energies have been recently studied by the HADES Collaboration at GSI. For more information on the development of the situation at lower energies see [15, 16].

The detailed studies show that the decays of hadrons in the final state of the interaction can be excluded as the source of the excess. Hence, the extra dileptons must be emitted at an earlier time during the collision event. The onset of the excess near twice the pion mass suggests that the $\pi\pi$ annihilation can be important source of additional dileptons. In the next Section we shall analyze this process in the framework of the kinetic theory. We shall follow the line of arguments presented by P. Koch in Ref. [17].

28.2 Dileptons from pion annihilation

The production rate of dileptons from $\pi^+\pi^-$ annihilation is defined by the expression that can be obtained from the kinetic-theory considerations similar to those presented in Sec. 9.2.1, see Ex. 29.1,

$$\begin{aligned} \frac{dN}{d^4x} = \sum_{\text{spins}} \int & dP_1 dP_2 dQ_1 dQ_2 \tilde{f}(p_1) \tilde{f}(p_2) \\ & \times (2\pi)^4 \delta^{(4)}(p_1 + p_2 - q_1 - q_2) \left| \mathcal{M}_{\pi^+\pi^- \to l^+ l^-} \right|^2 . \end{aligned} \tag{28.1}$$

Here p_1 and p_2 are the four-momenta of the annihilating pions, while q_1 and q_2 are the four-momenta of the outgoing leptons. The sum runs over the spins of the leptons. The quantities dP and dQ denote the integration measure

$$dP = \frac{d^3p}{(2\pi)^3 2\sqrt{\mathbf{p}^2 + m_\pi^2}}, \qquad dQ = \frac{d^3q}{(2\pi)^3 2\sqrt{\mathbf{q}^2 + m_l^2}}, \tag{28.2}$$

and $\tilde{f}(p)$ is the pion distribution function [2],

$$\tilde{f}(p) = \left[\exp \frac{\sqrt{\mathbf{p}^2 + m_\pi^2} - \mu_\pi}{T} - 1 \right]^{-1} . \tag{28.3}$$

The production rate of dileptons with fixed total four-momentum

$$q = p_1 + p_2 = q_1 + q_2$$

may be written as

$$\frac{dN}{d^4x d^4q} = \int dP_1 dP_2 \, \tilde{f}(p_1) \, \tilde{f}(p_2) \delta^{(4)}(q - p_1 - p_2) \left| \overline{\mathcal{M}}_{\pi^+\pi^- \to l^+ l^-} \right|^2 , \tag{28.4}$$

[2] In this Section we follow the original notation used in Ref. [17]. The distribution functions $\tilde{f}$ differ from the standard distribution functions introduced in Sec. 8.1 by a factor $(2\pi)^3$, namely, $f = \tilde{f}/(2\pi)^3$. This convention leads to simpler expressions in the cases where the equilibrium distribution functions appear in the formalism and $\hbar = 1$. We also note that the definition of integration measures (28.2) differs from the definition used in Chaps. 9 and 11 by a factor $2(2\pi)^3$.

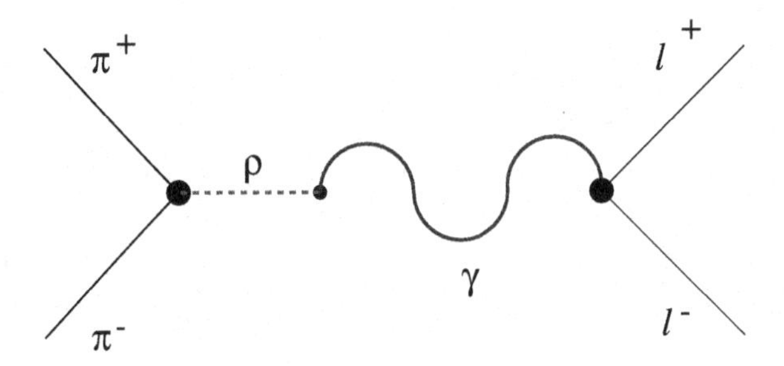

Fig. 28.2 Vector dominance model of dilepton production by pion annihilation

where

$$\left|\overline{\mathcal{M}}_{\pi^+\pi^-\to l^+l^-}\right|^2 \equiv \sum_{\text{spins}} \int dQ_1 dQ_2\ (2\pi)^4 \delta^{(4)}(q - q_1 - q_2)\left|\mathcal{M}_{\pi^+\pi^-\to l^+l^-}\right|^2. \quad (28.5)$$

28.2.1 *Vector dominance model*

The matrix element $\mathcal{M}_{\pi^+\pi^-\to l^+l^-}$ is usually calculated in the framework of the *vector dominance model* [18,19]. According to this model the electromagnetic interactions of hadrons are described by the intermediate coupling of hadrons to vector mesons. In this picture $\pi^+\pi^- \to l^+l^-$ annihilation can be understood as the formation of a vector meson by the annihilation of pions, and the subsequent decay of the vector meson into a lepton pair, see Fig. 28.2. The matrix element for this process has the structure

$$\mathcal{M}_{\pi^+\pi^-\to l^+l^-} = \sum_V g_{\pi\pi V}(p_1 - p_2)_\mu \frac{1}{m_V^2 - q^2 - i m_V \Gamma_V} \frac{e m_V^2}{g_V} \frac{1}{q^2}\, e L^\mu, \quad (28.6)$$

where m_V and Γ_V are the mass and width of the vector meson V, g_V is the parameter describing the coupling of the vector meson to the virtual photon, $1/q^2$ is the photon propagator, and $g_{\pi\pi V}$ is the effective coupling strength for the $\pi\pi V$ interaction. The leptonic current L^μ may be expressed in terms of the Dirac bispinors

$$L^\mu(q_1, s_1; q_2, s_2) = 2\ m_l\ \overline{u}(q_1, s_1)\ \gamma^\mu\ v(q_2, s_2), \quad (28.7)$$

where s_1 and s_2 denote the spin variables of the outgoing leptons [3]. Integration over momenta, with the fixed total four-momentum $q = q_1 + q_2$, and summation

[3]The structure of Eq. (28.7) follows from the application of the Feynman rules. In particular the factor $2m_l$ is required to change the "bosonic" normalization, $1/(2E)$, to the "fermionic" normalization, m/E.

over spins gives us the useful formula

$$L^{\mu\nu} \equiv \sum_{\text{spins}} \int dQ_1 dQ_2 \, (2\pi)^4 \delta^{(4)}(q - q_1 - q_2) L^\mu L^\nu$$
$$= \frac{1}{6\pi}\sqrt{1 - \frac{4m_l^2}{q^2}} \left(q^2 + 2m_l^2\right) \left(\frac{q^\mu q^\nu}{q^2} - g^{\mu\nu}\right). \tag{28.8}$$

In most of practical calculations, the sum over different vector meson states may be reduced to the single ρ-meson contribution (as other vector mesons are much heavier). In this case Eqs. (28.5), (28.6) and (28.8) give

$$\left|\overline{\mathcal{M}}_{\pi^+\pi^-\to l^+l^-}\right|^2 = g^2_{\pi\pi\rho}(p_1 - p_2)_\mu (p_1 - p_2)_\nu \frac{1}{g_\rho{}^2} \frac{m_\rho^4}{(m_\rho^2 - q^2)^2 + m_\rho^2 \Gamma_\rho^2} \frac{16\pi^2\alpha^2}{q^4} L^{\mu\nu}, \tag{28.9}$$

where we have introduced the fine structure constant $\alpha = e^2/(4\pi)$. The tensor product in Eq. (28.9) may be most easily done in the center-of-mass reference frame where $\mathbf{q} = \mathbf{p}_1 + \mathbf{p}_2 = 0$. A straightforward calculation gives

$$\left|\overline{\mathcal{M}}_{\pi^+\pi^-\to l^+l^-}\right|^2 = \frac{8\pi}{3}\left(\frac{g_{\pi\pi\rho}\alpha}{g_\rho}\right)^2 \frac{m_\rho^4}{(m_\rho^2 - q^2)^2 + m_\rho^2\Gamma_\rho^2}$$
$$\times \frac{\left(q^2 - 4m_l^2\right)^{\frac{1}{2}} \left(q^2 - 4m_\pi^2\right) \left(q^2 + 2m_l^2\right)}{q^5}. \tag{28.10}$$

Substitution of Eq. (28.10) into formula (28.4) yields the desired production rate of dileptons.

The last results can be used to compute the $\pi^+\pi^-$ annihilation cross section. It is defined by

$$\sigma_{\pi^+\pi^-\to l^+l^-} = \sum_{\text{spins}} \int dQ_1 dQ_2 \, (2\pi)^4 \, \delta^{(4)} \left(q - q_1 - q_2\right) \frac{\left|\mathcal{M}_{\pi^+\pi^-\to l^+l^-}\right|^2}{4F_{\text{i}}}, \tag{28.11}$$

where F_{i} is the invariant flux, see Eqs. (9.21) and (30.34),

$$F_{\text{i}} = \sqrt{\left(p_1 \cdot p_2\right)^2 - m_\pi^4} = \sqrt{q^2\left(\frac{q^2}{4} - m_\pi^2\right)}. \tag{28.12}$$

Since F_{i} is a function of q^2 only, we may simply write

$$\sigma_{\pi^+\pi^-\to l^+l^-} = \frac{\left|\overline{\mathcal{M}}(\pi^+\pi^- \to l^+l^-)\right|^2}{4F_{\text{i}}}, \tag{28.13}$$

and the explicit evaluation gives

$$\sigma_{\pi^+\pi^-\to l^+l^-} = \frac{4\pi}{3}\left(\frac{\alpha\, g_{\pi\pi\rho}}{g_\rho}\right)^2 |F_\pi(M)|^2 \times \frac{\left(M^2-4m_l^2\right)^{\frac{1}{2}}\left(M^2-4m_\pi^2\right)^{\frac{1}{2}}\left(M^2+2m_l^2\right)}{M^6}. \tag{28.14}$$

Here $M^2 = q^2$ is the dilepton invariant mass squared and

$$|F_\pi(M)|^2 = \frac{m_\rho^4}{\left(M^2-m_\rho^2\right)^2 + m_\rho^2\Gamma_\rho^2} \tag{28.15}$$

is the pion electromagnetic form factor. Equation (28.14) exhibits the resonance structure of the annihilation process.

28.2.2 *Effective decay width*

The production rate of dileptons can be expressed in terms of the decay width for the process $\rho \to l^+l^-$, see Fig. 28.3, and the formation rate for the reaction $\pi^+\pi^- \to \rho$. In both cases we take into consideration a virtual ρ-meson with the four-momentum q satisfying the condition $q^2 = M^2$. In the following we use the notation

$$q = (E_q, \mathbf{q}), \quad E_q = \sqrt{M^2 + \mathbf{q}^2}. \tag{28.16}$$

The width for the process $\rho \to l^+l^-$ is defined by the formula

$$\widehat{\Gamma}_{\rho\to l^+l^-} = \frac{1}{2E_q}\frac{1}{3}\sum_{i=1}^{3}\sum_{\text{spins}}\int dQ_1 dQ_2\, (2\pi)^4\delta^{(4)}(q-q_1-q_2)\left|\mathcal{M}^{(i)}_{\rho\to l^+l^-}\right|^2, \tag{28.17}$$

which follows from the general expression (30.29) and where the matrix element $\mathcal{M}^{(i)}_{\rho\to l^+l^-}$ has the form

$$\mathcal{M}^{(i)}_{\rho\to l^+l^-} = \epsilon^{(i)}_\mu(q)\frac{em_\rho^2}{g_\rho}\frac{1}{q^2}eL^\mu. \tag{28.18}$$

As usual we sum over the spins of outgoing leptons and average over three different polarizations of the decaying ρ-meson. The polarization vectors $\epsilon^{(i)}_\mu(q)$ satisfy the well-known relation

$$\sum_{i=1}^{3}\epsilon^{(i)}_\mu(q)\epsilon^{(i)}_\nu(q) = \frac{q_\mu q_\nu}{q^2} - g_{\mu\nu}. \tag{28.19}$$

A little algebra now shows that

$$\widehat{\Gamma}_{\rho\to l^+l^-} = \frac{4\pi\alpha^2 m_\rho^4}{3{g_\rho}^2 M^4 E_q}\left(1-\frac{4m_l^2}{M^2}\right)^{\frac{1}{2}}\left(M^2+2m_l^2\right). \tag{28.20}$$

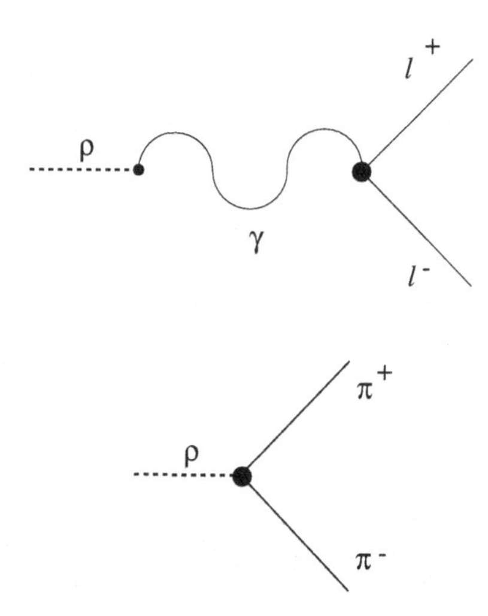

Fig. 28.3 Amplitudes for the decays $\rho \to l^+l^-$ and $\rho \to \pi^+\pi^-$.

In a similar way we compute the decay width for the processes $\rho \to \pi^+\pi^-$. In this case we have

$$\widehat{\Gamma}_{\rho\to\pi^+\pi^-} = \frac{1}{2E_q}\frac{1}{3}\sum_{i=1}^{3}\int dP_1 dP_2 \left[1+\tilde{f}(p_1)\right]\left[1+\tilde{f}(p_2)\right] \times (2\pi)^4\delta^{(4)}(q-p_1-p_2)\left|\mathcal{M}^{(i)}_{\rho\to\pi^+\pi^-}\right|^2, \tag{28.21}$$

where

$$\mathcal{M}^{(i)}_{\rho\to\pi^+\pi^-} = \mathcal{M}^{(i)}_{\pi^+\pi^-\to\rho} = g_{\pi\pi\rho}\,\epsilon^{(i)}_\mu(q)\,(p_1-p_2)^\mu. \tag{28.22}$$

The Bose enhancement factors in the definition (28.21) are important, since there are many pions in a hot hadronic system which is typically formed in the ultra-relativistic heavy-ion collisions. The large abundance of pions enhances the decay probability (28.21). On the other hand, in a dense pion system the opposite processes start to play a role. The creation of mesons in inelastic pion collisions becomes not negligible. In this case the *formation rate* of the ρ-meson (with a fixed polarization) due to pion annihilation is given by the formula, see Ex. 29.2,

$$\widehat{\Gamma}_{\pi^+\pi^-\to\rho} = \frac{1}{2E_q}\frac{1}{3}\sum_{i=1}^{3}\int dP_1 dP_2 \tilde{f}(p_1)\tilde{f}(p_2) \times (2\pi)^4\delta^{(4)}(q-p_1-p_2)\left|\mathcal{M}^{(i)}_{\pi^+\pi^-\to\rho}\right|^2. \tag{28.23}$$

Using Eqs. (28.1), (28.17) and (28.23) together with the identity

$$1 = \int dM^2 \int d^4q\,\delta\left(q^2-M^2\right)\delta^{(4)}\left(q-q_1-q_2\right) = \int dM^2 \int \frac{d^3q}{2E_q}\delta^{(4)}\left(q-q_1-q_2\right) \tag{28.24}$$

one finds the compact formula for the *three-momentum integrated lepton pair rate*

$$\frac{dN}{d^4x dM^2} = \int \frac{d^3q}{(2\pi)^3} \widehat{\Gamma}_{\rho \to l^+l^-} \frac{1}{\pi} \frac{3E_q \widehat{\Gamma}_{\pi^+\pi^- \to \rho}}{\left(m_\rho^2 - M^2\right)^2 + \left(m_\rho \Gamma_\rho\right)^2}. \tag{28.25}$$

We emphasize that the structure of Eq. (28.25) follows from the factorization of the matrix element (28.6) computed in the vector dominance model, see Ex. 29.3.

Equation (28.25) may be rewritten in a more conspicuous form. At first, it is important to realize that the *effective decay width* of the ρ-meson in medium should be given by the difference

$$\widehat{\Gamma}_\rho = \widehat{\Gamma}_{\rho \to \pi^+\pi^-} - \widehat{\Gamma}_{\pi^+\pi^- \to \rho}. \tag{28.26}$$

This definition takes into account a competition between decays and formation processes which are both important in a dense pion medium. Moreover, in thermal equilibrium, where the pion distributions are of the form (28.3), the two rates (28.21) and (28.23) are related,

$$\widehat{\Gamma}_{\rho \to \pi^+\pi^-} = \widehat{\Gamma}_{\pi^+\pi^- \to \rho} \exp\left(\frac{E_q - 2\mu_\pi}{T}\right). \tag{28.27}$$

This formula can be used to express the formation rate $\widehat{\Gamma}_{\pi^+\pi^- \to \rho}$ in terms of the effective decay width $\widehat{\Gamma}_\rho$

$$\widehat{\Gamma}_{\pi^+\pi^- \to \rho} = \widehat{\Gamma}_\rho \frac{1}{e^{(E_q - 2\mu_\pi)/T} - 1} \equiv \widehat{\Gamma}_\rho \, \tilde{f}_\rho(q). \tag{28.28}$$

In Eq. (28.28) one can recognize the equilibrium distribution function of the ρ-mesons with energy E_q and the chemical potential $\mu_\rho = 2\mu_\pi$. It appeared automatically in our approach, as a consequence of the assumption about thermal equilibrium. We still have to clarify the connection between $\widehat{\Gamma}_\rho$ and Γ_ρ appearing in the denominator of the ρ-meson propagator. As we shall see below, a consistent form of the meson propagator is achieved if we define

$$\Gamma_\rho = \frac{E_q}{m_\rho} \widehat{\Gamma}_\rho. \tag{28.29}$$

Finally, Eqs. (28.28) and (28.29) are used to rewrite the lepton pair rate (28.25) in the form

$$\frac{dN}{d^4x dM^2} = \int \frac{d^3q}{(2\pi)^3} \widehat{\Gamma}\left(\rho \to l^+l^-\right) \frac{1}{\pi} \frac{3E_q \widehat{\Gamma}_\rho}{\left(m_\rho^2 - M^2\right)^2 + \left(E_q \widehat{\Gamma}_\rho\right)^2} \tilde{f}_\rho(q). \tag{28.30}$$

This expression has a clear physical interpretation. It shows that the *indirect* dilepton production resulting from $\pi^+\pi^-$ annihilation in the ρ channel can be viewed as the *direct* decays of the ρ-mesons having the effective width which accounts for the formation and decay processes in the pion medium.

28.2.3 *Connection to field theory*

Description of pion annihilation into dileptons based on the kinetic theory is equivalent, in the case of equilibrium, to the approach based on the thermal field theory. In the latter case one derives the following formula [20]

$$\frac{dN}{d^4x dM^2} = \int \frac{d^3q}{(2\pi)^3} \widehat{\Gamma}_{\rho \to l^+ l^-} \left(2\rho_T + \rho_L\right) \tilde{f}_\rho\left(q\right), \tag{28.31}$$

where $\rho_{T,L}$ is the transverse *(T)* and the longitudinal *(L)* spectral function expressed by the appropriate components of the polarization tensor of the ρ-meson

$$\rho_{T,L} = \frac{1}{\pi} \frac{-\mathrm{Im}\Pi_{T,L}\left(q\right)}{\left(m_\rho^{(0)\,2} + \mathrm{Re}\Pi_{T,L}\left(q\right) - M^2\right)^2 + \left(\mathrm{Im}\Pi_{T,L}\left(q\right)\right)^2}. \tag{28.32}$$

The use of the explicit form for the decay width $\widehat{\Gamma}_{\rho \to l^+ l^-}$ gives us a compact expression for the dilepton pair rate (in the limit $m_l = 0$)

$$\frac{dN}{d^4x d^4q} = \left(\frac{\alpha\, m_\rho^2}{\pi\, g_\rho\, M}\right)^2 \frac{\left(2\rho_T + \rho_L\right)}{3} \tilde{f}_\rho\left(q\right). \tag{28.33}$$

For simplicity let us neglect the difference in the propagation of the transverse and longitudinal modes. In this case the imaginary part of the polarization function is directly related to the decay width

$$-\mathrm{Im}\Pi_{T,L}\left(q\right) = E_q \widehat{\Gamma}_\rho, \tag{28.34}$$

and we recover the formula derived from the kinetic theory, provided we make the identifications: $m_\rho^2 = m_\rho^{(0)\,2} + \mathrm{Re}\Pi_{T,L}\left(q\right)$ and $m_\rho \Gamma_\rho = E_q \widehat{\Gamma}_\rho$. If the transverse modes propagate in a different way from the longitudinal ones, the kinetic approach can be easily generalized in order to include such effects. One finds again the agreement between the kinetic and the field-theoretic approach.

So far we have considered only the $\pi^+\pi^-$ annihilation as a mechanism of extra dilepton production. Therefore, in order to obtain the agreement with the kinetic theory the polarization functions Π_T and Π_L, appearing in Eq. (28.32), should take into account only the effects of the interaction of the ρ-meson with pions. However, other processes contribute also to Π_T and Π_L, and they all have corresponding processes which contribute to the dilepton yield (for example, besides the $\rho\pi$ interaction one may include ρK scattering and the formation of s-channel resonances, like in the process $\rho\pi \to a_1$). We thus conclude that the more accurate is our calculation of the polarization function, the more processes are included as the source of the dilepton yield. The question remains which processes are dominant and this problem has been studied by many authors, for a review see [20].

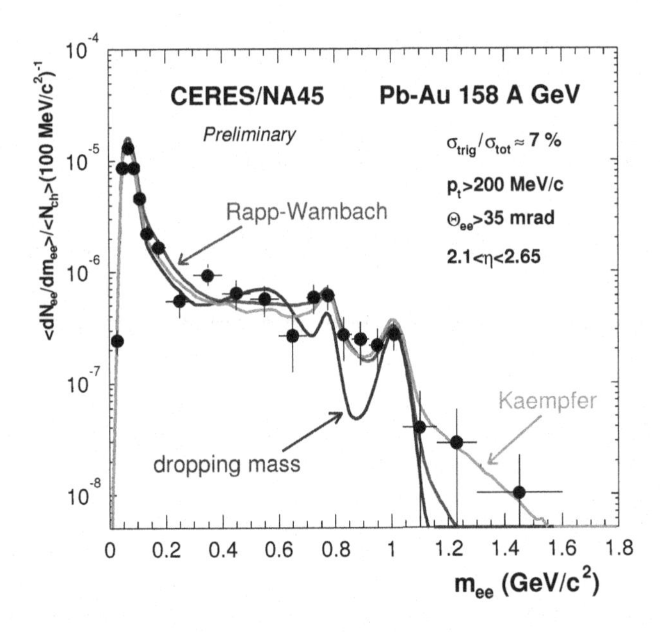

Fig. 28.4 Comparison of the CERES results [9] with the model calculations discussed in the text. Reprinted from [9] with permission from Elsevier.

28.3 Dropping masses and broadening widths

Low-mass dilepton data provided evidence for the emission of direct radiation from the hot and dense reaction volume created in central heavy-ion collisions. Several theoretical studies [17,21–24] showed that $\pi\pi$ annihilation represents the main contribution to the dilepton spectrum in the considered mass region. Different theoretical frameworks were used to calculate the dilepton yield: the *hydrodynamic models* with [25–27] or without [26,28,29] the phase transition, and the *transport models* [30–39] which explicitly propagate hadrons. All these calculations reproduce well the experimentally observed multiplicities and hadron spectra, however, they fail to reproduce the dilepton yield unless one assumes that ρ meson properties in dense matter are modified.

If the parameters used to describe the propagation of the ρ meson are the vacuum values, the calculations predict a resonance structure at around 780 MeV, which originates from $\pi\pi$ annihilation. Such a structure is absent in the data. The striking discrepancy can be solved by the assumption that meson properties in dense matter are different from those we measure in vacuum. Such modifications can wash out the resonance structure from the $\pi\pi$ annihilation displayed by Eq. (28.15). In the literature one considers usually two versions of the modifications. In the first version, the *Brown-Rho scaling* [40], one assumes that the meson masses decrease with the increasing density,

$$\frac{m^*}{m} \approx 1 - k_{\rm dm} \frac{\rho}{\rho_0}. \tag{28.35}$$

Here m^* is the effective in-medium mass, m is the vacuum mass, ρ is the baryon density of the medium, ρ_0 is the normal nuclear density, and $k_{\rm dm}$ is a positive parameter. In the second version the position of the mass peak does not change, however, the hadron widths increase in dense medium [20],

$$\frac{\Gamma^*}{\Gamma} \approx 1 + k_{\rm bw} \frac{\rho}{\rho_0}. \tag{28.36}$$

Here Γ^* is the effective in-medium width, Γ is the vacuum width, and $k_{\rm bw}$ is another positive parameter. The two options are known as scenarios of dropping masses and broadening widths, respectively.

The new CERES data discussed in Sec. 28.1 and especially the high resolution data of NA60 indicate the broadening of the resonance widths rather than the dropping of the in-medium masses, see Fig. 28.4.

Bibliography to Chapter 28

[1] T. Matsui and H. Satz, "J/psi Suppression by Quark-Gluon Plasma Formation," *Phys. Lett.* **B178** (1986) 416.

[2] **CERES** Collaboration, G. Agakishiev *et al.*, "Enhanced production of low mass electron pairs in 200 GeV/u S+Au collisions at the CERN SPS," *Phys. Rev. Lett.* **75** (1995) 1272–1275.

[3] **HELIOS** Collaboration, M. Masera, "Dimuon production below mass 3.1 GeV/c^2 in p W and S+W interactions at 200 A/GeV/c," *Nucl. Phys.* **A590** (1995) 93c–102c.

[4] **NA60** Collaboration, R. Arnaldi *et al.*, "First measurement of the rho spectral function in high- energy nuclear collisions," *Phys. Rev. Lett.* **96** (2006) 162302.

[5] **PHENIX** Collaboration, S. Afanasiev *et al.*, "Enhancement of the dielectron continuum in $\sqrt{s_{\rm NN}} = 200$ GeV Au+Au collisions," 0706.3034.

[6] **PHENIX** Collaboration, A. Adare *et al.*, "Enhanced production of direct photons in Au+Au collisions at $\sqrt{s_{\rm NN}} = 200$ GeV," 0804.4168.

[7] L. A. Winckelmann, H. Stoecker, W. Greiner, and H. Sorge, "Dielectron production in pp and pA collisions at 4.9 GeV," *Phys. Lett.* **B298** (1993) 22–26.

[8] L. A. Winckelmann, H. Sorge, H. Stoecker, and W. Greiner, "Baryon resonances: A primary $\rho \rightarrow$ lepton$^+$ lepton$^-$ source in p+p and p+d at 4.9 GeV," *Phys. Rev.* **C51** (1995) 9–11.

[9] **CERES** Collaboration, D. Miskowiec, "Collections of CERES results," *Nucl. Phys.* **A774** (2006) 43–50.

[10] **CERES** Collaboration, G. Agakichiev *et al.*, "e$^+$ e$^-$ pair production in Pb Au collisions at 158 GeV per nucleon," *Eur. Phys. J.* **C41** (2005) 475–513.

[11] P. Braun-Munzinger and J. Stachel, "Dynamics of ultra-relativistic nuclear collisions with heavy beams: An experimental overview," *Nucl. Phys.* **A638** (1998) 3–18.

[12] **DLS** Collaboration, G. Roche *et al.*, "First observation of dielectron production at the BEVALAC," *Phys. Rev. Lett.* **61** (1988) 1069–1072.

[13] C. Naudet *et al.*, "Threshold behavior of electron pair production in p Be collisions," *Phys. Rev. Lett.* **62** (1989) 2652.

[14] C. Ernst, S. A. Bass, M. Belkacem, H. Stoecker, and W. Greiner, "Intermediate mass excess of dilepton production in heavy ion collisions at BEVALAC energies," *Phys. Rev.* **C58** (1998) 447–456.

[15] **HADES** Collaboration, G. Agakichiev *et al.*, "Dielectron production in $^{12}C + {}^{12}C$ collisions at 2 A GeV with HADES," *Phys. Rev. Lett.* **98** (2007) 052302.

[16] **HADES** Collaboration, . F. Krizek *et al.*, "Inclusive dielectron production in Ar+KCl collisions at 1.76 A GeV studied with HADES," 0907.3690.

[17] P. Koch, "Low mass lepton pair production and pion dynamics in ultrarelativistic heavy ion collisions," *Z. Phys.* **C57** (1993) 283–304.

[18] J. J. Sakurai, "Currents and mesons," (University of Chicago Press, Chicago, 1969).

[19] C. Gale and J. I. Kapusta, "Vector dominance model at finite temperature," *Nucl. Phys.* **B357** (1991) 65–89.

[20] R. Rapp and J. Wambach, "Chiral symmetry restoration and dileptons in relativistic heavy-ion collisions," *Adv. Nucl. Phys.* **25** (2000) 1.

[21] K. Kajantie, J. Kapusta, L. D. McLerran, and A. Mekjian, "Dilepton emission and the QCD phase transition in ultrarelativistic nuclear collisions," *Phys. Rev.* **D34** (1986) 2746.

[22] K. Kajantie, M. Kataja, L. D. McLerran, and P. V. Ruuskanen, "Transverse flow effects in dilepton emission," *Phys. Rev.* **D34** (1986) 811.

[23] J. Cleymans, K. Redlich, and H. Satz, "Low mass dielectrons at LHC energy," *Z. Phys.* **C52** (1991) 517–526.

[24] B. Kampfer, P. Koch, and O. P. Pavlenko, "Kinetics of an expanding pion gas and low mass dilepton emission," *Phys. Rev.* **C49** (1994) 1132–1138.

[25] D. K. Srivastava, B. Sinha, and C. Gale, "Excess production of low mass lepton pairs in S+Au collisions at the CERN Super Proton Synchrotron and the quark - hadron phase transition," *Phys. Rev.* **C53** (1996) 567–571.

[26] J. Sollfrank *et al.*, "Hydrodynamical description of 200-A-GeV/c S+Au collisions: Hadron and electromagnetic spectra," *Phys. Rev.* **C55** (1997) 392–410.

[27] C. M. Hung and E. V. Shuryak, "Dilepton/photon production in heavy ion collisions, and the QCD phase transition," *Phys. Rev.* **C56** (1997) 453–467.

[28] R. Baier, M. Dirks, and K. Redlich, "Thermal dileptons from $\pi\rho$ interactions in a hot pion gas," *Phys. Rev.* **D55** (1997) 4344–4354.

[29] V. Koch and C. Song, "Dilepton production at SPS-energy heavy ion collisions," *Phys. Rev.* **C54** (1996) 1903–1917.

[30] G.-Q. Li, C. M. Ko, and G. E. Brown, "Enhancement of low mass dileptons in heavy ion collisions," *Phys. Rev. Lett.* **75** (1995) 4007–4010.

[31] G.-Q. Li, C. M. Ko, and G. E. Brown, "Effects of in-medium vector meson masses on low-mass dileptons from SPS heavy-ion collisions," *Nucl. Phys.* **A606** (1996) 568–606.

[32] W. Cassing, W. Ehehalt, and C. M. Ko, "Dilepton production at SPS energies," *Phys. Lett.* **B363** (1995) 35–40.

[33] W. Cassing, W. Ehehalt, and I. Kralik, "Analysis of the HELIOS-3 $\mu^+\mu^-$ data within a relativistic transport approach," *Phys. Lett.* **B377** (1996) 5–10.

[34] L. A. Winckelmann *et al.*, "Microscopic calculations of stopping and flow from 160-A- MeV to 160-A-GeV," *Nucl. Phys.* **A610** (1996) 116c–123c.

[35] R. Rapp, G. Chanfray, and J. Wambach, "Medium modifications of the rho meson at CERN SPS energies," *Phys. Rev. Lett.* **76** (1996) 368–371.

[36] R. Rapp, G. Chanfray, and J. Wambach, "Rho meson propagation and dilepton enhancement in hot hadronic matter," *Nucl. Phys.* **A617** (1997) 472–495.

[37] J. Murray, W. Bauer, and K. Haglin, "Low-mass dileptons at the SPS," *Phys. Rev.*

C57 (1998) 882–888.

[38] G.-Q. Li, C. M. Ko, G. E. Brown, and H. Sorge, "Dilepton production in proton nucleus and nucleus nucleus collisions at SPS energies," *Nucl. Phys.* **A611** (1996) 539–567.

[39] W. Cassing, E. L. Bratkovskaya, R. Rapp, and J. Wambach, "Probing the rho spectral function in hot and dense nuclear matter by dileptons," *Phys. Rev.* **C57** (1998) 916–921.

[40] G. E. Brown and M. Rho, "Scaling effective Lagrangians in a dense medium," *Phys. Rev. Lett.* **66** (1991) 2720–2723.

Chapter 29

Exercises to PART VIII

Exercise 29.1. *The $\pi^+\pi^-$ annihilation rate.*
Derive Eq. (28.1) starting from elementary kinetic considerations. *Answer:* The number of lepton pairs produced in $\pi^+\pi^-$ annihilation per unit time and unit volume is given by the expression analogous to Eq. (9.24),

$$dN = d^4x \int \frac{d^3p_1}{p_1^0} \frac{d^3p_2}{p_2^0} \frac{d^3q_1}{q_1^0} \frac{d^3q_2}{q_2^0} f(p_1)f(p_2)\; F_i\, q_1^0 q_2^0 \frac{d\sigma_{\pi^+\pi^-\to l^+l^-}}{d^3q_1 d^3q_2}. \tag{29.1}$$

Using the formula for the transition rate W, see Eq. (9.37), one finds

$$\begin{aligned} \frac{dN}{d^4x} = \int \frac{d^3p_1}{p_1^0} \frac{d^3p_2}{p_2^0} \frac{d^3q_1}{q_1^0} \frac{d^3q_2}{q_2^0}\, f(p_1)f(p_2) \\ \times \frac{1}{64\pi^2} \left|\mathcal{M}_{\pi^+\pi^-\to l^+l^-}\right|^2 \delta^{(4)}\left(p_1+p_2-q_1-q_2\right). \end{aligned} \tag{29.2}$$

This formula is equivalent to

$$\begin{aligned} \frac{dN}{d^4x} = \int \frac{d^3p_1}{2p_1^0(2\pi)^3} \frac{d^3p_2}{2p_2^0(2\pi)^3} \frac{d^3q_1}{2q_1^0(2\pi)^3} \frac{d^3q_2}{2q_2^0(2\pi)^3}\, \tilde{f}(p_1)\tilde{f}(p_2) \\ \times \left|\mathcal{M}_{\pi^+\pi^-\to l^+l^-}\right|^2 (2\pi)^4\delta^{(4)}\left(p_1+p_2-q_1-q_2\right), \end{aligned} \tag{29.3}$$

which agrees with Eq. (28.1) if the sum over spins is included.

Exercise 29.2. *Formation time.*
Derive Eq. (28.23) starting from elementary kinetic considerations. *Answer:* The number of the ρ-mesons produced in $\pi^+\pi^-$ annihilation in the phase space element

$V\Delta^3 q$ and within time Δt is again given by the expression analogous to Eq. (9.24),

$$\Delta N = V \Delta t \frac{\Delta^3 q}{q^0} \int \frac{d^3 p_1}{p_1^0} \frac{d^3 p_2}{p_2^0} \, f(p_1) f(p_2) \; F_i \, q^0 \frac{d\sigma_{\pi^+\pi^- \to \rho}}{d^3 q}. \tag{29.4}$$

Expressing the cross section by the matrix element one finds

$$\Delta N = V \Delta t \frac{\Delta^3 q}{2q^0 (2\pi)^3} \int \frac{d^3 p_1}{2p_1^0 (2\pi)^3} \frac{d^3 p_2}{2p_2^0 (2\pi)^3} \, \tilde{f}(p_1) \tilde{f}(p_2) \times (2\pi)^4 \delta^{(4)}(q - p_1 - p_2) \, |\mathcal{M}_{\pi^+\pi^- \to \rho}|^2. \tag{29.5}$$

To obtain the formation time we require that one meson is produced in the considered phase space element, $\Delta N = 1$. This condition requires also that the density of final states should be equal to unity, hence, we should have consistently $V\Delta^3 q/(2\pi)^3 = 1$. In this way we obtain

$$\frac{1}{\Delta t} = \frac{1}{2q^0} \int \frac{d^3 p_1}{2p_1^0 (2\pi)^3} \frac{d^3 p_2}{2p_2^0 (2\pi)^3} \, \tilde{f}(p_1) \tilde{f}(p_2) \times (2\pi)^4 \delta^{(4)}(q - p_1 - p_2) \, |\mathcal{M}_{\pi^+\pi^- \to \rho}|^2. \tag{29.6}$$

This quantity after averaging over different polarization states leads to the formation rate given by Eq. (28.23).

Exercise 29.3. *Factorization of the widths in the dilepton rate.*
Check the following identity

$$\left| \overline{\mathcal{M}}_{\pi^+\pi^- \to l^+ l^-} \right|^2 = \frac{1}{3} \sum_{i,j=1}^{3} \sum_{\text{spins}} \int dQ_1 dQ_2 \; (2\pi)^4 \delta^{(4)}(q - q_1 - q_2) \times \frac{\left| \mathcal{M}^{(i)}_{\pi^+\pi^- \to \rho} \right|^2 \left| \mathcal{M}^{(j)}_{\rho \to l^+ l^-} \right|^2}{(m_\rho - q^2)^2 + (m_\rho \Gamma_\rho)^2}. \tag{29.7}$$

Use it to derive Eq. (28.25).

Exercise 29.4. *Connection between the dilepton and virtual photon rates.*
Using the vector dominance model as an example show that the dilepton and the virtual photon rates are connected by the following relation

$$\frac{dN}{d^4 x d^4 q} = \frac{2\alpha}{3\pi} \frac{\left(1 - 4m_l^2/M^2\right)^{1/2} \left(1 + 2m_l^2/M^2\right)}{M^2} q_0 \frac{dN_{\gamma^*}}{d^3 q}. \tag{29.8}$$

Here dN/d^4xd^4q is the dilepton rate given by Eq. (28.1), while the virtual photon rate may be obtained from the formula

$$\frac{dN_{\gamma^*}}{d^4x} = \sum_{i=1}^{3} \int dQ\, dP_1\, dP_2 \tilde{f}(p_1)\tilde{f}(p_2) \\ \times (2\pi)^4 \delta^{(4)}(p_1 + p_2 - q) \left| \mathcal{M}^{(i)}_{\pi^+\pi^- \to \gamma^*} \right|^2 , \qquad (29.9)$$

where the matrix element has the form

$$\mathcal{M}^{(i)}_{\pi^+\pi^- \to \gamma^*} = \sum_V g_{\pi\pi V}(p_1 - p_2)_\mu \frac{1}{m_V^2 - q^2 - im_V\Gamma_V} \frac{em_V^2}{g_V} \epsilon^{(i)\mu}(q), \qquad (29.10)$$

In Eqs. (29.8)–(29.10) we use the notation $q^2 = M^2$.

PART IX
SUPPLEMENTS

Chapter 30

Cross Sections, Transition Rates, and Inclusive Distributions

In this Chapter we collect the elementary information about the *scattering theory*. The basic quantities such as the *cross section*, the *scattering matrix*, and the *transition rate* are defined. Following the popular treatment used in many textbooks, we consider the scattering processes in a simplified way, i.e., we put the colliding particles in a box and impose the periodic boundary conditions [1,2]. This is done instead of the more realistic but also more involved analysis of the wave packets.

The purpose of this Chapter is to present several issues that may be useful for better understanding of the ideas presented in the main parts of the book. For example, the expressions worked out below are used in Chap. 9, where the structure of the collision terms in the relativistic kinetic equations is analyzed.

We start our presentation with a simple geometrical picture of the cross section that is introduced in Sec. 30.1. Then, we gradually come to more complex notions presented in Secs. 30.2, 30.3, and 30.4. Section 30.5 is devoted to the discussion of the *inclusive cross sections*. We show how the inclusive cross sections are related to the *particle inclusive distributions*. This issue is rarely discussed in textbooks but its good understanding is important for the correct interpretation of various high-energy measurements.

30.1 Geometric interpretation of cross section

The scattering processes are commonly characterized by their cross sections. The value of the cross section, σ, determines the probability of a collision between the particles present in the two colliding beams. It may be interpreted as the effective transverse area of the colliding particles. To be more precise, let us assume that we have N_2 particles of the type "2", which are at rest, $v_2 = 0$, in the volume

$$V = dx\, dy\, dz = A\, dz. \tag{30.1}$$

These particles are bombarded by N_1 particles of the type "1", which move with the velocity $v_1 \neq 0$, see Fig. 30.1. The probability that a particle of the type "2" is hit by the particle of the type "1" is $N_2\sigma/A$. Hence, the total number of collisions may be calculated from the formula

$$\begin{aligned} d\nu &= N_1 N_2 \frac{\sigma}{A} \\ &= N_1 N_2 \frac{\sigma}{dx\, dy} \frac{dt}{dz} \frac{dz}{dt} \\ &= N_1 N_2 \frac{\sigma}{V}\, dt\, v_1 \\ &= V dt\, n_1^0 n_2^0 v_1 \sigma. \end{aligned} \tag{30.2}$$

Here we assumed that the volume V is swept by the particles of the type "1" going through the area A in the time interval dt. We introduced also the densities of particles defined as $n_1^0 = N_1/V$ and $n_2^0 = N_2/V$. Equation (30.2) may be used as a definition of the cross section. For the case $N_1 = N_2 = 1$ we get

$$\sigma = \frac{(d\nu/dt)}{(v_1/V)} = \frac{\text{collision rate}}{\text{flux of incoming particles}}. \tag{30.3}$$

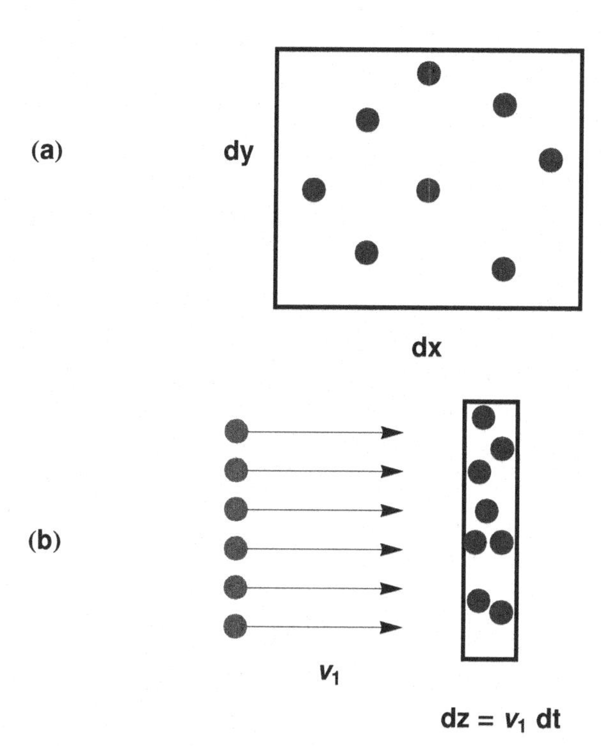

Fig. 30.1 **(a)** The view along the beam axis. In this way the projectile particles "see" the target particles. **(b)** The side view of the collision. In the laboratory frame the projectile particles move with velocity v_1.

Equation (30.2) is formulated in a special (i.e., laboratory) reference frame where the particles "2" are at rest. In order to generalize it to arbitrary reference frames we follow the treatment of Landau and Lifshitz [3] and write the formula for the number of collisions in the generally covariant form

$$d\nu = V dt\, n_1^0 n_2^0\, v_{\rm M}\, \sigma. \tag{30.4}$$

The particle densities have the form, see Eq. (8.54),

$$n_1^0 n_2^0 = \frac{n_1}{\sqrt{1-v_1^2}} \frac{n_2}{\sqrt{1-v_2^2}} = n_1 n_2 \frac{p_1^0 p_2^0}{m_1 m_2}, \tag{30.5}$$

where n_1 and n_2 are the densities of the particles "1" and "2" defined in their rest frames, hence n_1 and n_2 are Lorentz scalars. Since σ, $d\nu$, and $V dt$ are also scalars, the Lorentz structure of Eq. (30.4) requires that the quantity $v_{\rm M}$ is proportional to the inverse of the product of the energies $p_1^0 p_2^0$. In this case we may write

$$v_{\rm M} = \frac{p_1 \cdot p_2}{p_1^0 p_2^0} v_{\rm S}, \tag{30.6}$$

where $v_{\rm S}$ is a Lorentz scalar which is reduced to v_1 in the rest frame connected with the particles "2" (where for $v_2 = 0$ we have $\mathbf{p}_2 = 0$ and $p_1 \cdot p_2 = p_1^0 p_2^0$). In this frame v_1 may be calculated from the equation

$$p_1 \cdot p_2 = \frac{m_1}{\sqrt{1-v_1^2}} m_2, \tag{30.7}$$

which yields

$$v_1 = \sqrt{1 - \frac{(m_1 m_2)^2}{(p_1 \cdot p_2)^2}}. \tag{30.8}$$

Consequently, the quantity $v_{\rm S}$ may be obtained from the right-hand-side of Eq. (30.8). After simple manipulations, one obtains

$$\begin{aligned} d\nu &= V dt\, n_1^0 n_2^0\, v_{\rm M}\, \sigma = V dt\, n_1^0 n_2^0 \frac{p_1 \cdot p_2}{p_1^0 p_2^0} v_{\rm S}\, \sigma \\ &= V dt\, n_1^0 n_2^0 \frac{\sqrt{(p_1 \cdot p_2)^2 - (m_1 m_2)^2}}{p_1^0 p_2^0} \sigma \end{aligned} \tag{30.9}$$

and

$$v_{\rm M} = \frac{\sqrt{(p_1 \cdot p_2)^2 - (m_1 m_2)^2}}{p_1^0 p_2^0}. \tag{30.10}$$

We may now generalize our definition of the cross section, Eq. (30.3), to the arbitrary reference frames,

$$\sigma = \frac{(d\nu/dt)}{(v_{\rm M}/V)} = \frac{\text{collision rate}}{\text{flux of particles}}. \tag{30.11}$$

One can check that $v_{\rm M}$, known in the literature as the Møller velocity, may be expressed in terms of the three-velocities $\mathbf{v}_1$ and $\mathbf{v}_2$, see Ex. 10.8,

$$v_{\rm M} = \sqrt{(\mathbf{v}_1 - \mathbf{v}_2)^2 - (\mathbf{v}_1 \times \mathbf{v}_2)^2}. \tag{30.12}$$

In most cases, we deal with head-on collisions with the particles following parallel trajectories. In this case Eq. (30.12) is reduced to the simple expression,

$$v_{\rm M} = |\mathbf{v}_1 - \mathbf{v}_2|, \tag{30.13}$$

and the cross section may be defined as

$$\sigma = \frac{(d\nu/dt)}{(|\mathbf{v}_1 - \mathbf{v}_2|/V)}. \tag{30.14}$$

30.2 Scattering matrix

In the previous Section we introduced the cross section as the geometrical quantity that characterizes the probability that the two particles may collide with each other. Now we are going to give a more precise description of the collisions, specifying accurately the initial and final states of the colliding particles. Moreover, we shall take into account the possibility that a collision leads to the production of new particles, that have not been present in the initial state.

Generally speaking, in the analysis of scattering processes we are interested in a transition from the initial state i to the final state f. The state i describes the physical situation long before the collision, and the state f describes the situation long after the collision. A matrix containing the probability amplitudes for the transitions between the states i and f is called the *scattering matrix* S_{fi}. In general we assume that it has the following structure

$$S_{fi} = \delta_{fi} + T_{fi} = \delta_{fi} - i\,(2\pi)^\kappa\,\delta^{(4)}\,(P_f - P_i)\,M_{fi}, \tag{30.15}$$

where δ_{fi} is different from zero and equal to one if and only if the states i and f are exactly the same. The quantities P_f^μ and P_i^μ are the total four-momenta of particles in the states f and i, respectively.

The *transition matrix* T_{fi} defined by Eq. (30.15) describes non-trivial transitions from the initial state to the final state. Taking into account the energy and momentum conservation laws, we rewrite the transition matrix T_{fi} as a product of the delta function $\delta^{(4)}\,(P_f - P_i)$ and a new matrix M_{fi}. The value of the parameter κ is a matter of convention (for example, $\kappa = 1$ in the book by Weinberg [2], and $\kappa = 4$ in the book by Halzen and Martin [4]).

The scattering matrix may be used to calculate the probability of the transition from the state i to the state f. The standard problem in this case is, however, that the square of Eq. (30.15) containing the Dirac delta function is not well defined. This difficulty may be overcome by dealing with the wave packets instead of the plane waves (this is the way the real physical situation looks like) or by putting the system into the box with finite volume $V = L^3$. Below we shall follow the second method which gives the right answer after a few steps. The key observation in this case is that the values of the momenta are discrete and take the values

$$\mathbf{p} = \frac{2\pi}{L}(n_1, n_2, n_3), \tag{30.16}$$

where n_i are integers. Moreover, the Dirac delta function $\delta^{(3)}(\mathbf{p}-\mathbf{q})$ used to normalize the momentum states should be replaced by the Kronecker symbol $\delta_{\mathbf{p},\mathbf{q}}$,

$$\delta^{(3)}(\mathbf{p}-\mathbf{q}) \to \frac{1}{(2\pi)^3}\int_V d^3x\, e^{i(\mathbf{p}-\mathbf{q})\cdot\mathbf{x}} = \frac{V}{(2\pi)^3}\delta_{\mathbf{p},\mathbf{q}}\,. \tag{30.17}$$

The symbol $\delta_{\mathbf{p},\mathbf{q}}$ does not vanish and is equal to 1 if and only if the three integers representing $\mathbf{p}$ coincide with the three integers representing $\mathbf{q}$. In the similar way we replace the Dirac delta function expressing the energy conservation,

$$\delta(E_p - E_q) \quad \to \frac{1}{2\pi}\int_0^T dt\; e^{i(E_p-E_q)t} = \frac{T}{2\pi}\delta_{E_p,E_q}\,. \tag{30.18}$$

Here T plays a role of the time of the scattering process and the symbol δ_{E_p,E_q} does not vanish and is equal to 1 if and only if the two energies are equal, $E_p = E_q$.

The momentum states in the box $\left|p^{\rm box}\right\rangle$ are related to the continuum states $|p\rangle$ by the formula

$$\left|p^{\rm box}\right\rangle = \left[\frac{(2\pi)^3}{V}\right]^{1/2} |p\rangle\,. \tag{30.19}$$

We note that if the states $|p\rangle$ and $|q\rangle$ are normalized with the Dirac delta function (without any prefactors) then the corresponding states in the box are normalized to unity,

$$\langle q|p\rangle = \delta^{(3)}(\mathbf{p}-\mathbf{q}) \quad \longrightarrow \left\langle q^{\rm box}|p^{\rm box}\right\rangle = \delta_{\mathbf{p},\mathbf{q}}.$$

In this case, the properly normalized probability of the transition from the initial state $\left|i^{\rm \ box}\right\rangle$ (containing N_i particles with momenta $p_1^{\rm box}, \dots, p_{N_i}^{\rm box}$) to the final state $\left|f^{\rm \ box}\right\rangle$ (different from $\left|i^{\rm \ box}\right\rangle$ and containing N_f particles with momenta $q_1^{\rm box}, \dots, q_{N_f}^{\rm box}$) may be written as [1]

$$P(i \longrightarrow f) = \left|S_{fi}^{\rm \ box}\right|^2 = (2\pi)^{2\kappa}\left[\delta^{(4)}_{\rm box}(P_f - P_i)\right]^2 \left|M_{fi}^{\rm \ box}\right|^2. \tag{30.20}$$

With the help of Eqs. (30.17) and (30.18) we calculate the square of the delta function and the matrix element $\left|M_{fi}^{\rm \ box}\right|$. Afterwards, we come back to the continuum states and obtain

$$P(i \longrightarrow f) = (2\pi)^{2\kappa}\frac{T}{2\pi}\frac{V}{(2\pi)^3}\delta(E_f - E_i)\,\delta^{(3)}(\mathbf{P}_f - \mathbf{P}_i)\left[\frac{(2\pi)^3}{V}\right]^{N_f+N_i} \left|M_{fi}\right|^2. \tag{30.21}$$

[1] Note that in the special case of only two colliding particles this normalization agrees with the condition $N_{p_1} = N_{p_2} = 1$ that was used in the previous Section.

30.3 Differential transition rate

Strictly speaking, Eq. (30.21) is the *probability density* describing a transition between two discrete states. In order to get the probability we should multiply Eq. (30.21) by the density of final states dN_f. In this way one gets the *differential transition probability*

$$dP\,(i \longrightarrow f) = P\,(i \longrightarrow f)\; dN_f, \tag{30.22}$$

where

$$dN_f = \frac{Vd^3q_1}{(2\pi)^3} \cdots \frac{Vd^3q_{N_f}}{(2\pi)^3}. \tag{30.23}$$

Substituting (30.21) and (30.23) into Eq. (30.22) we find

$$dP\,(i \longrightarrow f) = T\,V^{1-N_i}\;\left|M_{fi}\right|^2 (2\pi)^{3N_i+2\kappa-4}\,\delta^{(4)}\,(P_f - P_i)\,d^3q_1 \ldots d^3q_{N_f}. \tag{30.24}$$

This formula may be used to calculate the *differential transition rate*, defined as the *transition probability per unit time*,

$$\begin{aligned} d\Gamma\,(i \longrightarrow f) &= \frac{dP\,(i \longrightarrow f)}{T} \\ &= V^{1-N_i}\;\left|M_{fi}\right|^2 (2\pi)^{3N_i+2\kappa-4}\,\delta^{(4)}\,(P_f - P_i)\;d^3q_1 ... d^3q_{N_f}\,. \end{aligned} \tag{30.25}$$

The structure of Eq. (30.25)) shows that the reaction rate is determined by the matrix element $\left|M_{fi}\right|$ and the volume of the accessible phase space element. To proceed further we note that the standard normalization of the relativistic plane waves contains a factor $(2\pi)^{3/2}\,\sqrt{2E_p}$. For example, the solutions of the Klein-Gordon equation may be taken in the form

$$\phi_p(t,\mathbf{x}) = \frac{1}{(2\pi)^{3/2}\,\sqrt{2E_p}} e^{-i(E_pt-\mathbf{p}\cdot\mathbf{x})}, \quad E_p = \sqrt{\mathbf{p}^2+m^2}. \tag{30.26}$$

In this case, it is convenient to introduce the invariant matrix $\mathcal{M}_{fi}$ defined by the equation

$$M_{fi} = \left[\frac{1}{(2\pi)^3\,2E\,(p_1)} \cdots \frac{1}{(2\pi)^3\,2E\,(p_{N_i})}\frac{1}{(2\pi)^3\,2E\,(q_1)} \cdots \frac{1}{(2\pi)^3\,2E\,(q_{N_f})}\right]^{1/2} \mathcal{M}_{fi}. \tag{30.27}$$

In this way Eq. (30.25) takes the form

$$\begin{aligned} d\Gamma\,(i \longrightarrow f) = \; & V^{1-N_i}\,\frac{1}{2E\,(p_1)} \cdots \frac{1}{2E\,(p_{N_i})}\,\left|\mathcal{M}_{fi}\right|^2 \\ & \times (2\pi)^{2\kappa-4}\,\delta^{(4)}\,(P_f - P_i)\,\frac{d^3q_1}{(2\pi)^3\,2E\,(q_1)} \cdots \frac{d^3q_{N_f}}{(2\pi)^3\,2E\,(q_{N_f})}. \end{aligned} \tag{30.28}$$

30.4 Differential decay rate and cross section

The case $N_i = 1$ describes the situation when a single particle with the four-momentum p_1 ($P_i = p_1$) is present in the initial state, hence Eq. (30.28) describes simply its decay rate

$$d\Gamma\,(i \longrightarrow f) = \frac{1}{2E\,(p_1)}\left|\mathcal{M}_{fi}\right|^2 (2\pi)^{2\kappa-4}\,\delta^{(4)}\,(P_f - P_i)\,\frac{d^3q_1}{(2\pi)^3\,2E\,(q_1)}\cdots\frac{d^3q_{N_f}}{(2\pi)^3\,2E\,(q_{N_f})}. \tag{30.29}$$

Introducing the concept of the invariant phase-space

$$dQ = (2\pi)^{2\kappa-4}\,\delta^{(4)}\,(P_f - P_i)\,\frac{d^3q_1}{(2\pi)^3\,2E\,(q_1)}\cdots\frac{d^3q_{N_f}}{(2\pi)^3\,2E\,(q_{N_f})} \tag{30.30}$$

we may write shortly

$$d\Gamma\,(i \longrightarrow f) = \frac{1}{2E\,(p_1)}\left|\mathcal{M}_{fi}\right|^2 dQ. \tag{30.31}$$

The total decay rate is obtained by integration over the available phase space and by summing over all possible decay channels

$$\Gamma(i) = \frac{1}{2E\,(p_1)}\sum_f\int\left|\mathcal{M}_{fi}\right|^2 dQ. \tag{30.32}$$

We note that the factor $1/E\,(p_1)$ introduces naturally the time dilation effect for the particles decaying in motion.

For $N_i = 2$ two particles collide and a certain number of the secondary particles (N_f) is produced. From our previous considerations, see Eq. (30.11), we know that in this case the transition (collision) rate divided by the flux of incoming particles yields the cross section. The flux has the structure

$$\Phi_i = \frac{v_{\mathrm{M}}}{V} = \frac{1}{V}\left|\frac{\mathbf{p}_1}{E\,(p_1)} - \frac{\mathbf{p}_2}{E\,(p_2)}\right| = \frac{1}{V}\frac{\sqrt{(p_1\cdot p_2)^2 - m_1^2m_2^2}}{E\,(p_1)\,E\,(p_2)}, \tag{30.33}$$

where v_{M} is the Møller velocity. Very often one introduces the invariant flux in the initial state

$$F_i = \sqrt{(p_1\cdot p_2)^2 - m_1^2m_2^2} \tag{30.34}$$

which leads to the compact expression

$$\Phi_i = \frac{F_i}{VE\,(p_1)\,E\,(p_2)}. \tag{30.35}$$

The cross section is then obtained from the formula

$$d\sigma\,(i \longrightarrow f) = \frac{d\Gamma\,(i \longrightarrow f)}{\Phi_i} = \frac{1}{4\sqrt{(p_1 \cdot p_2)^2 - m_1^2 m_2^2}} \left|\mathcal{M}_{fi}\right|^2 \times (2\pi)^{2\kappa-4}\,\delta^{(4)}\,(P_f - P_i)\,\frac{d^3q_1}{(2\pi)^3\,2E\,(q_1)} \cdots \frac{d^3q_{N_f}}{(2\pi)^3\,2E\,(q_{N_f})} \tag{30.36}$$

or, using the more compact notation, from the equation

$$d\sigma\,(i \longrightarrow f) = \frac{\left|\mathcal{M}_{fi}\right|^2}{4\sqrt{(p_1 \cdot p_2)^2 - m_1^2 m_2^2}}\, dQ = \frac{1}{4F_i}\left|\mathcal{M}_{fi}\right|^2 dQ. \tag{30.37}$$

Strictly speaking, the expression above is the differential cross section which specifies the probability of the production of particles with momenta contained in the phase space element $d^3q_1 ... d^3q_{N_f}$. The total cross section may be obtained by integration of Eq. (30.36).

30.5 Inclusive processes

Studying processes at very high energies it is necessary to distinguish between exclusive and inclusive experiments [5]. In the exclusive experiments we demand that certain particles are produced and no others. On the other hand, in the inclusive experiments we study the production of certain particles allowing that any other particles are also produced. Certainly, at low energies we typically deal with the exclusive reactions, controlling all (in practice a few) produced particles. Contrary, at very high energies the number of the produced particles becomes very large and only their subset may be studied in more detail. Below we discuss the structure of the inclusive cross sections and give them a natural physical interpretation.

30.5.1 *One-particle inclusive cross section*

At first, let us concentrate in more detail on the one-particle inclusive cross sections. The particles a and b collide producing a particle of the type c together with N other particles. The system of N particles is described shortly as X (note that among many different, in fact infinite, realizations of the state X there might be the case where N is zero and the case where there are particles of the same type as c appearing in X). In this case the differential transition rate is (compare Eq. (30.25) with $N_i = 2$ and $\kappa = 1$)

$$\Delta\Gamma_X\,(a\,b \longrightarrow c\,X) = \frac{1}{V}\,\left|M\,(a\,b\,|\,c\,X)\right|^2 (2\pi)^4 \delta^{(4)}\,(q + P_X - P_a - P_b)\,\Delta^3 q. \tag{30.38}$$

Here the subscript X denotes that we omitted the element of the phase space of the system X. The integration over X has the general form

$$\int dX = \sum_{n_1,\, n_2, \dots} \int d^3x_1 d^3x_2 \dots \,, \tag{30.39}$$

where x_1, $x_2, \dots$ are momenta and the sum over n_1, $n_2, \dots$ specifies which particles are included. To calculate the inclusive cross section we sum over all possible states X, hence the appropriate expression is

$$\begin{aligned} \Delta\sigma^{\text{incl}}\,(a\,b \longrightarrow c\,) &= \int dX\, \Delta\Gamma_X\,\left(a\,b \longrightarrow c\,X\right) \frac{V}{v_{\text{M}}} \\ &\equiv \int dX\; \Delta\sigma_X\,\left(a\,b \longrightarrow c\,X\right), \end{aligned} \tag{30.40}$$

or we may write

$$\begin{aligned} \frac{d\sigma^{\text{incl}}\,(a\,b \longrightarrow c\,)}{d^3q} &= \int dX \;\left|\left\langle cX|\hat{M}|ab\right\rangle\right|^2 \\ &= \int dX\, \left\langle ab|\hat{M}^{\,\dagger}|cX\right\rangle \left\langle cX|\hat{M}|ab\right\rangle . \end{aligned} \tag{30.41}$$

where we introduced the operator $\hat{M}$ defined by the matrix elements $M\,(a\,b\,|\,c\,X)$ multiplied by the appropriate coefficients that may be inferred from Eq. (30.38) and Eq. (30.40) [2].

The states $|cX\rangle$ are obtained by adding a particle c to the system X. This may be achieved with the help of the creation operator $a_c^\dagger\,(q)$ acting on $|X\rangle$,

$$\frac{d\sigma^{\text{incl}}}{d^3q}\,(a\,b \longrightarrow c\,) = \int dX\, \left\langle ab|\hat{M}^{\,\dagger}\, a_c^\dagger\,(q)\,|X\right\rangle \left\langle X|a_c\,(q)\,\hat{M}|ab\right\rangle . \tag{30.42}$$

Using the completeness of the states X, Eq. (30.42) may be rewritten in the form

$$\begin{aligned} \frac{d\sigma^{\text{incl}}}{d^3q}\,(a\,b \longrightarrow c\,) &= \left\langle ab|\hat{M}^{\,\dagger} a_c^\dagger\,(q)\, a_c\,(q)\,\hat{M}|ab\right\rangle \\ &= \left\langle \Phi\,\left|a_c^\dagger\,(q)\, a_c\,(q)\right|\,\Phi\right\rangle \\ &= \left\langle \Phi\,\left|N_c\,(q)\right|\,\Phi\right\rangle , \end{aligned} \tag{30.43}$$

where $N_c\,(q)$ is the number operator of the particles c, and the state Φ is a tremendously complicated multi-particle state obtained by the action of the operator $\hat{M}$ on the initial state $|ab\rangle$. The good news is, however, that we do not have to know any details about this state to realize that by the normalization of the inclusive cross section to the total cross section we obtain the average number of particles

[2] The dealing with the Dirac delta function is subtle. The appropriate transformations may be done if one applies again the periodic boundary conditions.

c produced in the collision. This may be seen from the formula defining the total cross section

$$\begin{aligned}\sigma^{\mathrm{tot}}(a\,b) &= \int dX \left\langle ab|\hat{M}^{\dagger}|X\right\rangle\left\langle X|\hat{M}|ab\right\rangle \\ &\equiv \int dX\; \sigma_X\,(a\,b \longrightarrow X) = \langle\Phi|\Phi\rangle\,,\end{aligned} \tag{30.44}$$

which combined with Eqs. (30.43) yields the very important result

$$\frac{1}{\sigma^{\mathrm{tot}}(a\,b)}\frac{d\sigma^{\mathrm{incl}}}{d^3q}(a\,b \longrightarrow c) = \frac{\langle\Phi\,|N_c\,(q)|\,\Phi\rangle}{\langle\Phi|\Phi\rangle} = \langle N_c\,(q)\rangle\,. \tag{30.45}$$

30.5.2 *Two-particle inclusive cross section*

Similar calculations may be performed in order to calculate the two-particle inclusive cross sections. In this case we detect two particles of the type c, denoted below as c_1 and c_2,

$$\begin{aligned}\Delta\sigma^{\mathrm{incl}}(a\,b \longrightarrow c_1c_2) &= \int dX\; \Delta\Gamma_X\,(a\,b \longrightarrow c_1c_2\,X)\,\frac{V}{v_{\mathrm{M}}} \\ &\equiv \int dX\; \Delta\sigma_X\,(a\,b \longrightarrow c_1c_2\,X)\end{aligned} \tag{30.46}$$

Repeating the steps known from the previous Section we get the formula

$$\begin{aligned}\frac{d\sigma^{\mathrm{incl}}(a\,b \longrightarrow c_1c_2)}{d^3q_1\,d^3q_2} &= \int dX\; \left|\left\langle c_1c_2X|\hat{M}|ab\right\rangle\right|^2 \\ &= \left\langle ab|\hat{M}^{\dagger}a_c^{\dagger}(q_1)\,a_c^{\dagger}(q_2)\,a_c(q_2)\;a_c(q_1)\,\hat{M}|ab\right\rangle \\ &= \left\langle\Phi\left|a_c^{\dagger}(q_1)\,a_c^{\dagger}(q_2)\,a_c(q_2)\;a_c(q_1)\right|\Phi\right\rangle,\end{aligned} \tag{30.47}$$

where we assumed that the first particle has the momentum q_1 and the second has the momentum q_2. Dividing the two-particle inclusive cross section by the total cross section we find

$$\frac{1}{\sigma^{\mathrm{tot}}(a\,b)}\frac{d\sigma^{\mathrm{incl}}}{d^3q_1d^3q_2}(a\,b \longrightarrow c_1c_2) = \frac{\left\langle\Phi\left|a_c^{\dagger}(q_1)\,a_c^{\dagger}(q_2)\,a_c(q_2)\;a_c(q_1)\right|\Phi\right\rangle}{\langle\Phi|\Phi\rangle}. \tag{30.48}$$

Equation (30.48) counts the number of pairs of the particles c with momenta q_1 and q_2. The integration of (30.48) over the momenta gives the average number of all such pairs, namely

$$\langle N_c\,(N_c-1)\rangle\,. \tag{30.49}$$

Note that the pairs which differ by ordering are counted separately here.

30.5.3 *Inclusive distributions*

The quantity defined by Eq. (30.45) and multiplied by the energy of the particle c is very often called shortly the *one-particle (invariant) inclusive distribution*

$$\mathcal{P}_1(q) = E_q \frac{1}{\sigma^{\rm tot}} \frac{d\sigma^{\rm incl}}{d^3q}. \tag{30.50}$$

Of course the multiplication by the energy is required to obtain the Lorentz invariant quantity. Similarly, Eq. (30.48) multiplied by the product of the energies of two particles of type c defines the *two-particle (invariant) inclusive distribution*

$$\mathcal{P}_2(q_1, q_2) = E_{q_1} E_{q_2} \frac{1}{\sigma^{\rm tot}} \frac{d\sigma^{\rm incl}}{d^3q_1 d^3q_2}. \tag{30.51}$$

According to the formulas worked out above, the one- and two-particle inclusive distributions satisfy the normalization conditions

$$\begin{aligned} \int \frac{d^3q}{E_q} \mathcal{P}_1(q) &= \langle N_c \rangle , \\ \int \frac{d^3q_1}{E_{q_1}} \frac{d^3q_2}{E_{q_2}} \mathcal{P}_2(q_1, q_2) &= \langle N_c(N_c - 1) \rangle . \end{aligned} \tag{30.52}$$

It is easy to generalize these definitions to the k-particle (invariant) inclusive distribution

$$\mathcal{P}_k(q_1, q_2, ..., q_k) \tag{30.53}$$

that satisfies the normalization condition

$$\int \frac{d^3q_1}{E_{q_1}} \frac{d^3q_2}{E_{q_2}} \cdots \frac{d^3q_k}{E_{q_k}} \mathcal{P}_k(q_1, q_2, \ldots, q_k) = \langle N_c(N_c - 1)(N_c - 2) \cdots (N_c - k + 1) \rangle . \tag{30.54}$$

Of course one may consider different types of the produced particles, hence the inclusive cross sections for the production of pions, kaons, protons and all other possible hadron species may be measured. It is also possible to consider the inclusive production of a group of particles. For example one may study the inclusive production of all charged particles. In Chap. 17 we discussed the pion correlation functions which are defined as the ratios of the one- and two-particle inclusive pion distributions. An interesting relation exists also between the event-by-event fluctuations and the inclusive distributions, as showed by Bialas and Koch in Ref. [6].

Bibliography to Chapter 30

[1] J. Bjorken and S. Drell, "Relativistic quantum mechanics," (McGraw-Hill, New York, 1964).
[2] S. Weinberg, "The Quantum Theory of Fields, Vol. 1: Foundations," (University Press, Cambridge, 1995).
[3] L. D. Landau and E. M. Lifshitz, "Classical Theory of Fields," (Butterworth-Heinemann, Oxford, 1980).

[4] F. Halzen and A. D. Martin, "Quarks and leptons: an introductory course in modern particle physics," (Wiley, New York, 1984).
[5] R. P. Feynman, "Very high-energy collisions of hadrons," *Phys. Rev. Lett.* **23** (1969) 1415–1417.
[6] A. Bialas and V. Koch, "Event by event fluctuations and inclusive distributions," *Phys. Lett.* **B456** (1999) 1–4.

Chapter 31

Useful Mathematical Formulae

31.1 Heaviside and Dirac functions

The Heaviside step function $\theta(x)$, also called the unit step function, is defined by the expressions:

$$\theta(x) = 1 \quad \text{for} \quad x > 0, \quad \theta(x) = 0 \quad \text{for} \quad x \le 0. \tag{31.1}$$

The Dirac delta function $\delta(x)$ is the distribution whose action on the function $f(x)$ is defined by the following formula

$$\int_{-\infty}^{\infty} \delta(x-a) f(x) dx = f(a). \tag{31.2}$$

The basic property of the Dirac delta function used in this book is

$$\delta\left(g(x)\right) = \sum_i \frac{\delta(x - x_i)}{|g'(x_i)|}, \tag{31.3}$$

where x_i are real roots of the function $g(x)$ and the prime denotes the derivative.

31.2 Modified Bessel functions $K_n(x)$

The second-order ordinary differential equation

$$x^2 \frac{d^2 w(x)}{d\,x^2} + x \frac{dw(x)}{dx} - \left(x^2 + n^2\right) w(x) = 0 \tag{31.4}$$

has a solution which is a linear combination of the modified Bessel functions of the first kind $I_n(x)$and the second kind $K_n(x)$ [1–3]

$$w(x) = c_1 I_n(x) + c_2 K_n(x). \tag{31.5}$$

The modified Bessel function $K_n(x)$ has the following integral representation

$$K_n(x) = \frac{\sqrt{\pi}\, x^n}{2^n\, \Gamma\left(n + \frac{1}{2}\right)} \int_1^{\infty} e^{-xt} \left(t^2 - 1\right)^{n - \frac{1}{2}} dt, \tag{31.6}$$

where $\Re(n) > -1/2$ and $\Re(x) > 0$. The importance of the functions $K_n(x)$ in the relativistic thermodynamics comes from the fact that many relativistic phase space integrals are reduced to the integrals of the form (31.6). Other integral representations were introduced as problems in Chap. 10, see Eqs. (10.7)–(10.11).

The Bessel functions $K_n(x)$ satisfy the following recurrence identities

$$K_n(x) = K_{n+2}(x) - \frac{2(n+1)}{x} K_{n+1}(x),$$
$$K_n(x) = K_{n-2}(x) + \frac{2(n-1)}{x} K_{n-1}(x), \tag{31.7}$$

which are used in the studies of the thermodynamic relations. The derivative of the Bessel function $K_n(x)$ can be expressed through other Bessel functions with different indices, for example

$$\frac{\partial K_n(x)}{\partial x} = -\frac{1}{2}\left(K_{n-1}(x) + K_{n+1}(x)\right), \tag{31.8}$$

this derivative can also be represented in other form

$$\frac{\partial K_n(x)}{\partial x} = -K_{n-1}(x) - \frac{n}{x} K_n(x) = \frac{n}{x} K_n(x) - K_{n+1}(x). \tag{31.9}$$

The series expansions of the modified Bessel functions $K_n(x)$ for $x \to 0$ and for $n = 1, 2, 3, 4$ are given by the formulas

$$K_1(x) = \frac{1}{x} + O(x^1),$$
$$K_2(x) = \frac{2}{x^2} - \frac{1}{2} + O(x^2),$$
$$K_3(x) = \frac{8}{x^3} - \frac{1}{x} + \frac{x}{8} + O(x^3),$$
$$K_4(x) = \frac{48}{x^4} - \frac{4}{x^2} + \frac{1}{4} - \frac{x^2}{48} + O(x^4). \tag{31.10}$$

The asymptotic expansion of the modified Bessel function $K_n(x)$ for $x \to \infty$ has the generic form

$$K_n(x) = e^{-x}\left(\sqrt{\frac{\pi}{2x}} + O\left(x^{-\frac{3}{2}}\right)\right). \tag{31.11}$$

31.3 Modified Bessel functions $I_n(x)$

The modified Bessel functions of the first kind may be defined as the integrals

$$I_n(x) = \frac{1}{\pi}\int_0^{\pi} e^{x\cos\phi}\cos(n\phi)d\phi. \tag{31.12}$$

In the special cases we have

$$I_0(x) = \frac{1}{\pi}\int_0^{\pi} e^{x\cos\phi} d\phi = \frac{1}{2\pi}\int_0^{2\pi} e^{x\cos\phi} d\phi, \tag{31.13}$$

$$I_1(x) = \frac{1}{\pi}\int_0^{\pi} e^{x\cos\phi}\cos(\phi)d\phi = \frac{1}{2\pi}\int_0^{2\pi} e^{x\cos\phi}\cos(\phi)d\phi. \tag{31.14}$$

We note that

$$I_1(x) = \frac{\partial}{\partial x} I_0(x). \tag{31.15}$$

The asymptotic expansion of the modified Bessel function $I_n(x)$ for $x \to \infty$ is

$$I_n(x) \approx e^x \sqrt{\frac{1}{2\pi x}}. \tag{31.16}$$

31.4 Cylindrical and light-cone coordinates

In the transition from the Cartesian to cylindrical coordinates one may use the following rules

$$\begin{aligned} \frac{\partial}{\partial x} &= -\frac{\sin\phi}{r}\frac{\partial}{\partial \phi} + \cos\phi \frac{\partial}{\partial r}, \\ \frac{\partial}{\partial y} &= \frac{\cos\phi}{r}\frac{\partial}{\partial \phi} + \sin\phi \frac{\partial}{\partial r}. \end{aligned} \tag{31.17}$$

Similarly, we find

$$\begin{aligned} \frac{\partial}{\partial t} &= \cosh\eta_{\|} \frac{\partial}{\partial \tau} - \frac{\sinh\eta_{\|}}{\tau}\frac{\partial}{\partial \eta_{\|}}, \\ \frac{\partial}{\partial z} &= -\sinh\eta_{\|} \frac{\partial}{\partial \tau} + \frac{\cosh\eta_{\|}}{\tau}\frac{\partial}{\partial \eta_{\|}}. \end{aligned} \tag{31.18}$$

Bibliography to Chapter 31

[1] M. Abramowitz and I. A. E. Stegun, "Handbook of mathematical functions with formulas, graphs, and mathematical tables," (Dover Publications, New York, 1964).
[2] http://mathworld.wolfram.com/ModifiedBesselFunctionoftheSecondKind.html.
[3] http://mathworld.wolfram.com/ModifiedBesselFunctionoftheFirstKind.html.

Index